TREATISE ON SOLID STATE CHEMISTRY

Volume 2
Defects in Solids

TREATISE ON SOLID STATE CHEMISTRY

Volume 1 • The Chemical Structure of Solids
Volume 2 • Defects in Solids
Volume 3 • Crystalline and Noncrystalline Solids
Volume 4 • Reactivity of Solids
Volume 5 • Changes of State
Volume 6 • Surfaces

TREATISE ON SOLID STATE CHEMISTRY

Volume 2
Defects in Solids

Edited by
N. B. Hannay
Vice President
Research and Patents
Bell Laboratories
Murray Hill, New Jersey

PLENUM PRESS • NEW YORK-LONDON

Library of Congress Cataloging in Publication Data

Hannay, Norman Bruce, 1921-
Defects in solids.

(His Treatise on solid state chemistry; v. 2)
Includes bibliographical references and index.
1. Crystals — Defects. I. Title.
QD478.H35 vol. 2 [QD921] 541'.042'1s [548'.842]
ISBN-13: 978-1-4684-0831-7 e-ISBN-13: 978-1-4684-0829-4
DOI: 10.1007/978-1-4684-0829-4 74-22197

Six-volume set: ISBN-13: 978-1-4684-0831-7

Softcover reprint of the hardcover 1st edition 1975

Published by Plenum Press, New York
A Division of Plenum Publishing Corporation
227 West 17th Street, New York, N.Y. 10011

United Kingdom edition published by Plenum Press, London
A Division of Plenum Publishing Company, Ltd.
4a Lower John Street, London, W1R 3PD, England

Foreword

The last quarter-century has been marked by the extremely rapid growth of the solid-state sciences. They include what is now the largest subfield of physics, and the materials engineering sciences have likewise flourished. And, playing an active role throughout this vast area of science and engineering have been very large numbers of chemists. Yet, even though the role of chemistry in the solid-state sciences has been a vital one and the solid-state sciences have, in turn, made enormous contributions to chemical thought, solid-state chemistry has not been recognized by the general body of chemists as a major subfield of chemistry. Solid-state chemistry is not even well defined as to content. Some, for example, would have it include only the quantum chemistry of solids and would reject thermodynamics and phase equilibria; this is nonsense. Solid-state chemistry has many facets, and one of the purposes of this *Treatise* is to help define the field.

Perhaps the most general characteristic of solid-state chemistry, and one which helps differentiate it from solid-state physics, is its focus on the chemical composition and atomic configuration of real solids and on the relationship of composition and structure to the chemical and physical properties of the solid. Real solids are usually extremely complex and exhibit almost infinite variety in their compositional and structural features.

Chemistry has never hesitated about the role of applied science, and solid-state chemistry is no exception. Hence, we have chosen to include in the field not only basic science but also the more fundamental aspects of the materials engineering sciences.

The central theme of the *Treatise* is the exposition of unifying principles in the chemistry, physical chemistry, and chemical physics of solids. Examples are provided only to illustrate these principles. It has, throughout, a chemical viewpoint; there is, perforce, substantial overlap with some areas of solid-state physics and metallurgy but a uniquely chemical perspective underlies the whole. Each chapter seeks to be as definitive as possible in its particular segment of the field.

The *Treatise* is intended for advanced workers in the field. The scope of the work is such that all solid-state chemists, as well as solid-state scientists and engineers in allied disciplines, should find in it much that is new to them in areas outside their own specializations; they should also find that the treatment of their own particular areas of interest offers enlightening perspectives.

Certain standard subjects, such as crystal structures, have been omitted because they are so well covered in many readily available standard references and are a part of the background of all solid-state scientists. Certain limited redundancies are intended, partly because they occur in different volumes of the series, but mainly because some subjects need to be examined from different viewpoints and in different contexts. The first three volumes deal with the structure of solids and its relation to properties. Volumes 4 and 5 cover broad areas of chemical dynamics in bulk solids. Volume 6 treats both structure and chemical dynamics of surfaces.

N.B.H.

Preface to Volume 2

Volume 1 of this Treatise was concerned with the chemical structure of solids. This volume treats disorder in solids in relationship to physical properties. Many of these properties stem from simple defects in crystals, i.e., isolated impurity atoms and the point defects, vacancies, and interstitials. In the case of mechanical properties, dislocations (line defects) are dominant. Control of defects in solids thus provides the key to control of properties in many practical solids. This volume examines this relationship in detail.

N. B. H.

Contents of Volume 2

Chapter 2

The Imperfect Solid—Color Centers in Ionic Crystals **133**

W. Beall Fowler

Chapter 3

The Imperfect Solid—Dielectric Properties **183**

G. E. Peterson

1

Electronic Structure and Spectra of Impurities in the More Ionic Crystals

Donald S. McClure
Department of Chemistry
Princeton University
Princeton, New Jersey

1. Introduction

The physics and chemistry of the solid state nearly always involve impurities or defects in crystals. It is impossible to get rid of them entirely, so they are present whether we want them or not. But a large class of phenomena and devices depend on impurity or defect behavior, so the control and understanding of these objects is very important.

This chapter is concerned mainly with the electronic structure and spectroscopy of defects and impurities in the more ionic crystals. Underlying the spectroscopy of these localized centers, however, should be an understanding of their local geometry and the electronic structure of the ground state. Electronic spectroscopy alone gives some, but not enough, of this information, and it must be supplemented by other methods such as electron spin resonance, electron-nuclear double resonance, nuclear quadrupole resonance, infrared and Raman spectra, and thermodynamic measurements. The great generality and the precise meaning of thermodynamic quantities give them a position of special significance in this field, as will be shown in detail in the first parts of the chapter.

The electronic states associated with defects and impurities are localized on a very few atoms, unlike the Bloch states of the perfect crystal. The localization is caused by a change in potential at one or a few lattice sites compared to the pure crystal and could be described in some cases as a perturbation of the band structure, as has been done in a famous paper by Koster and Slater.[1] On the other hand, the potential changes due to impurity substitution are caused by definite chemical differences between host and guest atom and are usually fairly large. Localized states entirely unrelated to host crystal band structure are introduced by transition metal impurities in non-transition metal compounds, for example. A useful classification of localized states can be based on the types of atomic orbitals available to electrons in impurity atoms, relative to those available in the host crystal.

The most localized impurity states are those of the $4f$ shell in rare earth atoms. Except for the splitting of degeneracies by a few hundred cm^{-1} and shifts of the same amount, the energy levels of the rare earth ions in ionic crystals are the same as those of the free atoms.[2] The $4f$ shell is also able to preserve its identity in semiconductors and metals.

As the number of radial nodes for a given l value increases, the orbital becomes more extended in space. Thus the $5f$ shell of the actinide series is more strongly perturbed than the $4f$ shell, and crystal field splittings are of the order of 1000 cm^{-1}.[3]

In the first transition series, Sc to Zn, where the 3d orbitals are being filled, the atomic degeneracies are split by 5000–25,000 cm^{-1} in crystals, while in the second and third series greater splittings are observed.[4]

The metals at the ends of the three transition groups have moderately stable $4s$, $5s$, or $6s$ orbitals because the d^{10} shells do not shield as well as p^6 shells of the pre-transition metals. Therefore the $4s$ and $4p$, $5s$ and $5p$, and $6s$ and $6p$ orbitals of post-transition metals are localized sufficiently well to be identified as atomiclike states when they occur as ions in crystals. The ns–np separation is considerably less than in the free atoms, however, and is strongly dependent on the environment.[5]

Even a bromide ion impurity in a NaCl lattice shows a distinct spectrum at lower energy than the band gap of NaCl, and thus it must have well-localized states. The $5p$ shell in this case is part of a highly electronegative atom. On the other hand, only weakly localized behavior is exhibited when electropositive metals such as potassium occur as impurities in NaCl.[6]

Another example of poorly localized impurity states is given by the levels resulting from the substitution of phosphorus in a germanium crystal. The wave functions of these states extend over many unit cells, and are best characterized as perturbed Ge conduction band levels.[7]

In the transition groups of elements, by which we mean both *d* and *f* series, both the initial and final states of a spectroscopic transition may be highly localized. At energies high enough, however, stronger transitions occur in which either the initial or final state is localized, but the other is not as well localized. The latter states are related to the band structure of the crystal, though they are modified by the presence of the impurity. The name "charge-transfer transition" is often given to these processes, implying that an electron is transferred from one atom to another, or to a group of other atoms.[8]

Closely related to charge-transfer transitions are the spectroscopic processes in "donor–acceptor pairs." These are two impurities, usually acting mutually as charge compensators in the crystal lattice, and an electron is transferred from one to the other. This type of system is the basis for phosphor behavior.[9]

Pairs of identical transition group atoms situated at nearest cation sites in a crystal display exchange interactions which have been related to the magnetic behavior of the corresponding transition metal compound. Nonidentical pairs, or a transition metal impurity in a transition metal crystal, also show interesting exchange effects, both in the ground and in excited states.[10]

Atomic or ionic vacancies act as donors or acceptors in crystals. The absence of an atom leaves weakly bound states which result in the occurrence of low-energy optical transitions. The most familiar of these is the anion vacancy, which becomes the *F* center when it captures an electron.[11]

Complexes formed between an *F* center and a transition metal or rare earth ion have been observed, and have interesting properties. For example Ce^{3+} + *F* center[12] in a CaF_2 crystal displays photochromic behavior. An *F* center with a neighboring alkali or alkali-earth ion different from the host cations is called an F_A center, and much has been learned from them about electrons near defects.[13]

A great deal of chemistry depends on which atomic orbitals in the range of valence energies are able to remain localized when the atom becomes part of a crystal or molecule. Our scheme for classifying localized states depends on being able to identify these localized valence orbitals. An illustration of the differences between the radial extent of different atomic orbitals is shown for the gadolinium atom in Figure 1.[14] Gadolinium has three valence electrons, one 5*d* and two 6*s*. As an impurity in a CaF_2 crystal in the presence of only a

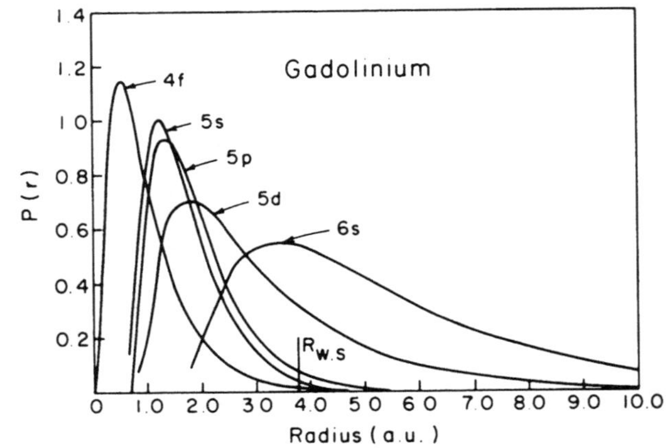

Fig. 1. The radial probability distribution $P(r) = \psi(r)^2 \cdot 4\pi r^2$ of electrons in the outer parts of the 4*f*, 5*s*, 5*p*, 5*d*, and 6*s* orbitals of atomic Gd. The radius of the Wigner–Seitz sphere (approximately one-half the nearest-neighbor distance in the metal) is shown for comparison.[10]

weak acceptor, it can exist in the divalent state with one electron in the 5*d* orbital and seven in the 4*f* orbitals. With a somewhat stronger acceptor it reaches the usual trivalent state when the 5*d* orbital is empty. The 4*f* shell is so localized that it has nearly the same magnetic and spectroscopic properties whether the atom is part of a metal or an ionic compound. Thus, as Figure 1 shows, the 4*f* orbital does not have significant amplitude at half the interatomic distance in the metal, while the 5*d* orbitals of neighboring metal atoms overlap strongly. In compounds, the 5*d* orbitals, if occupied, are strongly perturbed by the orbitals of the electronegative partners and the 6*s* orbitals are essentially completely mixed with them, giving the compound its ionic nature.

The radial extent of the orbitals relative to interatomic distances therefore gives a good indication of the chemical behavior of the valence electrons.

There is more to the story, however, because Figure 1 does not show the large kinetic energy associated with the *f* and *d* orbitals. Thus the 4*f* electrons are less tightly bound than the 5*s* and 5*p* electrons even though the 4*f* orbitals are on the average closer to the nucleus. Kinetic energy is manifested in the node patterns of *f* and *d* orbitals, which in turn leads to "crystal field splitting." The number of nodes is not affected by the introduction of the atom into a solid and it remains as an identifying feature of the nature of localized atomic orbitals.

The viewpoint of this chapter is perhaps unique and is definitely limited. Two recent books could be consulted for a coverage of other topics and for presentations having different emphasis.[15,16]

2. Incorporation of Impurities and Defects in Crystals

The incorporation of an impurity or other defect in a crystal involves both the movement of atoms and a change of electronic structure. There are examples of the extreme cases in which very little electronic reorganization takes place (K replacing Rb in RbBr) and in which a large amount of electronic reorganization occurs (P replacing Ge in a Ge crystal). In this section we hope to give some conception of the entire range of the possible effects of creating a defect on the geometric and electronic structure of a crystal. We will succeed mainly in displaying large areas of ignorance, but hopefully will make guessing a little less risky.

The incorporation of defects (including impurities) in a crystal may be carried out by equilibration with another phase. The concentration of the defects is controlled by the free energy for the process, and the equilibrium may be characterized by a standard free energy. The defect may be characterized by an internal energy relative to a standard state involving the perfect lattice and the other components. The most useful reference state is the gas-phase atoms since the energy of formation from this state represents the binding energy of the crystal. The difference between binding energies for imperfect and perfect crystal is the change in binding energy due to the defect. This binding energy could also be expressed as a change of lattice energy in the case of an ionic crystal, and then could be calculated by the standard theories of ionic crystals.

The binding energy of the defect found either experimentally or theoretically gives the ground-state energy relative to that of the pure crystal, and it may therefore be used to interpret spectroscopic transitions in the defect relative to the energy levels of the pure crystal. Since the defect wave function is limited in space, the spectroscopic transitions observed could involve an energy level of the pure crystal as an initial or final state, and a defect level as the other state. These transitions are called acceptor or donor charge-transfer transitions, as the defect acts either as an acceptor or donor relative to the lattice. The observation of such transitions fixes the energy levels of the defect relative to those of the crystal. These transition energies can be compared to the defect binding energies obtained from thermodynamic measurements. They are not the same but are related in complex ways which will be illustrated in later sections. As another example, the charge state, or oxidation–reduction potential, of a transition metal ion in a crystal is closely related to the heats of solution of the ion in its two charge states, and direct spectroscopic observation of the oxidation or reduction process is often possible. The comparison of thermal and optical energies of such processes has very important implications for solid-state chemistry and for chemistry in general.

There are not enough data of either type to make extensive comparisons possible. The next sections review what is known, and will suggest possibilities for further work.

2.1. Solubility and Heat of Solution of Impurities in Crystals

It is interesting to consider solubilities of impurity ions in crystals both to see what systems are available for experimental study and to consider the meaning of the corresponding heats of solution in terms of the electronic and atomic rearrangement which attends incorporation of an impurity into a lattice.

By solubility, one means the concentration of the solute in a solvent when an external phase containing solute has reached equilibrium with the solute–solvent system. Solubility is thus the quantity of interest for the practical man needing to know how much of A can get into B. The quantity of more theoretical interest is the heat or energy of solution. Since we are interested in the so-called point defects, we need to know the heat of the reaction in which one mole of guest G dissolves in an infinite amount of host H; this is the partial molar heat of solution of G in H at infinite dilution. This heat quantity, except for the small PV term, leads to the extra energy needed to replace a lattice atom by an impurity atom.

2.1.1. Heats of Solution and Substitution

For a host crystal HA one could write the chemical process of dissolving x moles of guest GA as

$$x\mathrm{GA}_{(s)} + (1 - x)\mathrm{HA}_{(s)} \rightarrow \mathrm{G}_x\mathrm{H}_{1-x}\mathrm{A}_{(s)} + \Delta H \qquad (1)$$

where ΔH is the heat of the reaction for x moles. When x is small,

$\Delta H/x$ is the partial molar heat of solution at infinite dilution. This quantity is rigorously expressed as

$$\left(\frac{\partial(\Delta H)}{\partial x}\right)_{x=0} = \lim_{x\to 0}\frac{\Delta H}{x} \equiv \chi \tag{2}$$

The abbreviation χ has been used by several writers on this subject.[17,18] It can be written in terms of the heats of formation of the reactants and product as

$$\begin{aligned}\chi &= \lim_{x\to 0}\frac{1}{x}[H(\mathrm{G}_x\mathrm{H}_{1-x}\mathrm{A}) - (1-x)H(\mathrm{HA}) - xH(\mathrm{GA})]\\ &= \lim_{x\to 0}\frac{H(\mathrm{G}_x\mathrm{H}_{1-x}\mathrm{A}) - H(\mathrm{HA})}{x} + H(\mathrm{HA}) - H(\mathrm{GA})\\ &= H_s + H(\mathrm{HA}) - H(\mathrm{GA})\end{aligned} \tag{3}$$

where H_s is the change in heat content of the HA lattice per mole of substituted guest

$$H_s = [H(\mathrm{soln}) - H(\mathrm{host})]/x = \chi - H(\mathrm{HA}) + H(\mathrm{GA}) \tag{4}$$

H_s is also the heat of the reaction

$$x\mathrm{GA}_{(s)} + \mathrm{HA}_{(s)} \longrightarrow \mathrm{G}_x\mathrm{H}_{1-x}\mathrm{A}_{(s)} + x\mathrm{HA}_{(s)} \tag{5}$$

The H(HA) are heats of formation from the elements in their standard thermodynamic states.

H_s can be called the heat of substitution. It is the change of the host crystal heat content resulting from the substitution of the impurity, per mole of impurity. Another interesting quantity is the "recovered lattice energy." If we look at the solution process as the destruction of one mole of guest lattice and the creation of a new, imperfect lattice, we want to know how much new lattice energy has been produced to compensate for the loss of the old. This is $H_s + H(\mathrm{HA})$ and it equals $\chi + H(\mathrm{GA})$. The value of H_s is the quantity of real interest and since H(HA), H(GA), and χ are obtainable from experiment, H_s can be found if these data are available. It can also be calculated from the ionic model of the crystal, and if the agreement with the experiment is good, other quantities of interest can be obtained reliably from the model. Thus our real interest in solubility is that it provides the final datum with which to derive the heat of substitution and the properties related to it. Figure 2 shows these quantities on an energy level diagram for the case of KCl in NaCl, and for NaCl in KCl, following a similar presentation of Ikeya *et al.*[19]

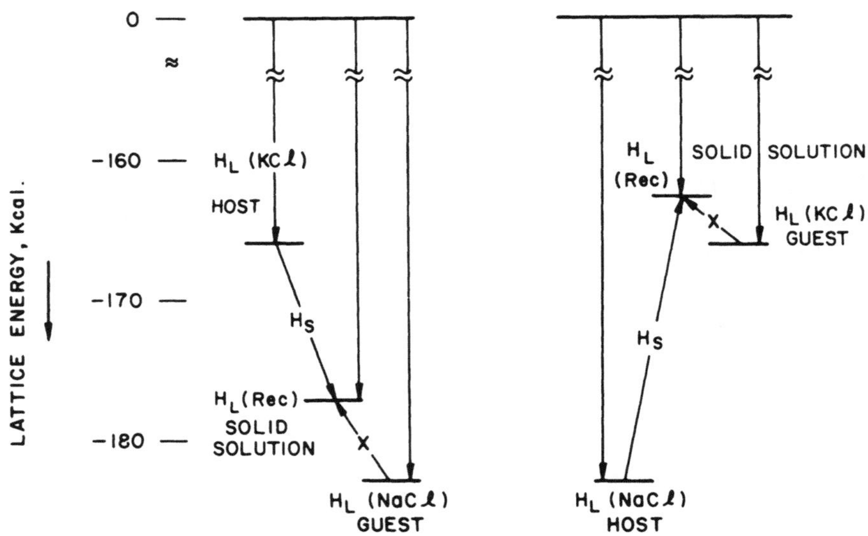

Fig. 2. Diagrams showing the relationships between heat of solution, heat of substitution, and recovered lattice energy for (left) NaCl in KCl, (right) KCl in NaCl. The total energy values are given as lattice energies.

The heat of substitution from Eq. (3) is expressed in terms of reaction (1) in which the compounds in their standard thermodynamic states participate. The alternative quantities of interest are the heat of substitution of atoms:

$$H_s^{\mathrm{a}} = \chi - H_B(\mathrm{HA}) + H_B(\mathrm{GA})$$

where H_B is the heat of bonding (or of formation of the crystal from neutral gas-phase atoms); and the heat of substitution of ions:

$$H_s^{\mathrm{i}} = \chi - H_L(\mathrm{HA}) + H_L(\mathrm{GA})$$

where H_L is the heat of formation of the lattice from gas-phase ions. The tabulated lattice energies are $U_L = -H_L$.

We will hardly ever have to consider the PV term in the enthalpy.

Not surprisingly, the alkali halides are the only materials for which there exist both a reasonable amount of data and of calculations. We must first distinguish between isovalent and nonisovalent (aliovalent) substitution. In the former case the mixed alkali halides have been the main focus of attention,[20–22] with a small amount of work on the silver halide–alkali halide solid solutions[23] and on alkali earth oxide solid solutions.[24] In the second case there has been some work on alkali-earth impurities in alkali halides, and it is reaching the stage where a systematic study of the results is possible.[18,25]

There has been a considerable amount of work on lattice defect calculations in the fluorite structure.[26] Most of this work is applied to the problem of the motion of defects or diffusion of impurities, and H_s is not known for many impurities.

2.1.2. Isovalent Substitution

In all of the theoretical work on ionic crystals the ions of the crystal are regarded as charged spheres interacting through their Coulomb fields and their repulsive potentials. Details of the electronic structure are thus ignored. Substitution of an impurity of the same charge causes local changes in the repulsive term, which results in displacements of neighboring ions. Since the latter were at centers of symmetry in the case of alkali halide crystals, the displaced atoms are moved away from high-symmetry points, and therefore acquire dipole moments.

There is no guarantee that the substituted ion will remain at the center of symmetry and indeed there are many cases known in which a small ion such as Li in replacing a larger ion does move away from the center. In $KCl:Li^+$ the displacement appears to be in the $\langle 111 \rangle$ direction by $0.2r_{KCl} = 0.63$ Å and the energy reduction is 0.03 eV, according to calculations of Quigley and Das[27] (also see Ref. 28). These energies have little effect on solubility, but the tunneling between equivalent $\langle 111 \rangle$ minima gives rise to many low-lying energy levels of spectroscopic interest. (See Ref. 29 for an example of recent experimental work.)

The lattice energy of the impure crystal is the sum of all the interaction terms at the new equilibrium positions. Since the latter are not known in advance, the calculation must be done iteratively until a minimum energy results. H_s is the difference between this energy and the formation energy of the pure crystal. These calculations are reviewed by Tosi.[30]

The additional quantities of interest which can come out of these calculations are the displacements of the near neighbors, the induced moments on the neighbors, the force system in the vicinity of the impurity, which can be used to calculate the altered vibrational spectrum, and others.

The most comprehensive and systematic theoretical investigation of the heats of solution among the alkali halides is that of Douglas.[20]

TABLE 1

Theoretical Heats of Solution of Alkali Halides in Each Other at Infinite Dilution Compared to Calorimetric Values, and Theoretical Heats of Substitution Derived from Heats of Solution

Solvent (host)	Solute	χ, kcal/mole of solute			H_s		$-H_L(298)$
		Theoret.	Calorim.	Difference	Ref. 20	Ref. 22	
KCl	RbCl	1.0	1.0[254]	0.0	6.73	5.1	165.8
RbCl	KCl	0.7	0.8[254]	−0.1	−5.03	−3.2	159.3
NaCl	KCl	6.0	5.5[255,256]	+0.5	21.74	21.8	182.6
KCl	NaCl	4.1	4.4[255,256]	−0.3	−11.61	−10.6	165.8
NaBr	KBr	5.2	4.6[255,256]	+0.6	19.23	—	173.6
KBr	NaBr	3.6	3.5[256]	+0.1	−10.45	—	158.5
NaI	KI	4.3	3.2[256]	+1.1	16.23	—	163.2
KI	NaI	2.9	2.3[256]	+0.6	−9.02	—	149.9
NaCl	NaBr	1.4	1.4[256,257]	0.0	9.83	—	182.6
NaBr	NaCl	1.2	1.3[256,257]	−0.1	−7.23	—	173.6
KCl	KBr	1.1	1.0[258–260]	+0.1	7.83	—	165.8
KBr	KCl	0.9	1.0[258–260]	−0.1	−5.84	—	158.5
NaBr	NaI	2.7	2.8[258–260]	−0.1	13.66	—	173.6
NaI	NaBr	2.3	2.0[256]	+0.3	−8.63	—	163.2
KBr	KI	2.1	1.5[256]	+0.6	10.96	—	158.5
KI	KBr	1.8	1.9[256]	−0.1	−7.09	—	149.9

Many parameters were varied during this study and all but the Cs-containing salts were studied (i.e., only the NaCl structure was studied). Douglas' paper contains a good discussion of his and other calculations which need not be repeated here, but Table 1 shows some of the results compared to the known calorimetric values of heats of solution. Calorimetric experiments can only be done with sufficient accuracy when the solubility exceeds a few percent, and so the cases for which comparison with theory is possible are those having heats of only a few kcal/mole. All the heats are positive and therefore solubility is promoted by the entropy. Computed heats of ≥ 15 kcal/mole correspond to solubilities of less than 0.01% just below the melting point. The calorimetric values are probably accurate to $\pm 20\%$, and the theoretical results shown in Table 1 are all within about 20% of the experimental value. This is remarkably good agreement and gives one confidence in the computed displacements, dipole moments, etc.

Values of H_s are also given in Table I (E_{total} in Douglas' notation) and a comparison with some values calculated independently by

Tosi and Doyama[22] is included. The agreement between the two calculations is very good. The values of H_s are positive when a larger ion replaces a smaller one and negative for the reverse case. The heat of formation of the lattice of the host crystal H_L is also given. The "recovered lattice heat" is $H_s + H_L$. When the solute is a larger ion this sum is smaller in absolute magnitude, showing that the crystalline solution has a smaller lattice energy than the host crystal. In the reverse case it is larger; but the increase of lattice energy is always less than the difference between host and guest crystal lattice energies so that a positive heat of solution results. In summary, small solute ions lead to a more stable lattice, large ones to a less stable lattice.

Douglas[20] has given the computed displacements of the neighbors of the impurity ion out to ten shells for both cation and anion replacement, extending earlier work by Dick and Das[31] and Tosi and Doyama.[22] The displacements of the nearest anion neighbors near a cation impurity resulting from Douglas' calculations are shown in Figure 3. The figure is a plot of $\Delta r/r_{\mathrm{H}}$ vs. $(r_{\mathrm{G}} - r_{\mathrm{H}})/r_{\mathrm{H}}$ where Δr is the displacement of the nearest anions, and r_{G} and r_{H} are guest and host crystal nearest-neighbor distances. The computed points are not shown on this graph since they fall very close to the lines given. For a larger ion replacing a smaller one the slope of the line is 0.50

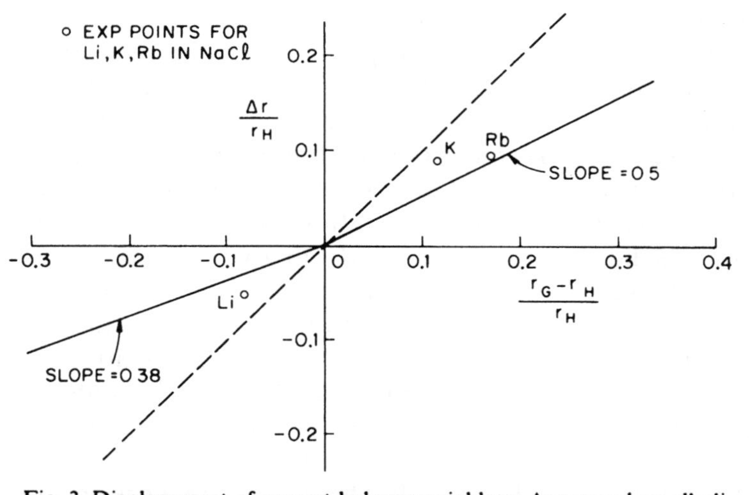

Fig. 3. Displacement of nearest halogen neighbors Δr around an alkali metal cation impurity as a function of $r_{\mathrm{G}} - r_{\mathrm{H}}$ ($r_{\mathrm{G}}, r_{\mathrm{H}}$ are the nearest cation–anion distances in guest and host alkali halide crystals). Solid line is a plot of calculations made by Douglas.[20] Dashed line of unit slope represents the rigid-sphere displacement. Points are Fukai's data.[32]

and for a smaller ion replacing a larger one it is 0.38. This difference reflects the fact that the repulsion force is very short ranged and nonlinear. The long-range Coulomb force has only a minor effect on the displacements and heats of substitution in the case of isovalent substitution. The dashed line of unit slope in Figure 3 is the maximum displacement, assuming rigid sphere behavior.

The three points in Figure 3 are experimental values of the displacement of the neighbors of Li, K, and Rb in NaCl obtained by Fukai.[32] These were obtained by analyzing the quadrupolar broadening of the ^{23}Na nuclear magnetic resonance signal in NaCl crystals. Because of the high symmetry at a cation site in pure NaCl there is no quadrupole field. The reduction in symmetry caused by impurities introduces quadrupole fields which interact with the nuclear quadrupole of ^{23}Na and cause the line broadening. The points in Figure 3 correspond to appreciably larger displacements than calculated, but the interpretation of the data is quite complex and it is not known how accurate they are.

A comparison of heat of solution in crystals with solution in water is interesting. If we dissolve NaCl in H_2O, the heat of solution χ is found to be very small. Therefore almost all of the lattice energy is recovered as heat of solution. Since the heat content of the host is small [$H_{vap}(H_2O) = 9.7$ kcal], H_s in this case nearly equals the lattice energy of the guest. On the other hand, if NaCl is dissolved in KCl, again the heat of solution is very small, and the recovered lattice energy is approximately equal to that of NaCl, but most of this energy goes into making the space in the KCl lattice, i.e., the recovered lattice energy is not H_s but H(KCl). Thus H_s is quite large for water solution, but relatively small for KCl, and the difference must largely be due to the much greater relaxation in water compared to the crystalline solid. For the case of NaCl in KCl the relaxation energy is only -1.4 kcal/mole,[22] while in water it is about 164 kcal/mole. About half of this last figure is attributable to the Cl^- ion, however.

Some calorimetric measurements of heats of formation of silver–alkali halide solutions by Kleppa and Meschel[23] should be mentioned. The values are close to those of the alkali halide solutions. The heats of solution at infinite dilution are AgCl : Na, 2.9; NaCl : Ag, 2.5; AgBr : Na, 2.6; NaBr : Ag, 1.9. These heats are more positive than those that correspond to the ionic radius difference, which is very small in this case.

There are also some calorimetric measurements of the heats of solution of alkaline earth oxides in each other.[24] The CaO–SrO and SrO–BaO systems exhibit complete intersolubility (at the temperature of the preparation, 1150°C) and correspondingly their heats of solution are very small, of the order of 1 kcal/mole. No value for CaO–BaO is given, though the solubility is limited and the heat of solution must be positive and high. The H_s values for CaO–SrO and SrO–BaO must therefore be just the difference between the lattice energies, ± 52 and ± 43 kcal/mole, respectively.

2.1.3. Aliovalent Substitution

The few known heats of solution of divalent halides in alkali halides have values of 15–45 kcal, or an order of magnitude higher than for the mixed alkali halides. Since an alkali-earth atom has an extra valence electron compared to an alkali, and the energy of promotion to the conduction band is prohibitive, the electron is captured in an alkali metal vacancy. Knowing the mechanism of incorporation, we write a typical such process as

$$xSrCl_2 + (1 - 2x)NaCl \longrightarrow Sr_xNa_{1-2x}Cl + \Delta H(\text{soln}) \qquad (6)$$

showing that two Na are removed for each Sr incorporated. The partial molar heat of solution is the heat of this reaction per mole of solute at infinite dilution:

$$\begin{aligned} \chi &= \lim_{x \to 0} \frac{\Delta H(\text{soln})}{x} \\ &= \lim_{x \to 0} \frac{1}{x}[H(Sr_xNa_{1-2x}Cl) - xH(SrCl_2) - (1 - 2x)H(NaCl)] \\ &= \lim_{x \to 0} \frac{H(\text{soln}) - H(NaCl)}{x} - H(SrCl_2) + 2H(NaCl) \qquad (7) \\ &= H_s - H(SrCl_2) + 2H(NaCl) \end{aligned}$$

Thus H_s is again the difference in enthalpy of formation between the impure and the pure lattice, per mole of impurity. In this case, however, there are two distinct imperfections for each molecule of added impurity; a Sr^{2+} ion replacing a Na^+ ion, and a Na^+ vacancy. It would be useful to separate H_s into components referring to the separate imperfections. Thus $H_s - H_{vac}$ is the change in enthalpy attributable to the Sr^{2+} and H_{vac} is that attributable to the vacancy. Figure 4 shows the energy of the solid solution on an energy level diagram, and also the lattice energy attributable to the Sr^{2+}.

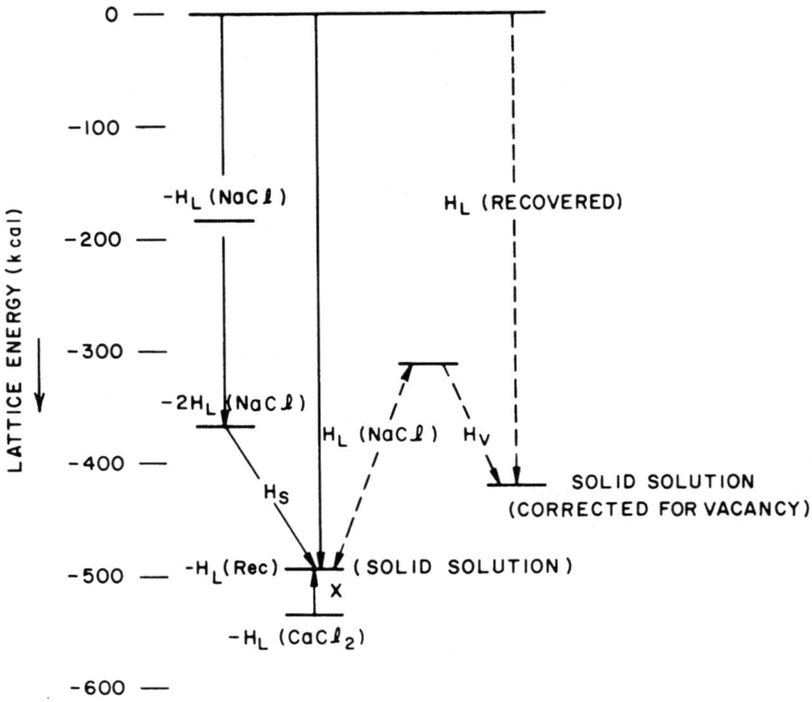

Fig. 4. Diagram showing relationships between heat of solution χ, heat of substitution H_s, heat of vacancy formation H_{vac}, and recovered lattice energy for the case of $CaCl_2$ in NaCl. The energies of host, guest, and solid solution are given on the diagram as lattice energies.

Both theoretical and experimental determinations of the solubility of divalent ions in alkali halides are more difficult than for isovalent substitution. Early calculations reviewed by Lidiard[17] gave negative values for χ, but they are surely positive. The experimental values are still uncertain by a factor of two, as discussed in a recent review by Barr and Lidiard.[25] Bowman[18,33] has recently been able to produce theoretical results which look very reasonable. Because of the limitations of the Born–Mayer theory, his work is limited to the cubic alkali-earth chlorides and fluorides dissolved in alkali chlorides and fluorides. The relaxation of the neighbors of the divalent impurity and of the vacancy are some of the interesting results of these calculations. Table 2 summarizes the experimental data and the theoretical results. Only the data considered the most likely to be correct are included, and the reader should consult Refs.

TABLE 2

Theoretical and Experimental Heats of Solution χ of Alkali Earth Halides in Alkali Halides, Theoretical Displacements $\Delta r/r_H$, Heats of Substitution H_s, and Recovered Lattice Energy H_L*

Host†	Guest	χ (exp), kcal/mole	χ (theoret) kcal/mole	$\Delta r/r_H$	$\frac{r_G - r_H}{r_H}$	H_s, kcal/mole	H_L (Rec) kcal/mole
NaCl	$CaCl_2$ (534)	23	—	—	0.028	−145	−435
−182.6	Sr (509)	42	23	0.007	0.100	−102	−393
	Ba (485)	46	35	0.054	0.160	−74	−365
	Cd (600)	29	—	—	0.00	−206	−496
	Vacancy	—	—	0.089	—	108	—
KCl	Sr (509)	33	31	−0.079	−0.018	−145	−415
−165.8	Ba (485)	39	40	−0.034	0.032	−114	−386
	Vacancy	—	—	0.083	—	104	—
RbCl	Sr (509)	33	32	−0.115	−0.067	−158	−421
−159.3	Ba (485)	—	39	−0.070	−0.018	−127	−390
	Vacancy	—	—	0.077	—	104	—
KBr	Ca (524)	17	—	—	−0.082	−192	−451
−158.5	Sr (488)	26	—	—	−0.017	−137	−396
	Ba (466)	43	—	—	0.029	−89	−346
	Cd (589)	29	—	—	−0.09	−233	−492
	Vacancy	—	—	—	—	(100)	—
NaF	Ca (626)	—	41	−0.053	0.023	−157	−501
−214.4	Sr (590)	—	61	−0.007	0.080	−101	−445
	Ba (555)	—	74	0.036	0.13	−52	−396
	Cd (662)	—	52	−0.055	0.00	−181	−525
	Pb (590)	—	63	0.003	0.066	−98	−442
	Vacancy	—	—	0.092	—	130	—

* The theoretical calculations are those of Bowman.[18,33] The experimental values are those quoted by Bowman, except for KBr, where values quoted by Barr and Lidiard are used.[25]

† Lattice energies of host are shown under formula for host crystal. Those of the guest are shown after formula for guest.

17 and 25 for a discussion of the measurements. Due to the very low solubilities, calorimetric methods are not feasible and the values obtained are taken from ionic conductivity data.

The displacements calculated for the neighbors of the impurity are of the same order as for the case of isovalent substitution. Comparing them with $r_G - r_H$, however, shows an interesting difference. The displacements for $r_G - r_H > 0$ are much smaller than $r_G - r_H$, and when $r_G - r_H < 0$ the displacements tend to be quite large. A plot of $\Delta r/r_H$ vs. $(r_G - r_H)/r_H$ similar to Figure 3 is shown in Figure 5.

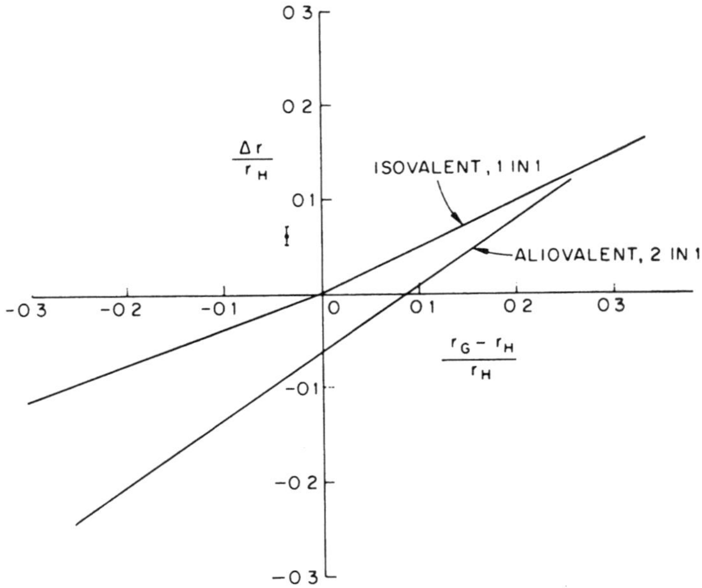

Fig. 5. Displacements of nearest halogen neighbors Δr around an alkaline-earth cation impurity as a function of $r_G - r_H$, as in Figure 3 (the case of divalent impurity in a monovalent lattice). Calculations of Bowman.[18] The lines of Figure 3 (monovalent in monovalent) are repeated for comparison. The one known example of monovalent in divalent (Na^+ in CaF_2) is shown as the plotted point.[34]

The data used there are for the systems in Table 2 whose heats of solution are known experimentally, but the displacements are the calculated ones. These fall on a straight line lying below the lines for isovalent substitution, also shown for comparison. This trend is evidently caused by the increased Coulomb attraction of the divalent ion over the monovalent one. A small, dipositive ion has both a larger Coulomb field and a smaller repulsion with which to cause displacements of the neighboring anions.

It is tempting to generalize from these results. A monovalent ion in a cubic MF_2 lattice might lie on a curve with a smaller slope and an intercept at about -0.1 on the $(r_G - r_H)/r_H$ scale. The only such example is the calculation by Franklin[34] of the distortion around a Na^+ ion at a Ca^{2+} site in CaF_2. This point is plotted in Figure 5. The corresponding heat of solution is $+32$ kcal/mole and the heat of substitution is $+344$ kcal/mole, excluding the energy of $V(F^-)$.

Since the heats of solution include a large, positive energy of vacancy formation (108 kcal/mole for NaCl), the 20–50 kcal positive heats of solution have been reduced by the considerable binding energy of the impurity. The value of H_s, which includes the positive energy of vacancy formation for the case of Sr in NaCl, is -94 kcal, showing that the mixed crystal has a deeper total energy than the pure crystal. The contribution of the Sr^{2+} ion to this stability is $H_s - H_{vac} = -202$, using Tosi and Doyama's value[22] of $H_{vac} = 108$. The lattice energy which would have been contributed by the undisturbed site is -182.6 and so the total lattice energy contributed by the impurity is -384.6. Bowman[33] has computed the energy necessary to remove a Sr^{2+} from the Na^+ site in NaCl leaving a relaxed vacancy behind. This value is 336 kcal. These last two numbers do not correspond to quite the same process, because of the effect of relaxation, but they should be of about the same magnitude, as they are found to be. Values of H_s and of the total lattice energy contributed by the impurity, H_L, calculated as for the above example, are also given in Table 2.

The values of the heats of substitution H_s shown in Table 2 are obtained from experimental heats of solution and lattice energies but the recovered lattice energy H_L is partly theoretical because of the use of the energy of an isolated vacancy. In the next section the experimental justification for the values of the vacancy energies will be discussed. The heats of substitution in Table 2 all have large, negative values (~ 100 kcal), whereas for isovalent substitution (Table 1) the values are much smaller and of either sign. The increase of H_s is an obvious consequence of the higher ionic charges. The recovered lattice energy is a large fraction of the host lattice energy, usually about 80%. It is not 100% because the dipositive impurity, while pulling in the nearest-neighbor anions, must also increase the anion to host cation distance in the next shell, thus raising the energy of many host lattice ions in this region.

2.1.4. Association of Impurity and Vacancy

A divalent ion at a cation site in an alkali halide has a Coulomb attraction for a cation vacancy, and at a short range other forces come into play. The heat of association is 8–15 kcal for the species with nearest-neighbor cation sites occupied by impurity and vacancy.[35] The formation of such a species aids solubility at low temperatures since the entropy favors association, and thereby the positive heat of solution is reduced. Barr and Lidiard review both the experimental and the theoretical situation in this area.[25]

In more general terms, an aliovalent ion and its charge compensator, be it vacancy, interstitial, or another impurity, will tend to associate. The extent of association depends both on the heat of association and the experimental conditions. Such pairs are often of major importance for some of the physical phenomena observed in impure crystals, and there will be a discussion of this later (Section 3.4.3; also see Refs. 36–39).

2.1.5. Impurities in the Fluorite Structure

Impurities and defects in the fluorite structure have attracted much attention.[40] The most work has been done on solution of YF_3 or trivalent rare earth fluorides in CaF_2. These form solid solutions up to very high proportions and the heat of solution must be quite small or even negative. Figure 6 shows the structure in the vicinity of a single trivalent ion where a fluoride interstitial F_I^- acts as the charge compensator. The energies of these interstitial ions will be discussed in Section 2.2, but because of the open structure of the CaF_2 crystal and an F^-–F^- attraction, their energy is reported to be actually negative, -60 to -80 kcal/mole.[26] This is in contrast to the positive energy needed to form the charge-compensating cation

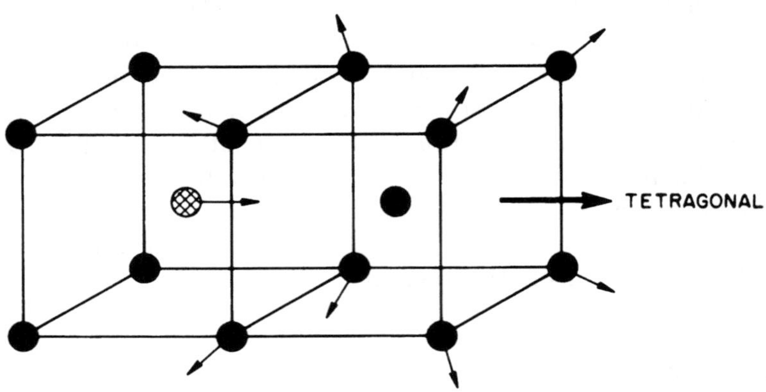

Fig. 6. Portion of the crystal structure of CaF_2 where Ca^{2+} ion has been replaced by a Ce^{3+} ion (diamond hatched), and an interstitial F^- ion occupies a normally vacant position in the neighboring cube of F^- ions. All other neighboring interstitial sites are empty, so the additional F^- ion gives rise to a tetragonal distortion whose axis is indicated. The small arrows show the directions of displacements as measured by ENDOR.[46] Each F^- is displaced by 0.16 Å and the Ce^{3+} by 0.16 Å.

vacancy in the alkali halides. Therefore, since the trivalent ion recovers a large fraction of its lattice energy, it is easily possible to have a negative heat of solution. If we write the solution process as

$$xYF_3 + (1 - x)CaF_2 \longrightarrow Y_xCa_{1-x}F_2F_x{}^I + \Delta H \qquad (8)$$

then the heat of solution at infinite dilution can be written as

$$\chi = [H_s - H(F^I)] + H(F^I) - H(YF_3) + H(CaF_2) \qquad (9)$$

where the term in brackets is the heat of substitution of the Y^{3+} ion, excluding the heat of formation of the interstitial. Using values of lattice energies and the interstitial energy (see next section), and assuming $\chi = 0$, the heat of substitution of the Y^{3+} ion would be about -600 kcal/mole, which is far higher than the values for divalent ions in alkali halides. The large numerical value is caused by the high value of the third ionization potential, which is greater than the sum of the first two. The ionic model is quite unrealistic for trivalent ions, since three electrons have no reason to leave the Y atom entirely, but it could nevertheless give values of H_s in agreement with the above. In fact, Franklin(34) has made a Born model calculation for Y in CaF_2 and obtains values of H_s in agreement with the above estimates. The calculated values of the displacements do not seem to be reliable, however.

Since the fluorite structure can dissolve such large amounts of MF_3, there are bound to be defects involving several M atoms. The structures of some of these are being studied by Cheetham *et al.*(41) and Catlow.(42)

Monovalent ions may also replace the Ca^{2+} ions in CaF_2 and on lattice sites they may act as charge compensators for trivalent ions. Oxide ions are often found as impurities on the fluorite lattice, where again they may act as charge compensators.

2.1.6. Experimental Determination of Displacements near Defects

The measurement of NMR line broadening caused by field gradients near impurities was mentioned earlier as a way to determine displacements near impurities. This method involves many details of the electronic structure around the impurity so the interpretation of the field gradient is not at all simple.(43) Perhaps only a SCF calculation, allowing the ions to relax to minimum energy positions, will be really satisfactory.

Another experimental method involves the use of electron-nuclear double resonance (ENDOR). It has been applied to the trivalent rare earth ions in CaF_2 for the case when the ion is in a cubic site[44] and for the case when it is near an interstitial F^- charge compensator.[45] The experiment depends on the measurement of the dipolar coupling of the electron of the paramagnetic rare earth to the nuclear dipoles of the fluoride ions in its vicinity. Since the dipole–dipole coupling depends in a known way upon distances and angles, the location of fluoride ions relative to a paramagnetic impurity can be deduced. The method works for the distant fluoride ions, which are not overlapped by the electronic orbital, but there is a much more complicated dependence on the position of the nearest neighbors where overlap is appreciable.

In spite of these difficulties, Baker *et al.*[45] were able to map out the displacements around a Ce^{3+} or Yb^{3+} at a Ca^{2+} site in CaF_2 when a F^- ion is in the nearest interstitial site acting as a charge compensator. The crystal structure at this defect is shown in Figure 6, and the displacements for the case of Ce^{3+} are given there. Most of the distortion is caused by the interstitial F^- ion (F_I^-), and its lattice F^- neighbors are pushed out by 0.16 Å or about 7% of the nearest-neighbor distance. The Ce^{3+} is attracted toward F_I^- and moves away from the cube center by about 0.16 Å[45]

The ENDOR method is powerful, but complex. A few other examples and a discussion of some difficulties with the method are given in Ref. 46.

When the concentration of the impurity ions is large enough the structure near impurity ions can sometimes be deduced from the changes in Bragg reflection intensities,[47] or from diffuse scattering intensities.[41]

2.1.7. Conclusion

The scarcity of solubility data makes the ionic model quite important for correlating and extrapolating what data there are, and in this section we have depended heavily upon it. In highly ionic compounds the model gives the heat of solution quite accurately, and the local displacements are probably given correctly. The very simple correlation of Δr with $r_G - r_H$ shown in Figures 3 and 5 can possibly be extended to other systems. There is only a rough correlation between heat of solution and $r_G - r_H$, since the other parameters of the ionic model are quite important for this property. In the off-center cases, such as Li in KCl, the calculations agree with the experimental measurements.[27] These successes make it a useful model, but it should not be considered as anything but a way to correlate and extrapolate experimental data. It does not tell much about the basic electronic structure around the impurity. Its deficiencies are revealed when one attempts to calculate field gradients or other electronic properties.[43] It can be used as a starting point for SCF calculations, but a lattice of neutral atoms can be used just as well.[48] The same point of view applies to the use of the ionic model in crystal field theory, where again the atoms of the crystal are treated as classical charged spheres.

In cases of aliovalent substitution the charge compensation mechanism is related to the lowest energy defect type and this is known for many crystals (see next section). The association of impurity and defect, discussed earlier, is understood in general. As we have seen, recent ionic model calculations of solubility[20] and defect energies[26] have been successful.

On the other hand, the solubility of impurities in the less ionic II–VI compounds of Zn and Cd is a more complicated and less understood subject. Calculations based on the ionic model apparently have not been attempted, and are not expected to be as successful as in the alkali halides or in alkaline earth oxides. A review of the situation has been given by Woodbury.[49] The small band gaps of these compounds make it possible to have electronic compensation (see Section 2.3). The formation of donor–acceptor complexes enhances solubility and is a more important mode of charge compensation than is the formation of native defects. For example, Cu is soluble only to the extent of 2×10^{-4} mole/mole in ZnS, but a continuous series of solid solutions between ZnS and $CuGaS_2$ may be formed.[50] We will return to these cases in Section 2.3.

It is surprising that heats of solution have not been measured for more crystal systems. Such data have a precise meaning and provide both a way to establish trends in a series of data, as has been done in this section, and accurate numbers with which to compare model calculations. Modern computer-based calculations of properties of ionic crystals are now reliable enough so that extensive comparisons of calculated and observed heats of substitution could lead to refinements of interionic potential functions, and ultimately to reliable displacements and polarizations of ions near an impurity. Reliable results of this type could serve in turn as the basis for electronic structure calculations near an impurity.

2.2. Energies of Defect Formation

Just as the heats of solution give some information about the electronic ground state of impurity ions, the energies of defect formation are useful for considering the electronic structure in the vicinity of a vacancy or interstitial. In the case of stoichiometric crystals the common types of defects are (1) Frenkel pairs, in which either a cation or an anion leaves the lattice site and goes into an interstitial position, (2) Schottky pairs, in which a cation and an anion vacancy appear together, and (3) antistructure defects, in which cation and anion interchange. In a nonstoichiometric lattice the defects do not have to appear in pairs, but some mechanism for preserving charge balance must be present. If a Na atom is removed from a NaCl crystal, the lattice is left charge neutral. The Na vacancy will have an electronic hole in its vicinity, and the hole may travel to another site or it may form a Cl_2^- molecule (a V_k center), but the total charge is conserved in whatever process occurs. The thermodynamic energy of vacancy formation, however, measures the actual formation energy of the vacancy plus the energy of subsequent electronic and ionic rearrangement.

In the alkali halides, where the most work has been done, the Schottky pair is the dominant type of defect and its heat of formation is usually measured by means of the ionic conductivity curve.(17) The defects present are not neutral atom vacancies but ionized vacancies, or the Schottky pair formed by removing a cation and an anion.

In fluorite structures, CaF_2, SrF_2, BaF_2, the anion Frenkel defect is the lowest energy type. These are fluoride ion vacancies accompanied by fluoride interstitials in the empty eight-coordinated sites.

The energies of anion and cation vacancies in alkali halides were calculated by Tosi and Doyama.(22) The parameters were determined by the same model with the same potentials as for the impurity calculations shown in Table 2. The nearest-neighbor displacements and dipole moment as well as the vacancy energy are shown in Table 3. The vacancy formation energy is split into a rigid lattice component and a relaxation component. The rigid lattice energy for forming a vacancy is nearly equal to the lattice energy per ion pair. The relaxation energy is one-third to one-half of the rigid lattice

TABLE 3
The Energies of Formation of Cation and Anion Vacancies in Alkali Halides*

Parameter†	NaCl		KCl		RbCl		NaF
	Cation	Anion	Cation	Anion	Cation	Anion	Cation‡
$\Delta r/r_H$	0.053	0.077	0.063	0.071	0.057	0.060	0.092
ν	−0.069	0.0057	−0.049	0.020	−0.045	0.027	−0.029
U_{relax}	75.6	65.5	64.0	59.0	59.5	56.5	—
$U_{r.l.}$	184	188	168	171	163	165	—
U_v	108	122	104	111	104	109	130
U_L	182		164		160		214
E_S (calc)	44.8		46.1		48.5		—
E_S (exp)	49.0		51.2		—		—

* From Tosi and Doyama.[22]
† $\Delta r/r_H$ is the displacement of a nearest neighbor from its lattice site divided by the cation–anion distance in the crystal. Positive values mean outward displacement. ν is the dipole moment induced on the nearest neighbors in units of electron-angstroms. A positive sign means the positive end of the dipole is outward. U_{relax} = relaxation energy; $U_{r.l.}$ = rigid lattice energy; U_v = energy of vacancy formation; U_L = lattice energy.
‡ The NaF column is from the work of Bowman.[18]

energy, far greater than for the case of impurity substitution. (Also see more recent results of Boswarva[51] and a careful study of the computational method by Boswarva and Lidiard.[52])

In order to compare these calculations to experiment, the energy to form a Schottky pair is calculated. This is

$$E_S = U_+ + U_- - U_L \tag{10}$$

where U_+ and U_- are the cation and anion vacancy energies and U_L is the lattice energy. Table 3 shows the calculated and observed values of E_S, and they are in reasonably good agreement with each other. A critical discussion of the experimental and theoretical results would be out of place here, but has been given by Barr and Lidiard[25] for most of the available examples. We will assume that the agreement between the observed and calculated Schottky pair energies justifies the serious consideration of the theoretically calculated quantities in Table 3.

It is interesting to note that the displacements around a vacancy are always outward. This is reasonable since the short-range repulsion acting on the ions surrounding the vacancy cannot do work on them over a great distance, while the attractive Coulomb forces of the crystal acting on these ions do not decrease greatly with displacement and therefore can do work over an appreciable distance. An anion vacancy containing an electron (F center) is presumably more compressed, since the cations are partially neutralized by the extra electron (see Chapter 2).

The experimental values of the anion Frenkel defect energy in alkaline earth fluorides are shown in Table 4. Recent calculations[26,53] have given results in good agreement with the experimental values, so that it is worth looking at the individual vacancy and interstitial energies resulting from these calculations. These are also shown in Table 4. The energy to form a Frenkel defect is quite small and this partly accounts for the high solubility of trivalent rare earth fluorides in CaF_2.

TABLE 4
Energies of Formation of Defects in Alkaline Earth Fluorides (kcal/mole)[26]

	Vacancy		Interstitial		Frenkel		
	Anion (calc)	Cation (calc)	Anion (calc)	Cation (calc)	Anion (calc)	Anion (exp)	U_L
CaF_2	135	538	−75	−343	60	62[261]	626
SrF_2	128	510	−73	−322	55	53	~590[34]
BaF_2	115	491	−71	−301	44	44[261]	555

The cation vacancy energy shown in Table 4 is 85–90% of the lattice energy; therefore the relaxation around the cation vacancy is very much smaller than is the case for the alkali halides (see Table 3), where the ratio U_v/U_L is about 0.6.

Up to this point we have ignored the problems of finding out what the nature of an experimentally observed defect is. There is not enough space to permit a review of this area, as it is a very large one. Henderson and Wertz[54] have reviewed the work on defects in alkaline earth oxides, mainly electron spin resonance work (also see Ref. 55). Jain[56] reviews divalent cation impurities in alkali halides. Much of the work on defects in CaF_2 has been done by Hayes and his group; some recent examples are Refs. 57–60. Some theoretical papers on defects in CaF_2 are Refs. 61–64. Examples of additional experimental work on CaF_2 are given in Refs. 65–70. Several papers have dealt with impurities and defects in perovskite-type crystals.[71–74] Smoluchowski and Dienes and their associates have contributed much to the theory of defects in alkali halides.[75–77] This cursory review will at least point to the vast literature on the properties of defects and impurities in crystals.

The defects formed in the less ionic binary compounds such as ZnS, CdTe, etc., can be studied by means of photoconductivity and other optical methods.[78] As a result, a great deal is known about the electronic energy levels associated with these defects. The poor photoconductive response of the highly ionic compounds renders a comparable study of defect energy levels much more difficult for them. On the other hand, the thermodynamics of the II–VI compounds is more complex and the ionic model cannot be used to extend the data as has been done with ionic compounds. Therefore, it is hard to find firm values of defect energies in the literature.

One of the complicating features of defect chemistry in the II–VI compounds is the involvement of electrons and holes. Thus the constituents of the crystal cannot be considered as charged spheres. This of course represents a partial breakdown of the ionic model of the crystal. Both the thermodynamics and electronic structure of these compounds are reviewed by Aven and Prener.[79]

In highly ionic crystals, electrons and holes are trapped in specific ways leading to defects which are unlike the Frenkel or Schottky disorder defects discussed earlier. These trapping mechanisms lead to the degradation of photoconductivity in ionic systems.[80] In the alkali halides an electron–hole pair created by radiation can apparently produce a pair of defects: an *F* center and an *H* center.[75,81] A hole on a halogen ion produced by light leaves a neutral halogen atom at a lattice site, but this is unstable and combines with a neighboring halogen ion to produce a molecule X_2^- at one of the halogen sites (*H* center). The electron is attracted to the halogen vacancy to become an *F* center. The *H* center may migrate or turn into a *V* center, an X_2^- molecule straddling two halogen sites. These types of centers can be described successfully using theoretical models based on ionic constituents.[82–84] They illustrate the fact that a crystal model consisting of charged spheres does not suffice to describe all the defects which even an "ionic" crystal can have. In the next section the interplay between electronic and atomic structures will be illustrated further.

2.3. Equilibria among Electrons, Impurities, and Defects

In the two preceding sections the role of the electrons in determining the properties of the defective crystal was suppressed because of the use of the Born ionic model, and because transition group atoms were not discussed. Defect chemistry becomes much more complex and interesting when electrons and holes in the conduction and valence bands are considered or when valence states of impurities or of the host cations can change.

A good introduction to defect chemistry is given by Van Gool,[85] and more elaborate and complete treatments are given by Kröger[86] and Brebrick.[87] The formalism for writing chemical equations involving defects and thermodynamic relations involving the Fermi level are presented in these places, and will be used here.

2.3.1. Redistribution of Electrons after Substitution

When considering substitution in ionic crystals one has the tendency to think in terms of ionic impurities, or the creation of ionic defects. This is an inconvenient point of view with compounds such as CdS or GaAs, and nonsense for the elemental semiconductors. In order to have a picture of defect behavior useful for all crystals, one should think in terms of substituting or removing neutral atoms. The subsequent rearrangement of the electrons may then be considered separately. The electronic reorganization after the formation of a defect is the main subject of this section.

The removal of a neutral atom and its replacement with a neutral isovalent atom should cause relatively small changes in the electronic structure of the crystal. Some examples of the spectra of the mixed alkali halides (Section 3.3.3) will illustrate this point. Another kind of example is the substitution of a divalent transition metal atom for Mg or Zn in a crystal such as $KMgF_3$ or $ZnCl_2$. In these cases the 4*s* band (conduction band) of the crystal is only slightly perturbed, but a new set of unfilled 3*d* levels appears at the impurity site (Section 3.2).

With heterovalent substitution a different number of valence electrons are present at the impurity site in the host crystal, and some way must be found to accommodate this change. A well-known example is the substitution of P for Si in a silicon crystal, where phosphorus brings in an extra electron. In this case the electron is weakly bound in the Coulomb field of the P atom, and it occupies an extended orbital made up chiefly of conduction band wave functions of the Si lattice. At normal temperatures, therefore, Si becomes an *n*-type semiconductor.

There are not any strict analogs in highly ionic systems of the case, such as P in Si, in which the extra electron goes into the conduction band. Consider, for example, why NaCl when doped with Ca does not become an *n*-type semiconductor. The extra Ca electron would have to be promoted to the conduction band of NaCl, at an energy cost equal to the band gap of NaCl, 10 eV. A much less energetic way to accommodate the extra electron is to create a Na vacancy. The extra Ca electron in effect fills the electronic vacancy in the Cl neighbors of the lost Na atom. The Ca atom, in accommodating to the local valence requirements, has become an ionized donor, in the present point of view, and the Na vacancy an ionized acceptor.

Now considering a less ionic host crystal, such as ZnS, it is possible that both modes of disposing of the extra electron of a donor atom could be operative. A normally trivalent atom, such as Al, in ZnS might be expected to act as a donor, like P in Si, and with a band gap of only 3.5 eV the promotion energy is within the range of defect energies. The question in this case is, Does the extra electron go into the conduction band of ZnS or is it energetically more economical to create a defect which accommodates the charge? This question will be answered quantitatively in the next part of this section.

One might think that a sharp distinction could be made between the behavior of electrons in transition and in non-transition metal atoms upon substitution in a lattice, and to a certain extent this can be done. The neutral $3d$ transition metal atoms always have $3d^n4s^2$ as the lowest, or nearly lowest, configuration, and they tend to supply the $4s$ electrons and as many $3d$ electrons to the more electronegative nonmetal atoms of the host crystal as may be necessary to satisfy the valence shells of these atoms. In a situation where fewer valence electrons are needed the d shell can absorb another electron. So instead of promoting electrons to the conduction band, or taking them out of the valence band (e.g., Ga in Si), the d shell acts as a buffer for the electron concentration, tending to level it in accord with host crystal valence requirements.

As an example of this behavior, when manganese is substituted at a silicon site in silicon, the deficiency in its valence shell seems to be made up by the transfer of two $3d$ electrons to the $4s$, $4p$ shell so that the tetrahedral bonding can be completed.[88] A species Mn^{+} containing two $3d$ electrons and a species Mn^{2-} containing five $3d$ electrons have been observed; both of these have a completed valence shell. In order to be present in appreciable amounts, they require a charge compensator, X^{-} and X^{2+}, respectively. The $3d^3$ species does not require a charge compensator, but was probably not detected in the spin resonance experiments which led to the discovery of the other two because it would have a degenerate ground state.

The buffering action of the $3d$ shell is also illustrated by the behavior of Mn in an interstitial site of the Si lattice. Ludwig and Woodbury[88] found Mn with a d^7 shell, and it appears that the two valence electrons have been demoted into the d shell. The reason is that the valence structure of Si is satisfied, and the Mn valence electrons are not needed. The Si, being more electronegative than Mn, determines the valence structure, while the Mn makes the best response possible by swallowing its own valence electrons.

An example of the reverse case, that of having a highly electronegative atom at an interstitial site, is interstitial fluorine in CaF_2. Fluorine never exists as an atom since it is a powerful electron acceptor and it does either of two things: (1) It removes an electron from a substituted trivalent atom at a lattice site, such as Ce or Yb; it is then called a charge compensator; or (2) it disrupts the lattice by forming a F_2^- molecule, or "V_K center" as mentioned earlier in Section 2.2. In either action it displays its high electronegativity.

Returning to transition metal atoms, we want to make another point about the valence state. Both Cr and Al are considered trivalent atoms, and when they substitute for Zn in ZnS there are the two options for getting rid of the extra electron, as discussed in the previous paragraph. With Cr, however, we could say there is a third option: demoting an electron into the d shell. However, Cr atom has the $3d^4\,4s^2$ ground configuration anyway, so the free atom is "in the divalent state." The normal chemical trivalent state exists because d^3 forms a more stable chemical ground state than d^4. The reason for this behavior is given by crystal field theory (Section 3). The valence state of a metal is determined by the number of its easily removed electrons. In this example one of these is a d electron. So in one sense the sharp distinction between transition and non-transition metals is blurred. On the other hand, the d shell is always distinct and recognizable in atoms in crystals and therefore the number of d electrons present is the best way to characterize a transition metal atom in a solid. Chromium in ZnS can be present as Cr^{3+}, Cr^{2+}, and Cr^{+}, but it is better to think of these species as d^3, d^4, and d^5.

2.3.2. Some Quantitative Examples

A comprehensive quantitative treatment of the interplay between impurities, defects, and electrons is inherent in the work of Kröger.[86] Much remains to be done, however, to estimate the many equilibrium constants and the energies and entropies of formation of defects so that this work can correlate data and have predictive value. So far, very little has been done to relate the ionic crystal calculations discussed in Sections 2.1 and 2.2 with the wider class of problems being discussed in this section. There also are obvious relationships between the chemical oxidation–reduction potentials of transition metal atoms in solids and optical charge-transfer spectra, but these relationships are only beginning to be recognized and used.

Instead of presenting the algebra of a complete treatment of the electron–impurity–host crystal equilibria, as is done by Kröger,[86] some examples will be discussed. Mandel[89] has addressed himself to the question, Will the electrons from a donor atom in a given host crystal occupy the conduction band or will they be trapped at defect sites? A question which is not asked is, What is the charge state of the donor atom? It is assumed to be a completely ionized single donor. In the next example impurities having several charge states will be considered.

2.3.2.1. The Fate of Electrons Introduced by Donor Atoms

Consider a binary compound MN into which donor atoms D have been introduced. The vapor pressures of M and N, P_{M} and P_{N}, are fixed at a temperature T, and we suppose the donor species to have a fixed concentration [D] in the solid, and negligible vapor pressure at T. As in the examples Ca in NaCl or Al in ZnS, we assume that there is only one kind of acceptor level, an M vacancy. The donor is supposed to have a very low ionization energy and to produce electrons in the conduction band at concentration n, some of which fall into acceptors at concentration $[V_{\mathrm{M}}^{-}]$. The conservation of charge gives

$$[\mathrm{D}^{+}] = n + [V_{\mathrm{M}}^{-}] \tag{11}$$

The equilibrium within the crystal which we need to consider is

$$e + V_{\mathrm{M}} \longrightarrow V_{\mathrm{M}}^{-} \tag{12}$$

and its equilibrium constant can be written

$$[V_{\mathrm{M}}^{-}]/n[V_{\mathrm{M}}] = (1/N_c)\exp[(E_c - E_A)/kT] \tag{13}$$

where N_c is the density of states at the bottom of the conduction band, E_c is the energy of the conduction band, and E_A is the energy of the acceptor level in the M vacancy. Here $E_c - E_A$ is minus the ionization energy of V_M^-. Eliminating $[V_M^-]$ between (11) and (12) gives

$$[D^+]/n = 1 + ([V_M]/[N_c]) \exp[(E_c - E_A)/kT]$$
$$= 1 + ([V_M]/[N_c]) \exp[(E_g - E_A')/kT] \quad (14)$$

where E_g is the band gap energy and E_A' is the energy of the acceptor level above the valence band. The vacancy concentration V_M is controlled by adjusting the external pressure through the equilibrium

$$M_{(s)} \rightarrow V_M + M_{(g)} \quad (15)$$

where $M_{(s)}$ means the perfect metal lattice and $M_{(g)}$ means metal atoms in vapor phase. Then we have

$$V_M = (RTN_M/P_M) \exp(\Delta S_M/k) \exp(-\Delta H_M/kT) \quad (16)$$

where N_M is the number of metal atom sites per unit volume. Mandel's question reduces to asking whether or not the value of P_M can ever be made high enough to reduce $[V_M]$ to a small value. If so, then $[D^+]/n$ can approach unity, and the electrons go into the conduction band. When (16) is substituted into (14) one sees two important and opposing factors; the ratio $N_M/N_c \approx 1000$, favoring small values of n; and in the exponent $E_g - \Delta H_M$, which must be small enough to result in a negative exponent in order to overcome N_M/N_c if n is to have a high value. The heat of vacancy formation ΔH_M is not known accurately, but should vary from one crystal to another approximately as the cohesive energy E_{coh} of the crystal. Thus it was found that crystals for which $E_g/E_{coh} > 1$ always form traps (e.g., KCl), and those for which it is less than 0.5 do not (e.g., GaAs). These latter crystals act as semiconductors.

This example emphasizes the host crystal. The conditions under which the impurity atom actually donates an electron to the lattice is another but equally interesting question. Atoms with filled, stable inner shells having one more valence electron than the host atom being replaced are the D^+ of the previous example (Al in ZnS, Cd in KCl). They are sure to act as donors. (This does not mean there cannot be localized excitonlike states in the vicinity of these atoms. Such states exist, but the electron is usually ionized at moderate temperatures.) Atoms having incomplete inner shells, such as Ce at a Ca site in CaF_2, can exist in several charge states, depending on the chemical potential of the surrounding metal vapor. In ionic crystal notation we could have Ce^{2+} or Ce^{3+} or Ce^{4+}.

2.3.2.2. The Ratio of Ce^{2+} to Ce^{3+} in CaF_2

The reduction of Ce^{3+} at a Ca site in CaF_2 can be accomplished by exposing the crystal to a vapor pressure of Ca metal at about 600°C. Ce^{3+} is cerium with one $4f$ electron. The ground state of Ce^{2+} in CaF_2 is derived from the $4f$, $5d$ configuration.[90] Normally Ce^{3+} has a F_I^- charge compensator in a nearby interstitial site. The chemical reduction equation can be written

$$Ce_x^{III}Ca_{1-x}F_2F_x^{I} \longrightarrow Ce_x^{II}Ca_{1-x}F_2 + \tfrac{1}{2}xF_{2(g)} + \Delta H_1 \qquad (17)$$

where the solid solution of Ce^{3+} is shown as evaporating away its interstitial fluorine, but retaining an electron to produce Ce^{2+} in the lattice. The value of ΔH_1 can be written down in terms of the heats of solution discussed in Section 2.1. For the "reactant" solution

$$Ce_x^{III}Ca_{1-x}F_2F_x^{I} \longrightarrow xCeF_3 + (1-x)CaF_2 - \chi^{III}$$

and for the product

$$xCe^{II}F_2 + (1-x)CaF_2 \longrightarrow Ce_xCa_{1-x}F_2 + \chi^{II}$$

We subtract these from Eq. (17) to get

$$xCeF_3 \longrightarrow xCeF_2 + \tfrac{1}{2}xF_{2(g)} + \Delta H_2 \qquad (18)$$

and $\Delta H_2 = \Delta H_1 + \chi^{III} - \chi^{II}$.

The enthalpy of reaction (18) is just the difference between the standard heats of formation of CeF_2 and CeF_3: $\Delta H_2 = H(CeF_2) - H(CeF_3)$, and these are known, or can be estimated. Thus the enthalpy for the loss of the interstitial fluorine and the reduction of the metal, Eq. (17), is

$$\Delta H_1 = H(CeF_2) - H(CeF_3) - \chi^{III} + \chi^{II} \qquad (19)$$

Both χ^{III} and χ^{II} are expected to be small, as suggested by the results of Sections 2.1.1 and 2.1.5. Now, $H(CeF_3) = -416$ kcal/mole[91] and $H(CeF_2)$ is taken to be the same as for CaF_2, -290 kcal/mole, so we have $\Delta H_1 \cong 126$ kcal/mole at 298°K.

The actual chemical process uses Ca metal vapor, so the ΔH_1 will be reduced by half the heat of formation of CaF_2 and will be -19 kcal. There does not seem to be any experimental test of this figure, but its sign and value are reasonable considering the chemical process which occurs.

There is a spectroscopic analog of the oxidation process in which Ce^{2+} is converted to Ce^{3+} with an electron in the conduction band.

With light of energy greater than about 1.5 eV, divalent rare earth-doped CaF_2 becomes photoconducting.[92] The process is

$$Ce^{2+} \longrightarrow Ce^{3+} + e_{cr} + \Delta H(\text{spect}) \tag{20}$$

or in the solid solution notation

$$Ce_x{}^{II}Ca_{1-x}F_2 \longrightarrow Ce_x{}^{III}Ca_{1-x}F_2 + e_{cr} \tag{21}$$

This process differs from the chemical one in that Ce^{III} is uncompensated by any lattice defect and there is an electron in the conduction band instead of $F_{2(g)}$ as a reactant. The relation between the two can be shown by adding the following reactions to Eq. (17):

$$F^-_{(g)} \longrightarrow F_I{}^- \qquad H(F_I{}^-) \tag{22}$$

$$e_{(g)} + \tfrac{1}{2}F_{2(g)} \longrightarrow F^-_{(g)} \qquad H = -63.5 \text{ kcal/mole} \tag{23}$$

$$e_{(g)} \longrightarrow e_{cr} \qquad -A \tag{24}$$

where A is the electron affinity of the crystal. The sum of (21)–(24) equals the reverse of the process shown in Eq. (17), and therefore $-\Delta H_1 = \Delta H(\text{spect}) + H(F_I{}^-) - A - 63.5$. In Section 2.1, Table 4, $H(F_I{}^-)$ is given as -75 kcal. Therefore, inserting the value of ΔH_1, we have

$$\Delta H(\text{spect}) = A + 12.5 \text{ kcal} \tag{25}$$

The observed $\Delta H(\text{spect}) \cong 35$ kcal. We do not know A, but a value of 1 eV as given by this calculation is reasonable. There are, however, some other factors to take into account. The relaxation energy of the Ce atom in the crystal when the electron is removed could be appreciable. This energy does not appear in the spectroscopic measurement, so $\Delta H(\text{spect})$ should be larger than is given by Eq. (25). But in view of our ignorance of A, it is not possible to check on the effects of relaxation.

Calculations involving thermochemical cycles in the solid state such as we have just presented usually end with one or more vital quantities not being accurately known. In the above example the two heats of solution were estimated to be zero, a calculated value for the energy of interstitial fluoride was used, and the electron affinity of the crystal was not known to better than 1 V. Such results are unsatisfactory in some respects, but they do at least put reasonable limits on some interesting and important quantities.

Actual studies of charge conversion in CaF_2 have been made by Merz and Pershan.[93] They did not study this system at thermodynamic equilibrium, however, but instead measured the activation energy for the thermal reoxidation of $CaF_2:Re^{2+}$ produced by X irradiation. They showed that the activation energy measured was that for hole motion; the holes are self-trapped (V_K centers) at low temperatures but move at an appreciable rate above 140°K.

The two examples given here should provide some hint of the link between the electrical, optical, and thermodynamic quantities which characterize a defect solid. Other chapters in these volumes and works referenced here cover these areas more systematically. It is a fact, however, that the changes described by thermodynamic quantities differ in essential ways from those described by spectroscopic quantities. The latter can usually be described by single-particle transitions, while the former are many-particle changes. This difference tends to keep these areas apart. The following sections of this chapter will be largely devoted to spectroscopy, but the link to thermochemical quantities will be mentioned whenever possible.

3. Electronic States of Impurity Ions in Crystals

A theoretical description of electronic levels of defects in crystals is necessary for understanding and correlating spectroscopic and other observations. A theoretical explanation may aim at various levels, from simply relating one piece of data, to another to purporting to solve the Schrödinger equation for the system. Our main interest in this chapter is to establish a first-order description of the electronic states of many different types of defects in crystals. We will not attempt a detailed treatment by rigorous quantum mechanical theories.

The principal ways to learn about the electronic structure of defects are the various methods of spectroscopy. Electrical and thermodynamic measurements are also important. The emphasis here will be on electronic spectroscopy since it gives the first-order description of the electronic structure most directly.

The transition metal and rare earth ions permit one to obtain a wealth of spectroscopic information, far more than for the other classes of impurity ions mentioned in the introduction. The description of the inner shell transitions of these ions by way of crystal field theory is also much more complete and accurate than for other types of transitions. This unevenness in our knowledge is represented in this chapter by the long section on crystal field spectra compared to shorter sections on descriptions of other types of optical spectra. We begin this section with a brief discussion of spectroscopy in the solid state.

3.1. The Absorption and Emission of Radiation

3.1.1. Absorption Spectra

The optical properties of a crystal can be characterized by an index ellipsoid whose three orthogonal principal axes are the three independent refractive indices at a given wavelength.[94] Light can travel through the crystal in such a way that only one of these refractive indices acts on it. Absorption spectra are done this way in order to simplify the interpretation of the data. The experiment is done by cutting a crystal in the form of a plate parallel to one of the principal planes of the index ellipsoid, and then sending in light at normal incidence polarized along one or the other of the principal indices. Two such plates are needed for crystals having less than tetragonal symmetry in order to find the three absorption coefficients $\alpha_x, \alpha_y, \alpha_z$. For each direction we have

$$dI = \alpha_x I \, dl \tag{26}$$

leading to the equation for the optical density

$$\rho \equiv \log(I_0/I) = l\alpha_x/2.303$$

Since an optical density greater than four cannot ordinarily be measured, this equation shows the practical limit of crystal thickness for a given absorption coefficient.

The absorption coefficient can be written as $\alpha_x = N\sigma^x$, where σ^x is the cross section in cm^2 and N is the number of absorbing sites per cm^3. The cross section is a function of the transition energy E and is given by Dexter[95] as

$$\sigma^x_{m\to k}(E) = \frac{4\pi^2 e^2}{c\hbar} x^2_{mk} E_{mk} S^x_{mk}(E) \frac{1}{n_x}\left(\frac{E^x_{\text{eff}}}{E_0{}^x}\right)^2 \tag{27}$$

where x_{mk} is a component of the transition moment vector:

$$x_{mk} = \int U_m{}^* \sum_i x_i U_k \, d\tau \tag{28}$$

and the U are the steady-state wave functions of the initial and final states. In (27) $S^x_{mk}(E)$ is a shape function, which integrates to unity over an electronic absorption region, $\int S^x(E)\, dE = 1$, and contains significant information about the transition. Examples of the shapes of absorption bands are given in Figures 7–9, and will be discussed later. n is the refractive index and $(E^x_{\text{eff}}/E_0{}^x)$ is an internal field correction.

This last term shows one of the difficulties in trying to distinguish rigorously between the absorbing center and the rest of the crystal. $E_0{}^x$ is the average electric field in the crystal due to the x-polarized light wave which is causing the transition, while E_{eff}^x is the field effective at the center where the transition occurs. This effective field includes the reaction field due to the polarization produced in the medium by the change in the charge distribution of the center making the transition. For a very diffuse center, or for very weak transitions, $E_{\text{eff}}/E_0 = 1$, and for strong dipole transitions the Lorentz factor is appropriate: $1 + (n^2 - 1)/3$. For less clear-cut cases one does not know what to do except set it equal to one. This correction would disappear, however, if the crystal were treated as if it were a large molecule, the only rigorously correct way to view the impurity in a

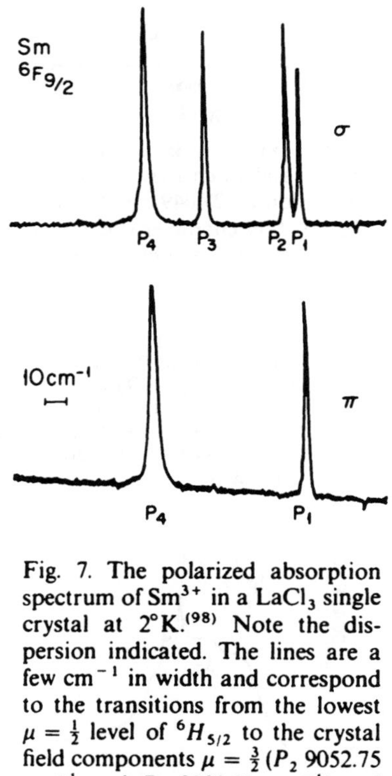

Fig. 7. The polarized absorption spectrum of Sm^{3+} in a $LaCl_3$ single crystal at 2°K.[98] Note the dispersion indicated. The lines are a few cm^{-1} in width and correspond to the transitions from the lowest $\mu = \frac{1}{2}$ level of ${}^6H_{5/2}$ to the crystal field components $\mu = \frac{3}{2}$ (P_2 9052.75 cm^{-1} and P_3 9088.21 cm^{-1}) and $\mu = \frac{5}{2}$ (P_1 9047.57 cm^{-1} and P_4 9114.12 cm^{-1}) of ${}^6F_{9/2}$. This is a typical weak $4f \rightarrow 4f$ spectrum. The f number of the line P_2, for example, is 8.7×10^{-8}.[98]

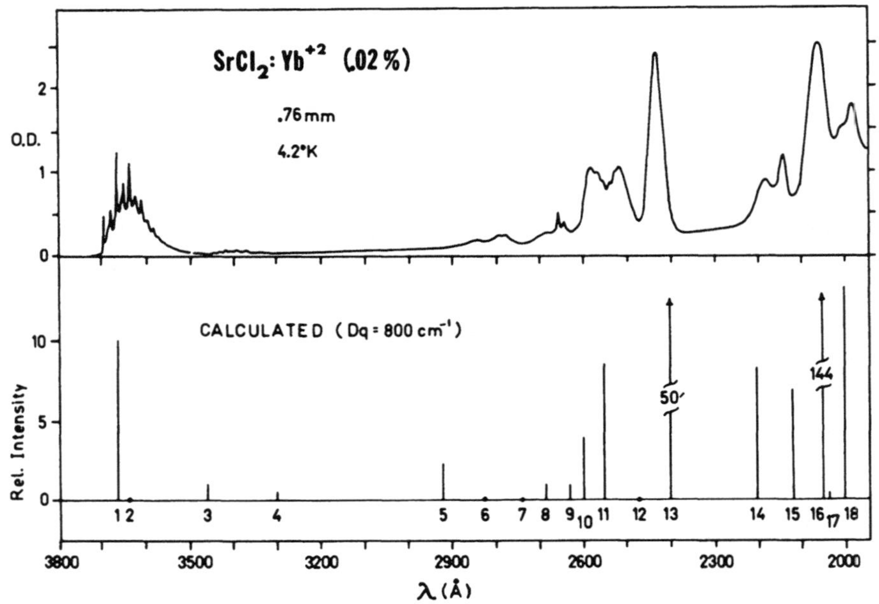

Fig. 8. Absorption spectrum of Yb^{2+} in $SrCl_2$ single crystal at 4.2°K. Below are shown the calculated energies and intensities of the 18 allowed transitions. The origin band of the first transition at 27,049 cm^{-1} is only a few cm^{-1} wide. The vibrational spacing of 210 cm^{-1} is characteristic of the $SrCl_2$ lattice. This is a typical $4f \rightarrow 5d$ spectrum having an f number for band 1, for example, of 10^{-2}, and a total f of about 0.2.[102]

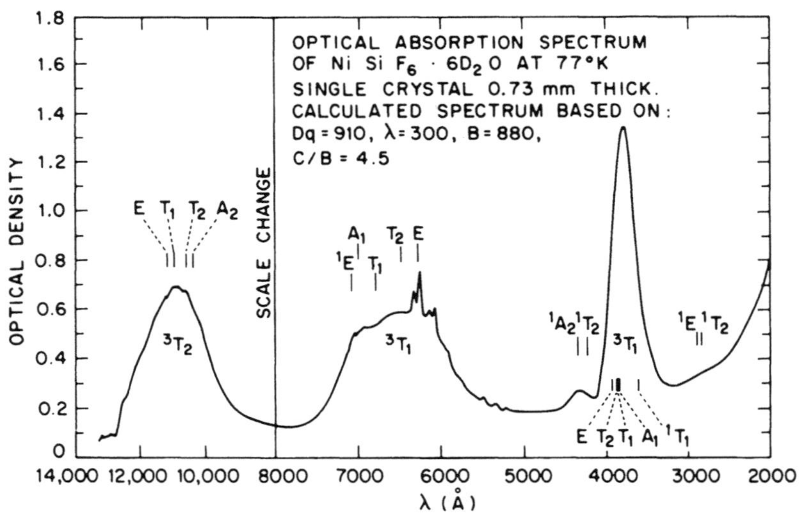

Fig. 9. Absorption spectrum of single crystal $NiSiF_6 \cdot 6H_2O$ at 77°K. (The long-wavelength part is of $NiSiF_6 \cdot 6D_2O$.) The level positions calculated with crystal field theory parameters are also shown. The f number of the band at 6500 Å and 4.2°K is 2.7×10^{-4}. This is a typical $3d \rightarrow 3d$ spectrum.[101]

crystal. An example of a case in which the medium had to be treated explicitly is the pseudo-quadrupole transition in rare earth ions in crystals or solutions.[96]

The factor $1/n_x$ appears because the energy density of radiation in the crystal is greater by a factor of n^2 than in vacuum but the photon flux is reduced by a factor of n; upon dividing by n, we attribute a smaller cross section to the center, and the derived transition moment can be compared with that of the analogous absorber in a vacuum or with the transition moments of similar centers in different crystals.

A quantity of immediate interest derivable from the absorption cross section is the electronic transition moment x_{mk}. By the use of polarized light the absorption coefficients in the directions permitted by crystal optics[94] may be found. If the absorption region corresponding to the electronic transition of interest can be identified, then the transition moment for a particular polarization can be obtained from experiment as follows:

$$\bar{x}_{mk} = \left[n_x \left(\frac{E_0{}^x}{E^x_{\text{eff}}} \right)^2 \frac{\hbar c}{4\pi^2 e^2} \int \frac{\sigma^x_{mk}(E)\, dE}{E} \right]^{1/2} \tag{29}$$

(The shape function disappears if x_{mk} is considered constant throughout the absorption band.) The quantity $\bar{x}_{mk}$ obtained from experimental data through Eq. (29) can be compared to its theoretical value from Eq. (28). Table 5 gives examples of various types of localized transitions in impurities in crystals. It is arranged in order of f number and shows that this quantity can range over ten orders of magnitude. The corresponding transition dipole moment matrix element x_{mk} is also given. These examples show that different types of transitions have very different absorption strengths. Observed absorption strengths provide one important way to identify an absorption feature.

A quantity often quoted is the oscillator strength, although the transition moment is more directly related to theory and is available from experiment through Eq. (29). The oscillator strength f is

$$f_{mk} = (2m/\hbar)\omega_{mk} \langle x_{mk} \rangle^2 \tag{30}$$

for a one-dimensional oscillator. If the expression for x^2_{mk} from Eq. (29) is inserted and if the E in the denominator of the integral is set equal to an average energy expressed as $\hbar\omega_{mk}$, then the f number can be derived from experimental data as follows:

$$f_{mk} = n \left(\frac{E_0}{E_{\text{eff}}} \right)^2 \frac{mc^2}{e^2} \int \sigma^x \, d\bar{\nu} \tag{31}$$

where $\bar{\nu}$ is now the frequency expressed in wave numbers.

TABLE 5

Dipole Transition Moment, f-Number, and Bandwidth for Several Types of Localized Transitions in Crystals*

System and transition†	Transition characteristics‡	Average frequency, cm^{-1}	Bandwidth, cm^{-1}	f Number	Transition moment, Å	Reference
$RbMnF_3$, Mn^{2+}, 2°K $^6A_{1g} \rightarrow {}^4T_{1g}(I)$	$3d \rightarrow 3d$, MD zero phonon line	18,221	3	2.5×10^{-11}	6.5×10^{-6}	142
$CaF_2:Dy^{2+}$, 4°K $^5I_8(T_2) \rightarrow {}^5I_7(T_1)$	$4f \rightarrow 4f$, MD line	4,239.8	0.07	1.0×10^{-7}	0.0008	262
$LaCl_3:Sm^{3+}$, 4°K $^6H_{5/2}(\frac{1}{2}) \rightarrow {}^6F_{9/2}(\frac{3}{2})$	$4f \rightarrow 4f$, MD line	9,052.8	2	8.7×10^{-8}	0.0005	98
$Al_2O_3:Cr^{3+}$ (ruby), 4°K $4A_2 \rightarrow {}^2E(\bar{E})(R_1)$	$3d \rightarrow 3d$ ED spin-forbidden line	14,421.12	0.2	7.5×10^{-7}	0.0009	263
$RbMnF_3$, Mn^{2+}, 2°K $^6A_{1g} \rightarrow {}^4T_{1g}(I)$	$3d \rightarrow$ ED magnon sideband	18,285	25	4.7×10^{-9}	8.8×10^{-5}	142
$Al_2O_3:Cr^{3+}$ (ruby), 4°K $^4A_2 \rightarrow {}^4T_2$	$3d \rightarrow 3d$, ED spin, symmetry allowed	18,450	2500	3.9×10^{-4}	0.026	264
$CaF_2:Ce^{3+}$, $-F_I^-$, 4°K $^2F_{5/2}(G, E_{3/2}) \rightarrow {}^2E\,(E_{3/2})$	$4f \rightarrow 5d$, ED allowed band	33,000	1000	4.3×10^{-3}	0.063	265
$KMgF_3:Mn^{2+}$, 4°K $^6A_{1g} \rightarrow {}^6A_{1g}$	$3d \rightarrow 4s$, ED vibronic band	60,600	3300	6.3×10^{-4}	0.017	191
$KMgCl_3:Ni^{+2}$, 77°K $^3A_{2g} \rightarrow {}^3T_{1g}(p)$	$3d \rightarrow 3d$ ED vibronic band	20,400	1300	2.2×10^{-4}	0.018	104
$Cs_2ZrCl_6:Ir^{4+}$, 4°K $^2T_{2g}(E'') \rightarrow {}^2T_{1u}(U_u')$	hal $\rightarrow 5d$, ED CT	19,500	800	0.11	0.41	114
Fe(II)Fe(III)$(CN)_6$ First band	Mixed valence ED	14,000	6000	0.7	1.2	266

* The formula for the one-dimensional oscillator strength is used; data in the literature have been corrected to follow this formula. ε is the molar decadic extinction coefficient $= (1/cl)\log(I_0/I)$, c in moles/liter, l in cm, $\tilde{\nu}$ in cm^{-1}. $F = 1.30 \times 10^{-8} \int \varepsilon\, d\tilde{\nu}$, $\mu^2 = 3.95 \times 10^{-4} \int \varepsilon\, d\nu/\nu\ (Å^2)$.

† The ground $\rightarrow$ excited state designations are given. See references for details of these designations.

‡ ED = electric dipole, MD = magnetic dipole.

In case the transition is caused by the magnetic field of the light wave, the transition moments to replace Eq. (28) are of the type

$$y_{mk}(\text{magnetic}) = \int U_m \sum_i \frac{L_{yi} + 2S_{yi}}{2mc} U_k \, d\tau \tag{32}$$

The internal field correction should be $(H_{\text{eff}}/H_0)^2$, but the refractive index appears in the same places, properly corrected for polarization, which is specified by the direction of the *electric* vector. This internal field correction is usually set equal to unity.

Both the transition moment and the spectral shape function give important information about the initial and final states of the spectroscopic transition.[97] Their product occurs in the expression for the transition between two vibrational electronic states. The electronic vibrational wave functions are expressed as products according to the Born–Oppenheimer approximation:

$$\psi_{ev} = U(r_i, q_j^0)V(q_j) \tag{33}$$

where r_i represent electron coordinates and U is evaluated for a fixed configuration q_j^0 of the nuclei, V is the vibrational wave function, and q_j denotes the vibrational coordinates. The transition moment between two electronic states m and k with vibrational levels α and β can be written

$$\begin{aligned} x_{mk,\alpha\beta} &= \int U_m{}^* V_{m\alpha} \sum (x_i + q_i) U_k V_{k\beta} \, d\tau_e \, d\tau_v \\ &= \int U_m{}^* \sum x_i U_k \, d\tau_e \int V_{m\alpha} V_{k\beta} \, d\tau_v = x_{mk} S_{mk\alpha\beta} \end{aligned} \tag{34}$$

where the "Condon approximation" made in the second step implies that the electronic transition moment is independent of the nuclear configuration in the vicinity of the configuration q_j^0. When summed over all initial states α and final states β the shape function $S_{mk\alpha\beta}$ gives unity.

The band shape shown in Figure 7 for a $4f \longrightarrow 4f$ transition in Sm^{3+} [98] is characteristic of an electronic transition in which almost no vibrations are excited in the surrounding lattice. The finite linewidth of several cm^{-1} has not been interpreted, but may be caused by interaction with low-frequency lattice waves. In this case the shape function is very narrow, and α and β would refer to different wave vectors of the phonon branch of the host crystal.

An example of somewhat stronger phonon sideband structure is the fluorescence spectrum of Cr^{3+} in MgO.[99] The band shape has been interpreted by Sangster,[100] who shows that S_{mk} closely resembles the density-of-states function of the MgO lattice.

Figure 9 shows the very broad $3d \rightarrow 3d$ bands of Ni^{2+}, in which it is not possible to identify the origin band at all. Even the vibrational structure is too complex to resolve and analyze.[101]

Figure 8 shows several band types in Yb^{2+} in $SrCl_2$, a $4f \rightarrow 5d$ transition. The origin of the first of these bands is very sharp ($\sim 1\ cm^{-1}$ half-width) and several kinds of sidebands forming a progression of at least five members each can be seen. The band shape function in this case involves several quanta of a few specific vibrations rather than the entire density of vibrational states, even though these special vibrational modes are heavily excited.[102]

The expression for the electronic transition moment, Eq. (34), implies selection rules which can be briefly stated as: The product of the representations of the wave functions U_m and U_n must contain the representation of the transition moment operator x. If x is the electric dipole operator, the transitions occur between states of opposite parity. The behavior of x_{mk} toward other symmetry operations can be determined from tables of representations of point groups.[103] Table 6 gives the often used cubic representations.

If the wave function explicitly involves electron spin, there are several cases to consider: (1) Often the spin angular momentum will be the same in the initial and final states and the selection rule on spin

TABLE 6
Character Table for the Groups T_d and O*

T_d / O	E / E	$8C_3$ / $8C_3$	$3C_2$ / $3C_2$	$6S_4$ / $6C_4$	$6\sigma_d$ / $6C_2'$	Representations of some tensor components
A_1	1	1	1	1	1	$xx + yy + zz$
A_2	1	1	1	−1	−1	
E	2	−1	2	0	0	$xx + yy - 2zz, xx - yy$
T_1	3	0	−1	1	−1	x, y, z in O; $R_x R_y R_z$
T_2	3	0	−1	−1	1	x, y, z in T_d; xy, xz, yz

Spinor representations

	E	R	$8C_3$	$8RC_3$	$3C_2 + 3RC_2$	$6C_4$	$6RC_4$	$6C_2' + 6RC_2'$
$E_{1/2}$	2	−2	1	−1	0	$\sqrt{2}$	$-\sqrt{2}$	0
$E_{5/2}$	2	−2	1	−1	0	$-\sqrt{2}$	$\sqrt{2}$	0
G	4	−4	−1	1	0	0	0	0

* $O_h = O \times i$; E = identity; C_3 = threefold axis; C_4 = fourfold axis; $C_2 = C_4^2$; C_2' = twofold (110) axes; R = rotation by 2π.

is $\Delta m_s = 0$. The orbital part of the transition moment integral obeys the normal selection rules. (2) If there is a change in spin angular momentum, the transition is forbidden in a first approximation, and the wave functions will have to be expanded to include the effects of spin–orbit coupling in order to find nonvanishing moments. (3) If there is an even number of electrons, the symmetry of the complete wave function is the spin–orbit product symmetry, which is found by the reduction of the product of the representations of the spin part and the orbit part. The overall selection rules (including spin–orbit effects) are determined by the product of these symmetries for the initial and final states. (4) If there is an odd number of electrons, the representations will fall into the spinor classes of the group, and the expanded symmetry group must be used to determine the selection rules.[103] The extra spinor representations for the cubic group are shown in Table 6.

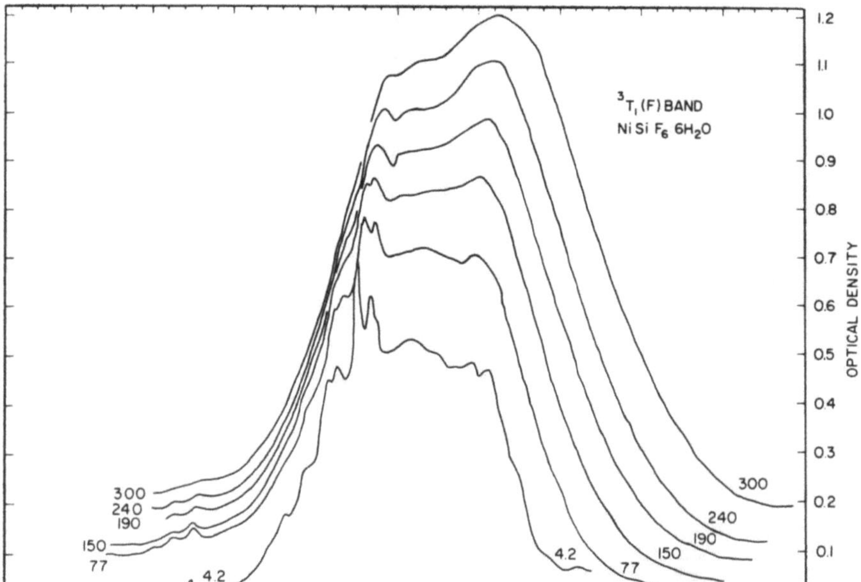

Fig. 10. Temperature dependence of the $^3A_2 \rightarrow {}^3T_1$ absorption band of $NiSiF_6 \cdot 6H_2O$. Its intensity more than doubles between 4.2 and 300°K, but does not accurately follow a $\coth(h\nu/2kT)$ law. Nevertheless, the strong temperature dependence shows that the intensity is vibronically induced.[101]

Transitions occurring between the states of the same atomic configuration, such as $3d \rightarrow 3d$ or $4f \rightarrow 4f$, are forbidden as electric dipole transitions because both states of the transition have the same parity (see Figures 7 and 9). In a crystal the parity may be destroyed by the surroundings of the ion, but the transitions are still quite weak compared to fully allowed transitions between states of opposite parity in the atom, such as $3d \rightarrow 4f$, as several examples in Table 5 show. If the environment of the impurity ion is centrosymmetric, then the transition only becomes allowed through the effect of odd-parity vibrations.[104] Since the vibrational displacement increases with quantum number, these types of transitions are stronger, the higher

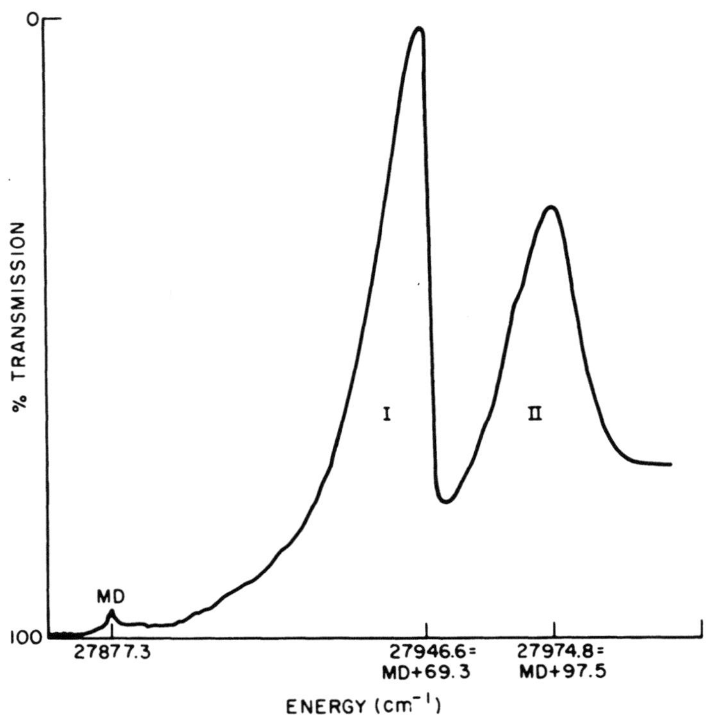

Fig. 11. The magnetic dipole origin (MD) and two spin-wave sidebands of the transition $^6A_{1g} \rightarrow {}^4T_{2g}$ (II) at 27,877.3 cm^{-1} in $RbMnF_3$ single crystals at 2°K.[142]

the temperature. The absorption coefficient then has the following temperature dependence:

$$\alpha(T) = \alpha(0)\coth(h\nu/2kT) \tag{35}$$

The presence of a temperature effect is another important diagnostic criterion as to the type of transition being observed. A typical temperature-dependent band is shown in Figure 10.

Another way for forbidden transitions to become allowed is by way of exchange coupling between pairs of like atoms.[105–107] For example, in MnF_2 the spin-forbidden $3d \rightarrow 3d$ transitions are several orders of magnitude more intense than in isolated Mn ions, such as Mn in $KMgF_3$ (Table 5). These transitions originate on a spin-wave sideband, a feature which can be recognized in an antiferromagnetic crystal by its response to a magnetic field[108] and by its shape.[109] An example is shown in Figure 11.

3.1.2. Modulation Spectroscopy

Absorption spectra are often less informative than one would like, and in order to extract more information about excited states, one must use some method for perturbing the spectrum and measuring the response to this perturbation. The application of electric or magnetic fields, usually accompanied by ac detection methods, has made spectroscopy into an entirely new field during the past ten years. Many aspects of this revolution have been reviewed by Cardona.[110] Magnetic-circular dichroism, now one of the most widely used of these methods, has been reviewed by Buckingham and Stephens,[111] Schatz and McCaffery,[112] and Stephens.[113]

In magnetic-circular dichroism (MCD) one measures the quantity $\alpha_L - \alpha_R$, the difference between the absorption coefficients for left and right circularly polarized light induced by a magnetic field directed along the path of propagation of the light. The measurement can only be done in isotropic or uniaxial materials. The MCD measured in a material has a magnitude and sign, a magnetic field dependence, and a temperature dependence. These variables provide criteria which, in connection with the absorption spectrum, enable assignments of symmetry of ground and excited states to be made. An example from Schatz's group[114] is shown in Figure 12, the beginning of the charge-transfer spectrum of Ir^{4+} in Cs_2ZrCl_6.

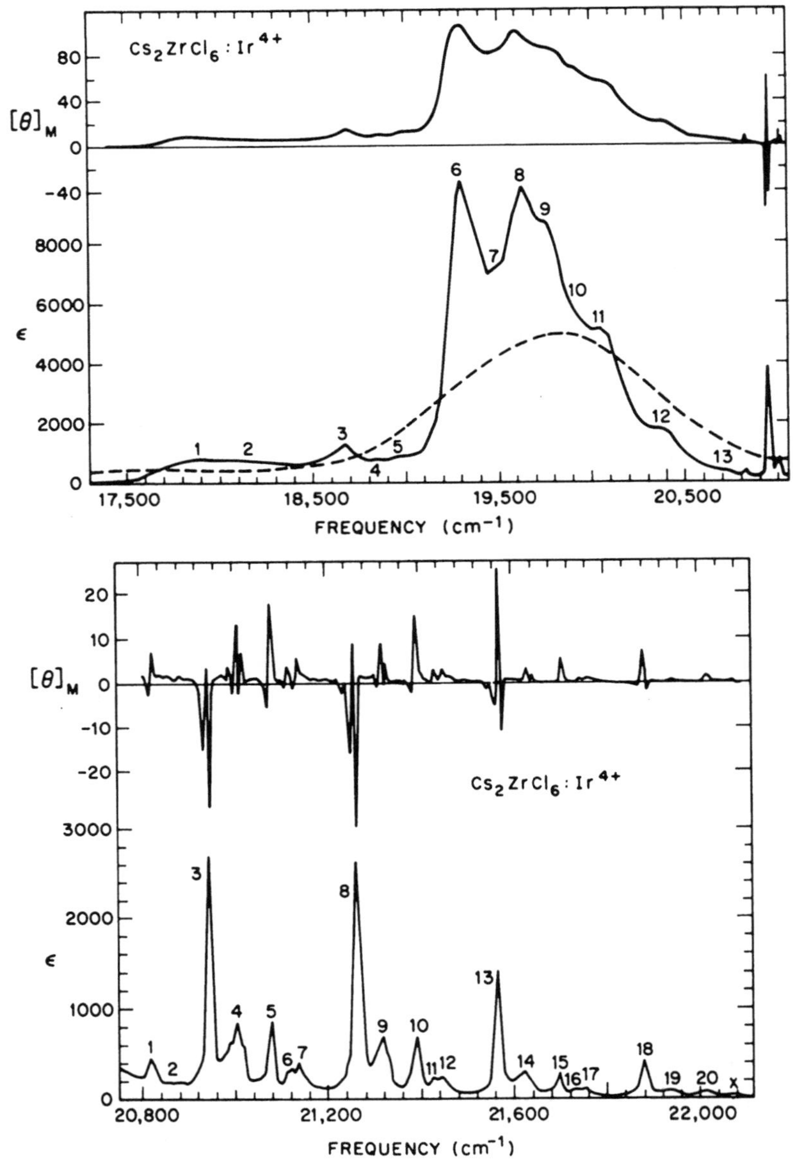

Fig. 12. The absorption and magnetic-circular dichroism spectrum of $Cs_2ZrCl_6:Ir^{4+}$ single crystal at about 5°K. Three transitions (at least) are shown, and they are identified as follows (upper section): (1, 2)$E_g''(^2T_{2g})$ (ground state) ⟶ $U_g'(^2T_{1g})$, a weak, parity-forbidden transition (see Figure 36 and Ref. 195 for notation); (3) ⟶ $^2E_g''(^2T_{1g})$; (6–13) ⟶ $U'(^2T_{1u})$, a strongly allowed transition. In the lower part of the figure, the spectral region shows the parity-forbidden transitions to $U_g'(^2T_{2g})$ and $E''(^2T_{2g})$. The MCD data have eliminated some of the competing assignments of these spectra.[114]

3.1.3. Emission Spectra

Radiation may be emitted from an excited state of an ion in a crystal either spontaneously or in response to a radiation field already present.[115] An excited state may also decay nonradiatively by interacting with the vibrational levels of the crystal. The radiative emission probability for a transition from upper level i to lower level j is

$$dn_i/dt = n_i A_{ij} + (n_i - n_j) B_{ij} U(\nu_{ij}) \tag{36}$$

where n_i and n_j are the number of atoms in levels i and j per unit volume, $U(\nu_{ij})$ is the radiation density, and A_{ij} and B_{ij} are coefficients of spontaneous and induced emission, respectively. The usual situation with low-density radiation is that only the first term is effective. Then A_{ij} can be thought of as the first-order rate constant for the radiative decay process. The reciprocal lifetime of the upper state is

$$\sum_j A_{ij} = 1/\tau_i \tag{37}$$

The A_{ij} and B_{ij} are related to each other, and therefore the spontaneous emission rate $i \longrightarrow j$ is proportional to the absorption cross section for the process $j \longrightarrow i$. The oscillator strength and lifetime are approximately related as

$$f = 1.5/\nu^2 \tau \tag{38}$$

where ν is in cm^{-1}, τ in seconds.

The B_{ij} is either an absorption or an emission coefficient depending on whether $n_i < n_j$ or $n_i > n_j$. If $n_i \approx 0$, then the usual conditions for absorption spectroscopy prevail:

$$dn_i/dt = -n_j B_{ij} U(\nu_{ij}) = -dn_j/dt \tag{39}$$

This equation leads to Eq. (1), Beer's law, with $\sigma = hB/\lambda$.

On the other hand, if $n_i > n_j$, then the system is in a condition to exhibit forced emission and instead of an absorption coefficient there is a gain coefficient, $(n_i - n_j)hB/\lambda$ cm^{-1}.

In a practical laser device the excited-state population n_i is produced by an intense light source (or in semiconductor lasers, an electric current), called the pump source. When the rate of forced emission exceeds the spontaneous rate, then laser action is possible, i.e., when

$$(n_i - n_j) B_{ij} U(\nu_{ij}) > n_i A_{ij} \tag{40}$$

In a laser cavity one or a few particular radiation modes, that is, one frequency, polarization, and direction of propagation, are reinforced above the others, whereas in spontaneous emission there is no such selection.

Radiationless loss of energy from excited states may often compete with radiation. The spontaneous emission quantum yield is the ratio of the radiative rate to the sum of all loss rates:

$$\Phi = k_r/(k_r + k_n) \tag{41}$$

where k_r and k_n are the radiative and nonradiative decay rates ($k_r = \sum_j A_{ij}$). For ions in crystals the nonradiative rate depends on the coupling between the electronic and the vibrational modes of the crystal. The smaller the interval between a pair of electronic levels, the greater the mixing caused by vibrations, and the more efficient the radiationless process. The intervals between successive pairs of excited levels is usually much less than the interval between the ground state and the lowest level, so it is often found that emission

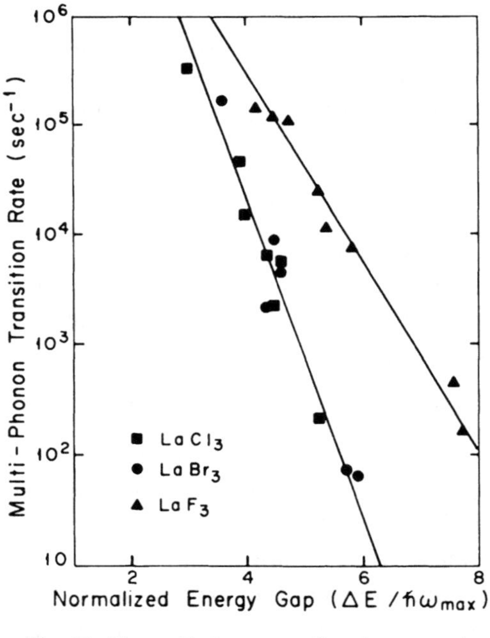

Fig. 13. Nonradiative rates of various transitions of rare earth ions in lanthanum halide host crystals plotted as a function of the energy gap ΔE (energy to next lowest level) divided by the maximum phonon frequency of the lattice $\hbar\omega_{max}$.[116]

occurs from only the lowest of the excited states. The quantum efficiency is usually greater for large energy intervals between the ground and first excited states, than for small ones.

The nonradiative transition rates k_n in rare earth ions in $LaCl_3$ and similar host crystals were measured by Moos and his collaborators, and reviewed by Moos.[116] Figure 13 from this review shows the measured radiationless rates from an excited level to the next lower level as a function of the energy gap between the levels. The abscissa is actually $\Delta E(\text{gap})/\hbar\omega_{max}$, where ω_{max} is the highest optical frequency of the lattice. The data for many different ions in $LaCl_3$ and $LaBr_3$ fall on one straight line on this logarithmic plot. The data for ions in LaF_3 fall on another line; but the crystal structure of this host is different. These results show that radiationless transitions are caused by the conversion of electronic energy into the energy of several phonons. The fastest conversions are those requiring the smallest number of phonons, i.e., the smallest electronic energy gaps.

3.2. Electronic Structure and Spectroscopy of the Transition Group Ions

The theoretical methods for discussing spectra and electronic structures of the *d* and *f* transition group atoms in solids do not differ in principle from those used to treat molecules; at present the self-consistent field method is the closest we can come to the true solution of the Schrödinger equation for any of these systems. Since our interest here is to interpret the observed energy level scheme rather than to reproduce it à priori, we will spend more space discussing methods which specifically describe the inner shell levels of transition group ions. These are called crystal field theory, or ligand field theory, or molecular orbital theory, in order of increasing detail in the treatment of the atoms surrounding the transition metal atom. These methods have been discussed in many other places and therefore will be only briefly described here. The principal sources for further treatments are Refs. 4, 103 and 117–119.

3.2.1. Crystal Field Theory

A "crystal field" is a model potential acting on the electrons of an atom or molecule embedded in a crystal. It is a static potential which has been constructed so as to give the best account of the energy levels of the embedded atom. The static potential is a substitute for some very complicated electronic motions and the fact that it is highly successful is of enormous importance. "Crystal field theory," as it is called, ranks along with the band theory of solids and the Hückel-type theories of molecules in its simplicity, accuracy, and wide range of application. These "theories" are actually models since they depend on parameters which must be constructed by means of other calculations or obtained from experiment. They have only a tenuous relation to ab initio methods. It is safe to say, however, that these less fundamental descriptions are far more useful and correlate much more data than have attempts to solve the Schrödinger equation by self-consistent fields or other approximations.

When applied to ions in crystals, crystal field theory is used to deal with the inner *d* or *f* shells of transition group atoms. The theory can be carried out at several different levels of complexity, depending on the purpose for which it is being used. The lowest form could be called the pictorial method, and it will be presented without apology, since it always gives the right answers to an important group of questions. Ascending from this level, one can work up to the solution of high-order secular equations involving electron–electron interactions, spin–orbit interaction, exchange and magnetic effects, and so on. Applications at this level involve detailed interpretations of spectroscopic data, for which the power of the method is fully exploited.

3.2.1.1. One Electron in a Cubic Crystal Field

Crystal field theory is used to predict the magnitude and direction of the shifts and splittings of *d* and *f* levels of atoms in solids. Table 7 shows the *d* wave functions in the central field approximation in two kinds of quantization; the first is suitable in an axial magnetic field and is of the familiar $|lm_l\rangle$ form; the second is adaptable to a cubic field such as would be found at a lattice site in many types of crystals. We will dwell on the latter since it has wide application to transition metal atoms in crystals.

There are two types of *d* wave functions in a cubic field, represented by the set (u, v) and (ξ, η, ζ). The symmetry operations of a cube (Table 6) convert u and v into linear combinations of each other and also interconvert ξ, η, ζ; but no such symmetry operation ever transforms a member of one set into any combination of the other set. The u, v set is called the *e* type, since it belongs to the E (or Γ_3) representation of the cubic point group; and the other set is called t or t_2, since it

TABLE 7

3d Wave Functions: $u_{3d}(r, \theta, \varphi) = R_{3d}(r)\Theta_{lm}(\theta)\Phi_m$

In free atom

l	m_l	Θ_{lm}	Φ	Dirac notation $\|lm_l\rangle$
2	2	$(\sqrt{15}/4)\sin^2\theta$	$e^{i2\varphi}\sqrt{2\pi}$	$\|22\rangle$
2	1	$-(\sqrt{15}/2)\cos\theta\sin\theta$	$e^{i\varphi}/\sqrt{2\pi}$	$\|21\rangle$
2	0	$(\sqrt{5}/2\sqrt{2})(2\cos^2\theta - \sin^2\theta)$	$1/\sqrt{2\pi}$	$\|20\rangle$
2	-1	$(\sqrt{15}/2)\cos\theta\sin\theta$	$e^{-i\varphi}/\sqrt{2\pi}$	$\|2-1\rangle$
2	-2	$(\sqrt{15}/4)\sin^2\theta$	$e^{-2i\varphi}/\sqrt{2\pi}$	$\|2-2\rangle$

In cubic field

$$u = |20\rangle = (5/16\pi)^{1/2}(3z^2 - 1)\ R_{3d}(r)$$
$$v = (1/\sqrt{2})(|22\rangle + |2-2\rangle) = (15/16\pi)^{1/2}(x^2 - y^2)\ R_{3d}(r)$$
$$\xi = (i/\sqrt{2})(|21\rangle + |2-1\rangle) = (15/4\pi)^{1/2}yz\ R_{3d}(r)$$
$$\eta = -(1/\sqrt{2})(|21\rangle - |2-1\rangle) = (15/4\pi)^{1/2}zx\ R_{3d}(r)$$
$$\zeta = -(i/\sqrt{2})(|22\rangle - |2-2\rangle) = (15/4\pi)^{1/2}xy\ R_{3d}(r)$$

belongs to the T_2 (or Γ_5) representation. The characters of the representations are shown in Table 6.

The pictorial method of applying crystal field theory begins with d-orbital diagrams such as those of Figure 14. These diagrams follow from the e, t_2 orbitals of Table 7. Even in a solid, the d-orbital shapes are not badly distorted, since the largest force acting on the electron is still the central field. Therefore it is legitimate to consider the surroundings of the atom as a perturbation, and indeed this is what is done in the mathematical treatments of crystal field theory. Figure 15 shows an e- and a t-type orbital in a rectangular coordinate system so that the locations of atoms in fourfold, sixfold, and eightfold cubic coordination can be shown. These coordination numbers correspond to tetrahedral, octahedral, and cubic arrays of anions around the central ion. Since the anions are negatively charged, they repel the electron in the d orbital. Since in the e orbital the electron density is highest along the axes x, y, z, the anions at $\pm x$, $\pm y$, $\pm z$ in octahedral coordination strongly repel an electron in an e orbital, while an electron in a t orbital, which has its nodes in the coordinate planes, is not as strongly repelled. Thus in octahedral coordination the d level of an atom is split into a lower t and an upper e level. On the other hand, in cubic or tetrahedral coordination the splitting is inverted, since the anions lie in [111] directions rather than in [100] directions. The cubic field splitting normally lies in the range from 2000 to 20,000 cm^{-1}.

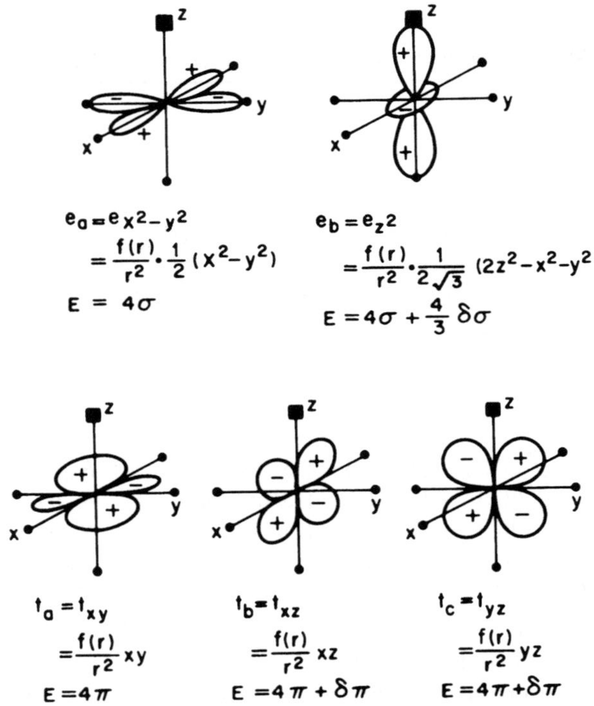

Fig. 14. Diagrams showing the spatial distribution of the *d* orbitals relative to an octahedral array of ligands. A special ligand on the *z* axis is marked and it illustrates the fact that a different ligand perturbs each *d* orbital in a characteristic way. The analytical form of the orbital is given. The energy of each orbital is given in terms of σ and π antibonding parameters, and in terms of the change in these parameters for the special ligand, $\delta\sigma$ and $\delta\pi$.[268]

The doubly degenerate *e* and triply degenerate *t* levels can be further split by departures from cubic symmetry, such as would be caused by having two different types of anions in the coordination sphere or a vacancy nearby in the lattice. Figure 14 shows how a single different anion can split the *e* and *t* levels. The *u* partner of the *e* orbital pair, which has the angular dependence $3z^2 - 1$, obviously interacts more strongly with a *z* perturbation than does the *v* orbital $(x^2 - y^2)$; if the perturbation corresponds to increased antibonding, then $E(u) > E(v)$. Similarly the *xz* and *yz* orbitals of the *t* set interact more strongly than does the *xy* orbital, and the splitting pattern would be an upper pair of levels and a lower single one. The splitting

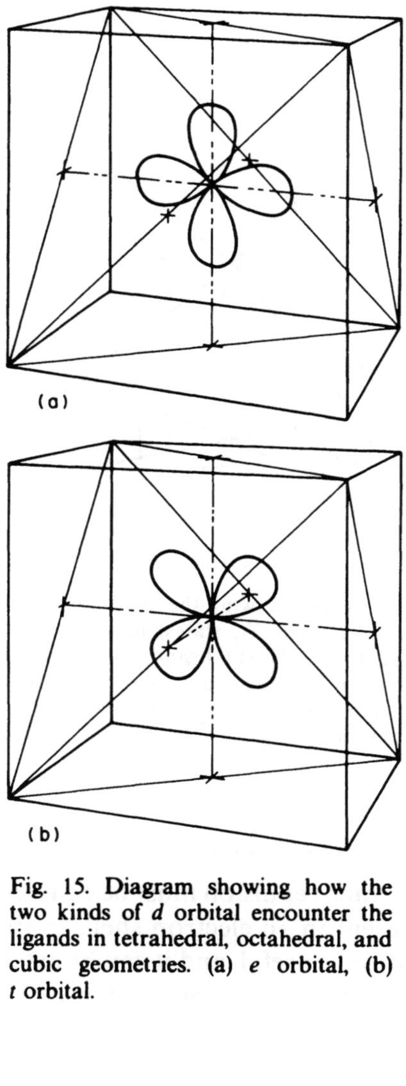

Fig. 15. Diagram showing how the two kinds of *d* orbital encounter the ligands in tetrahedral, octahedral, and cubic geometries. (a) *e* orbital, (b) *t* orbital.

diagram corresponding to an axial perturbation is shown in Fig. 16. The levels there are labeled by the representations of the point group C_{4v}, and some possible spectroscopic transitions and their polarizations are also shown. The transitions are electric dipole since the center of symmetry has been removed by the perturbation. The splitting diagram would look the same, however, with two perturbing centers at $\pm z$, and then the transitions would be magnetic, since the center of symmetry would remain.

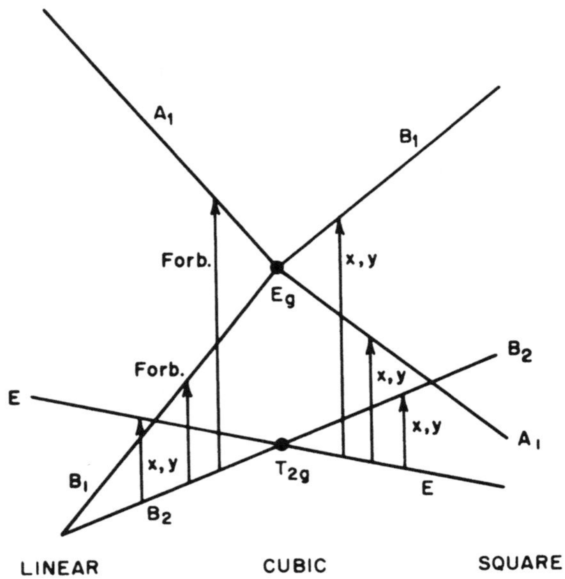

Fig. 16. Diagram showing the splitting of e and t orbitals (or E_g and T_{2g} states) in an octahedron perturbed by a different ligand, as in Figure 14. The extremes of square and linear coordination are also shown as would be achieved by symmetric distortions involving the polar and equatorial ligands.

3.2.1.2. Several Electrons in a Crystal Field

The interelectronic repulsion must be considered when more than one electron is present in an electron shell. In free atoms the idea of electron configuration is useful, and it can be extended easily to atoms in crystal fields. The idea of this approximation is to neglect the interelectronic repulsion at first, and to form basis states consisting of products of single-electron wave functions. Later, interelectronic repulsion is computed in this basis set so as to give the energies of the many-electron states. Initially, however, a great deal of insight about crystal field energy levels can be obtained by simply considering the crystal field configurations available in a given case and their relationships to the states of the free atom.

The crystal field configurations can be described as in atoms by simply giving the number of electrons in each type of orbital. The notation for a configuration in a cubic field is $t^m e^n$, where there are $m + n$ electrons in the d shell, and m are in t and n in e cubic field orbitals. For one electron the energy difference between t^1 and e^1 is called Δ or $10Dq$. This same parameter determines the energy difference between successive configurations such as $t^m e^n$ and $t^{m-1}e^{m+1}$.

In more formal terms, we can write the crystal field Hamiltonian as a sum of one-electron operators:

$$H_c = e \sum_i V(r_i) \tag{42}$$

where i sums over electrons, r_i is the radius vector of an electron, and V is the one-electron potential. A wave function belonging to a configuration, say t^2e, for example, can be written as a product of two of the t-type orbitals and one of the e type. Some of these are $\eta\zeta u$, $\eta\zeta v$, $\xi\eta u$, $\xi\eta v$, $\zeta\xi u$, and $\zeta\xi v$. In our present approximation these have the same energy, but when interelectronic repulsion is included there will be several energy levels. In using the configuration concept, one assumes that the latter splitting is small or at most comparable to the separation of different configurations.

Using the crystal field Hamiltonian, Eq. (42), and any one of the energetically equivalent wave functions of a configuration, we can write the configuration energy as, for example,

$$\begin{aligned} E(t^2e) = \langle \eta\zeta u|H_c|\eta\zeta u\rangle &= [\langle \eta|V|\eta\rangle + \langle \zeta|V|\zeta\rangle + \langle u|V|u\rangle] \\ &= 2E_t + E_e \end{aligned}$$

Since $E_e - E_t = 10Dq$ and since the center of gravity of a pair of levels which split apart under a perturbation is fixed (i.e., $3E_t + 2E_e = 0$), we find $E_e = 6Dq$ and $E_t = -4Dq$. Thus $E(t^2e) = -2Dq$. The energies of all d-electron configurations can be written down easily and are shown in Figure 17. The next question is, What states can arise from each configuration, and, primarily, which configuration contains the ground state?

The ground states of the free atoms of the first transition series follow from Hund's rule, which states: The lowest state of a configuration has the highest spin multiplicity, and if there are several such states, it is the one with highest angular momentum. Thus for d^3 there are two quartet states ^4F and ^4P having maximum possible multiplicity, and ^4F is the ground state. Hund's rule holds for the lower oxidation states of transition metal ions in crystals. When applied to the problem of determining which crystal field configuration gives the ground state, one finds that only a limited number of configurations can have maximum multiplicity; others cannot contain high-spin states without violating the Pauli principle. For the d^3 case we can ask which configurations may contain quartets and thus be derivable from the free-atom ground state. It is seen that t^3, t^2e, and te^2 can contain quartet states, but that e^3 cannot. The lowest configuration in an octahedral field is t^3 $[3 \times (-4Dq)]$ and this is the one which contains the $^4A_2$ ground state.

In a tetrahedral field the te^2 configuration would contain the lowest of the quartet states, but e^3 is the lowest energy configuration. Therefore in weak crystal fields where atomic forces dominate, a state of te^2, $^4T_1$, becomes the ground state, while in stronger crystal fields the lowest state of e^3, which can only be a doublet, would become the ground state. This is an example of a spin change brought about by increasing the crystal field strength. Another way to put it is that spin angular momentum is "quenched" by a strong crystal field.

These facts about configurations are brought out in Figure 17. The energy order is appropriate to an octahedral field, but when inverted it would be correct for a cubic or tetrahedral field. The maximum spin multiplicity $(2S + 1)$ of the states in each configuration is indicated by a number in brackets and one can see that the Hund rule ground state cannot be the state of lowest crystal field energy near the middle of the transition series. One consequence of this fact, which was pointed out earlier, is that with increasing crystal field energy, the spin multiplicity of the ground state shifts to a lower

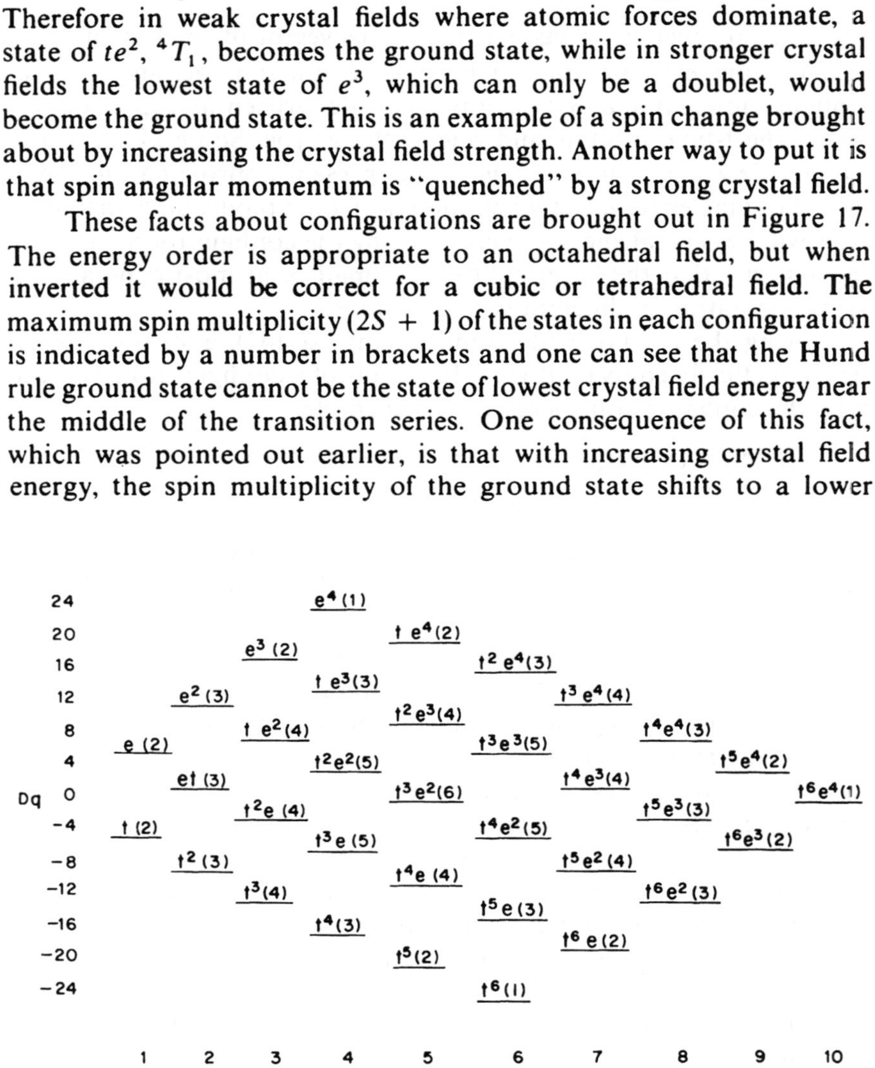

Fig. 17. The possible configurations t^ne^m of d electrons and their maximum spin multiplicity.

value. Thus Fe^{2+}, d^6, can have two lower multiplicities compared to its normal quintet state in hydrated crystals or fluoride host crystals. Drickamer and Frank[120] have shown that the spin change can be induced by application of high pressure, whereby the crystal field strength is increased beyond the critical value for the change.

The *central field* configurations are just as useful in a solid as in the free atom, but the splitting into *crystal field* configurations is a useful concept only in "strong" crystal fields. Usually, the $4f$ orbitals of a rare earth ion are split by amounts which are very small compared to interelectronic repulsion, while d orbitals are split by amounts comparable to or greater than interelectronic repulsion. For rare earths, therefore, it is better to use the free atom as the starting point for the calculation, and this is called the "weak field" scheme. For transition metal atoms the weak or the strong field scheme may be used.

In order to go beyond the configuration energies or the strong field levels we must see how the states of a configuration arise. Only a simple example will be given here. In the two-electron configuration *et* there are six ways to place two electrons into the two *e* and three *t* orbitals, leading to six states. By applying symmetry operations, however, one finds that two triply degenerate states actually result. Furthermore, since the two orbitals are different, there is no Pauli exclusion, and the two states may be singlet or triplet. Thus the *et* configuration gives rise to the states ${}^1T_1$, ${}^1T_2$, ${}^3T_1$, ${}^3T_2$. These states are split apart by the interelectronic repulsion. This splitting is illustrated in Figure 18. Other examples are more complex, but the complete theory and its results have been worked out, and are presented in Ref. 117.

The weak field and the strong field schemes are simply two extreme basis sets, and in practice no system conforms accurately to either one. Instead real systems could be described as cases of intermediate coupling, some being closer to one extreme than the other. Formally it does not matter which basis set is chosen so long as all the matrix elements of the crystal field Hamiltonian are calculated and used to build up the secular equation, because the roots of the latter are independent of a unitary transformation of the secular equation. The energy levels of an atom can be calculated and plotted as a function of the ratio of a crystal field splitting parameter to an interelectronic repulsion parameter as this ratio goes from zero to infinity. Such a diagram illustrates the correlation between the two extreme basis sets.

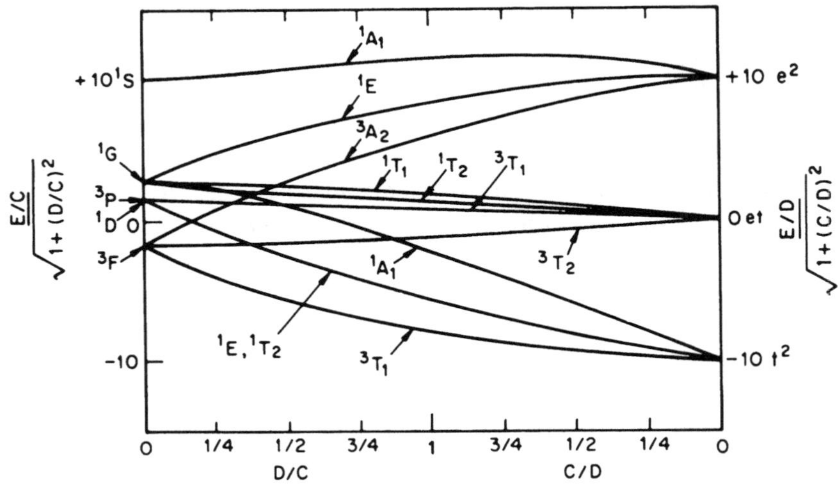

Fig. 18. Correlation diagram showing relation between energy levels of d^2 in free atom and in strong crystal field. D is the cubic crystal field parameter and C is the Racah electrostatic repulsion parameter.[118]

Figure 18 is such a correlation diagram for two d electrons in an atom such as V^{3+}. Energy is measured in units of $(D^2 + C^2)^{1/2}$, where D is the crystal field parameter Dq, and C is one of the Racah parameters for interelectronic repulsion [Eq. (49)]. Several general points are illustrated by this diagram. Note that lines corresponding to states of the same symmetry do not cross. They result from the solution of a secular equation involving all states of the given symmetry. These states do not belong to a single configuration when the parameter ratio C/D is greater than zero. Those states that are the only one in a symmetry class belong to a single configuration for all values of the parameter ratio, for example, ${}^3T_2$ and ${}^3A_2$ in Figure 18.

The correlation diagram also shows the important position of the free atom energy levels and wave functions in crystal field theory. The left side of the diagram corresponds to the levels of the free atom.

The use of the ratio D/C (or C/D) in plotting Figure 18 makes the actual value of C unimportant for the diagram. In fact, the values of B and C do change with increasing crystal field,[121] but usually decrease only 10–20% below the free atom value. That this happens in a particular case can be determined only by fitting the spectrum with parameters Dq, B, and C. The parameter values are known for many cases, so that for any new system their probable range is known and spectral fitting is usually straightforward, once the correlation diagram is known. (See Table 10 for B and C values.)

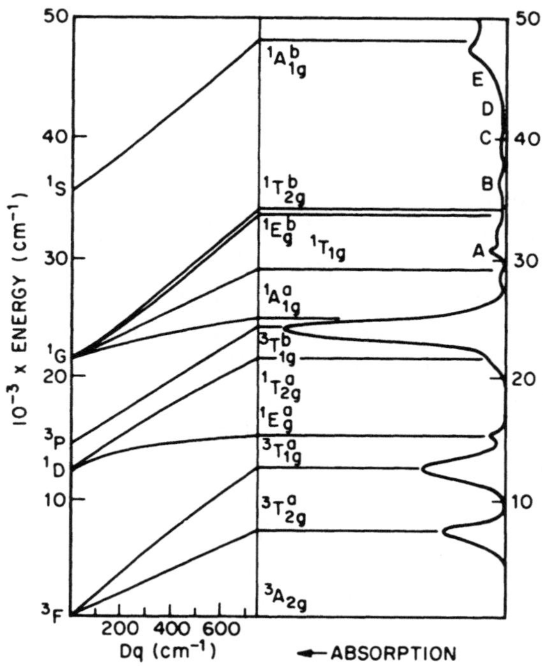

Fig. 19. Ligand field energy level diagram for Ni^{2+} compared to the spectrum of $KNiF_3$. The weak bands marked A–E are ion pair excitations.[122]

An example of a spectral assignment is shown in Figure 19 for Ni^{2+} in $KNiF_3$.[122] The d^8 system fits the diagram for d^2 when the crystal field splitting is inverted, but the electrostatic splitting is left the same. The resulting diagram for the energy levels proceeding from zero crystal field to a value of Dq which fits the spectrum is shown on the left of the figure, while the spectrum is shown on the right. The fit was accomplished with $Dq = 740\ \text{cm}^{-1}$, and values of the Slater–Condon parameters F_2 and F_4 which were slightly smaller than the free atom values. The excellent agreement between theory and experiment settles the assignments of the gross spectral features.

More spectral detail may be resolved by going to low temperatures, studying magnetic effects, or using magnetic-circular dichroism, uniaxial stress, etc. One can then uncover the spin–orbit splitting, Jahn–Teller effects, exchange splittings, etc. The gross spectral assignments discussed in this section, however, are necessary for supplying the wave functions needed for the interpretation of these finer details.

3.2.1.3. The Free Atom

A qualitative and semiquantitative discussion of crystal field theory suffices for a rough understanding of spectra, but to use the theory fully requires a careful mathematical development beginning with the theory of the free atom. The observed atomic levels must be accurately represented by the wave functions and energies of the Hamiltonian of the free atom H_F. Recently, Ferguson[123] has given a discussion of atomic energy level calculations for the 3d atoms, and earlier extensive discussions were given by Griffith[103] and Condon and Shortley.[124] These authors are mainly interested in fitting the Slater–Condon theory as given in Ref. 124 to experimental data. Watson[125] has carried out extensive calculations of wave functions for the 3d series by the restricted Hartree–Fock method and has computed the parameters of Slater–Condon theory. Ferguson[123] has shown how these results compare with experiment.

The principal source of data for the free atom energy levels is Moore's compilation.[126] The data for the rare earths are still scarce. There is a good review of what was known up to 1967 in Dieke's book,[2] and the theory of rare earth spectra is discussed in Wybourne's book.[127]

The central field Hamiltonian of the free atom can be written as

$$H_F = H_0 + H_e + H_s$$
$$= \sum_{i=1} \left(\frac{\hbar^2}{2m} \nabla_i^2 - eU_i \right) + e^2 \sum_{i<j}^{k} \left| \frac{1}{r_{ij}} \right| + \sum_{i}^{k} \zeta_i l_i s_i \qquad (43)$$

The first term, H_0, contains the electron kinetic energy and the central field attraction. This term gives rise to the shell structure of the atom (each shell being specified by the quantum numbers n and l). Its wave functions are products of k one-electron functions, $\prod_a^k u_a(r_i, \theta_i, \phi_i)$ ($a = nlm_l sm_s$), since the operators in H_0 are one-electron operators. The u function is a product of functions of each coordinate of the electron:

$$u_a(r_i, \theta_i, \phi_i) = R_{nl}(r_i)\Theta_l^m(\theta_i)e^{im\phi_i}\sigma_i(m_s) \qquad (44)$$

where the notation is explained with reference to 3d functions in Table 7. The quantum numbers are $nlm_l sm_s$. The central field U_i is not a Coulomb field except for hydrogen and helium, because it represents an effective potential including the nucleus and all electron shells other than the one to which the u_a belongs. The Hamiltonian H_F therefore incorporates the shell structure in H_0. The term H_e then represents specifically the electrostatic interactions of electrons in a shell. The electrostatic interaction with electrons of other shells is taken into account by the value of U_i. The term H_s represents the spin–orbit interaction. The parameter ζ_i also refers to electrons in one shell and the presence of electrons in other shells is accounted for in its value. It is normally determined empirically from multiplet splittings.

The outer, partially filled *nd* or *nf* shells are the ones for which a theory is needed. Then the most important term in H_F is H_e, the interelectron repulsion in the *nd* or *nf* shell. Only for the heavier rare earths does H_s become comparable to H_e. Under these conditions the wave functions of $H_0 + H_e$ are constructed by making eigenfunctions of orbital angular momentum L and its projection M_L, using techniques described in Ref. 124.

The two-electron wave functions have the form

$$|LM_L\rangle = \sum_{m_{l_1}, m_{l_2}} \langle l_1, l_2, m_{l_1}, m_{l_2}|LM\rangle |l_1 m_{l_1}\rangle |l_2 m_{l_2}\rangle \tag{45}$$

with $m_{l_1} + m_{l_2} = M_L$. This is an example of a vector coupling formula in which the vector coupling coefficients $\langle l_1, l_2, m_{l_1}, m_{l_2}|LM\rangle$, often written as 3-$j$ symbols,[128] join the various one-electron wave functions $|l_1 m_{l_1}\rangle$ into two-electron wave functions. These coefficients are tabulated.[128] When there are more than two electrons, or when the spin is introduced, the coupling formula becomes more involved. A considerable amount of work has gone into clarifying these coupling problems, and it is expertly reviewed in works by Judd[129] and Wybourne.[127] Examples of coupled two-electron wave functions are shown in Table 8 for ^3F and ^3P of d^2. The crystal field transform of these functions is also shown for later use. The matrix elements of H_e in this basis reduce to integrals over the one-electron functions:

$$\int u_a(1)u_b(2)(e^2/|r_{12}|)u_c(1)u_d(2)\, dv_1\, dv_2 \tag{46}$$

When $1/|r_{12}|$ is expanded into a multipole series the integrals can be separated into a sum of products over radial and angular factors:

$$\langle ab|\frac{e^2}{r_{12}}|cd\rangle = \delta(ms_a ms_c)\,\delta(ms_b ms_d)\,\delta(m_{la} + m_{lb}, m_{lc} + m_{ld}) \times \sum_{k=0}^{\infty} c^k(l_a m_{la}; l_c m_{lc})c^k(l_d m_{ld}; l_b m_{lb})R^k(ab; cd) \tag{47}$$

TABLE 8

Two d Electrons in Free Atom and in Cubic Field, Using Notation of Table 7

^3F State of free atom		Cubic components of ^3F	
$\|LM\rangle = \sum \|m_{l_1}\rangle\|m_{l_2}\rangle\langle l_1 l_2 m_{l_1} m_{l_2}\|LM\rangle$		$\|\Gamma, \gamma\rangle$	
33	$\|2\rangle\|1\rangle$	A_2	$\|uv\rangle$
32	$\|2\rangle\|0\rangle$	$T_2\ \zeta$	$\|\zeta u\rangle$
31	$\|1\rangle\|0\rangle(\sqrt{2/5}) \pm \|2\rangle\|-1\rangle(\sqrt{3/5})$	$T_2\ \eta$	$-\|\eta\sqrt{3}v + u)/2\rangle$
30	$\|1\rangle\|-1\rangle(2/\sqrt{5}) + \|2\rangle\|-2\rangle(1/\sqrt{5})$	$T_2\ \xi$	$\|\xi(\sqrt{3}v - u)/2\rangle$
3−1	$\|0\rangle\|-1\rangle(\sqrt{2/5}) + \|1\rangle\|-2\rangle(\sqrt{3/5})$	$T_1\ \alpha$	$(2/\sqrt{5})\|\zeta\eta\rangle + (1/\sqrt{5})\|\xi(v + \sqrt{3}u)/2\rangle$
3−2	$\|0\rangle\|-2\rangle$	$T_1\ \beta$	$(2/\sqrt{5})\|\zeta\xi\rangle - (1/\sqrt{5})\|\eta(v - \sqrt{3}u)/2\rangle$
3−3	$\|-1\rangle\|-2\rangle$	$T_1\ \gamma$	$(2/\sqrt{5})\|\xi\eta\rangle + \|\zeta v\rangle$

^3P State of free atom		Cubic component of ^3P	
$\|11\rangle$	$\|2\rangle\|-1\rangle(\sqrt{2/5}) - \|10\rangle\|0\rangle(\sqrt{3/5})$	$T_1\ \alpha$	$(1/\sqrt{5})\|\zeta\eta\rangle - (2/\sqrt{5})\|\xi(v + \sqrt{3}u)/2\rangle$
$\|10\rangle$	$\|2\rangle\|-2\rangle(2/\sqrt{5}) - \|1\rangle\|-1\rangle(1/\sqrt{5})$	$T_1\ \beta$	$(1/\sqrt{5})\|\zeta\xi\rangle + (2/\sqrt{5})\|\eta(v - \sqrt{3}u)/2\rangle$
$\|1-1\rangle$	$\|1\rangle\|-2\rangle(\sqrt{2/5}) - \|0\rangle\|-1\rangle(\sqrt{3/5})$	$T_1\ \gamma$	$(1/\sqrt{5})\|\xi\eta\rangle - (2/\sqrt{5})\|\zeta v\rangle$

	States of V^{3+} SC theory	Exp.	$F_2 = 1360$ $F_4 = 100$	$F_2 = 1336$ $F_4 = 94.1$ $\alpha = 124$
^3F	$F_0 - 8F_2 - 9F_4$	0	0	0
^1D	$F_0 - 3F_2 + 36F_4$	10,540	11,300	10,193
^3P	$F_0 + 7F_2 - 84F_4$	12,980	12,900	12,959
^1G	$F_0 + 4F_2 + F_4$	17,970	17,300	17,970
^1S	$F_0 + 14F_2 + 126F_4$	·	31,650	

The c^k are numbers tabulated in Refs. 117 and 124. They arise from the integrations over angles. The R^k are radial integrals which cannot be given accurate theoretical values because the radial wave functions are not known accurately. They are therefore considered to be parameters. Fortunately only three parameters, called F_0, F_2, F_4, are needed to describe the interelectronic repulsion in the d shells and four (F_0, F_2, F_4, F_6) in the f shells, and of these F_0 is not needed when only the energy level splittings of the configuration are wanted. The F_k are, for example,

$$F_k(3d;3d) = \frac{1}{D_k} e^2 \int_0^\infty \int_0^\infty [R_{3d}(r_1)]^2 [R_{3d}(r_2)]^2 \frac{r_<^k}{r_>^{k+1}} r_1^2 r_2^2 \, dr_1 \, dr_2 \tag{48}$$

where $D_0 = 1$, $D_2 = 49$, $D_4 = 441$; $r_<$ is the smaller of r_1 and r_2, $r_>$ is the greater, and R_{3d} is a $3d$ radial one-electron function. Integrals of this type, called the Slater–Condon parameters, are normally taken from experiment. Watson[125] has calculated them for the $3d$ series using SCF functions, and gets values about 20% higher than the experimental ones.[123]

A different notation for the interelectronic repulsion parameters, due to Racah, is used in much the current literature on transition group atoms. For d electrons it is

$$A = F_0 - 49F_4; \qquad B = F_2 - 5F_4; \qquad C = 35F_4 \tag{49}$$

The theoretical energies of the terms of the d^n configurations are given in Table 9. This tabulation is useful both to show how the energy levels are distributed and to see how large the thermodynamic stabilization is (Section 3.2.6). The values of B and C for free ions are

TABLE 9

Theoretical Energies of Terms of d^n Configurations for $n \leq 5$*

d^2	d^3
$^3F = A - 8B$	$^4F = 3A - 15B$
$^3P = A + 7B$	$^4P = 3A$
$^1G = A + 4B + 2C$	$^2H = {}^2P = 3A - 6B + 3C$
$^1D = A - 3B + 2C$	$^2G = 3A - 11B + 3C$
$^1S = A + 14B + 7C$	$^2F = 3A + 9B + 3C$
	$^2D = 3A + 5B + 5C$
	$\pm (193B^2 + 8BC + 4C^2)^{1/2}$
d^4	d^5
$^5D = 6A - 21B$	$^6S = 10A - 35B$
$^3H = 6A - 17B + 4C$	$^4G = 10A - 25B + 5C$
$^3G = 6A - 12B + 4C$	$^4F = 10A - 13B + 7C$
$^3F = 6A - 5B + 5\frac{1}{2}C$	$^4D = 10A - 18B + 5C$
$\pm \frac{3}{2}(68B^2 + 4BC + C^2)^{1/2}$	$^4P = 10A - 28B + 7C$
$^3D = 6A - 5B + 4C$	$^2I = 10A - 24B + 8C$
$^3P = 6A - 5B + 5\frac{1}{2}C$	$^2H = 10A - 22B + 10C$
$\pm \frac{1}{2}(912B^2 - 24BC + 9C^2)^{1/2}$	$^2G = 10A - 13B + 8C$
$^1I = 6A - 15B + 6C$	$^2G' = 10A + 3B + 10C$
$^1G = 6A - 5B + 7\frac{1}{2}C$	$^2F = 10A - 9B + 8C$
$\pm \frac{1}{2}(708B^2 - 12BC + 9C^2)^{1/2}$	$^2F' = 10A - 25B + 10C$
$^1F = 6A + 6C$	$2D' = 10A - 4B + 10C$
$^1D = 6A + 9B + 7\frac{1}{2}C$	$2D = 10A - 3B + 11C$
$\pm \frac{3}{2}(144B^2 + 8BC + C^2)^{1/2}$	$\pm 3(57B^2 + 2BC + C^2)^{1/2}$
$^1S = 6A = 10B + 10C$	$^2P = 10A + 20B + 10C$
$\pm 2(193B^2 + 8BC + 4C^2)^{1/2}$	$^2S = 10A - 3B + 8C$

* The relative energies of the terms of d^{10-n} are the same as those of d^n.

TABLE 10

Free-Ion Values of *B* and *C*, According to Electron Configuration, for Gaseous Ions

	Ion	*B*	*C*		Ion	*B*	*C*
$3d^2$	Ti^{2+}	718	2629	$4d^3$	Mo^{3+}	610	—
	V^{3+}	861	4165	$4d^6$	Rh^{3+}	720	—
	Cr^{4+}	1039	4238	$4d^7$	Rh^{2+}	620	4002
$3d^3$	Sc^{+}	480	—	$4d^8$	Pd^{2+}	683	2620
	V^{2+}	766	2855	$5d^2$	Os^{6+}	780	—
	Cr^{3+}	918	3850	$5d^3$	Re^{4+}	650	—
	Mn^{4+}	1064	—		Ir^{6+}	810	—
$3d^4$	Cr^{2+}	830	3430	$5d^4$	Os^{4+}	700	—
	Mn^{3+}	1140	3675	$5d^6$	Ir^{3+}	660	—
$3d^5$	Mn^{2+}	960	3325		Pt^{4+}	720	—
$3d^6$	Fe^{2+}	1058	3901	$5d^8$	Pt^{3+}	600	—
	Co^{3+}	1100	—				
$3d^7$	Co^{2+}	971	4366				
$3d^8$	Ni^{2+}	1041	4831				

given in Table 10. These values have to be reduced slightly for application to ions in crystals (see next section).

The results of this analysis are a series of energy levels and wave functions to be compared to experiment. Table 8 shows some of the $|LS\rangle$ wave functions in Russell–Saunders notation, e.g., ^3F, etc., and the corresponding energy in terms of the parameters for the d^2 configuration. The experimental levels of V^{3+} are also shown. If the values $F_2 = 1360\,\text{cm}^{-1}$ and $F_4 = 100\,\text{cm}^{-1}$ are taken, then the energy values of the fourth column are found. These are in fairly good agreement with the experiment. If one adds a term $\alpha L(L+1)$, an empirical term discovered by Trees,[130] then the fit is better, taking $F_2 = 1336$, $F_4 = 94.1$, and $\alpha = 124$ (see column 5). In many cases the α parameter gives a significant improvement, both for the free ion and for the ion in a crystal.[123] It was explained by Racah as being a result of configuration interaction.[131,132]

The Slater–Condon parameters have been found for the first transition elements by fitting spectra to theory.[133,134] For the rare earths, where there are many more levels, it is better to fit the parameters to the known levels for a given ion and use these to assign the other levels of that ion, rather than attempt a general fitting of all spectra.[135]

The spin–orbit coupling H_s has the effect of splitting the $|L, S\rangle$ state into its components, specified by total angular momentum J.[136]

A level such as ^3F of d^2 then splits into $^3F_2$, $^3F_3$, $^3F_4$, an array of levels called a multiplet. In V^{3+}, ^3F spans 200 cm^{-1} and in Ni^{2+}, 2200 cm^{-1}. In the latter case the off-diagonal spin–orbit matrix elements which, for example, connect $^3F_2$ with $^1D_2$, are large enough to change the multiplet pattern appreciably. This mixing of singlet and triplet states is also significant in that it increases the intensity of transitions between states of differing multiplicity. The values of the one-electron spin–orbit parameter are given in Ref. 137. They are important in diagnosing fine structure and intensity in spectra of ions in crystals.

3.2.1.4. Atom in a Crystal Field

Having discussed a workable theory of the free atom, let us see how this can be applied to the atom in a crystal field. Experience with the data shows that the F_k in crystals are very slightly reduced below their free ion values, of the order of 1% in the case of the rare earths, and of the order of 10–20% for the 3*d* transition group. The structure of the theory is not badly upset by these changes, so the Slater–Condon theory of the free atoms becomes the basic theory of atoms in crystals. It provides the basis set for calculating the crystal field energies, and it can be used to give the electrostatic energies. The reason for the relatively small changes in the electrostatic repulsion values is that they are determined largely by the inner parts of the wave function, and although chemical binding in the crystal field may change the outer parts of the wave function drastically, the inner parts of *nd* or *nf* wave functions are not greatly affected. It is remarkable that even when the crystal field splittings are of the same order of magnitude as the atomic level separations, the free-atom level scheme is still easily recognizable. Figure 19 shows such a relationship for Ni^{2+} in $KNiF_3$.

The reduction of symmetry in passing from free ion to crystal causes the free ion terms to split into substates of lower degeneracy. For example, ^3F in a cubic field splits into $^3A_2$, $^3T_2$, and $^3T_1$. Since ^3P is $^3T_1$ in a cubic field, it interacts with the $^3T_1$ component of ^3F. Thus there is an off-diagonal crystal field matrix element connecting ^3F and ^3P. The calculation of these matrix elements is discussed in the next section after the crystal field potential is examined.

The two limiting forms of crystal field theory discussed earlier can be described more fully as follows.

1. The term weak crystal field is applied when $H_e \gg H_c$. Under these conditions the interelectron repulsions determine the states of

TABLE 11

Energy Levels in Octahedral and Tetrahedral Complexes[121]*

d^2	t_2^2	$^3T_1$	$7.5B - 3Dq - \frac{1}{2}(225B^2 + 100Dq^2 + 180DqB)^{1/2}$
		^1E	$-8Dq + 9B + 2C - 6B^2/10Dq$
		$^1T_2$	$-8Dq + 9B + 2C - 12B^2/10Dq$
		$^1A_1$	$-8Dq + 18B + 5C - 108B^2/10Dq$
	t_2e	$^3T_2$	$+2Dq$
		$^3T_1$	$7.5B - 3Dq + \frac{1}{2}(225B^2 + 100Dq^2 + 180DqB)^{1/2}$
	e^2	eA_2	$+12Dq$
d^3	t_2^3	$^4A_2$	$-12Dq$
		^2E	$-12Dq + 9B + 3C - 50B^2/10Dq$
		$^2T_1$	$-12Dq + 9B + 3C - 24B^2/10Dq$
		$^2T_2$	$-12Dq + 15B + 5C - 176B^2/10Dq$
	t_2^2e	$^4T_2$	$-2Dq$
		$^4T_1$	$7.5B + 3Dq - \frac{1}{2}(225B^2 + 100Dq^2 - 180DqB)^{1/2}$
	t_2e^2	$^4T_1$	$7.5B + 3Dq + \frac{1}{2}(225B^2 + 100Dq^2 - 180DqB)^{1/2}$
d^4	t_2^4	$^3T_1$	$-16Dq + 6B + 5C - 64B^2/10Dq$
		^1E	$-16Dq + 12B + 7C - 82B^2/10Dq$
		$^1T_2$	$-16Dq + 12B + 7C - 208B^2/10Dq$
		$^1A_1$	$-16Dq + 21B + 10C - 436B^2/10Dq$
	t_2^3e	^5E	$-6Dq$
	$t_2^2e^2$	$^5T_2$	$+4Dq$
d^5	t_2^5	$^2T_2$	$-20Dq + 15B + 10C - 140B^2/10Dq$
	t_2^4e	$^4T_1$	$-10Dq + 10B + 6C - 26B^2/10Dq$
		$^4T_2$	$-10Dq + 18B + 6C - 38B^2/10Dq$
	$t_2^3e^2$	$^6A_1$	0
		$^4A_1$	$10B + 5C$
		^4E	$10B + 5C$
		$^4T_2$	$13B + 5C + \cdots$
		^4E	$17B + 5C$
		$^4T_1$	$19B + 7C - \cdots$
		$^4A_2$	$22B + 7C$
d^6	t_2^6	$^1A_1$	$-24Dq + 5B + 8C - 120B^2/10Dq$
	t_2^5e	$^3T_1$	$-14Dq + 5B + 5C - 70B^2/10Dq$
		$^3T_2$	$-14Dq + 13B + 5C - 106B^2/10Dq$
		$^1T_1$	$-14Dq + 5B + 7C - 34B^2/10Dq$
		$^1T_2$	$-14Dq + 21B + 7C - 118B^2/10Dq$
	$t_2^4e^2$	$^5T_2$	$-4Dq$
	$t_2^3e^3$	^5E	$+6Dq$
d^7	t_2^6e	^2E	$-18Dq + 7B + 4C - 60B^2/10Dq$
	$t_2^5e^2$	$^4T_1$	$7.5B - 3Dq - \frac{1}{2}(225B^2 + 100Dq^2 + 180DqB)^{1/2}$
	$t_2^4e^3$	$^4T_2$	$+2Dq$
		$^4T_1$	$7.5B - 3Dq + \frac{1}{2}(225B^2 + 100Dq^2 + 180DqB)^{1/2}$
	$t_2^3e^4$	$^4A_2$	$+12Dq$
d^8	$t_2^6e^2$	$^3A_2$	$-12Dq$
		^1E	$-12Dq + 8B + 2C - 6B^2/10Dq$
		$^1A_1$	$-12Dq + 16B + 4C - 108B^2/10Dq$
	$t_2^5e^3$	$^3T_2$	$-2Dq$
		$^3T_1$	$7.5B + 3Dq - \frac{1}{2}(225B^2 + 100Dq^2 - 180DqB)^{1/2}$
		$^1T_2$	$-2Dq + 8B + 2C - 12B^2/10Dq$
		$^1T_1$	$-2Dq + 12B + 2C$
	$t_2^4e^4$	$^3T_1$	$7.5B + 3Dq + \frac{1}{2}(225B^2 + 100Dq^2 - 180DqB)^{1/2}$

* The terms $nB^2/10Dq$ are calculated for $C = 4B$.

the many-electron system and the free atom wave functions are used. The crystal field is then a small perturbation and the crystal field matrix elements are of the form $\langle \alpha SLJM|H_c|\alpha SL'J'M'\rangle$, where free atom wave functions are represented by their quantum numbers.

2. The term strong crystal field is applied when $H_c \gg H_e$. Now the crystal field splitting is large compared to the interelectronic repulsion. The Hamiltonian is $H_0 + H_c$ for this limiting case. Its two terms are one-electron operators, since H_e has been left out. The real one-electron wave functions (u, v; ξ, η, ζ) form a basis set with which to express the crystal field matrix. Recalling Eq. (42), the crystal field matrix elements can be expressed as sums of one-electron matrix elements, and the rules for forming these sums are the same as for one-electron operators in atomic problems.[124]

Actual examples of the evaluation of matrix elements in the weak field limit are given in Dieke's book[2] and the best place to see how the strong field scheme works in detail is the book by Sugano *et al.*[117]

The energy levels for d electrons in cubic fields have been obtained by both methods. Table 11 lists the energies of the most commonly observed states in terms of the crystal field parameter Dq and the Racah parameters B and C. The free-ion ground term (Table 9) as obtained by Slater–Condon theory has been subtracted. The ground term of the crystal field state is below the ground term of the free atom, and this energy can be interpreted as a thermodynamic stabilization of the ion in the crystal (see Section 3.6).

3.2.1.5. The Crystal Field Potential

The emphasis on the wave functions up to this point has been intentional. The potential of the crystal is best thought of as a model potential to be determined from experimental data and therefore it is defined by its matrix elements. Working in this spirit, the crystal field potential for n electrons is written

$$V = \sum_{i}^{n} V_i = \sum_{i=1}^{n} \sum_{k,q} A_{kq} Y_k^q(\theta_i \varphi_i) \tag{50}$$

where the V_i are single-electron potentials, the A_{kq} are constants (or depend only on r_i) describing the field, and $Y_k^q(\theta_i\varphi_i)$ is a spherical harmonic whose arguments are the angular coordinates of electron i. In this expansion the first term, $k = q = 0$, is spherically symmetric, since Y_0^0 = const, and subsequent terms have more and more angular dependence. As k increases by one unit an additional node in angle is added. Here q designates the orientation of the node. The derivation of a potential of this form from a point charge model will be given in a later section, but the form does not depend on any model, as shown especially by Newman.[138]

A matrix element of V_i, a one-electron component of V, may be written as

$$\int U_i{}^* V_i U_i{}' \, d\tau_i$$

$$= \langle nlm | V | nlm' \rangle$$

$$= \sum_{k,q} \int |R_{nl}(r_i)|^2 A_{kq} r_i{}^2 \, dr_i \int Y_l^{m*} Y_k{}^q Y_l^{m'} \sin\theta \, d\theta \, d\varphi$$

$$= \sum_{k,q} \bar{A}_{kq} (l \| Y_k \| l) \begin{pmatrix} l & k & l \\ -m & q & m' \end{pmatrix} (-1)^{l-m} \tag{51}$$

where in the last line the Wigner–Eckart theorem has been used. We could just as well have used the $c_k(lm; l'm')$ coefficients to give the value of the integral over spherical harmonics, but the reduced matrix elements $(l \| Y_k \| l)$ and the 3-j coefficients $\left(\begin{smallmatrix} l & k & l \\ -m & q & m' \end{smallmatrix}\right)$ are more commonly used. The reader unfamiliar with the Wigner–Eckart theorem should read Ref. 139. A generalization of this theorem to the point groups is also very useful, and is described fully by Sugano *et al.*[117] The 3-j coefficients are tabulated in Ref. 128 and the reduced matrix elements in Ref. 140. The value of the last form of Eq. (51) only becomes apparent when dealing with many-electron states. In that case one has wave functions $|L, M\rangle$, as exemplified in Table 8, where L is a vector sum of l_i, and $M = \sum m_i$. The matrix element $\langle LM | V | LM' \rangle$ has the same form as Eq. (51), but the reduced matrix element $(L \| \sum_i (Y_k)_i \| L)$ is different. The latter, however, are tabulated in Ref. 140, and a great deal of labor can be avoided if one is familiar with these tables.

Adjustments to the values of the $\bar{A}_{kq}$ of Eq. (51) are made until the model potential fits the data. One is then interested either in the trends in the $\bar{A}_{kq}$ and how they relate to other data, or in getting a refined wave function of the atom in a crystal potential in order to calculate other quantities. Examples of both procedures will be given.

First, however, it is desirable to point out some simplifications. Most crystal field potentials have symmetry, and the total number of terms in Eq. (51) will be reduced as a result. The triangular rule reduces the number of k which can have matrix elements with a given l to $k = 0, 2, 4, \ldots, 2l$. The combination of these two results leaves a cubic field potential with the terms V_0^0, V_4^0, V_4^4, V_6^0, V_6^4 when the m value is measured on the C_4 symmetry axis, and in addition there are relationships between V_4^0 and V_4^4 and between V_6^0 and V_6^4. The spherical term V_0^0 only shifts the levels. Thus only two potential constants for a cubic field are needed:

$$V_i = \sum_i D_4[Y_4^0 + (5/14)^{1/2}(Y_4^4 + Y_4^{-4})] + D_6[Y_6^0 - (7/2)^{1/2}(Y_6^4 + Y_6^{-4})] \tag{52}$$

Both of the terms affect f electrons, but only the first affects d electrons. (One should be careful to realize that D_4 is not the same constant for d electrons as it is for f electrons.)

Even if the symmetry is not high, it is often a great advantage to assume a higher symmetry, and to express the potential in terms of deviations from the high symmetry. For example, the potential of an ion in the corundum lattice is roughly cubic, but has a component D_{3h}, and a smaller part C_{3v}, and smaller yet C_3. So a perturbation series could be arranged with a potential of the form:

$$V = V(O_h) + \alpha V(D_{3d}) + \beta V(C_{3v}) + \gamma V(C_3) \tag{53}$$

where $1 > \alpha > \beta > \gamma$.

3.2.1.6. The Point Charge Model

The potential of a set of point charges acting on an electron is

$$V = \sum_j \sum_i (e_j/|R_j - r_i|) \tag{54}$$

where R_j is the vector from the origin to the ligand charge site, r_i is the electron position vector, $r_i - R_j$ is the distance of the electron from the ligand, and e_j is the charge of the site. The expansion of V in a series of spherical harmonics leads to Eq. (50) with the A_{kq} expressed as

$$A_{kq} = A'_{kq}(r^k/R_j^{k+1})e_j \tag{55}$$

for a set of symmetrically equivalent points R_j. When the radial average for the d or f shell is taken the constant becomes

$$A_{kq} = (A'_{kq}/R_j^{k+1})\langle r^k \rangle e_j \tag{56}$$

where

$$\langle r^k \rangle = \int_0^\infty |R_{nl}(r)|^2 r^k r^2 \, dr \tag{57}$$

This expression is not rigorously correct because the integration should not be carried beyond $r = R$ unless $\langle r^k \rangle$ is replaced by $\langle R^k/r^{k+1} \rangle$. But $\langle r^k \rangle$ should be considered a parameter anyway, so its theoretical expression is not of great significance.

The value of the point charge model is that it gives an easily visualized potential. For a single electron the constant in the cubic field potential of Eq. (52) for an octahedron of point charges is

$$D_4(\text{oct}) = (7\pi/3)e_j r^4/R^5 \tag{58}$$

and when it is converted to Cartesian coordinates the fourth-power potential expression becomes

$$V_4(\text{oct}) = (35e_j/4R)(r/R)^4(x^4 + y^4 + z^4 - \tfrac{3}{5}) \tag{59}$$

where x, y, and z are electron direction cosines. It is not difficult to use this expression with the wave functions given in Table 7 to obtain the crystal field matrix elements by direct integration. Probably the best easy reference to the details of the point charge model is an article by Hutchings.[141] A much deeper discussion with application to spectra is provided by the work of Newman.[138]

The point charge model makes the prediction that Dq varies as R^{-5}. Drickamer and Frank have measured this distance dependence by observing spectra under high hydrostatic pressure.[120] Ions substituted in corundum all followed the R^{-5} law, but ions substituted in MgO all had stronger dependences, as shown in Figure 20. This

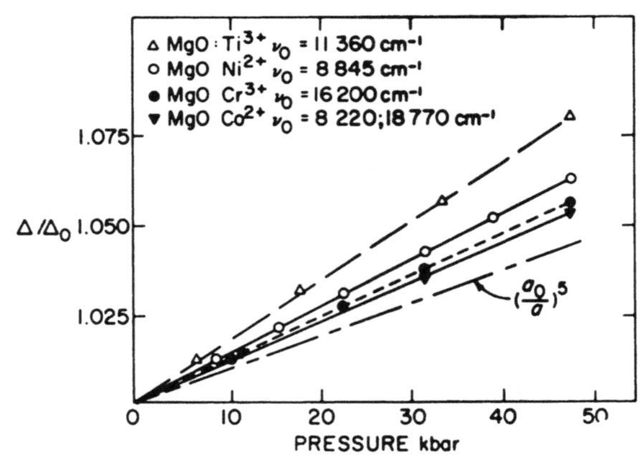

Fig. 20. Shifts of transition energies of impurity ions in MgO plotted as a function of hydrostatic pressure. The shifts are compared to $(a_0/a)^5$, the shift expected from the point charge model of crystal field theory.[120]

partial agreement does not necessarily support the ionic model, since a strong distance dependence would be exprected on any other model, for example, the molecular orbital model. Nevertheless, the distance dependences given by the point charge model are a good first approximation even for the more complicated uniaxial strains.[142]

3.2.1.7. Tensor Operators

After one has mastered the basic ideas of crystal field theory he must take another step in order to be able to handle real problems in a professional way. One often has to deal with many-electron problems, and it is difficult to handle these with elementary methods. Instead the tensor algebra methods should be used.

These methods can be illustrated by reference to Eqs. (50) and (51), where the problem of finding matrix elements of the crystal field was discussed. Equation (50) could be rewritten

$$V = \sum_{k,q} A_{kq} T_q^{(k)} \tag{60}$$

where $T_q^{(k)}$ transforms like $Y_k^q(\theta_i \varphi_i)$ and is called a tensor operator. The definitions of the tensor operator are sometimes given in terms of the quantities $C_k^q(\theta_i \varphi_i)$ where $C_k^q = (4\pi/2k+1)^{1/2} Y_k^q$, but they always have the same transformation properties. These properties permit one to use the Wigner–Eckart theorem to simplify the matrix elements. Thus, even though $T_q^{(k)}$ is a many-electron operator, its matrix elements can be reduced as was done in Eq. (51) for a one-electron matrix element.

Other operators, such as the spin–orbit operator, can be expressed in tensor operator form, and the benefits of the Wigner–Eckart theorem can also be obtained for these matrix elements.

The extent of the simplifications possible with this method can be illustrated by the reduction of a crystal field matrix element for a rare earth ion. Its free ion states are described by quantum numbers $\alpha LSJM$, where α represents all specifications of the wave function other than its angular momenta. The matrix element of $T_q^{(k)}$ is

$$\begin{aligned} &\langle \alpha LSJM | T_q^{(k)} | \alpha' L' S J' M' \rangle \\ &= (-1)^{J-M} \begin{pmatrix} J & k & J' \\ -M & q & M \end{pmatrix} \langle \alpha LSJ \| T^{(k)} \| \alpha' L' S J' \rangle \end{aligned} \tag{61}$$

where the second factor is the 3-j symbol used in Eq. (51) and the third factor is a reduced matrix element. The latter can be reduced further because the crystal field operator does not depend on spin.

This means that we should uncouple L and S from J. The result of this process is a formula for the reduced matrix element:

$$\langle \alpha LSJ \| T^{(k)} \| \alpha' L'SJ' \rangle = (-1)^{S+k+J+L'}[(2J+1)(2J'+1)]^{1/2} \times \begin{Bmatrix} L & J & S \\ J' & L' & k \end{Bmatrix} \langle \alpha SL| \, \| T^{(k)} \| \, |\alpha' SL' \rangle \quad (62)$$

The factor in curly brackets is called a 6-j symbol and is tabulated,[128] and the last factor is a doubly reduced matrix element, and these, too, have been tabulated.[140] Thus the angular dependence of any matrix element of the crystal field can be obtained from tables.

In the example just given the spherical rotation group (R_3) was used. A parallel treatment using the cubic group was developed by Tanabe and Sugano.[117] Their method deals with d-electron configurations in a strong cubic field, $t^m e^n$. Many-electron problems are treated by an adaptation of the tensor operator methods. A valuable systematization of these methods for molecular symmetry groups in general was provided by Griffith.[143] His methods have been used extensively by Day and his co-workers to parametrize not only ligand field spectra[144] but charge-transfer spectra as well.[145]

3.2.2. Molecular Orbital Theory

The next step in model building beyond crystal field theory is the molecular orbital theory. The symmetry aspects of both theories are the same and therefore they have some of the same parameters. The tensor operator formalism can be used equally well in both. While the essential feature of molecular orbital theory is the inclusion of several atoms rather than a single one as in crystal field theory, the crystal field parameters determined from experiment already contain many-atom contributions. In using molecular orbital theory, we explicitly describe these many-atom aspects.

The only practical formulation of molecular orbital theory is the so-called "linear combination of atomic orbitals" or LCAO method.[146] In this form the theory is formally identical to tight binding theory used to describe band structure.[147] In the present context it can be used to describe the excitations of transition metal atoms in crystals and the electronic structure in the vicinity of less localized defects such as a vacancy or F center.

The general method, illustrated for a pair of atoms a and b, is to form a wave function

$$\psi = C_a\varphi_a + C_b\varphi_b \tag{63}$$

as a linear combination of atomic orbitals φ_a and φ_b. Then, using the Hamiltonian h, whose form can be specified later, we must have

$$h\Psi = e\Psi \tag{64}$$

where e is an energy eigenvalue. Inserting Ψ from Eq. (63) and multiplying through by φ_a or φ_b and integrating gives the two equations

$$\begin{aligned} C_a(e_a - e) + C_b(\beta - Se) &= 0 \\ C_b(\beta - Se) + C_b(e_b - e) &= 0 \end{aligned} \tag{65}$$

where $e_a = (\varphi_a|h|\varphi_a)$, $e_b = (\varphi_b|h|\varphi_b)$, $\beta = (\varphi_a|h|\varphi_b)$, and $S = (\varphi_a|\varphi_b)$. These two equations each give a value of C_a/C_b; in order for them to give the same value, e must be restricted to the values given by the determinantal equation

$$\begin{vmatrix} e_a - e & \beta - Se \\ \beta - Se & e_b - e \end{vmatrix} = 0 \tag{66}$$

If the values of e given by Eq. (66) are inserted in (65) and the normalization condition

$$C_a^2 + 2SC_aC_b + C_b^2 = 1 \tag{67}$$

is used, then the individual values of C_a and C_b can be found. There are two values of e which fit the consistency condition, and therefore two sets of C_a, C_b. These define two molecular orbitals (MO) which are mutually orthogonal.

The algebra of MO theory becomes somewhat involved, as even this simple example would show if the results were fully written out. For applications to transition metal atoms in crystals, however, some simplifications can be made because one of the coefficients will be much smaller than the other, and the overlap integral S is small. An added complication, however, is that an entire group of neighboring atoms must be taken into account. For example in an octahedral complex MX_6^{n-} the six neighbors are equivalent by symmetry and all have the same overlap with M. The partial LCAO's involving only the X atoms can be grouped into symmetry-adapted combinations and then treated as single orbitals as in Eq. (63).

The symmetry-adapted LCAO MO's for an octahedral array of closed-shell atoms around a transition metal atom are shown in Table 12. The atomic basis functions are the *ns* and *np* orbitals of the ligands and the 3*d*, 4*s*, and 4*p* orbitals of the first row transition metal atom. Analogous orbital arrays could be constructed for other

TABLE 12
Orbitals for Octahedral Array of Atoms around a Central Atom*

Symmetry	Central atom	Outer atoms (ligands)	Combined orbital
A_{1g}	s	$(1/\sqrt{6})(\sigma_x - \sigma_{\bar{x}} + \sigma_y - \sigma_{\bar{y}} + \sigma_z - \sigma_{\bar{z}})$	$\sigma a_{1g}, sa_{1g}$
E_g	$de = (2x^2 - y^2 - z^2)$,	$(1/2\sqrt{3})[2(\sigma_x - \sigma_{\bar{x}}) - (\sigma_y - \sigma_{\bar{y}} + \sigma_z - \sigma_{\bar{z}})]$	$de_g, \sigma e_g$
	$(y^2 - z^2)$	$\frac{1}{2}[\sigma_y - \sigma_{\bar{y}} - \sigma_z + \sigma_{\bar{z}}]$	
T_{1u}	$p = (x, y, z)$	$\frac{1}{2}(\sigma_z + \sigma_{\bar{x}})$, etc.	$pt_{1u}, \pi t_{1u}, \sigma t_{1u}$
		$\frac{1}{2}[(x_z + x_{\bar{x}}) + (x_y + x_{\bar{y}})]$, etc.	
T_{1g}		$\frac{1}{2}[(x_z - x_{\bar{z}}) - (z_x - z_{\bar{x}})]$, etc.	πt_{1g}
T_{2u}		$\frac{1}{2}[(x_z + x_{\bar{z}}) - (x_y + x_{\bar{y}})]$, etc.	πt_{2u}
T_{2g}	$dt = (xz, \text{etc.})$	$\frac{1}{2}[(x_z - x_{\bar{z}}) + (z_x - z_{\bar{x}})]$, etc.	$\pi t_{2g}, dt_{2g}$

* The outer atom orbitals are explained in Figure 22. In the A_{1g}, E_g, and T_{1u} classes there is more than one orbital in the symmetry class, and the correct "combined orbitals" are mixtures. The descriptions σa_{1g}, sa_{1g} are no longer appropriate after mixing, and they are called $1A_{1g}$, $2A_{1g}$, etc. The coefficients of the central atom and ligand orbitals in the combined orbital are determined in principle by the variation method.

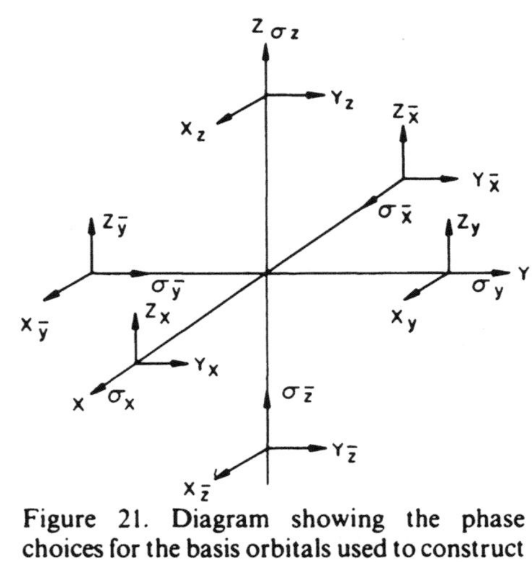

Figure 21. Diagram showing the phase choices for the basis orbitals used to construct molecular orbitals of an octahedral array of ligands.[118]

coordination numbers and types of atoms. The phase conventions for setting up the basis orbitals must be carefully defined, and Figure 21 shows the choices upon which Table 12 is based. A first good guess as to the order of energy of these orbitals can be based on the reasonable assumptions that $E_s < E_p$ and $E_\sigma < E_\pi$ and finally the order of energies within, say, the set of π orbitals can be determined from the node patterns.[118] Figure 22 shows the node patterns of the orbitals of Table 12. The greater the number of nodes, the higher is the orbital

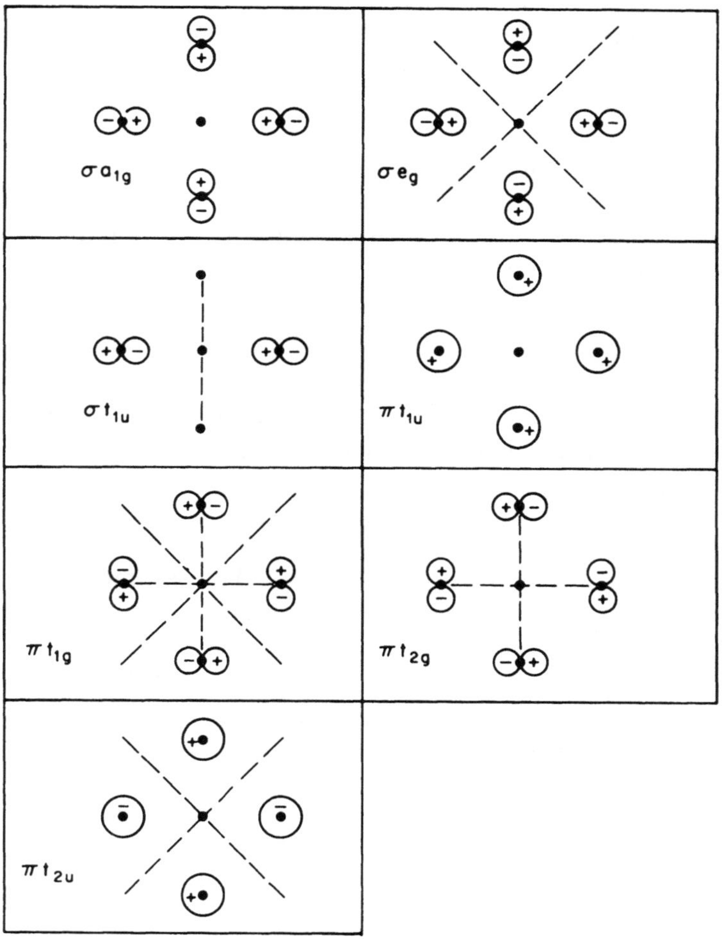

Fig. 22. Nodal patterns of the ligand orbitals of an octahedral array. The ligands are here supposed to have three equivalent p orbitals, so that the partners of degenerate orbitals look the same as the ones shown, but differ in orientation.[118]

energy. The highest energy ligand orbital should be the one labeled πt_{1g} since it is nonbonding with respect to the d shell and has the maximum number of nodes between ligands. This should be the highest filled orbital and the one most easily excited in an electronic transition. Since it has even parity, as do the d orbitals, a $\pi t_{1g} \rightarrow nd$ transition is expected to be weak. This feature has been observed, for example, in the charge-transfer spectrum of $IrCl_6^{2-}$. A weak band at the beginning of the spectrum is identified with this forbidden transition (see longest wavelength band of Figure 12). An energy level diagram giving a reasonable order of levels in an octahedral complex is shown in Figure 23. The actual order of levels has been established experimentally for a number of octahedral second and third row transition metal halide complexes, and will be discussed in Section 8.

The linear combinations of the d and ligand orbitals of Table 12 may be described as bonding or as antibonding. The bonding orbitals are largely localized on the ligands, while the antibonding orbitals are nearly pure d orbitals. The $T_2\xi$ orbitals, for example, are

$$\begin{aligned} \Psi(T_2\xi, b) &= N_b^{-1/2}[\gamma_\pi \xi_d + \xi_l] \\ \Psi(T_2\xi, a) &= N_a^{-1/2}[\xi_d - \lambda_\pi \xi_l] \end{aligned} \tag{68}$$

where γ_π and λ_π are positive and are small compared to unity. When these two orbitals are made orthogonal to each other one finds the relationship between γ and λ:

$$\lambda = (\gamma + S)(1 + \gamma S) \tag{69}$$

where S is the overlap integral: $S = \langle \xi_d | \xi_l \rangle$. The parameter γ is called the covalency parameter. A detailed discussion of covalency in this form is given in Ref. 117, and will not be pursued here.

When molecular orbital theory is used to describe the effects of bonding on d orbitals it applies to the parameters used to fit optical spectra of the $d \longrightarrow d$ type, to ESR spectra, and to charge-transfer spectra.

For the first of these applications a truncated form of the theory is adequate for many purposes.[148] The wave function of a d electron can be expressed as

$$\Psi = C_d \varphi_d + C_l \varphi_l \tag{70}$$

where φ_d and φ_l are the d and ligand orbitals in a row of Table 12 and C_d and C_l are the MO coefficients to be determined. Since ψ is mainly a d orbital, $C_d \gg C_l$.

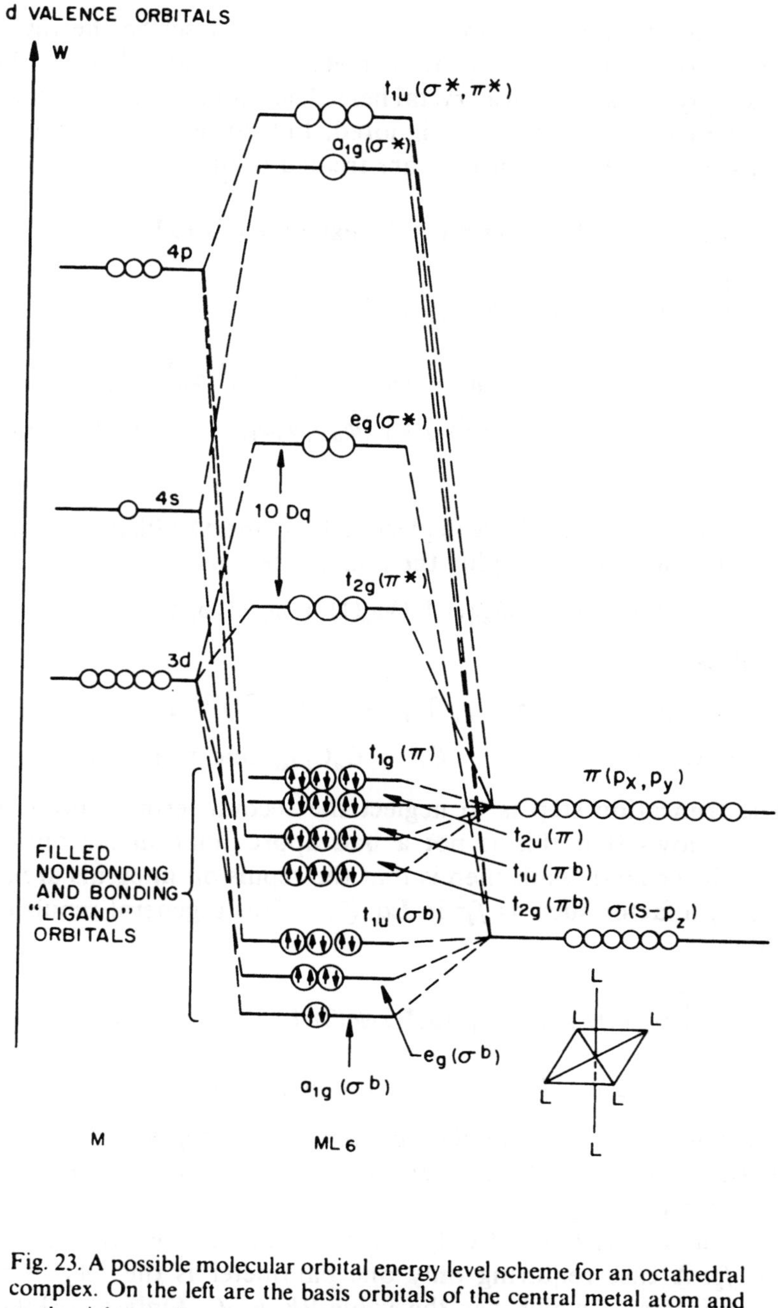

Fig. 23. A possible molecular orbital energy level scheme for an octahedral complex. On the left are the basis orbitals of the central metal atom and on the right are the π and σ orbitals of the ligands, shown as if they did not interact and split. In the center is the energy level diagram of the complex, showing a possible orbital occupation.[146]

The Hamiltonian we want to use here will describe the inter-electronic repulsion in the transition metal atom and will treat the ligands simply as a source of a crystal field. The details of the electron–ligand interaction are therefore ignored, but since the C_d and C_l remain as parameters, the ligands are treated in more detail than in crystal field theory.

An electron repulsion term can be expanded as follows:

$$\int \psi^*(1)\psi(1)(e^2/r_{12})\psi^*(2)\psi(2)\,dv_1\,dv_2$$

$$\equiv [\psi^2, \psi^2] = C_d[\varphi_d{}^2, \varphi_d{}^2] + 2C_d{}^2C_l{}^2[\varphi_d{}^2, \varphi_l{}^2] + 4C_d{}^2C_l{}^2[\varphi_d\varphi_l\varphi_d\varphi_l] + 4C_d{}^3C_l[\varphi_d{}^3, \varphi_d\varphi_l] \quad (71)$$

where terms in $C_l{}^3$ and $C_l{}^4$ are neglected. Next the Mulliken approximation is introduced to simplify the mixed integrals:

$$\varphi_d{}^*(1)\varphi_l(1) = \tfrac{1}{2}S_{dl}[\varphi_d{}^*(1)\varphi_d(1) + \varphi_l{}^*(1)\varphi_l(1)] \quad (72)$$

The result is

$$[\psi^2, \psi^2] = [\varphi_d{}^2, \varphi_d{}^2]f_d{}^2 + [\varphi_d{}^2, \varphi_l{}^2]f_l{}^2C_d{}^2 \quad (73)$$

where $f_d = C_d{}^2 + C_dC_lS_{dl}$, $f_l = C_l{}^2 + C_dC_lS_{dl}$, and $f_d + f_l = 1$.

The truncation step is to neglect the second term in Eq. (73). Lohr[148] shows that this is not a bad approximation for optical spectra. The general result then is that the repulsion integrals are all reduced by a factor such as $f_d{}^2$ in Eq. (73). Thus a general repulsion integral is

$$\int \psi_i{}^*(1)\psi_j(1)(e^2/r_{12})\psi_k{}^*(2)\psi_l(2)\,dv_1\,dv_2$$

$$\equiv [ij, kl] = (\alpha A + \beta B + \gamma C)[f_if_jf_kf_l]^{1/2} \quad (74)$$

where the first factor is the electron repulsion integral in terms of Racah parameters for the free atom and the second factor is the reduction factor.

In a cubic environment only two kinds of f factors are needed, f_{t2} and f_e. A case requiring only one parameter is the ${}^4A_2 \rightarrow {}^2E$ transition in Cr^{3+}, which has the value $9B + 3C$. Only t_2 orbitals are involved here. Using the B and C values of Table 10 which fit the free atom, the transition energy would be 19,800 cm^{-1}. In Al_2O_3:Cr it is observed at 14,400, giving a value for f_{t2} of 0.85. Such a value is in agreement with the assumptions of the theory, which required $C_d \gg C_l$.

Spectral assignments can usually be made even when the B and C parameters are uncertain to 20% or so. A single reduction factor can then be determined by fitting a spectral interval. One generally finds reductions of the intervals of the order of 10–30% from the value expected for the free ion, as in the previous example. With more careful fitting of several intervals one can get both f_{t2} and f_e. A general review with many illustrations is given by Lever.[121]

The introduction of the reduction factors at this point is fairly empirical. All we have really done is to give a plausible explanation for the reduced electrostatic repulsion parameters and a way to get them from the spectra. They should be obtainable from a more rigorous theory and should relate to other experimental observables, as we go on to investigate.

In applications to spin resonance spectra the orbital angular momentum matrix elements between molecular orbitals must be found. These quantities give the orbital contributions to the g factor. In the literature of this subject,[149] the ratio of the MO matrix element to the free atom matrix element is called a reduction factor, but it is not the same factor as the one defined for the reduction of the electrostatic repulsion integral. It must contain contributions from the angular momentum on the ligands and cannot be truncated in the way done for the electrostatic integrals. In fact, the truncation step introduces an error which only cancels out in spectroscopic intervals. Therefore these two types of reduction factors are only distantly related at the present level of theory.

The nuclear spins of the ligands, particularly F^- ions, interact with the electron spin of the transition metal atom to give extra line splitting in ESR spectra. This "transferred hyperfine interaction" can also be observed in NMR spectra. It can be interpreted directly with the LCAO model so as to give a measurement of the mixing of d and ligand orbitals. Shulman *et al.*[150] carried out the experimental and theoretical work on a number of transition metal ions octahedrally coordinated by fluorides.[150] Values of λ and γ of Eq. (68) were found; for example, in $KNiF_3$, $\gamma_s = 0.04$, $\lambda_s = 0.12$, and $\gamma_\sigma = 0.34$, $\lambda_\sigma = 0.34$. These numbers represent a large amount of covalency. They also imply a value of C_d [Eq. (70)] of 0.90 and $f_d = 0.84$. Since this is a reasonable f_d value, the hyperfine data are consistent with the Racah parameter reduction values. In a simple model theory this is gratifying and gives us confidence in using it. Extensive comparisons of this type have not been made, however.

Thornley[151] has obtained the MO coefficients from the two kinds of experimental data, g-factors and transferred hyperfine interaction for the IrX_6^{2-} complexes. The values from the two sources do not agree very well. The paper, however, gives a good presentation of the method.

Hayes' group has studied the binding in the 3*d* transition group fluorides[152] following earlier work of Tinkham.[153] Covalent bonding in the rare earths was considered by Jorgensen *et al.*[154] and Burns and Axe.[155] Ligand effects on spin–orbit coupling have been worked out by Misetich and Buch.[156]

There is a continuing effort to develop semiempirical molecular orbital models which are easy to apply and which can handle several kinds of data consistently with a single set of parameters. A good discussion of the problems in achieving this goal is given by Dahl and Ballhausen[157] and Ballhausen and Gray.[158] Examples of recent efforts are given in Refs. 159 and 160.

3.2.3. Self-Consistent Field Theory

Dissatisfaction with the oversimplified models used to relate portions of the experimental data and a desire to investigate electronic structure more deeply have led a number of investigators[161–163] to work with better approximations to the solution of the Schrödinger equation.

In one recent study the method of the self-consistent field in several approximations was applied to the calculation of spectral properties of the NiF_6^{4-} cluster.[161] The positions of several excited states were calculated with fair accuracy. The wave functions give reasonable covalency parameters and the transferred hyperfine parameters are fairly close to experiment. These calculations were done by open-shell techniques so that the SCF orbitals for each excited state could be found and the changes of the MO coefficients from one state to another can be followed. An interesting result is the decrease of *d*-orbital occupation of about 0.03 electron when the atom makes the transition ${}^3A_{2g}(t^6e^2) \longrightarrow {}^3T_{2g}(t^5e^3)$. This change is consistent with the increase of σ compared to π bonding for a $t \longrightarrow e$ transition.

The type of information obtained in these calculations is highly detailed, and only in a few particular ways can it be combined so as to lead to values of observables. One hopes that the good agreement with experimental values of the observables justifies believing the rest of the picture. In the present example[161] the authors could comment on the validity of the molecular orbital picture and could discuss atomic *d*-orbital expansion, *s–p* hybridization on the fluorines, and many other details. These are important matters for the understanding of electronic structure in general, so these detailed calculations on a particular example have much wider significance than might appear at first sight. They are so difficult, lengthy, and expensive to do, however, that for the present we can expect that only a few systems will be investigated in this way.

3.2.4. Extended Hückel Calculations

A popular approximation of the SCF method is a highly simplified molecular orbital approach called the extended Hückel method. It is derived from the early treatments of organic molecules. The structure of the method is the same as molecular orbital theory, so that the roots of a secular equation must be found; these are the molecular orbital energies. The matrix elements are entirely empirical, and are determined by various sets of rules, depending on what works best in a particular instance. A good discussion and critique of the method is given by Ballhausen and Gray.[158]

The method has been applied to spectra of many inorganic complexes, beginning with a paper on MnO_4^- by Wolfsberg and Helmholtz,[164] and it has since been developed by many authors.[165–167] Recently Messmer and Watkins have used the method to describe the energy levels of an impurity in a solid,[168] the case of nitrogen at a silicon site in silicon. The results were encouraging enough to show that the method should be applied to other systems. Messmer and Watkins' paper contains a good review of the method and a survey of recent work with it.

3.2.5. Crystal Field Spectra

The spectra caused by tightly bound *d* or *f* electrons for which crystal field theory gives a good description can be classified as crystal field spectra. There are a number of recent reviews of these spectra, so our discussion here will be brief.

3.2.5.1. $f \rightarrow f$ Spectra

The prime reference for $4f \rightarrow 4f$ spectra of the rare earths is Dieke's book,[2] where theoretical and experimental progress is reviewed. In recent years there has been some progress on the analysis of the spectra of the free atoms and ions of the rare earth elements, so it has become possible to compare levels of these ions in the free and crystal states. It is still true, however, that most of what we know about the free ion spectra is derived from the spectra of crystals. The trivalent ion spectra have been analyzed almost completely for several host crystal systems, such as $LaCl_3:RE^{3+}$. A very careful study of the fitting of these spectra to crystal field models has been made by Newman and his co-workers, and has been reviewed by Newman.[138]

Rare earths normally are trivalent ions in crystalline hosts, but the entire series can be obtained in the divalent state,[169] and Ce, Pr, and Tm are obtained as tetravalent ions. The $f \rightarrow f$ spectra in these other valence states have been partially investigated.[170]

3.2.5.2. $f \rightarrow d$ Spectra

The $4f \rightarrow 5d$ transitions of trivalent rare earths occur at energies above most of the $4f \rightarrow 4f$ transitions.[171] The $5d$ orbitals are quite well shielded by the $4f$ electrons, so that as nuclear charges are added through the rare earth series the binding energy of the $5d$ electrons changes very little. Therefore, the $4f \rightarrow 5d$ transition energies through the series reflect the $4f$ binding energies almost exclusively. The transitions are from the lowest of the $4f$ states and therefore, as the number of electrons increases and the configuration splitting increases, the $4f \rightarrow 5d$ transition energy increases. This trend continues up to the half-filled shell (Gd^{3+}, $4f^7$), but the eighth electron cannot be exchange-stabilized and so the transition energy decreases abruptly at Tb^{3+}, $4f^8$. Then the same trend is repeated in the second half of the series. The spectra due to these transitions occur at energies from 4 to 8 eV.[171]

In the divalent rare earths, however, the $5d$ level is much lower and the $4f \rightarrow 5d$ spectra occur in the visible or infrared, as shown in Figure 24. Indeed, there are four ions, La^{2+}, Ce^{2+}, Gd^{2+}, and Tb^{2+}, having a $5d$ orbital occupied in the ground state. These four have interesting photochromic behavior as impurities in CaF_2.[12]

The details of the spectra of these mixed configurations have been interpreted in only a few cases.[90,172–174] An example of one

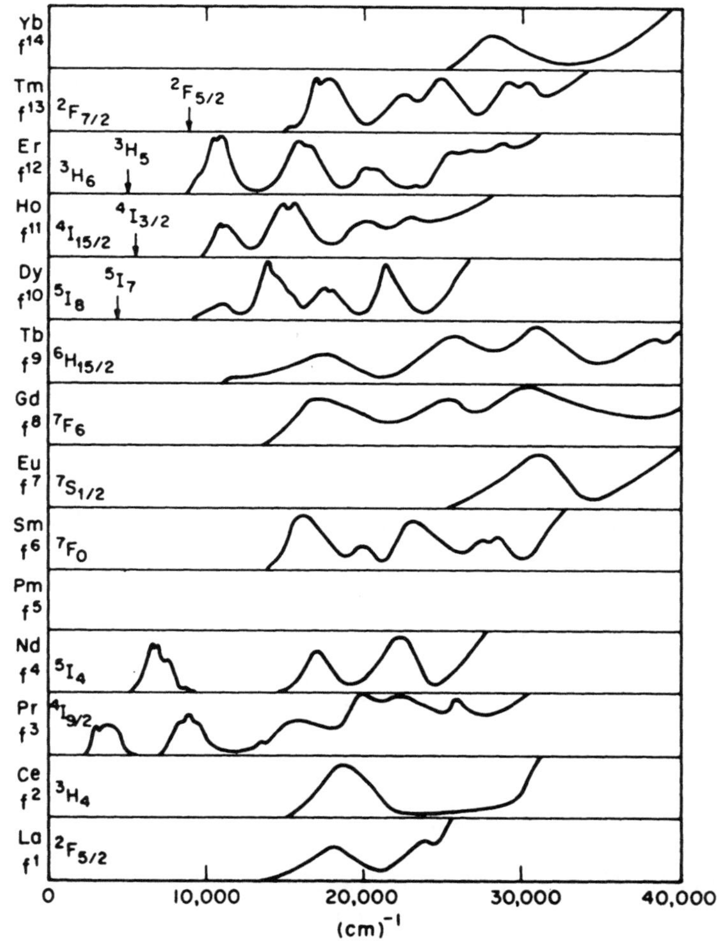

Fig. 24. Low resolution spectra of the divalent rare earths in CaF_2. The broad bands are the 4f ⟶ 5*d* transitions, and a few of the 4*f* ⟶ 4*f* transitions are marked with arrows. The spectra for La, Ce, Gd, and Tb are not those of the eight-coordinated divalent ions but correspond to a complex of the rare earth at a Ca^{2+} site and a F^- vacancy containing one or two electrons (see Ref. 12).[169]

is shown in Figure 25, where the absorption of Ce^{2+} in CaF_2 is given.[90] In Ce^{2+} the large crystal field splitting of the 5*d* electron is enough to force a state of 4*f* 5*d* below all states of $4f^2$. The spectrum therefore shows 5*d* ⟶ 4*f* absorption transitions. The states of the $4f^2$ configuration are very clearly seen. In the early survey, as

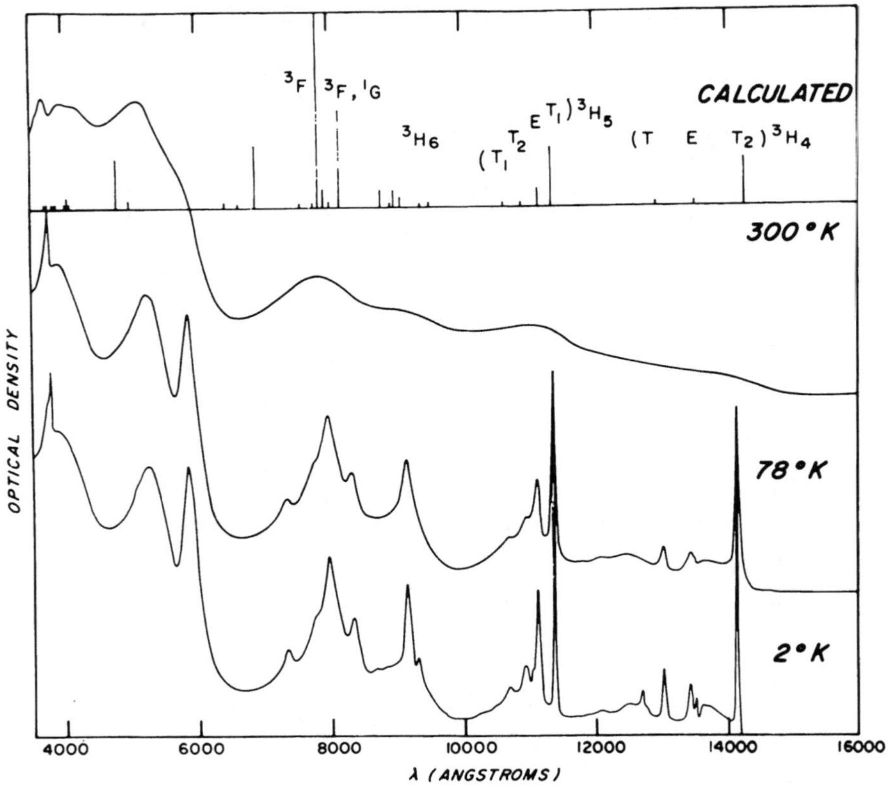

Fig. 25. The spectrum of Ce^{2+} in CaF_2 showing the levels of the excited $4f^2$ configuration. The ground state is a T_2 state of $4f\,5d$.[90]

illustrated in Figure 24, the spectra of the ions La^{2+}, Ce^{2+}, Ga^{2+}, and Tb^{2+} were not actually being observed; instead the spectra shown in Figure 24 for these ions are the spectra of a defect consisting of the trivalent ion with an *F* center nearby.[12] Only in the case of Ce^{2+} has the true divalent ion spectrum been observed (Figure 25).

3.2.5.3. *d* → *d* Spectra

A recent review of the spectroscopy of 3*d* transition metal ions in crystals has been written by Ferguson.[123] Since crystal field effects are much larger than in *f*–*f* spectra and the number of atomic states is smaller, more emphasis is placed on the way the transition metal ion interacts with the crystal. The importance of the Jahn–Teller effect in spectral analysis will be pointed out in the next section and in Section 3.3.2. The molecular orbital theory of covalency has been developed largely to deal with 3*d* spectra.

By far the largest amount of work has been done with 3*d* transition metal ions in octahedral coordination. Here the $d \longrightarrow d$ transitions are parity forbidden, and the spectra only become observable through vibronic coupling, the effect of low-symmetry fields, or exchange interactions with other open shell atoms. The much more intense spectra of tetrahedrally coordinated atoms (noncentrosymmetric) have been studied in connection with inorganic complexes and also for ions in II–VI compounds (see Section 3.3.2 and Ref. 175). Some of the spectra of the 3*d* transition series ions have also been observed in the eightfold cubic coordination of a Ca^{2+} site in CaF_2,[176] where again the transitions are forbidden. The theories of spectra in these three coordination geometries are the same; only the magnitudes of the cubic splitting parameters differ. From the point charge model the relationships are Dq(octahedral) = $-(9/4)Dq$(tetrahedral) = $-(9/8)Dq$(cubic), and these are fairly well followed in the known examples. The change of sign between octahedral and tetrahedral has important spectral as well as thermodynamic consequences[177] (Section 3.2.6).

There has been considerable interest recently in the two commonest five-coordinated geometries (square pyramid and trigonal bipyramid) but these shapes occur only in complexes in solution.[178] Other geometries normally occurring are distortions of octahedral coordination. As Figure 16 shows, these distortions range from linear to square planar if C_4 symmetry is maintained. A charge compensator near an ion in a crystal reduces the symmetry and causes spectral splittings.[179]

The spectra of the 4*d* and 5*d* transition metal ions have been investigated in recent years. Quite often the spectra of these ions are ligand $\longrightarrow$ metal charge transfer rather than $d \longrightarrow d$, because high oxidation states of the 4*d* and 5*d* groups occur commonly, but only recently have these assignments been definitely established.[180] For Ir^{4+} in a Cl^- environment, for example, the strong charge-transfer transitions apparently occur at energies lower than those of the first $d \longrightarrow d$ transitions. The latter are so weak compared to the charge-transfer bands that it is very difficult to identify them.

3.2.5.4. Spectroscopy of $RbMnF_3$

An example of a recent application of crystal field theory is a study of $RbMnF_3$ under uniaxial stress.[142] This crystal is geometrically cubic even at the lowest temperatures. Since it is antiferromagnetic, below 80° K each ion is subjected to an exchange field whose direction can be controlled by uniaxial stress or by an external magnetic field. There were two uses for crystal field theory in this investigation. One was to generate serviceable wave functions to help explain the details of the spectral structure, and another was to rationalize the stress response and Jahn–Teller energies found from experiment.

The wave functions were found for the Mn^{2+} ion, whose low-lying states arise from the $3d^5$ cubic configurations, t^3e^2, t^4e, and t^2e^3. The Tanabe–Sugano matrices[117] were solved for reasonable values of the cubic crystal field parameter, $10Dq$, and the electrostatic repulsion energies B and C. The values which best fit the six lowest energy absorption bands, shown in Figure 26, were found, and then the wave functions could be determined in the form

$$({}^4T_{1g}I) = \alpha_1{}^4T_{1g}(t^4e) + \beta_1{}^4T_{1g}(t^3e^2) + \gamma_1{}^4T_{1g}(t^2e^3) \qquad (75)$$

where α_1, β_1, and γ_1 are coefficients which determine the fraction of the ${}^4T_{1g}$ type wave function of each configuration such as t^4e in the state ${}^4T_{1g}$(I). Similar sets of coefficients were obtained for the other T_{1g} states and for states of other symmetries. The ${}^4T_{1g}$(I) state is the one corresponding to the first spectral absorption band shown in Figure 26. Other states are also shown in this figure. The good fit of the theory to the experimental bands gives one some confidence that

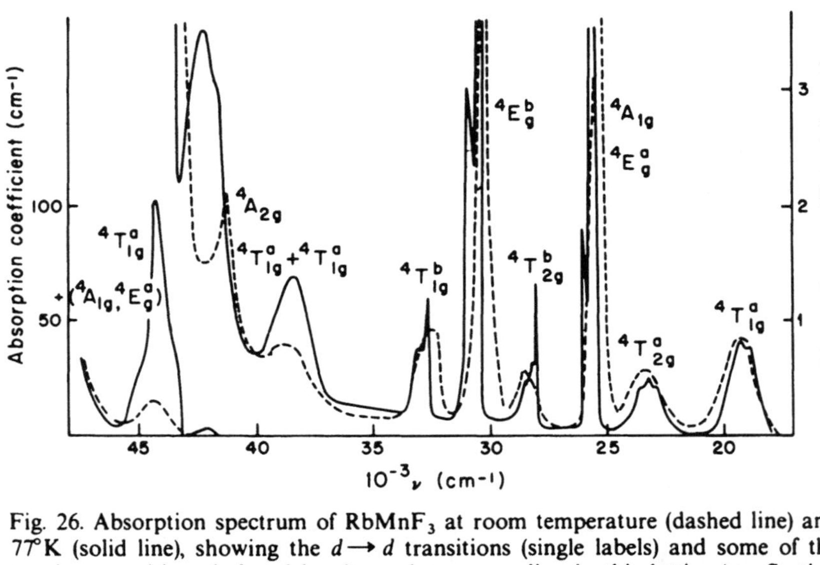

Fig. 26. Absorption spectrum of $RbMnF_3$ at room temperature (dashed line) and 77°K (solid line), showing the $d \rightarrow d$ transitions (single labels) and some of the two-ion transitions induced by the exchange coupling in this lattice (see Section 3.4).[269]

the wave functions should be adequate for the calculation of other properties.

The effects of the exchange field, Jahn–Teller distortion, spin–orbit coupling, uniaxial stress, and electron–phonon coupling were all of interest in this study. Therefore the wave functions for each state were to be acted upon by a rather complicated Hamiltonian which could describe all these effects. A $^4T_{1g}$ state has three orbital and four spin components, a total of 12 substates. Therefore each spectroscopic multiplet is described by a 12 × 12 Hamiltonian matrix. Its roots give the positions of the 12 substates. Since the ground state is $^6A_1$, and the exchange field forces the $M_s = -5/2$ component lowest, the only excited states observed in the spectra will be ones containing some fraction of $M_s = -3/2$ character. Only the lowest three will be observed unless the spin–orbit coupling is very large and the Jahn–Teller distortion small. An analysis of the $^4T_{2g}$(II) state is shown in Figure 27. Five multiplet components are calculated to have enough intensity to be seen, and from the stress behavior, four are identified in approximately the right positions, as shown in the figure. The rest

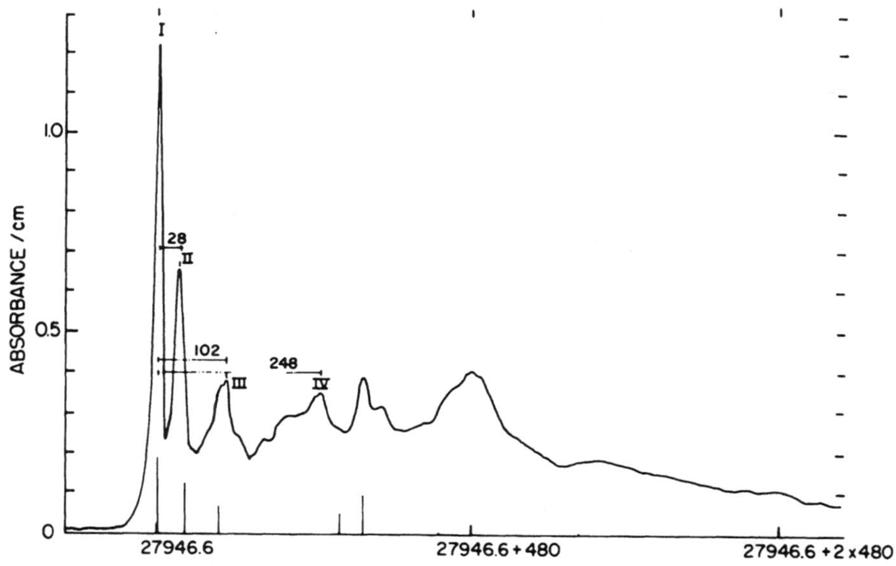

Fig. 27. The $^6A_{1g}$–$^4T_{2g}$(II) transition in $RbMnF_3$. The vertical lines below the spectrum are the calculated energies of the multiplet components; the heights of the lines are the computed intensities. Only five components are expected to be observable. The dotted line shows the computed magnitude of the LO_3 phonon sideband.[142]

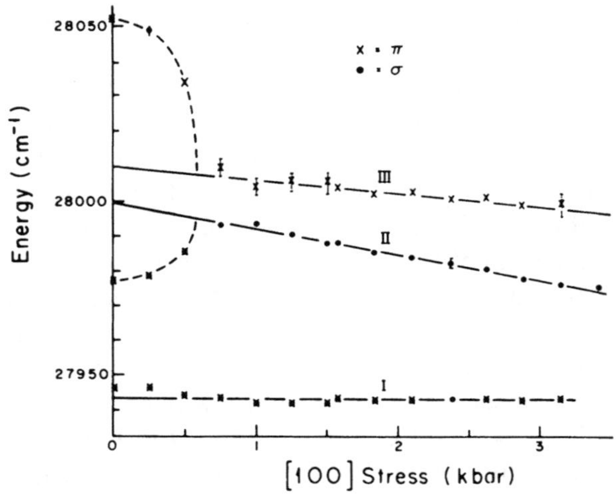

Fig. 28. Plot of the energies of the first three spin-wave sidebands of Fig. 27 as a function of stress. The spin reorientation effects below 0.5 kbar and the stress response above are shown.[142]

of the structure is due to phonons, and has been analyzed. The effect of stress applied along a cubic axis is shown in Figure 28 for the lowest three substates of ${}^4T_g$(II). The most remarkable observation is the drastic change of the multiplet level positions induced by spin reorientation. The $RbMnF_3$ crystal undergoes a spin reorientation from the ⟨111⟩ direction to the ⟨100⟩ direction when the applied ⟨100⟩ stress is greater than 0.5 kbar. The orbital states, however, change very little with the application of this amount of stress. The result is that the spin–orbit coupling changes and the multiplet components shift. The theoretical calculations reproduce this effect quite well for the three spin directions investigated.

3.2.6. Thermodynamic Properties

The effects of the *d* shell on thermodynamic properties is important for understanding lattice energies and heats of solution and for the interpretation of charge-transfer spectra.[181] In a Born cycle calculation of the lattice energy the transition metal ion is prepared in the gas phase in its ground state. Unlike non-transition series elements, these ground states normally are degenerate. When the crystal forms, the degeneracy is split and the lowest energy component becomes the ground state. The energy difference between this lowest component and the energy of the unsplit ionic state is the crystal field stabilization energy δH, which must be added to the lattice energy as usually calculated. Figure 29 is an energy diagram illustrating the formation of NiF_2 via the Born cycle, showing the magnitude of the crystal field stabilization energy δH relative to the other energies. Although δH is very small relative to the lattice energy, it is not so small compared to the heat of formation from atoms, or from the elements, as also shown in Figure 29. (The diagram also illustrates how artificial the lattice energy is as a measure of crystal stability. Most of the lattice energy is used to overcome the ionization energy, and the magnitude of the lattice energy does not necessarily indicate

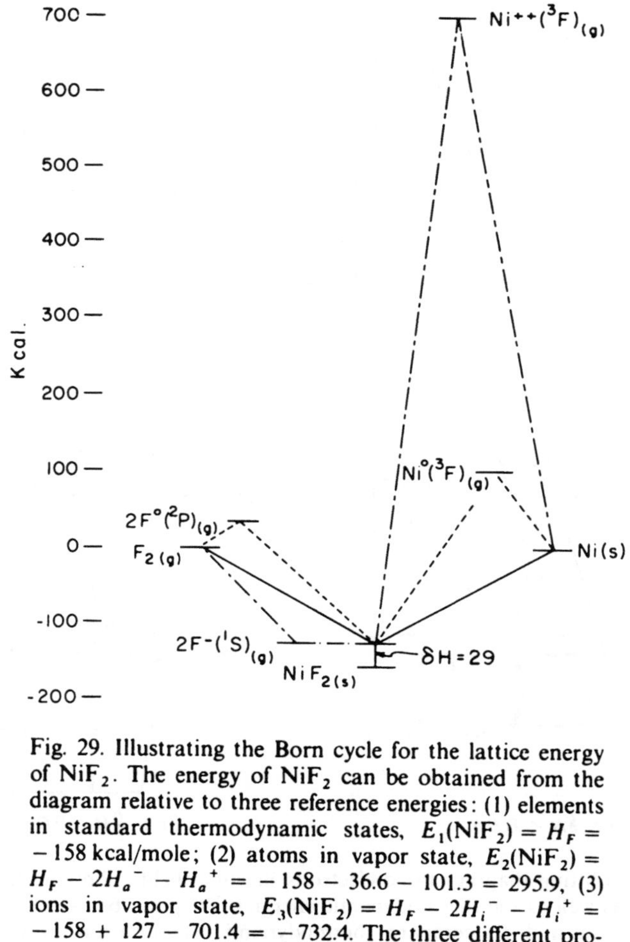

Fig. 29. Illustrating the Born cycle for the lattice energy of NiF_2. The energy of NiF_2 can be obtained from the diagram relative to three reference energies: (1) elements in standard thermodynamic states, $E_1(NiF_2) = H_F = -158$ kcal/mole; (2) atoms in vapor state, $E_2(NiF_2) = H_F - 2H_a^- - H_a^+ = -158 - 36.6 - 101.3 = 295.9$, (3) ions in vapor state, $E_3(NiF_2) = H_F - 2H_i^- - H_i^+ = -158 + 127 - 701.4 = -732.4$. The three different processes are indicated by the three kinds of lines.

chemical stability, the distance below zero on the diagram.) Since δH varies from zero to as much as 60 kcal through the transition series ions, it causes significant and apparently erratic changes from one member to the next.

Figure 30 shows the energy level splitting diagrams for the states of highest multiplicity of the first row transition metal ions in cubic crystal fields. The crystal field stabilizations δH are indicated. The diagrams can also be used to fit spectral data and thereby to derive the cubic splitting parameter Dq.

When the lattice or hydration energies of the transition metal compounds are plotted against atomic number it is found that there is not a smooth variation as would be expected from ionic radius

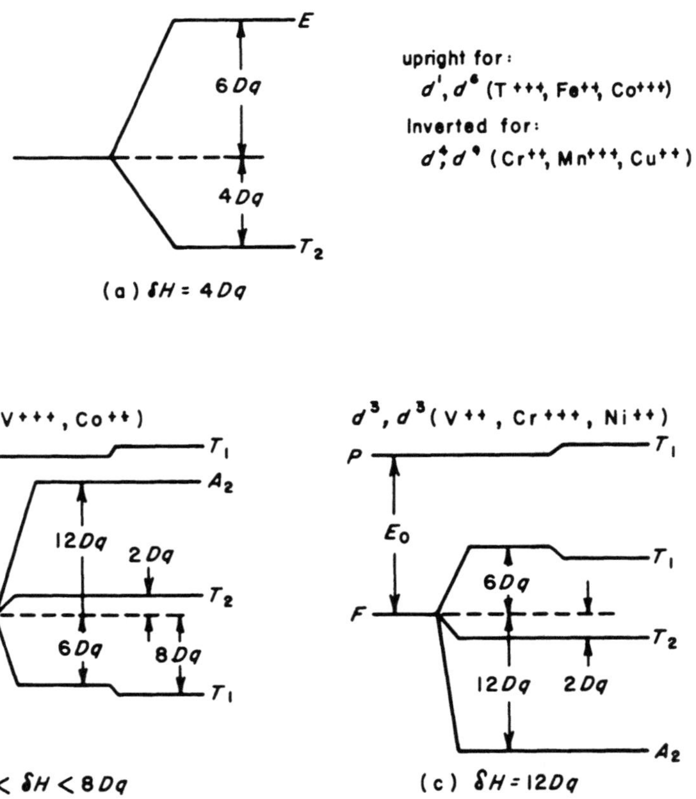

Fig. 30. Splitting diagram for d-electrons in cubic fields, showing the optical energies as well as the thermodynamic energies. They can be used to derive Dq from spectra and to compare thermodynamic stabilization to the optical values.[181]

considerations, but instead a double-humped curve. This type of curve is illustrated in Figure 31 for the hydration energies. When the spectroscopically determined δH values are subtracted the expected smooth variation is revealed, showing that the crystal field effect is indeed present. This subtraction is also shown in the figure.

The crystal field stabilization is large enough to mask other expected trends in properties throughout a transition series. Both the chemical oxidation potentials, $M^{2+} \longrightarrow M^{3+}$, and donor-type charge-transfer spectra should follow the variation of the third ionization potential with atomic number. This trend is only poorly represented in the oxidation potential data, but when corrections are made for the crystal field stabilization of both the M^{2+} and the M^{3+} ions the data are accurately represented by the trend in the third ionization potential.[181] (Also see next section.)

The extra stabilization due to the *d* shell is also implicit in Table 11 (instead of Fig. 30). Numerical results can be obtained by substituting parameter values. Table 13 gives such values for octahedral and tetrahedral coordination.

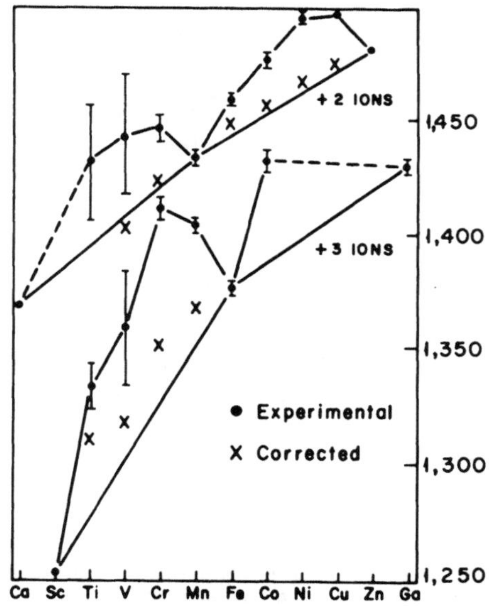

Fig. 31. Heats of hydration of transition metal ions. The values of δH derived from spectroscopic data by the use of Figure 30 are subtracted from each hydration energy, indicated by a ×. Straight-line interpolation curves are also shown.[181]

One notes from Figure 30 or Table 13 that certain ions in a particular coordination geometry have high values of δH; for example, Cr^{3+} on octahedral sites or Co^{2+} on tetrahedral sites. These tendencies influence crystal structure[182] and site choices of impurities.[177] As an example, Cr^{3+} is never found at tetrahedral (or A) sites in spinels, AB_2O_4, since it has the highest preference of any ion for octahedral or B sites. Ni^{2+} has the highest preference of any divalent ion for octahedral sites, but according to the numbers in Table 13, Cr^{3+} should drive Ni^{2+} onto the tetrahedral sites. In fact, the formula for the spinel nickel chromite is $NiCr_2O_4$, with Ni^{2+} on a tetrahedral site, rather than $Cr(NiCr)O_4$. Nickel gallate is $Ga(NiGa)O_4$, since gallium has no crystal field site preference.[177] (But note the ion size effect.[183])

TABLE 13
Crystal Field Theory Data for Transition Metal Ions*

Number of *d* electrons	Ion	Free ion ground state*	Octahedral field ground state†	Tetrahedral field ground state	*Dq*, cm^{-1} oct.	*Dq*, cm^{-1} tetr.	Stabilization, kcal oct.	Stabilization, kcal tetr.	Oct. site preference energy, kcal/mole
1	Ti^{3+}	^2D	$^2T_{2g}$	^2E_g	2030	900	23.1	15.4	7.7
2	V^{3+}	^3F	$^3T_{1g}$	$^3A_{2g}$	1800	840	41.5	28.7	12.8
3	V^{2+}	^4F	$^4A_{2g}$	$^4T_{1g}$	1180	520	40.2	8.7	31.5
	Cr^{3+}	^4F	$^4A_{2g}$	$^4T_{1g}$	1760	780	60.0	13.3	46.7
4	Cr^{2+}	^5D	^5E_g	$^5T_{2g}$	1400	620	24.0	7.0	17.0
	Mn^{3+}	^5D	^5E_g	$^5T_{2g}$	2100	930	35.9	10.6	25.3
5	Mn^{2+}	^6S	$^6A_{1g}$	$^6A_{1g}$	750	330	0	0	0
	Fe^{3+}	^6S	$^6A_{1g}$	$^6A_{1g}$	1400	620	0	0	0
6	Fe^{2+}	^5D	$^5T_{2g}$	^5E_g	1000	440	11.4	7.5	3.9
	Co^{3+} §	^5D	$^1A_{1g}$	^5E_g	—	780	45	26	19
7	Co^{2+}	^4F	$^4T_{1g}$	$^4A_{2g}$	1000	440	17.1	15.0	2.1
8	Ni^{2+}	^3F	$^3A_{2g}$	$^3T_{1g}$	860	380	29.3	6.5	22.8
9	Cu^{2+}	^2D	^2E_g	$^2T_{2g}$	1300	580	22.2	6.6	15.6
10	Zn^{2+}	^1S	$^1A_{1g}$	$^1A_{1g}$	0	0	0	0	0

* The data given are:
1. Number of *d* electrons.
2. Transition metal ions.
3. Free-ion Russell–Saunders ground term (spin-orbit coupling is neglected in the term designation).
4. The standard group theory symbol for the ground state of the octahedrally coordinated ion in a solid (using nomenclature of Eyring *et al.*[267])
5. The group theory symbol for the tetrahedrally coordinated ionic ground state.
6. *Dq* values for octahedral hydrates of the ions.
7. *Dq* calculated for tetrahedral coordination.

8, 9. The thermodynamic stabilization in octahedral or tetrahedral fields.

10. The octahedral site preference, or the difference between columns 8 and 9.

§ The octahedral site stabilization of Co^{3+} was estimated from the heat of hydration increment caused by the crystal field, and the tetrahedral site stabilization was taken to be the same as for Cr^{3+}.

3.2.7. Oxidation–Reduction Potentials in a Transition Series

The members of the transition groups of elements usually have several charge or oxidation states in crystals and it is interesting to consider the energy changes from one such state to another. In a change such as

$$M^{2+} \longrightarrow M^{3+} + e \qquad \Delta E \tag{76}$$

(where the electron may be taken up by an acceptor) there will be a crystal field stabilization for both M^{2+} and M^{3+}, but the major energy change should be the third ionization potential. Thus

$$\Delta E = I_3 - \delta H(M^{3+}) + \delta H(M^{2+}) + C \tag{77}$$

where C is a constant to be determined for a particular series; I_3 represents the binding energy of the electron in the free atom. Figure 32

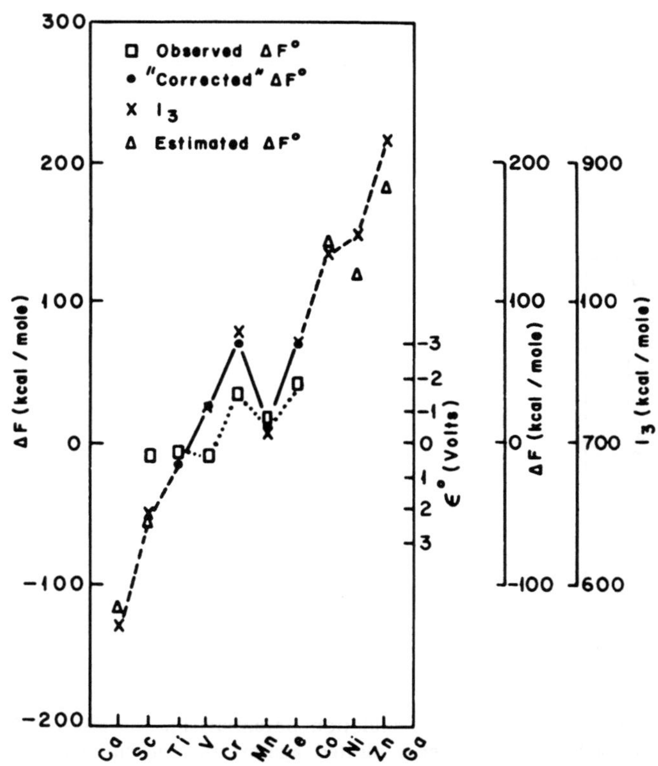

Fig. 32. The $M^{2+} \longrightarrow M^{3+}$ oxidation potentials of the first transition series ions. The corrected ΔF^0 values are the observed ones minus the crystal field correction, $\delta H(3^+) - \delta H(2^+)$. These values follow the third ionization potential very closely.[181]

shows such an analysis of the II–III oxidation potentials of the transition metal ions.[181]

In the rare earth and actinide series, where the electron binding energy is not known, the electrostatic splitting of the configuration can be expressed in terms of Slater–Condon theory, and the variation of the ground-state electron binding energy with atomic number is obtained.[169] In these two f series the crystal field contributions are small so the δH values are not as important.

This same equation, but with a different value of C, gives a reasonable description of the positions of charge-transfer absorption bands. Also, the relative location of the $4f \rightarrow 5d$ band positions of the rare earths is given by the Slater–Condon theory of the $4f$ configuration splitting, since the $5d$ binding energy is small and changes little with atomic number.[184] It is therefore pleasingly simple and

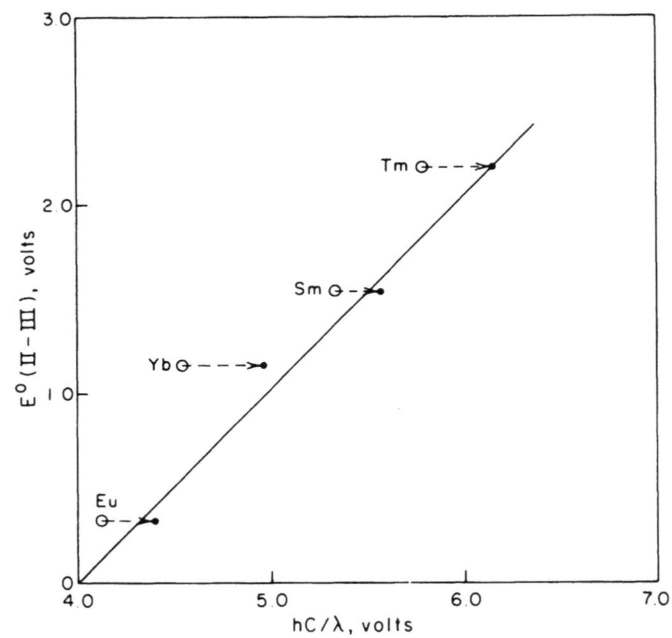

Fig. 33. The II → III oxidation potentials for the lanthanide hexachloride complexes plotted against the measured charge transfer band energy. The points × are measured values and the arrow tips represent values of the CT band energy that have been increased by 0.04q eV, where q is the number of f electrons in the divalent ion. Thus the charge transfer band energy for those MCl_6^{3-} ions whose oxidation potentials cannot be measured could be used to find the oxidation potentials, and this has been done.[186]

remarkably accurate to correlate chemical oxidation–reduction potentials, charge-transfer energies, and interconfigurational transition energies with each other. These correlations help fill in missing data points and provide a critique of new data.

Many examples of such correlations are given in a series of papers by Nugent *et al.*[185,186] and one example is shown in Figure 33. In this figure the energy values of the lowest charge-transfer band for MCl_6^{3-}, where M = rare earth, are plotted against the II–III oxidation potentials. The charge-transfer band is chloride to metal, so the metal goes from M^{3+} to M^{2+} in the spectroscopic process. The uncorrected values lie close to a straight line of unit slope, showing that the chemical and spectroscopic processes must be nearly the same. If a correction proportional to the number of electrons is added to the charge transfer energies, amounting to 0.04 eV per electron, the points lie accurately on a line of unit slope. This correction accounts for changes proportional to atomic number, which could be the differences between the effect of the lanthanide contraction on the chemical versus the optical process. Nugent *et al.*[185,186] have used such plots to help analyze and extend the chemical and spectroscopic observations on the actinide elements. It has been possible to make accurate estimates of oxidation potentials and to calculate those for which experimental data are not available. Although most of this work has been carried out with a view toward liquid solution chemistry, it applies equally well to solid solutions.

3.2.8. Charge-Transfer Spectra

There are high-energy optical transitions involving the 3*d* transition metal ions which are too intense and occur at too high an energy to be assigned as $3d \rightarrow 3d$ transitions. These are often assignable to ligand $\rightarrow$ metal charge-transfer transitions. An example is shown in Figure 34, where the spectra of the complex ions NiX_4^{2-} (as evaporated films of the tetramethylammonium halide salts) are given.[145] These spectra show a regular progression toward lower energies in the order Cl > Br > I. The reason for the assignment ligand $\rightarrow$ metal is now clear: It takes more energy to remove an electron from Cl^- than from Br^- and more from Br^- than from I^-; hence the spectroscopic energy is highest for Cl^- and lowest for I^- for halogen $\rightarrow$ metal transfer.

There is now a large and growing body of literature on charge-transfer spectra of inorganic complex ions having halide and oxide ligands, much of which has been reviewed by Jorgensen.[187] The

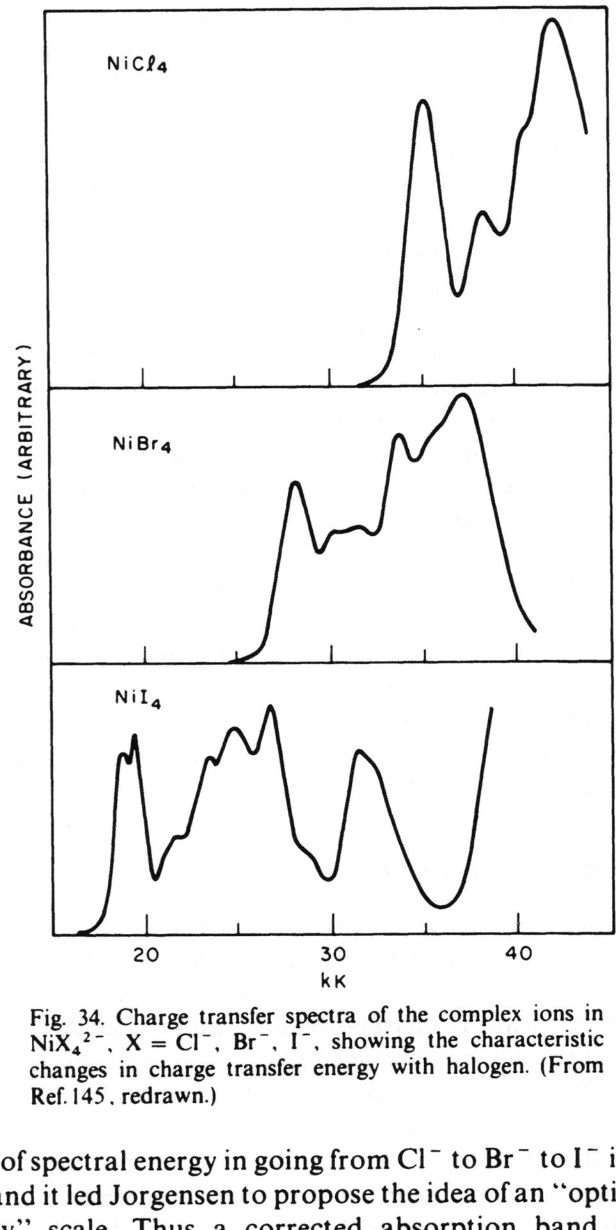

Fig. 34. Charge transfer spectra of the complex ions in NiX_4^{2-}, X = Cl^-, Br^-, I^-, showing the characteristic changes in charge transfer energy with halogen. (From Ref. 145, redrawn.)

decrease of spectral energy in going from Cl^- to Br^- to I^- is a familiar feature, and it led Jorgensen to propose the idea of an "optical electronegativity" scale. Thus a corrected absorption band position is given by

$$\sigma_{corr} = 30{,}000\ \text{cm}^{-1} \times [\chi_{opt}(X) - \chi_{opt}(M)] \qquad (78)$$

The correction will be for the spin pairing energy (a few thousand

cm^{-1}) to be discussed later and for the crystal field state of the electron. The values χ_{opt} for halides are: F^- 3.9, Cl^- 3.0, Br^- 2.8, I^- 2.5, and they roughly parallel Pauling's electronegativity scale. Values can be assigned to the metals as well, taking account of their coordination number.[187] Equation (78) gives a good first approximation to the position of the charge-transfer bands of a large number of systems.[188]

There has been some work on the corresponding spectra of impurity ions in crystals, although not enough to correlate the behavior of a large number of systems. Tippins[189] has shown that the spectra of trivalent transition metal ions in corundum, observed in the vacuum-ultraviolet region, correspond to oxygen-to-metal electron transitions, because the energy of these transitions decreases as one goes from Ti^{3+} to Co^{3+}. This decrease is expected because of the increasing electron affinity of the metal (increasing third ionization potential). In semiconductor language, the metal ions act as acceptors, and the electron comes from the valence band.

There have been few examples reported of donor activity on the part of the metal ion for these highly ionic systems. Anderson[92] observed photoconductivity in CaF_2 doped with divalent rare earth ions. The photocurrent is undoubtedly an electron current. Therefore these highly reduced ions are acting as would be expected: They lose an electron to the conduction band and return to their normal trivalent state. The internal transitions in the divalent rare earths are $4f \rightarrow 5d$. The excited $5d$ levels apparently overlap the states of the conduction band because the photoconductivity threshold of $\sim$1–2 eV is quite low in the manifold of $f^{n-1}d$ states. Therefore practically all of these states are unstable with respect to "internal Auger" processes.[190]

Recently, Sabatini[191] has shown that new data obtained by him on the vacuum-ultraviolet spectra of divalent transition metal ions in $KMgF_3$, MgF_2, and CaF_2 prove that the metal ion acts as a donor. This behavior of the metal in fluoride lattices is in sharp contrast to the behavior of the same metals in lattices containing the other halides referred to in earlier paragraphs. In fluoride lattices the transition energy is well above the highest $3d \rightarrow 3d$ internal transition but could correspond to $3d \rightarrow 4s$. But $4s$ should be quite delocalized and it may be correct to consider the transition as a true metal-to-conduction band process. Because of the small overlap between $3d$ impurity ion states and $4s$ or band states, these transitions are weak, relative to ligand $\rightarrow$ metal-charge transfer transitions, but are still stronger than $3d \rightarrow 3d$ transitions.

3.2.8.1. Theory of Charge-Transfer Transitions

Charge-transfer transitions of impurity ions in crystals are probably quite localized on the central atom and a few of its near neighbors. Since a charge separation is always involved, the electron and hole are attracted by a Coulomb field, and at least the lowest charge-transfer states are bound states for this reason. The simplest and most useful theoretical description of these states is the molecular orbital, LCAO description (Section 3.3.2). It is possible to parametrize this theory and determine the parameters from the spectra.[145] Theoretical estimates of the parameters can also be made.

The simplest examples of charge–transfer processes are those in which there is only one open shell in each state of the transition: That is, the metal atom (acceptor) must have a single hole in an otherwise filled shell in the ground state, and in the excited state the hole transfers to the ligand orbitals. This special case will be discussed first.

(a) *The Case of One Open Shell.* The halide complexes of tetravalent iridium have strong absorption bands in the visible region.

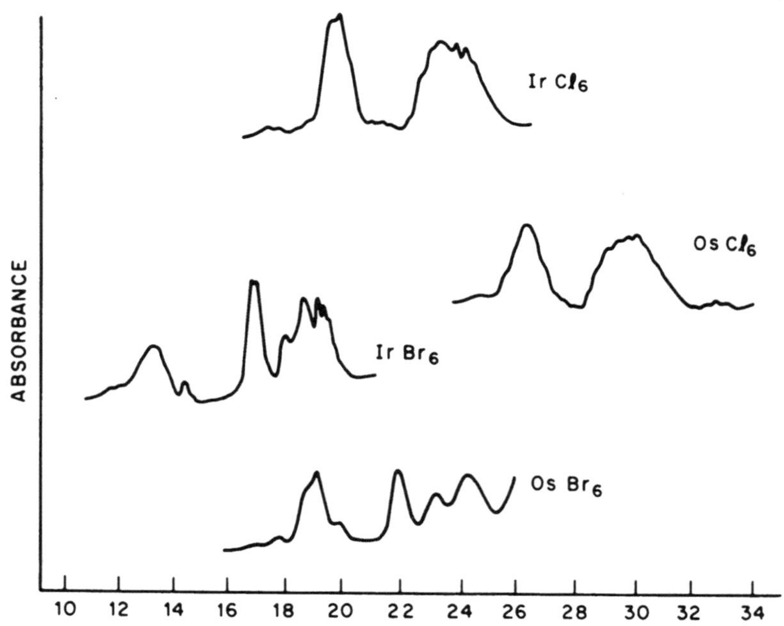

Fig. 35. Charge transfer spectra of Ir^{4+} and Os^{4+} halides, showing both halide shift and metal ion shift. The higher energy for the lower atomic number Os^{4+} ion shows that the ions are acting as electron acceptors in these transitions. (Ref. 145, redrawn.)

Figure 35 shows the $IrCl_6^{2-}$ and $IrBr_6^{2-}$ spectra,[192] and from the fact that the bromide comes at a markedly lower energy (7000 cm^{-1}) these bands are reasonably ascribed to charge transfer. The comparison with osmium also supports this assignment, since osmium, having one less nuclear charge, has its corresponding bands at considerably higher energies than iridium ($\sim$6000 cm^{-1}).

The details of the spectra of these complexes have been worked out in a series of beautiful papers by Schatz and his co-workers.[114, 193–196] Figure 12 shows the first three absorption regions of $IrCl_6^{2-}$ in Cs_2ZrCl_6 at low temperatures. Both the absorption and the magnetic circular dichroism spectra are given.[114]

The Ir^{4+} ion has a $5d^5$ shell, and in a very strong crystal field the lowest term is $^2T_2(t^5)$. Because of the greater radial spread of the $5d$ orbital relative to the $3d$, the ions of the third transition group experience large crystal fields and their ground states are derived from the lowest configurations (see Figure 17). The acceptor-type charge-transfer transition places an electron in the t^5 shell, filling it and leaving a hole in the ligand orbitals. Thus there is only one open shell in the ground and in the excited state.

As shown in Figure 23, the most weakly bound of the ligand molecular orbitals is the πt_{1g} orbital, whose nodal pattern shows it to be antibonding between all ligands (Figure 22). The transition $\pi t_{1g} \rightarrow dt_{2g}$ is parity forbidden and therefore weak. The weak band shown in Figure 12 near 17,000 cm^{-1} is attributed to this transition.[193,194] The next region is very strong and, according to Schatz *et al.*,[193] corresponds to the $t_{1u} \rightarrow dt_{2g}$ transition of Figure 23. This transition is out of order on the energy diagram constructed from the node patterns (Figure 22), yet the assignment based on the use of magnetic-circular dichroism (MCD) seems quite certain.[193–195] The next two transitions, from the t_{2u} and t_{2g} π levels, are in the expected order. The theory of these level orders has been considered by Jorgensen and Preetz,[197] who show that the πt_{1u} orbital energy relative to πt_{2u} can be quite variable from one complex to another, and that the πt_{1g} and πt_{2u} separation should be rather constant.

An energy level diagram of the $IrBr_6^{2-}$ complex[195] showing the spin–orbit splitting is given in Figure 36. This is a state diagram, not an orbital diagram as is Figure 23, and it represents the spectroscopic band positions. The spin–orbit product notation is that of Griffith.[103] Next to each symmetry symbol is given the theoretical value of the C term of MCD and then the dipole strength. The agreement in sign and magnitude between the observed and calculated C values as well

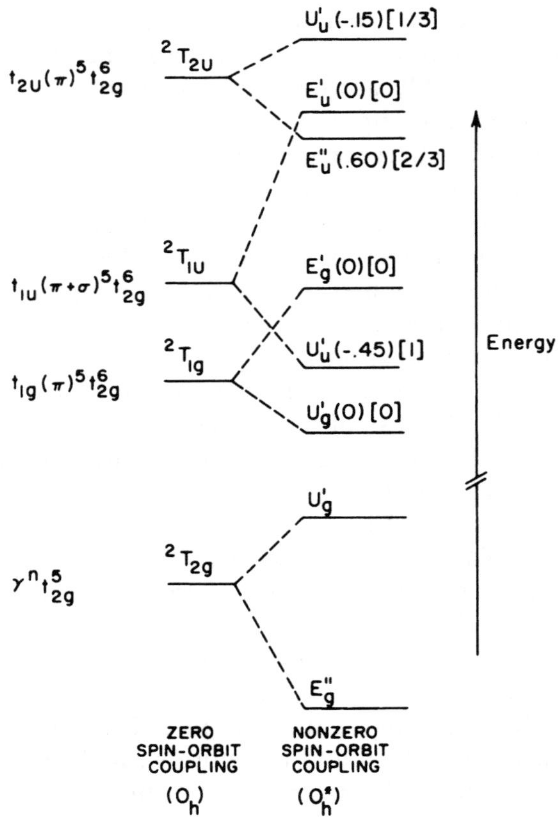

Fig. 36. Energy level diagram for charge transfer transitions of $IrBr_6^{2-}$ in Cs_2ZrBr_6. The lowest level E_g'' is the ground state. The orbital parentage is $^2T_{2g}$, and U_g', E_g'' are spin–orbit components. This diagram is fairly well verified by experiment. The theoretical value of the C term in the MCD spectrum is shown in parentheses, and the theoretical dipole strength is shown in square brackets.[195]

as the agreement with the D value confirm the spectroscopic assignments. Those states having $D = 0$ are observed to have vibronically induced spectra, showing both the expected temperature dependence [Eq. (35)] and the spectral structure.[195] The general appearance of the low-resolution spectra of Figure 35 is in agreement with this diagram. When the ligands are Cl^- instead of Br^- the spin–orbit splitting of the excited states is about one-fourth as great, and high-resolution spectra, such as is shown in Figure 12, are necessary to be able to make assignments.

(b) *More Than One Open Shell.* The OsX_6^{2-} spectra of Figure 35 look remarkably similar to those of IrX_6^{2-}. This is surprising because the ground configuration of Os^{4+} is t^4 and the acceptor-type excited states must belong to configurations $(L)^5(dt_{2g})^5$. For each ligand hole, say, for example, $(\pi t_{2u})^5$, there will be many states, as is shown by multiplying out the representations $t_{2u} \times t_{2g}$, giving eight states, and when spin–orbit coupling is included there are 15 states. The selection rules, however, permit only two of these states to have high intensities, the same number as in the corresponding configuration of IrX_6^{2-}. This remarkable simplification applies to the other configurations of OsX_6^{2-} and is commonly observed in other complexes.[145,192]

The electrostatic interaction splits the nine orbital components, but the magnitude of this splitting is not known. It is presumed to be small,[114] since the $5dt_2$ and $np\pi$ orbitals are localized on different atoms, and furthermore the $np\pi$ hole is distributed over six halogen atoms.

In recent work the magnitude of an electrostatic splitting in $Mo^{4+}(Cl)_6$ has been found.[198] For the $t_{1u} \rightarrow t_{2g}$ excitation the $^3T_{2u}$–$^5T_{2u}$ splitting is close to 515 cm^{-1}.

As a result of the work on the charge-transfer transitions, the order of orbital levels (averaged spin–orbit levels) has been found to be that implied by Figure 36.

3.2.8.2. Donor-Type Charge-Transfer Transitions

The description of donor-type transitions has not progressed as far as that for acceptor transitions. In Sabatini's work on transition metal impurities in metal fluoride hosts[191] the donor assignment is verified by the dependence of the transition energy on atomic number of the metal. Figure 37 shows the absorption bands of V^{2+}, Cr^{2+}, Mn^{2+}, Fe^{2+}, Co^{2+}, and Ni^{2+} in $KMgF_3$, along with the calculated position of the onset of absorption. Since the lattice is the same for all of the ions, the only quantities which vary along this series are those associated with the metal ion. For a charge-transfer process these are the third ionization potential I_3, the crystal field stabilization of the ground state Δ_1, and that of the excited state Δ_2. So we should have

$$\Delta E(\text{transition}) = I_3 + \Delta_1 - \Delta_2 + C \qquad (79)$$

where C is a constant determined from one data point. The fit is quite good, as shown in Figure 37. An equally good fit is provided,

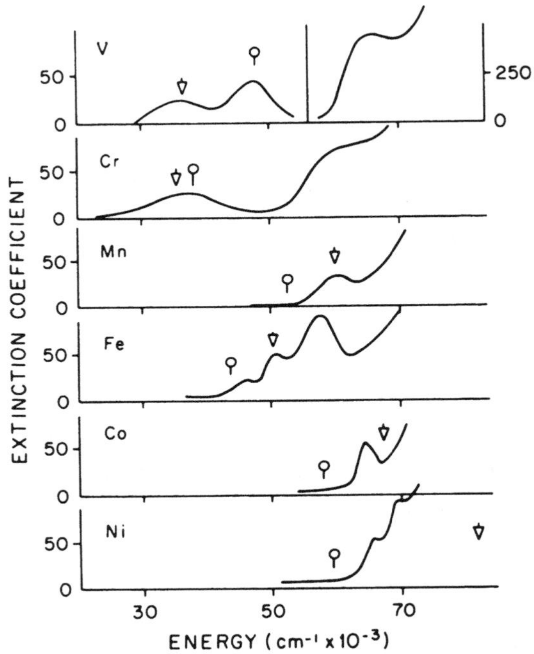

Fig. 37. Ultraviolet absorption bands of the $3d$ transition metal ions in $KMgF_3$ host crystal. These spectra represent preliminary data. Arrows show energies calculated for donor-type charge transfer transitions, using Mn^{2+} to determine C in Eq. (79). Circles show energy of $3d \rightarrow 4s$ transition in free atom, corrected for crystal field effects.[191]

however, by the energies of the $3d \rightarrow 4s$ transitions of the free divalent ions; and in this case no adjustable constant is necessary. These data are also shown in Figure 37.

The detailed properties of these spectra are in agreement with the donor or $d \rightarrow s$ assignment. Their oscillator strengths are low for charge-transfer transitions, varying from 10^{-4} to 10^{-3}, while for the acceptor-type transitions in $FeCl_4^-$, $f = 0.1$–0.3. The low value of oscillator strength is understandable for a $3d \rightarrow 4s$ assignment, since the transition is parity forbidden. In fact the oscillator strength is temperature dependent; for $KMgF_3$:Co it follows Eq. (35) with $\omega = 320\ cm^{-1}$. Thus local metal–ligand vibrations are necessary to break the parity restriction on the transition.

There are many questions to be answered by future investigations of these donor transitions. The nature of the wave function of the excited state would be one of the important ones to clarify.

3.3. Electronic Structure and Spectroscopy of Non-Transition Group Impurities

3.3.1. Post-Transition Elements

The elements occurring at the ends of the *d* transition groups have valence shell electrons which are more strongly bound than in the analogous pre-transition element. For example, the Cd 5*s* electron has a binding energy of 11.1 eV, while in Sr it is only 6.2 eV. The reason is that the nuclear charge is less well shielded by *d* electrons than by *p* electrons. The s^2 shell is therefore relatively stable in the post-transition metals, and, for example, in Sn or Tl, there are two stable oxidation states, 2 and 4 for Sn; 1 and 3 for Tl. The s^2 shell is always filled in the lower of these oxidation states.

When these elements in their lower oxidation state occur as impurities in crystals, one can observe excitations from the localized *s* shell to higher levels of the impurity–crystal system. Even these higher levels are somewhat localized. They are approximately described as the atomic $s^2 \longrightarrow sp$ transitions[199] (see Ref. 118, pp. 511–523), as is supported by the observation of the correct number of bands. The excited states are $^3P_0$, $^3P_1$, $^3P_2$, $^1P_1$. Atomic selection rules require the $^1S_0 \longrightarrow {}^1P_1$ to be the only allowed transition, but since spin–orbit coupling in these heavy elements is strong, $^1S_0 \longrightarrow {}^3P_1$ is also permitted. Thus two T_{1u} bands are expected and are observed, as shown in Figure 38. The $^1S_0 \longrightarrow {}^3P_2$ transition is also seen in absorption in some cases as a very weak band,[200] and $^1S_0 \longrightarrow {}^3P_0$ is probably responsible for some emission phenomena.

The atomic $^1S_0 \longrightarrow {}^1P_1$ transition occurs at a much higher energy than its supposed analog in crystals. Recalling that the $3d \longrightarrow 4s$ transitions in the transition series metals occur quite near the free atom values (Section 3.2.8.2), one is tempted to conclude that the 4*s* level is less perturbed by the crystal than is the 4*p* level, and that 4*p* is strongly stabilized by crystal forces. In fact, it must become mixed into conduction band states of the host crystal having the appropriate local symmetry, T_{1u} in cubic crystals. In this sense, the $s^2 \longrightarrow sp$ transitions must have an appreciable component of donor-type charge transfer.

The principal interest in these spectra in recent years has been the understanding of the Jahn–Teller distortion, which has an effect on both the absorption and the emission spectra. The T_{1u} excited states can interact with the nontotally symmetric local cluster modes of e_g and t_{2g} symmetry as well as with the symmetric a_g mode of the

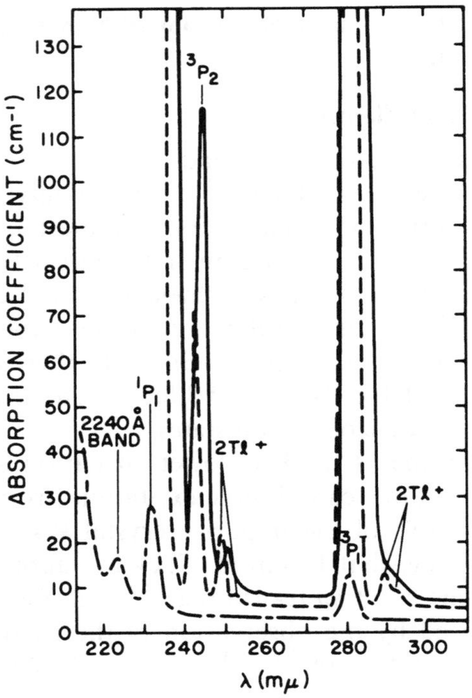

Fig. 38. Absorption spectrum of Tl^+ in KI. (·- -: lowest curve) Tl^+ concentration 1.2×10^{-4} mol %; 4°K. Only the strongest bands, namely $^3P_1$, $^1P_1$, and the 2240-Å bands are visible. (---) Tl^+ concentration 3.7×10^{-2} mol %; 77°K. The $^3P_2$ band appears strongly, and evidence of bands due to pairs of neighboring Tl^+ ions is present. (— —) Same concentration for solid curves, but at 4°K. The $^3P_2$ band is considerably weaker than at 77°K and four bands due to Tl^+ pairs are resolved.[200]

octahedral environment for the case of an ion like Tl^+ in KCl. Cho[201] has done a good model calculation of the band shape, allowing the excited state to interact linearly with these modes, and using the Franck–Condon principle to determine the transition probability. The calculations explain qualitatively the weak triplet structure observed in these absorption bands as being due to the ground–excited state vibrational overlap distribution with the multiple potential surfaces of the upper state. Fukuda[202] has obtained and analyzed some of the experimental data on these spectra and has attempted to find the coupling constants used in the theoretical description. The problem becomes more complex at this point, however, since it appears that configurational interaction between different states is a function of the Jahn–Teller distortion, and that quadratic coupling is important.

A recent study of the response of these bands to high pressure[203] shows, for the case of In^+ in alkali halides, the expected increase in the band splitting, consistent with the Jahn–Teller effect, but no such increase was observed for Sn^{2+} in KBr, etc. It was surmised that the elastic constants in the vicinity of Sn^{2+} are much higher than in the bulk of the crystal, so that no appreciable displacements could have been caused by the pressure.

These systems are particularly interesting since the electrons involved are moderately delocalized, intermediate between *d* electrons and band electrons. Their study will undoubtedly lead to a better understanding of electronic wave functions in solids. Some recent references will illustrate the type and extent of the work being done in this area.[201–212]

3.3.2. Impurities in II–VI Compounds

The compounds ZnS, CdTe, etc. formed from a column IIb and a column VI element are considered to be less ionic than the alkali halides and the a-column oxides. The b-column elements occur at the ends of the *d* transition groups and the *s* shells of these elements are much more strongly bound than those of the analogous pre-transition element. This increased binding energy for valence electrons is the reason for the increased covalency or reduced ionicity of post- compared to pre-transition metal compounds. Impurities in these compounds induce electronic properties which are desirable for many applications. The most important phosphor systems are based on ZnS and CdS,[213,214] and important photoconductors are based on CdS.[78]

The vast literature on these systems is summarized in articles in Aven and Prener's volume,[79] and the only reason to include the present section is to contrast these host crystals with the more ionic ones which have been discussed.

Highly localized *f* or *d* states are formed by the rare earth and transition group elements, as shown by ESR and optical absorption and emission spectra of a fair number of examples.[175,215,216] Since the coordination is tetrahedral in these hosts when the impurity is at a lattice site, the values of Dq and other crystal field parameters are smaller than in the octahedrally coordinated ionic compounds, but they are comparable to the values observed in oxides or halides for this same coordination number.

The first indications were therefore that the f and d shells in II–VI hosts are not appreciably different from those in other, more ionic hosts. It was natural to wonder, however, if in some of these hosts the d shell would become strongly delocalized. The orbital reduction factors (see Section 3.2.2) of 0.7–0.9 deduced from the optical or ESR spectra of transition metal atoms in crystals are interpretable in terms of the transfer of some of the d electron to the ligands. In one case a reduction factor of 0.46 was reported, namely for Cu in ZnO,[217] and it was explained in terms of MO theory. Thus it might be expected that reduction factors between zero and one could be found if one looked hard enough.

The reduction factors are found in ESR spectroscopy from the deviation of the g-factor from the value it would have if the d orbitals were free atom orbitals. For Cu^{2+} in a tetrahedral site in hexagonal ZnO the ground state is $^2T_2$ with a slight trigonal splitting. The reduction factor is the matrix element of orbital angular momentum: $k_{11} = -\langle t_2 1 | 1_z | t_2 1 \rangle$, where the component $\langle t_2 1 |$ is a trigonal basis state with orbital angular momentum -1 along the trigonal axis for the case of a pure d orbital. Thus $k_{11} = 1$ for an unperturbed d orbital, but 0.46 for Cu in ZnO.

There is nothing wrong with the measurement of the reduction factor, but its interpretation in terms of extensive delocalization has been shown to be wrong.

Yamaguchi and Kamimura[218] have shown that the Jahn–Teller effect is responsible for part of the reduction of the g-value, and that the part attributable to covalency is not appreciably different from the normal values found in other compounds. Thus the d electrons are largely localized on the copper atom. So far as reduction factors are concerned, it is indeed true that values from zero to one can be found, but this has more to do with Jahn–Teller effects than with electron delocalization. Because of the occurrence of these two factors, it is an involved process to interpret experimental data so as to separate the two contributions.

In ZnS : Ni there is a band in the blue region of the spectrum above all the d–d bands, and because of its high intensity and position, it must be a charge-transfer band. This is shown in Figure 39, from Weakliem.[175]

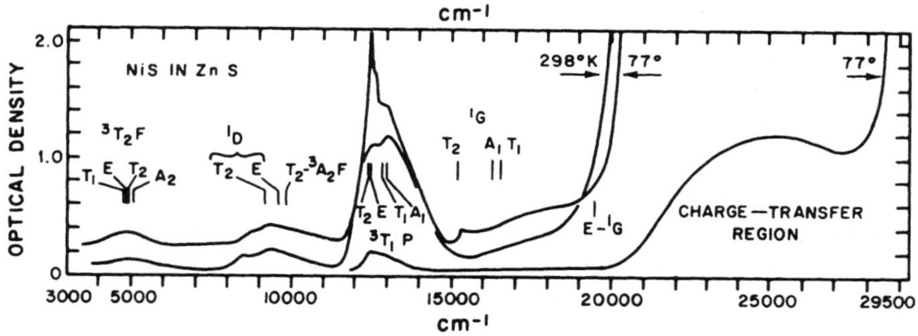

Fig. 39. Absorption bands of Ni^{2+} in ZnS showing *d–d* bands and a charge-transfer band.[175]

Kazanskii and Ryskin[219] have shown that the excited state of this transition corresponds to having an a_1 hole in the S^{2-} lattice and a $^2T_2$ Ni^+ ion. Thus it is an acceptor-type transition. ZnS:Co shows a strong absorption region at higher energy than Ni, in agreement with the acceptor assignment. Kazanskii and Ryskin[220] claim that the hole is released into the valence band when the optical excitation is high enough above the onset of the absorption band, i.e., by 0.17 eV or more. The released hole was trapped by Cu^+ ions in their experiments and this act was detected by the observation of infrared luminescence of the Cu^{2+} so formed via the $^2T_2 \longrightarrow {}^2E$ *d–d* band.

The mobility of the hole implied in the previous example and the high mobility of electrons lead to the classes of phenomena for which these host crystals are well known, luminescence[221] and photoconductivity.[78] The oxidation and reduction processes occurring in transition metal impurities are important in the mechanisms of energy storage, trapping, luminescence, and quenching of luminescence.

The most versatile impurity in ZnS is copper. Whereas Co and Ni are moderately soluble in their divalent states, copper is not, as mentioned in Section 2.3, and prefers to enter the lattice as a monovalent ion with a charge compensator. The latter is usually Cl at a S site or a trivalent ion, say Al at a Zn site. Since the copper and the compensator tend to associate in the crystal, the defect is a dipolar pair which has the special electronic properties associated with copper as an impurity. Thus copper is not analogous to the other transition metals except in the rare case when it occurs at a Zn site with no compensator nearby.

The driving force tending to convert copper from a d^9 ion to a d^{10} ion is the exchange energy gained by adding an electron to the d^9 shell. Thus the free atom Cu^+ prefers d^{10} to d^9s by nearly 3 eV. In the less electronegative environments of the II–VI compounds, copper can exist as the d^{10} ion with a hole in the valence band, or as a d^9 ion with a completed valence shell, i.e., with the hole transferred to the *d* shell; while in the more electronegative environments of oxide or halide the d^9 configuration would be preferred. In the presence of an electron donor, therefore, copper in II–VI compounds accepts an electron so that both the *d* shell and the valence shell are filled. Note that in compounds the electrons of the valence shell around the copper are chiefly bound by the electronegative atoms such as sulfur, although the 4*s* orbital of copper provides the extra binding energy distinguishing its behavior from that of the monovalent alkali metals. Copper has one less nuclear charge than the zinc atom it replaces, so the binding energy of the valence electrons, largely sulfur electrons, in its vicinity is slightly decreased compared to unsubstituted sites.

A similar reduction of binding energy occurs when S is replaced by group Vb elements, P, As, and Sb, which also have one less effective nuclear charge than the atom replaced. But when Zn is replaced by Ga (or S by Cl) there is one extra nuclear charge and these centers bring in an extra electron, which is not needed in the valence structure of the crystal, and can be donated. The Cu–Cl pair thus has a charge distribution which can be loosely described as Cu^-–Cl^+ when in its lowest state. The charge shift lowers the energy of the system by nearly the band gap energy of 3.7 eV (in ZnS), thereby completing the tetrahedral bonding around each impurity, and if the two impurity atoms are closely associated, up to another $\frac{1}{2}$ eV of energy may be released. The electronic structure in the ground state thus approximates the band structure of the pure crystal. The crystal structure in the vicinity of a near-neighbor pair is shown in Figure 40, although its precise geometry is not known.

The large reduction of energy resulting from the electron transfer reduces the heat of solution and increases the solubility of the donor–acceptor pair. (See p. 21.)

When considering the electronic excitations of the pair one must include the electrons shared by the surrounding atoms. Figure 40 shows that the Cu is surrounded by electronegative S atoms and that Cl is surrounded by electropositive Zn atoms. The electronic excitation Cu^-–$Cl^+ \longrightarrow Cu^\circ$–$Cl^\circ$ could be better described as an electron being excited from S orbitals to Zn orbitals. But this is none other

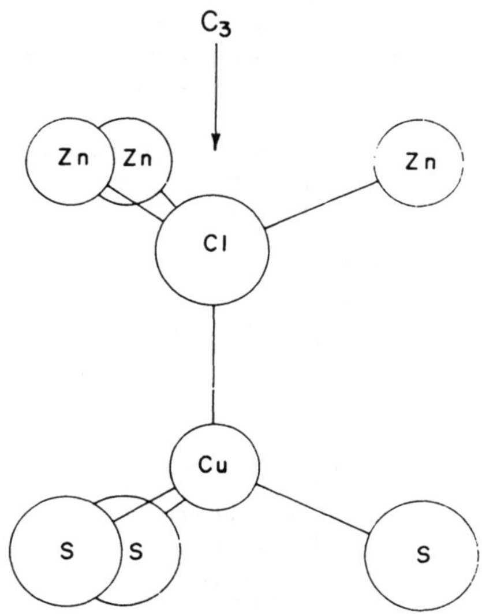

Fig. 40. Crystal structure in the vicinity of a Cu–Cl pair in ZnS.

than the band gap transition. Actually, since the binding energy of the electrons around the Cu atom is less, and that of the electrons around the Cl atom greater, the transition of the dipolar pair is at a somewhat lower energy than in the pure crystal.

A molecular orbital description of the electrons in the vicinity of the dipolar pair gives a reasonable idea of the relationship between the localized transition and the band to band transition. As shown in Figure 40, the dipolar pair has C_{3v} symmetry. Table 14 shows the molecular orbitals which can be constructed by using the tetrahedral bonding orbitals near the Cu and Cl, classified according to the representations of C_{3v}.

An energy diagram shown in Figure 41 is constructed using the following principles: (1) The greater the atomic number, the deeper the binding energy for a given type of orbital. (2) For a given type of atom, the MO with the least number of nodes is the lowest. (3) The A–E splitting is comparable to the bandwidth, either conduction or valence band, depending upon whether the state is one of the filled or one of the empty orbitals in the ground state. (4) The gap between filled and unfilled orbitals is comparable to the band gap; the highest filled orbital is the sulfur E orbital corresponding to the Γ_{15} band state at the Γ point in the zone of the pure crystal,[222] and the lowest unfilled orbital is the zinc A orbital, corresponding to the Γ_1 state of

TABLE 14
Molecular Orbitals for the Eight-Atom System $S_3CuClZn_3$

Basis orbitals*		
Sulfur sp^3	$S(A_1) = (\sigma_1 + \sigma_2 + \sigma_3)/\sqrt{3}$	A_1
	$S(E_1) = (2\sigma_1 - \sigma_2 - \sigma_3)/2\sqrt{3}$	E_1
	$S(E_2) = (\sigma_2 + \sigma_3)/\sqrt{2}$	E_2
Copper 4s	$Cu(A_1) = s_c$	A_1
Chlorine $sp^3\sigma$	$Cl(A_1) = (g_\alpha + g_\beta + g_\gamma)/\sqrt{3}$	A_1
	$Cl(A_1, \delta) = g_\delta$	A_1
	$Cl(E_1) = (2g_\alpha - g_\beta - g)/2\sqrt{3}$	E_1
	$Cl(E_2) = (g_\beta - g_\gamma)/\sqrt{2}$	E_2
Zinc 4s	$Zn(A_1) = (s_1 + s_2 + s_3)/\sqrt{3}$	A_1
	$Zn(E_1) = (2s_1 - s_2 - s_3)/2\sqrt{3}$	E_1
	$Zn(E_2) = (s_2 - s_3)/\sqrt{2}$	E_2
Mixed orbitals in order of increasing energy†		
$1A_1$	$Cl(A_1, \delta) + \mu Cu(A_1)$	
$2A_1$	$Cl(A_1) + \varepsilon Zn(A_1)$	
$3A_1$	$S(A_1) + \nu Cu(A_1)$	
$1E$	$Cl(E) + \varepsilon Zn(E)$	
$2E$	$S(E)$	
$4A_1$	$Zn(A_1) - \varepsilon Cl(A_1)$	
$3E$	$Zn(E) - \varepsilon Cl(E)$	
$5A_1$	$Cu(A_1) - \mu Cl(A_1, \delta)$	

* σ_1, etc., are sulfur $sp^3\sigma$ orbitals directed toward Cu.
g_α, etc., are Cl $sp^3\sigma$ orbitals directed toward Zn.
g_δ is a Cl orbital directed toward Cu.
s_c and s_1, etc., are Cu and Zn 4s orbitals.
† μ, ν, ε are small numbers.

the conduction band. The Cu and Cl orbitals are not strongly involved, but affect the transition energy through their effect on the sulfur and zinc electron binding energies. The Cl atom increases the ability of the Zn to bind an electron in the conduction band, leading to a shallow donor level, and the Cu atom decreases the ability of the S atoms to bind electrons in the valence band, leading to a shallow acceptor level.

The energy levels on the diagram of Figure 41 are filled through level $2E$. The transition $2E \longrightarrow 4A_1$ should correspond to a transition somewhat lower than the band gap and be polarized perpendicular to the Cu–Cl axis. The reverse process should correspond to the green emission band of Cu–Cl-doped ZnS phosphors. The polarization of a similar defect pair, Cl–V_{Zn}, has been shown by Koda and Shinoya[223] to be polarized perpendicular to the Cl–V_{Zn} axis. Therefore it appears that this simple energy diagram rationalizes the main facts. No use has been made of the copper d orbitals, so in this case it does not behave as a transition metal atom. On the other hand, the orbitals shown in Table 14 only form a crude basis for the problem. In a

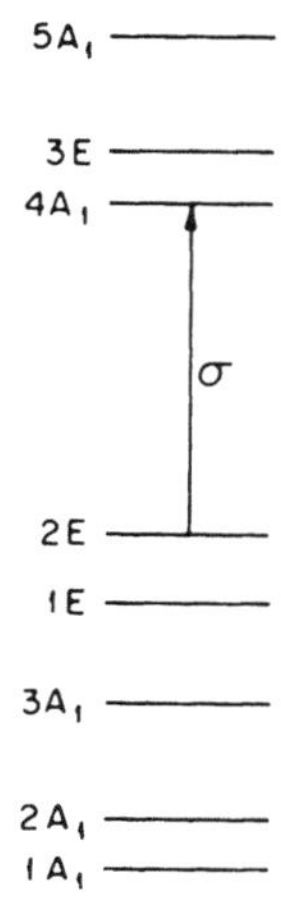

Fig. 41. Molecular orbital diagram of a Cu–Cl pair in ZnS.

continuation of the treatment one would need to consider the mixing of orbitals having the same symmetry, and the mixing of configurations caused by electron–electron interaction. No such detailed calculations have been made.

This example illustrates one of the least localized types of electronic transitions in solids. Analogous donor–acceptor pair transitions also are observed in III–V compounds, GaP, etc., and will be discussed briefly in Section 3.4.3.

It may seem surprising that the IIb–VI compounds have many properties in common with the III–V compounds, and indeed with the elemental group IV semiconductors, while the IIA–VI compounds (e.g., MgO, CaS) are much more similar to the ionic alkali halides. The greater *s*- and *p*-electron binding energy of the b-group metals is enough to make tetrahedral bonding preferable to ionic bonding, and therefore to stabilize the zincblende structure over the sodium chloride structure. The smaller band gap of the b-group compounds results in more possibilities for impurity and defect geometries since electrons are less localized. Thus many of the localized transition metal centers are photosensitive[219,224,225] and nontransition impurities are donors, acceptors, or compensated pairs.[226,227]

3.3.3. Spectra of the Substituted Alkali Halides

An interesting illustration of the conditions for the localization of an electronic excitation is the comparison of the spectra of mixed alkali halides when on the one hand the composition of the halide lattice is being changed, and on the other when the metal lattice is

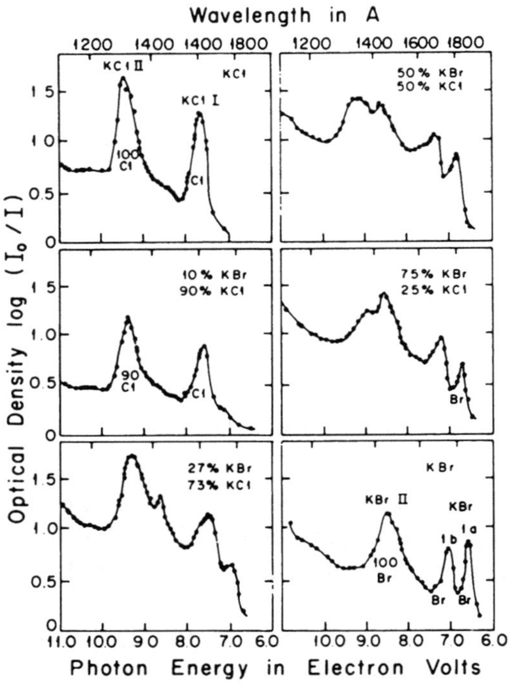

Fig. 42. Absorption spectra of thin films of KCl–KBr mixed crystals for various compositions. Note that the spectra of the mixtures are nearly a superposition of the spectra of the pure materials.[228]

being changed. An example of the first is Mahr's study[228] of the KCl–KBr system shown in Figure 42. It is easily seen that to a first approximation the spectra of the mixtures are superpositions of the spectra of the pure components with only small band shift. In marked contrast to this behavior is that of the KCl–RbCl system, shown in Figure 43, taken from the work of Nakai *et al.*[229] The spectrum of the mixed system is an average of those of the pure systems, and the bands shift nearly linearly with composition.

Onodera and Toyozawa[230] label the two types of mixed crystal spectra as persistence and amalgamation type. (Also see the recent work of Ref. 231.) The persistence type, of which halogen-substituted alkali halides are examples, show spectra which seem to be due to the excitations of the two halogen ions, e.g., Cl^- and Br^- in KCl–KBr mixtures. They merely change intensity with concentration. The second type shows only one spectral structure which shifts in energy

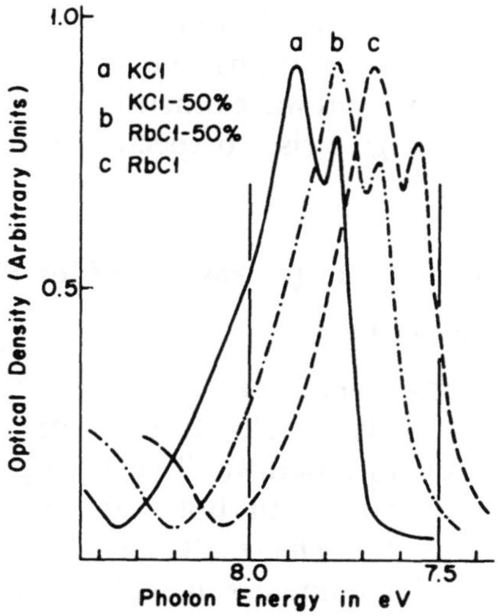

Fig. 43. Spectra of RbCl, KCl, and 0.5RbCl:0.5KCl films, showing shift instead of superimposed spectra in the mixture.[229]

with changes of composition. The alkali-metal-substituted alkali halides and the Ge–Si alloys are examples.

When the difference of the atomic binding energies $E_A - E_B$ of two atoms A and B is large compared to the bandwidth the electron states tend to be localized, and two separate spectra will be observed which are characteristic of atoms A and B. When $E_A - E_B$ is small compared to the bandwidth the transfer integral from one atom of the crystal to another is large, and the crystal does not act like an imperfect one; the spectrum resembles an average of the spectra of the two pure systems.

Returning to the alkali halides, we can see qualitatively why the halogen impurity produces its own localized excitation to a good approximation. Its neighbors are alkali atoms and the impurity halide electrons are moderately well localized in the ground state, but in the excited state the excitation spreads over the alkali atoms as it would in a band-to-band transition. Therefore transitions from either the impurity or the host halogen atoms begin on well-localized and energetically different levels, and end on delocalized levels at nearly the same energy in each case. An alkali impurity acts as an electron acceptor in an optical transition, but does not have a highly localized orbital available. Thus the final state is a mixture of orbitals spread out over many metal atoms and the impurity orbital contributes only a part of the final wave function. It is clear therefore that alkali impurities should give rise to spectra different from those of halide impurities.

3.4. Electronic Structure and Spectroscopy of Ion Pair Systems

3.4.1. Exchange-Coupled Pairs

When two impurity metal ions are in nearby lattice sites their inner shells may be weakly coupled to each other either directly or through a "superexchange pathway" involving the anions. Such systems have been studied with both optical and electron spin resonance methods.[232–234] The results have been used to determine exchange parameters for comparison with the parameters derived from the magnetic behavior of the pure compounds of the interacting ions.[10]

This field of study obviously impinges on the enormous area of magnetism, and also on the large area of inorganic chemistry occupied with the synthesis and study of polynuclear metal complexes.[235]

A large number of observations on both ground- and excited-state exchange interactions can be correlated by using the empirical formula for the exchange energy:

$$\mathcal{H}_{\text{ex}} = JS_a \cdot S_b \tag{80}$$

where J is the exchange energy parameter and S_a and S_b are the spins of the interacting ions. The value of J is 4.5 cm^{-1} for the 180° pair V–F–V in $KMgF_3:V^{2+}$, 50 cm^{-1} for V–O–V in MgO, and 240 cm^{-1} for the nearest-neighbor Cr^{3+} in Al_2O_3, to quote a few examples from the extensive compilation in the review by Owen and Harris[234] for exchange between ground-state ions. These interactions are antiferromagnetic, that is, the lowest state of the pair has zero total spin.

This form of interaction can be justified on theoretical grounds when the two ions are in nondegenerate states.[236] However, it is a special case of a more general form of interaction:

$$\mathcal{H}_{\text{ex}} = \sum_{i,j=1}^{n} J_{ij} s_i^a \cdot s_j^b \tag{81}$$

Here the J_{ij} is the exchange constant for the interaction between orbitals i and j on atoms a and b, respectively, and s_i^a and s_j^b are the spin operators for the electrons. This Hamiltonian is obviously better able to describe the interaction between ions in arbitrary states of excitation. It has been used by Huang[237] and Van der Ziel,[232] for example, to describe the interactions between Cr^{3+} pairs in which one or both members are electronically excited.

Going beyond these semiempirical expressions, the exchange constants can be calculated in terms of the parameters of a molecular orbital theory. This has been done most successfully by Freeman,[238] who applied it to the excited states of MnF_2.

When spin–orbit coupling is strong it is no longer justifiable to use the isotropic interaction operators $s_i^a \cdot s_j^b$. Various forms of anisotropic interaction then come into play. There is a comprehensive review of these by Levy.[239]

In the optical spectra of pairs the total spin $S = S_a + S_b$ of the pair is a good quantum number if spin–orbit coupling is not too large. The selection rule $\Delta S = 0$ is fairly well obeyed in some examples. One such example is that of the Cr^{3+} pairs in $LaAlO_3$, shown in Figure 44, from Van der Ziel's paper.[232] This figure shows the absorption spectrum in the region of the nearest-neighbor pairs, Cr–O–Cr, with the angle nearly 180°, and separation of 3.79 Å. There is strong superexchange interaction between the Cr atoms, so the ground state has $S = 0$ $(^4A_2 + {}^4A_2)$. The single-ion lines are the strong pair at 13,624.5 and 13,629.9 cm^{-1} (the "R" lines), corresponding to the ^2E excited state, split by the trigonal field of the distorted perovskite structure. The separation of $S = 0$ and $S = 1$ in the ground state corresponds to the separation of the weak lines 13,633 (marked 0-1) and the strong line 13,562 (marked 1-1) in Figure 44. A J value of about 70 cm^{-1} fits the three levels $(\frac{3}{2} + \frac{3}{2})$ of the ground state.

The excited states near the R lines correspond to singly excited pairs, $^4A_2 + {}^2E$. The excitation could be on either ion, and there must therefore be an energy corresponding to the transfer of excitation. In a pure crystal this transfer would lead to a bandwidth or a "Davydov splitting."[240] In the pair spectrum it represents the energy difference between the symmetric and antisymmetric wave functions, such as

$$\Psi_a(^4A_2)\Psi_b(^2E_\varepsilon) \pm \Psi_a(^2E_\varepsilon)\Psi_b(^4A_2) \qquad (82)$$

In the example we have been discussing this energy difference amounts to about 130 cm^{-1}, which is of the same order of magnitude as the Davydov splitting in Cr_2O_3.[240]

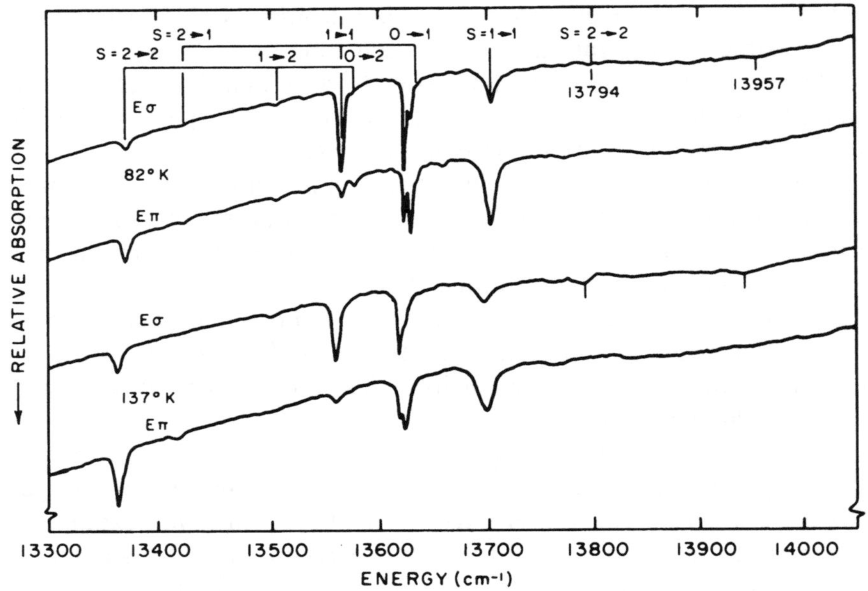

Fig. 44. Polarized absorption spectra at 82 and 137°K of Cr^{3+} pairs in $LaAlO_3$. The assignments of spin levels in the transitions are shown. The R lines ($^4A_2 \rightarrow {}^2E$) of single Cr^{3+} ions are seen in the center of the figure. The effect of temperature on the $S = 2$ level of the ground state can be clearly seen in several of the transitions.[232]

When there is exchange coupling between ions it is possible for one quantum of light to produce two (or more) quanta of ionic excitation. With weak exchange coupling these excitations are nearly the sums of individual ionic excitation energies. For a pair of ^2E Cr^{3+} ions the spin may be zero or one, and there are four states of each multiplicity. Three of the $S = 0 \rightarrow S = 0$ and one or two of the $S = 1 \rightarrow S = 1$ transitions have been identified in $LaAlO_3{:}Cr^{3+}$.

In concentrated materials, for example, $RnMnF_3$,[241] pair transitions are very prominent. These doubly excited levels occur at high energies and therefore may be mixed with charge-transfer states. These bands are shown in Figure 26.

Ferguson[242] has identified an electron transfer band of a pair of ions, Mn^{2+} and Cu^{2+}, in a host crystal of $KZnF_3$. The interpretation of these bands is probably fairly complex, and no definite conclusions as to the nature of the excited-state wave functions can be drawn as yet.

Studies of ion pair states are obviously of great interest both for the elucidation of ground-state magnetic properties and of band structure in transition metal compounds.

3.4.2. Mixed Valence Systems

When two ions of different valence states but of the same metal occur at neighboring sites in a crystal there is a possibility that an electron transfer will occur, interchanging the identity of the oxidized and reduced forms. If the surroundings of the two ions have the same chemical constitution, the only geometric feature by which they could be distinguished would be the bond lengths between metal and ligands. The transfer of an electron would also have to be accompanied by an interchange of bond lengths. In order for the two ions to be distinguishable, the electron would have to be localized sufficiently so that the difference in bond lengths could be sustained. Robin and Day have reviewed mixed valence compounds[243] and have a simple classification based on the degree of localization of the valences.

In partially reduced TiO_2 crystals the evidence is that the oxide lattice remains intact, but that Ti atoms move into interstitial positions, where they act as shallow donors. The lattice Ti ions share the extra electrons when the temperature is high enough to ionize the donors, and the crystal acts like a metal. In this case the Ti ions on the lattice are not distinguishable. These are called class III compounds in Robin and Day's classification.

On the other hand, the "vanadium bronzes," $Na_xV_2O_5$, which should contain both V(IV) and V(V), seem to behave as semiconductors, and therefore the valences are trapped. It is not known how much the Na ions help to localize the 3*d* electrons on the V(V) lattice, but it is not likely that all V(V) sites are strictly equivalent except for bond lengthening with electron localization. These are called class II systems.

The class I systems are those for which the different valence sites have different coordination numbers. They act as if there were only minor interactions between the two different ions. There would be a large energy change if the valences were to interchange. This would not be true in the other cases.

The delocalization of electrons in excited states is expected to be greater than in the ground state, and therefore the optical spectra of compounds with mixed valences could, and often do, show new bands which are associated with electron transfer. In fact, color changes are often the most striking and immediate evidence for the existence of mixed valences. The darkening of TiO_2 when it is reduced, the black color of Fe_3O_4, and the blue of reduced WO_3 are familiar examples.

In addition to Robin and Day's review, there is an interesting pair of articles by Allen and Hush[244] and Hush[245] where some useful models and relationships between mixed valence spectra and chemical reactivity are discussed.

Oxides of the transition metals nearly always display effects due to mixed valences because a change of stoichiometry can be accommodated by a change in the occupation number of the *d* shell. Then, not only is there the question of the localization of the valence, but also the problems of crystal structure and overall electronic structure and the resulting physical properties. All of these matters are discussed by Goodenough.[246]

3.4.3. Donor–Acceptor Pairs

In Section 3.3.2 we discussed the electronic state of the Cu–Cl pair in ZnS. This is an example of a donor–acceptor pair. The donor–acceptor pair is one of the most significant concepts in the spectroscopy of II–VI compounds. Since its introduction by Prener and Williams in 1956[247] there has been substantial experimental evidence for the widespread occurrence of these pairs. In the formation of a doubly doped crystal of a compound such as ZnS, impurities having opposite formal charge must tend to come together because of their Coulomb attraction. When the temperature at which diffusion ceases is high and when the dielectric constant is large the degree of association is small, and the members of the pair will be distributed over a wide range of distances of separation. In the II–VI and III–V compounds a broad range of pair separations is observed. The electronic energy of excitation of such a pair is higher the more strongly bound it is, i.e., the smaller the distance between donor and acceptor. The simplest expression for the excitation energy is

$$h\nu = E_G - (E_D + E_A) + e^2/\varepsilon r \tag{83}$$

where E_G is the band gap energy, E_D and E_A are the energies of donor and acceptor from the nearest band edges, ε is the dielectric constant, and r is the separation of the pair. The Coulomb energy term is the attraction for the excited electron by the center it has left, and not the thermodynamic donor–acceptor attraction. Because of the large value of ε in these compounds, the donor and acceptor electron and hole orbits are very large and therefore there is appreciable transition probability for large separations. The transition probability depends

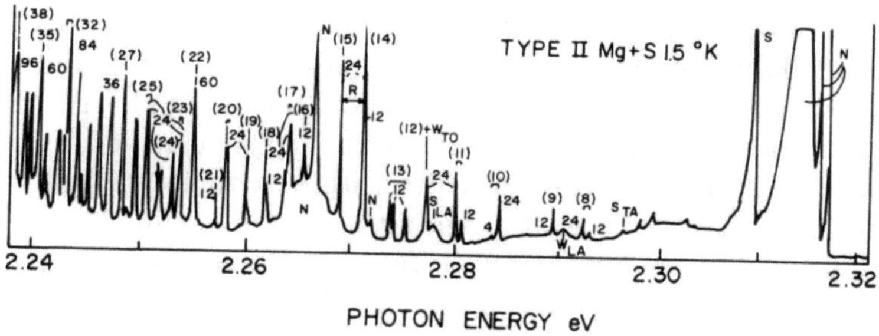

Fig. 45. Part of the photoluminescence spectrum of GaP doped with Mg and S, showing lines due to the donor–acceptor pairs. The bracketed integers give the shell number of the pair. The unbracketed integers give the number of pairs in the shell or subshell.[253,270]

on the hole–electron overlap and therefore falls exponentially with separation:

$$W(r) = W(0)\exp(-2r/r_B) \tag{84}$$

where r_B is a Bohr radius in the solid, $r_B = r_H\varepsilon$.

Whether to use the static or the electronic dielectric constant in Eq. (83) depends on the orbit period τ_0 relative to the lattice vibration period τ_l.[248] If $\tau_0 > \tau_l$, then the lattice can respond to the instantaneous electronic position, and one should use ε_0, the static dielectric constant. But if $\tau_l > \tau_0$, only the electronic polarization can respond, and one should use ε_∞, the high-frequency dielectric constant.

The reality of the donor–acceptor pairs was shown most dramatically by the line emission spectra of GaP by Thomas and Hopfield[249,250] (Figure 45). By assuming a random distribution of intrapair separations, it was possible to assign emission lines to almost all of the possible pairs out to 50th or 100th neighbors. The fit to Eq. (83) was quite good. In II–VI compounds, line emission pair spectra have only been observed in ZnSe[251] and in CdS.[252]

The work in this area has recently been reviewed by Dean.[253]

4. Conclusions

This survey of the electronic structure and spectra of impurities and defects in solids has been necessarily brief and incomplete. One could conclude from the material presented, however, that a first-order understanding of most impurity spectra in solids has been achieved. That is, we have semiquantitative ideas about the wave functions of the ground and excited states of these systems. For the *f–f* of *d–d* spectra of the transition groups these ideas are fairly refined, making useful quantitative calculations possible (as illustrated in Section 3.2.5.4). In the last few years acceptor-type charge-transfer spectra of transition metal ions have been sharply clarified, as shown in Section 3.2.8. The spectra of post-transition elements are understood in principle and a good deal is known about the Jahn–Teller effects in their excited states, but their wave functions have not been described very well. The localized states in the II–VI compounds are similar to those in covalent semiconductors and therefore are amenable to the effective mass approximation; deeper studies of these wave functions are needed, however.

The pair systems discussed in Section 3.4 provide the conceptual route from impurity ions to neat solids; exchange-coupled pairs lead to magnetic materials; mixed valence systems to substances having metal–insulator transitions; and donor–acceptor pairs to ternary compounds. Obviously, an understanding of pair systems can lead to an understanding of the properties of the neat systems and provide information for the design of new materials.

We began with a discussion of the incorporation of impurities in solids, treating the problem at first at the classical level of the ionic theory of solids. A later section gave some examples of the broader problems of equilibria involving impurities, defects, and electrons. We have not, however, integrated Sections 2 and 3 very well; this means that a complete theory of the electronic states of an impure solid has not been presented, nor has such a goal been achieved by anyone. In our present state of knowledge we have had to focus on isolated details of impurity levels in order to reach a first-order level of understanding. As progress continues we will be better able to treat impurities and host crystal as a single system and approach the ultimate level of understanding.

References

1. G. F. Koster and J. C. Slater, Wave functions for impurity levels, *Phys. Rev.* **95**, 1167–1176 (1954).
2. G. H. Dieke, *Spectra and Energy Levels of Rare Earth Ions in Crystals*, Interscience Publishers, New York (1968).
3. C. K. Jorgensen, *Lanthanides and 5f Elements*, Academic Press, New York (1970).
4. C. J. Ballhausen, *Introduction to Ligand Field Theory*, McGraw-Hill, New York (1962).

5. Ch. Lushchik, Survey of luminescence in alkali halide crystals, *J. Luminescence* **1,2**, 594–609 (1970).
6. T. Murata and Y. Nakai, Optical absorption of KCl·KBr solid solutions, *J. Phys. Soc. Japan* **23**, 904 (1967); N. Nagasawa, H. Nakagawa, and Y. Nakai, Absorption edge in solid solutions of alkali halides, *J. Phys. Soc. Japan* **24**, 356–362 (1968).
7. W. Kohn, in *Solid State Physics* (F. Seitz and D. Turnbull, eds.), Vol. 5, pp. 257–320, Academic Press, New York (1958).
8. C. K. Jorgensen, Chemical bonding from absorption spectra, in *Solid State Physics* (F. Seitz and D. Turnbull, eds.), Vol. 13, pp. 275-462, Academic Press, New York (1962).
9. F. E. Williams, Theoretical basis for solid state luminescence, in *Luminescence of Inorganic Solids* (P. Goldberg, ed.), pp. 254–277, Academic Press, New York (1966).
10. H. J. Zeiger and G. W. Pratt, *Magnetic Interactions in Solids*, Clarendon Press, Oxford (1973).
11. W. B. Fowler, Electronic states and optical transitions of color centers, in *The Physics of Color Centers* (W. B. Fowler, ed.), pp. 53–179, Academic Press, New York (1968); Jordon J. Markham, F-Centers in alkali halides, Supplement 8, *Solid State Physics* (F. Seitz and D. Turnbull, eds.), Academic Press, New York (1966).
12. C. H. Anderson and E. S. Sabisky, EPR studies of photochromic CaF_2, *Phys. Rev. B* **3**, 527–536 (1971).
13. F. Lüty, F_A Centers in alkali halide crystals, in *The Physics of Color Centers* (W. B. Fowler, ed.), Academic Press, New York (1968), pp. 181–242.
14. A. J. Freeman and R. E. Watson, Theoretical investigation of some magnetic and spectroscopic properties of rare earth ions, *Phys. Rev.* **127**, 2058–2075 (1962).
15. C. P. Flynn, *Point Defects and Diffusion*, Clarendon Press, Oxford (1972).
16. J. H. Crawford and L. Slifkin (eds.), *Point Defects in Solids*, Vol. 1, *General Ionic Crystals*, Plenum Press, New York (1972).
17. A. B. Lidiard, in *Handbuch der Physik*, Vol. 20, pp. 246–349 Springer-Verlag, Berlin (1957).
18. R. C. Bowman, Jr., Theoretical solution enthalpies of divalent impurities in the alkali halides, *J. Chem. Phys*, **59**, 2215–2223 (1973).
19. M. Ikeya, N. Itoh, and T. Suita, Distribution coefficients of various impurities in alkali halides, *Japan. J. Appl. Phys.* **7**, 837–845 (1968).
20. T. B. Douglas, Calculated heats of dilute solid solution among the alkali halides other than cesium, *J. Chem. Phys.* **45**, 4571–4585 (1966).
21. D. L. Fancher and G. R. Barsch, Lattice theory of alkali halide solid solutions. I. Heat of formation, *J. Phys. Chem. Solids* **30**, 2503–2516 (1969).
22. M. P. Tosi and M. Doyama, Vacancies and monovalent cation impurities in the alkali halides, *Phys. Rev.* **151**, 642–648 (1966).
23. O. Kleppa and S. V. Meschel, Heats of formation of solid solutions in the systems (Na–Ag)Cl and (Na–Ag)Br, *J. Phys. Chem.* **69**, 3531–3537 (1965).

24a. G. V. Flidlider, T. V. Kovtunenko, E. V. Kisselva, and A. A. Bundel, *Russ. J. Phys. Chem*, **40**, 1329–1331 (1966).

24b. V. S. Urusov, Theory of heats of formation of solid solutions of alkaline-earth metal oxides, *Russ. J. Phys. Chem.* **43**, 1703–1705 (1969).

25. L. W. Barr and A. B. Lidiard, Defects in ionic crystals, in *Physical Chemistry—An Advanced Treatise*, Vol. 10 (Eyring, Henderson, and Jost, eds.), Academic Press, New York (1970).
26. C. R. A. Catlow and M. J. Norgett, Shell model calculations of the energies of formation of point defects in alkaline earth fluorides, *J. Phys. C* **6**, 1325–1339 (1973).
27. R. J. Quigley and T. P. Das, Theory of off center displacements in solid solutions of ionic crystals, *Phys. Rev.* **164**, 1185–1199 (1967); Dependence of KCl:Li^+ and KBr:Li^+ off-center exchange on lattice parameter, *Phys. Rev.* **177**, 1340–1345 (1969).
28. G. K. Pandey and D. K. Shukla, Off-center direction of certain atomic impurities in alkali halide matrices, *Phys. Rev. B* **5**, 3362–3364 (1972).
29. M. S. Li, M. de Souza, and Fritz Lüty, Off-center Cu^+ ions in potassium halides studied with ionic thermocurrents, *Phys. Rev. B* **7**, 4677–4682 (1973).
30. M. P. Tosi, Cohesion of ionic solids in the Born model, in *Solid State Physics* (F.Seitz and D. Turnbull, eds.), Vol. 16, pp. 1–120, Academic Press, New York (1964).
31. B. G. Dick and T. P. Das, Lattice deformations in alkali-halide solid solutions, *Phys. Rev.* **127**, 1053–1062 (1962).
32. Y. Fukai, Lattice distortion around impurity ions in alkali halide crystals. I. Calculation of lattice distortion comparison with X-ray data of lattice parameter changes, *J. Phys. Soc. Japan* **18**, 1413–1421 (1963).
33. R. C. Bowman, Jr., Solubility of divalent impurities in the alkali halides, *Phys. Rev. Lett.* **26**, 1239–1241 (1971).
34. A. D. Franklin, Born model calculation of enthalpies of solution, *J. Am. Ceram. Soc.* **50**, 648–652 (1969).
35. L. A. Vredvoe, C. R. Chilver, and F. K. Fong, Dipolar impurities in alkali halides, in *Progress in Solid State Chemistry* (H. Reiss, ed.), Vol. 5, pp. 431–495, Pergamon Press, New York (1971).
36. D. Roy and A. K. Ghosh, Entropy of formation of vacancy pairs in alkali-halides, *J. Phys. Chem. SS* **4**, 941 (1971).
37. B. Henderson, J. E. Wertz, T. P. D. Hall, and R. D. Dowsing, An EPR study of vacancy-impurity association in MgO, *J. Phys. Chem. SS* **4**, 107–116 (1971).
38. F. Z. Gilfanov, L. D. Livanova, M. S. Orlov, and L. Solov, Effect of the activator conc. on the degree of association of impurity centers in the system SrF_2:Gd^{3+}, *Soviet Phys—Solid State* **11**, 1779–1782 (1970).
39. G. K. Miner, T. P. Graham, and G. T. Johnston, Effect of a Ce^{3+} codopant on the Gd^{3+} EPR spectrum of SrF_2 at room temperature, *J. Chem. Phys.* **57**, 1263–1270 (1972).
40. W. Hayes (ed.), *The Physics of Fluorite Compounds* Clarendon Press, Oxford, to be published.
41. A. K. Cheetham, B. E. F. Fender, and M. J. Cooper, Defect structure of calcium floride containing excess anions. I. Bragg scattering, *J. Phys. C: Solid State Phys.* **4**, 3107–3121 (1971).
42. C. R. A. Catlow, Defect clusters in doped fluorite crystals, *J. Phys. C* **6**, L64–L70 (1973).

43. D. F. Ikenberry and T. P. Das, Nuclear quadrupole effects in alkali-halide solid solutions, *Phys. Rev.* **184**, 989–998 (1969).
44. U. Ranon and J. S. Hyde, Electron-nuclear double-resonance and electron-paramagnetic-resonance analysis of the yterbium–fluorine superhyperfine interaction in CaF_2: Yb^{3+}, *Phys. Rev.* **141**, 259–274 (1966).
45. J. M. Baker, E. R. Davies, and J. P. Hurrell, Electron nuclear double resonance in calcium fluoride containing Yb^{3+} and Ce^{3+} in tetragonal sites, *Proc. Roy. Soc. A* **308**, 403–431 (1968).
46. J. M. Baker, E. R. Davies, and T. R. Reddy, Detailed mapping of atomic positions using electron nuclear double resonance (ENDOR), *Contemp. Phys.* **13**, 45–59 (1972).
47. S. C. Moss and R. F. Newnham, The chromium position in ruby, *Z. Krist.* **120**, 359–363 (1964).
48. R. C. Chaney, E. F. Lafon, and C. C. Lin, Energy band structures of lithium fluoride crystals by the method of tight binding, *Phys. Rev. B* **4**, 2734–2741 (1971).
49. H. H. Woodbury, Diffusion and solubility studies, in *Physics and Chemistry of II–VI Compounds* (M. Aven and J. S. Prener, eds.), North-Holland Publishing Co., Amsterdam (1967), pp. 224–264.
50a. E. F. Apple, Investigations in the $CuGaS_2$–ZnS and $AgGaS_2$–ZnS systems, *J. Electrochem. Soc.* **105**, 251–255 (1958);
50b. B. Tell, J. L. Shay, and H. M. Kasper, Electrical properties, optical properties and band structure of $CuGaS_2$ and $CuInS_2$, *Phys. Rev. B* **4**, 2463–2471 (1971).
51. I. M. Boswarva, Further calculations of the energies of formation of Schottky defects in NaCl structure ionic crystals, *J. Phys. C* **5**, L5–L8 (1972).
52. I. M. Boswarva and A. B. Lidiard, Energy of formation of Schottky defects in ionic crystals, *Phil. Mag.* **16**, 805–826 (1967).
53. M. J. Norgett, A Born model calculation of the energies of vacancies, interstitials and inert gas atoms in calcium fluoride, *J. Phys. C: Solid State* **4**, 298–306 (1971).
54. B. Henderson and J. E. Wertz, Defects in alkaline earth oxides, *Adv. Phys.* **17**, 749–855 (1968).
55. P. Weightman, D. E. Dugdale, and L. V. Holroyd, An EPR study of Mn^{2+} in single crystals of BaO, *J. Phys. C* **4**, 3292–3296 (1971).
56. V. K. Jain, Behavior of divalent cation impurities in alkali halide crystals, *Phys. Stat. Sol. b* **44**, 11–28 (1971).
57. W. Hayes and P. H. S. Smith, Paramagnetic resonance in ground and optically excited states of Tm^{+2} in alkaline earth fluoride, *J. Phys. C: Solid State* **4**, 840–851 (1971).
58. W. Hayes, D. L. Kirk, and G. P. Summers, The self-trapped hole and the self-trapped excitation in alkaline earth fluorides, *Solid State Commun.* **7**, 1061–1664 (1969).
59. Y. K. Voronko, V. V. Osiko, and I. A. Shcherbakov, Optical centers and the interaction of Yb^{3+} ions in cubic fluorite crystals, *Soviet Phys.—JETP* **29**, 86–90 (1969).
60. T. C. Ensign and N. E. Byer, Correlated electron paramagnetic resonance and optical study of CdF_2:Er^{3+}. I. C_{2v} Local-site symmetry, *Phys. Rev. B* **6**, 3227–3239 (1972).

61. M. J. Norgett and A. M. Stoneham, The self-trapped hole in alkaline earth fluorides: I. Static properties, *J. Phys. C* **6**, 229–237 (1973); II. Hopping motion, *J. Phys. C* **6**, 238–246 (1973).
62. J. M. O'Hare, Model for locally compensated cubic sites of tripositive rare earth ions in flourite crystals, *J, Chem. Phys.* **57**, 3838–3843 (1972).
63. A. D. Franklin, Born-model calculations of defect energies in CaF_2, *J. Phys. Chem. Solids* **29**, 823–841 (1968).
64. M. S. Norgett, A Born model calculation of the energies of vacancies, ion interstitials and inert gas atoms in strontium and barium fluorides, *J. Phys. C* **4**, 1284–1288 (1971).
65. G. G. P. van Gorkom, A new Mn^{2+} Centre in CaF_2, *J. Phys. Chem. Solids* **31**, 905–912 (1970).
66. H. Bernhardt, P. Görlich, and G. Kötitz, Foreign ions with d^1-states in alkaline earth flouride crystals, *Phys. Stat. Sol. (a)* **6**, 479–486 (1971).
67. R. S. Singh, D. W. Galipean, and S. S. Mitra, *U*-Band transition energies in alkaline-earth fluorides, *J. Chem. Phys.* **52**, 2341–2344 (1970).
68. J. Wagner and S. Mascarenhas, Ionic-thermocurrent study of rare-earth CaF_2 and SrF_2, *Phys. Rev. B* **6**, 4867–4873 (1972).
69. I. T. Jacobs, G. D. Jones, K. Zdansky, and R. A. Satten, Electron-phonon interaction effects in the spectra of hydrogenated, deuterated, and tritiated crystals of calcium and strontium fluorides containing cerium, *Phys. Rev. B* **3**, 2888–2910 (1971).
70. M. R. Brown, H. Thomas, J. S. S. Whiting, and W. A. Shand, Experiments on Er^{3+} in SrF_2. I. Fluorescence quantum efficiencies and lifetimes, *J. Chem. Phys.* **50**, 881–890 (1969).
71. S. E. Stokowski and A. L. Schawlow, Spectroscopic studies of $SrTiO_3$ using impurity-ion probes, *Phys. Rev.* **178**, 457 (1969).
72. J. J. Krebs and R. K. Jack, Electron nuclear double resonance and electron spin resonance evidence for a Fe^{3+} (K^+ vacancy) center in iron-doped $KZnF_3$, *Phys. Rev. B* **5**, 3499–3505 (1972).
73. Kazumi Horoi, Masanori Arokawa, and Yasuharu Shimogu, EPR and ^{19}F ENDOR investigations in KMF_3: V^{2+} , in *J. Phys. Soc. Japan* **33**, 565 (1972).
74. J. J. Davies and K. Horai, The hyperfine interactions of fluorine ions in the vicinity of Cr^{3+}: Endor measurements, *J. Phys. Chem. (Solid State)* **4**, 671–681 (1971).
75. R. Smoluchowski, O. W. Lazareth, R. D. Hatcher, and G. J. Dienes, Mechanism of point-defect formation in ionic crystals, *Phys. Rev. Lett.* **27**, 1288–1290 (1971).
76. G. J. Dienes, R. D. Hatcher, O. W. Lazareth, B. S. H. Royce, and R. Smoluchowski, Theory of Mn^0 centers in NaCl, *Phys. Rev. B* **7**, 5332–5335 (1973).
77. G. J. Dienes and R. Smoluchowski, Interstitial halogens in alkali halides, *J. Phys. Chem. Solids* **27**, 611–612 (1966).
78. R. H. Bube, *Photoconductivity of Solids*, Wiley, New York (1960).
79. M. Aven and J. S. Prener (eds.), *Physics and Chemistry of II–VI Compounds* North-Holland Publishing Co., Amsterdam (1967).
80. G. Huggett and K. Teegarden, Intrinsic photoconductivity in the alkali halides, *Phys. Rev.* **141**, 797–802 (1966).

81. B. S. H. Royce, The creation of point defects in alkali halides, in *Progress in Solid State Chemistry* (H. Reiss, ed.), Vol. 4, pp. 213–243, Pergamon Press (1967); W. Känzig and T. V. Woodruff, The electronic structure of an *H*-center, *J. Phys. Chem. Solids* **9**, 70–92 (1958).
82. A. N. Jette, T. L. Gilbert, and T. P. Das, Theory of the self-trapped hole in the alkali halides, *Phys. Rev.* **184**, 884–894 (1969).
83. D. Ikenberry, A. N. Jette, and T. P. Das, Theoretical calculation explaining negative transferred hyperfine constants of the V_K center, *Phys. Rev. B* **1**, 2785–2789 (1970).
84. A. N. Jette and T. P. Das, Theory of the self trapped hole in CaF_2, *Phys. Rev.* **186**, 909–925 (1969).
85. W. Van Gool, *Principles of Defect Chemistry of Crystalline Solids*, Academic Press, New York (1966).
86. F. A. Kröger, *The Chemistry of Imperfect Crystals*, North-Holland Publishing Co., Amsterdam (1964).
87. R. F. Brebrick, Non-stoichiometry in binary semiconductor compounds $M_{1-\delta}N_{1+\delta}$, in *Progress in Solid State Chemistry* (H. Reiss, ed.), Vol. 3, pp. 213–264, Pergamon Press, Oxford (1967).
88. G. W. Ludwig and H. H. Woodbury, Electron spin resonance in semiconductors, in *Solid State Physics* (F. Seitz and G. Turnbull, eds.), Vol. 13, pp. 223–304, Academic Press, New York (1962).
89. G. Mandel, Self-compensation-limited conductivity in binary semiconductors. I. Theory, *Phys. Rev.* **134A**, 1073–1079 (1964); G. Mandel, F. F. Morehead, and P. R. Wagner, Self-compensation-limited conductivity in binary semiconductors. III. Expected correlation with fundamental parameters, *Phys. Rev.* **136A**, 826–832 (1964).
90. R. C. Alig, Z. J. Kiss, J. P. Brown, and D. S. McClure, Energy levels of Ce^{2+} in CaF_2, *Phys. Rev.* **186**, 276–284 (1969).
91. *Selected Values of Chemical Thermodynamic Properties*, Circular 500, National Bureau of Standards, U. S. Govt. Printing Office (1952).
92. C. H. Anderson, unpublished report, RCA Laboratories; also in *The Physics of Fluorite Compounds* (W. Hayes, ed.), Clarendon Press, Oxford, to be published.
93. J. L. Merz and P. S. Pershan, Charge conversion of irradiated rare-earth ions in calcium fluoride. I. *Phys. Rev.* **162**, 217–235 (1967); II. Thermoluminescent spectra, *Phys. Rev.* **162**, 235–247.
94. M. Born and E. Wolf, *Principles of Optics*, 3rd ed., Ch. 14, Pergamon Press, New York (1965).
95. D. L. Dexter, Theory of optical properties of imperfections in non-metals, in *Solid State Physics* (F. Seitz and D. Turnbull, eds.), Vol. 6, pp. 353–411, Academic Press, New York (1958).
96. C. K. Jorgensen and B. R. Judd, Hypersensitive pseudoquadruple transitions in lanthanides, *Molecular Phys.* **8**, 281–290 (1964).
97. K. K. Rebane, *Impurity Spectra of Solids*, Plenum Press, New York (1970).
98. F. Varsanyi and G. H. Dieke, Infrared absorption of rare-earth chlorides, *J. Chem. Phys.* **36**, 835–840 (1962).
99. A. M. Glass, Reactions between vacancies and impurities in magnesium oxide. I. Cr^{3+} -Ion impurities, *J. Chem. Phys.* **46**, 2080–2091 (1967).

100. M. J. L. Sangster, Calculation of the phonon sideband of the *R*-line in MgO:Cr^{3+}, *Phys. Rev.* **36**, 254–258 (1972).
101. D. S. McClure, The optical spectrum of $NiSiF_6 \cdot 6H_2O$, in RCA Laboratories Report to Air Force Cambridge Research Laboratory, Contract No. AF19(604)-5541, June 1, 1961.
102. T. S. Piper, J. P. Brown, and D. S. McClure, *fd* and $f^{13}d$ Configurations in a crystal field, and the spectrum of Yb^{++} in cubic crystals, *J. Chem. Phys.* **46**, 1353–1358 (1967).
103. J. S. Griffith, *The Theory of Transition Metal Ions*, Cambridge University Press, London (1961).
104. J. Brynestad, H. L. Yakel, and G. P. Smith, Temperature dependence of the absorption spectrum of nickel (II)-doped $KMgCl_3$ and the crystal structure of $KMgCl_3$, *J. Chem. Phys.* **45**, 4652–4664 (1966).
105. L. L. Lohr and D. S. McClure, Optical spectra of divalent manganese salts. II. The effect of interionic coupling on absorption strength, *J. Chem. Phys.* **49**, 3516–3521 (1968).
106. L. L. Lohr, Jr., Spin-forbidden electronic excitation in transition metal complexes, *Coordination Chem. Rev.* **8**, 241–259 (1972).
107. J. Feruguson, M. J. Guggenheim, and Y. Tanabe, Exchange effects in the electronic absorption spectrum of Mn (II) on perovskite fluorides, *J. Appl. Phys.* **36**, 1046–1049 (1951).
108. P. G. Russell, D. S. McClure, and J. W. Stout, Zeeman splitting of narrow lines in crystalline MnF_2, FeF_2, and CoF_2. Evidence for spin-wave excitations, *Phys. Rev. Lett.* **16**, 176–178 (1966).
109. D. D. Sell, R. L. Greene, and R. M. White, Optical exciton-magnon absorption in MnF_2, *Phys. Rev.* **158**, 489–510 (1967).
110. M. Cardona, Modulation spectroscopy, in *Solid State Physics*, Suppl. 11 (F. Seitz, D. Turnbull, and H. Ehrenreich, eds.), Academic Press, New York (1969).
111. A. D. Buckingham and P. J. Stephens, Magnetic optical activity, *Ann, Rev. Phys. Chem.* **17**, 399–452 (1966).
112. P. N. Schatz and A. J. McCaffery, The Faraday effect, *Quart. Rev. (Chem. Soc. London)* **23**, 552–584 (1969).
113. P. J. Stephens, Theory of magnetic circular dichroism, *J. Chem. Phys.* **52**, 3489–3516 (1970).
114. S. B. Piepho, J. R. D. Dickenson, J. A. Spencer, and P. N. Schatz, High resolution magnetic circular dichroism spectrum of Cs_2ZrCl_6:Ir^{4+}, *J. Chem. Phys.* **57**, 982–990 (1972).
115. B. A. Lengyel, *Introduction to Laser Physics*, Wiley, New York (1966).
116. H. W. Moos, Spectroscopic relaxation processes of Rare earth ions in crystals, *J. Luminescence* **1,2**, 106–121 (1970).
117. S. Sugano, Y. Tanabe, and H. Kamimura, *Multiplets of Transition-Metal Ions in Crystals*, Academic Press, New York (1970).
118. D. S. McClure, Electronic spectra of molecules and ions in crystals. Part II, in *Solid State Physics* (F. Seitz and D. Turnbull, eds.), Vol. 9, Academic Press, New York (1959).
119. C. K. Jorgensen, *Modern Aspects of Ligand Field Theory*, North-Holland Publishing Co., Amsterdam (1971).

120. H. G. Drickamer and C. W. Frank, *Electronic Transitions and the High Pressure Chemistry and Physics of Solids*, Chapman and Hall, London (1973).
121. A. B. P. Lever, *Inorganic Electronic Spectroscopy*, Elsevier, Amsterdam 1968).
122. J. Ferguson, Electronic absorption spectrum of nickel II in cubic perovskite fluorides. III. The ultraviolet spectrum, *Aust. J. Chem.* **21**, 323–330 (1968).
123. J. Ferguson, Spectroscopy of 3*d* complexes, in *Progress in Inorganic Chemistry* (S. Lippard, ed.), Vol. 12, pp. 159–293, Interscience, New York (1970).
124. E. U. Condon and G. H. Shortley, *The Theory of Atomic Spectra*, The Cambridge University Press (1935).
125. R. E. Watson, *Iron Series Hartree–Fock Calculations* Technical Report No. 12, June 15, 1959, Solid State and Molecular Theory Group, Massachusetts Institute of Technology, Cambridge, Massachusetts.
126. C. E. Moore, *Atomic Energy Levels*, Circular of the National Bureau of Standards No. 467; Vol. I, 1949; Vol. II, 1952; Vol. III, 1958.
127. B. G. Wybourne, *Spectroscopic Properties of the Rare Earths*, Interscience, New York (1965).
128. M. Rotenberg, R. Bivens, N. Metropolis, and J. K. Wooten, Jr., *The 3-j and 6-j Symbols*, The Technology Press, Massachusetts Institute of Technology, Cambridge, Massachusetts (1959).
129. B. R. Judd, Effective operators for configurations of equivalent electrons, *Adv. Chem. Phys.* **14**, 91–127 (1969).
130. R. E. Trees, Configuration interaction in Mn II, *Phys. Rev.* **83**, 756–760 (1951); Term values in the $3d^5\,4s$ configuration of Fe III, *Phys. Rev.* **84**, 1089–1091 (1951).
131. G. Racah, $L(L+1)$ Correction in the spectra of the iron group, *Phys. Rev.* **85**, 381–382 (1952); R. E. Trees and C. K. Jorgensen, Parameters α and β in the spectra of the iron group, *Phys. Rev.* **123**, 1278–1280 (1961).
132. K. Rajnak and B. G. Wybourne, Configuration interaction effects in l^N configurations, *Phys. Rev.* **132**, 280–290 (1963).
133. J. Hinze and H. H. Jaffe, Slater–Condon parameters from spectral data, *J. Chem. Phys.* **38**, 1834–1847 (1963).
134. D. A. Brown, Discussion of Slater parameters for elements of the first long period, *J. Chem. Phys.* **28**, 67–70 (1958).
135. K. Rajnak and W. F. Krupke, Energy levels of Ho^{+3} in $LaCl_3$, *J. Chem. Phys.* **46**, 3532–3542 (1967).
136. A. Pasternak and Z. B. Goldschmidt, Spin-dependent interaction in the $3d^N$ configurations of the third spectra of the iron group, *Phys. Rev. A* **6**, 55–68 (1972).
137. B. W. N. Lo, K. M. S. Saxena, and S. Fraga, Complete spin–orbit interaction in many-electron atoms, *Theoret. Chim. Acta (Berl.)* **25**, 97–102 (1972).
138. D. J. Newman, Theory of lanthanide crystal fields, *Adv. Phys.* **20**, 197–256 (1971).
139. M. Tinkham, *Group Theory and Quantum Mechanics*, McGraw-Hill, New York (1964).
140. C. W. Nielson and G. F . Koster, *Spectroscopic Coefficients for the p^n, d^n and f^n Configurations*, MIT Press, Cambridge, Massachusetts (1963).
141. M. T. Hutchings, Point-charge calculations of energy levels of magnetic ions in crystalline electric fields, in *Solid State Physics* (F. Seitz and D. Turnbull, eds.), Vol. 16, pp. 227–273, Academic Press, New York (1965).

142. E. Solomon and D. S. McClure, A comparison of the Jahn–Teller effect in four triply degenerate states of Mn^{++} in $RbMnF_3$, *Phys. Rev. B* **9**, 4690–4718 (1974).
143. J. S. Griffith, *The Irreducible Tensor Method for Molecular Symmetry Groups*, Prentice-Hall, Englewood Cliffs, New Jersey (1962).
144. B. D. Bird, E. A. Cooke, P. Day, and A. F. Orchard, Derivation and testing of a molecular orbital description of ligand field spectra, *Phil. Trans., Roy. Soc. (London)* **276**, 277–339 (1974).
145. B. D. Bird and P. Day, Analysis of the charge transfer spectra of some first-transition-series tetrahalide complexes, *J. Chem. Phys.* **49**, 392–403 (1968).
146. H. B. Gray and C. J. Ballhausen, *Molecular Orbital Theory*, W. A. Benjamin, New York (1964).
147. J. C. Slater, *Quantum Theory of Molecules and Solids*, Vol. 2, McGraw-Hill, New York (1965).
148. L. L. Lohr, Optical spectra of divalent manganese salts. I. Energy levels for cubic and lower-symmetry complexes, *J. Chem. Phys.* **45**, 3611–3622 (1966).
149. A. Abragam and B. Bleaney, *Electron Paramagnetic Resonance of Transition Ions*, Clarendon Press, Oxford (1970).
150. R. G. Shulman and S. Sugano, Covalency effects in $KNiF_3$. I. Nuclear magnetic resonance studies, *Phys. Rev.* **130**, 506–511 (1963); K. Knox, R. G. Shulman, and S. Sugano, Covalency effects in $KNiF_3$. II. Optical studies, *Phys. Rev.* **130**, 512–516 (1963); S. Sugano and R. G. Shulman, Covalency effects in $KNiF_3$. III. Theoretical Studies, *Phys. Rev.* **130**, 517–530 (1963).
151. J. H. M. Thornley, The magnetic properties of $(IrX_6)^{2-}$ complexes, *J. Phys. C (Proc. Phys. Soc)* **1**, 1024–1037 (1968).
152. T. P. P. Hall, W. Hayes, R. W. H. Stevenson, and J. Wilkins, Investigation of the bonding of iron-group ions in fluoride crystals. I, *J. Chem. Phys.* **38**, 1977–1984 (1963); II. *J. Chem. Phys.* **39**, 35–39 (1963).
153. M. Tinkham, Paramagnetic resonance in dilute iron group fluorides. I. Fluorine hyperfine structure, *Proc. Roy. Soc. (London) A* **236**, 535–548 (1956); II. Wave functions of the magnetic electrons, *Proc. Roy. Soc (London) A* **236**, 549–563 (1956).
154. C. K. Jorgensen, R. Pappalardo, and H. H. Schmidtke, Do the "ligand field" parameters in lanthanides represent weak covalent bonding?, *J. Chem. Phys.* **39**, 1422–1430 (1963).
155. G. Burns and J. P. Axe, Covalent bonding effects in rare earth crystal fields, in *Optical Properties of Ions in Crystals* (H. M. Crosswhite and H. W. Moos, eds.), pp. 53–71, Interscience, New York (1967).
156. A. A. Misetich and T. Buch, Gyromagnetic factors and spin-orbit coupling in ligand field theory, *J. Chem. Phys.* **41**, 2524–2529 (1964).
157. J. P. Dahl and C. J. Ballhausen, Molecular orbital theories of inorganic complexes, *Advan. Quantum Chem.* **4**, 170–226 (1968).
158. C. J. Ballhausen and H. B. Gray, Electronic structure of metal complexes, in *Coordination Chemistry* (A. Martell, ed.), Vol. 1, pp. 3–83, A.C.S. Monograph 168, Van Nostrand-Reinhold, New York (1971).

159. A. Dutta-Ahmed and E. A. Boudreaux, Investigations in semi-empirical molecular orbital methods, *Inorg. Chem.* **12**, 1590–1597 (1973).
160. H. M. Gladney and A. Veillard, Limited basis set Hartree–Fock theory of NiF_6^{4-}, *Phys. Rev.* **180**, 385–395 (1969).
161. T. F. Soules, J. W. Richardson, and D. M. Vaught, Electronic structure and spectrum of the NiF_6^{4-} cluster: Results of calculations based on self-consistent-field models, *J. Chem. Phys.* **3**, 2186–2204 (1971).
162. J. Demunck and A. Veillard, Electronic structure of the tetracyanonickelate $Ni(CN)_4^{2-}$ and nickel carbonyl $Ni(CO)_4$. An ab-initio LCAO-MO-SCF calculation, *Theoret. Chim. Acta (Berl.)* **28**, 241–265 (1973).
163. A. Trautwein and F. E. Harris, Molecular orbital structure, Mössbauer isomer shift and quadrupole splitting in iron complexes, *Theoret. Chim. Acta (Berl.)* **30**, 45–58 (1973).
164. M. Wolfsberg and L. Helmholtz, The spectra and electronic structure of the tetrahedral ions MnO_4^-, CrO_4^{2-} and ClO_4^-, *J. Chem. Phys.* **20**, 837–843 (1952).
165. R. Hoffman, An extended Hückel theory. I. Hydrocarbons. *J. Chem. Phys.* **39**, 1397–1412 (1963).
166. M. D. Newton, F. P. Boer, and W. N. Lipscomb, Molecular orbital theory for large molecules. Approximations of the SCF LCAO Hamiltonian matrix, *J. Am. Chem. Soc.* **88**, 2353–2360 (1966).
167. G. Blyholder and C. A. Coulson, Basis of extended Hückel formalism, *Theoret. Chim. Acta (Berl.)* **10**, 316–324 (1968).
168. R. P. Messmer and G. D. Watkins, Molecular orbital treatment for deep levels in semiconductors: Substitutional nitrogen and the lattice vacancy in diamond, *Phys. Rev. B* **7**, 2568–2590 (1973).
169. D. S. McClure and Z. J. Kiss, A survey of the optical spectra of the divalent rare earth ions in cubic crystals, *J. Chem. Phys.* **39**, 3251–3257 (1963).
170. Z. Kiss, C. H. Anderson, and R. Orbach, Zeeman effect studies of the $^5I_7$–$^5I_8$ Transition in CaF_2:Dy^{+2}, *Phys. Rev.* **137A**, 1761–1766 (1965); Z. Kiss and H. A. Weakliem, Stark effect of $4f$ states and linear crystal field in BaClF:Sm^{+2}, *Phys. Rev. Lett.* **15**, 457–460 (1965).
171. M. Schlesinger and J. Szazurek, $4f \rightarrow 5d$ Transition studies of some rare-earth ions in CaF_2, *Phys. Rev. B* **8**, 2367–2368 (1973); Eugene Loh, Ultraviolet-absorption spectra of europium and ytterbium in alkaline earth fluorides, *Phys. Rev.* **184**, 348–352 (1969).
172. R. C. Alig, Optical spectrum of Yb^{+2} in $SrCl_2$: nephelauxetic effects, *Bull. Am. Phys. Soc. II* **19**, 258 (1974).
173. H. A. Weakliem, Electronic interactions in the $4f^n5d$ configuration of Eu^{2+} in crystals, *Phys. Rev. B* **6**, 2743–2748 (1972).
174. R. C. Alig, R. C. Duncan, Jr., and B. K. Mokross, Reduced $4f$–$5d$ electrostatic interaction of Tm^{2+} in $SrCl_2$, *J. Chem. Phys.* **59**, 5837–5841 (1973).
175. H. A. Weakliem, Optical spectra of Ni^{2+}, Co^{2+}, and Cu^{2+} in tetrahedral sites in crystals, *J. Chem. Phys.* **36**, 2117–2140 (1962).

176a. R. Stahl-Brada and W. Low, Optical spectrum of Co^{++} in the cubic crystalline field of CaF_2, *Phys. Rev.* **113**, 775–780 (1959).

176b. H. A. Weakliem, Systematic study of the crystal field of CaF_2, RCA Laboratories Report to Office of Naval Research, NONR 4133(00), August 20, 1964

177. D. S. McClure, The distribution of transition metal cations in spinels, *J. Phys. Chem. Solids* **3**, 311–317 (1957).
178. M. Norgett, J. Thornley, and L. Venanzi, The visible and ultraviolet spectra of trigonal bipyramidal complexes of d^6-, d^7-, and d^8-metal ions, *J. Chem. Soc. A* **1967**, 540–544.
179. W. M. Fairbank, Jr. and G. K. Klauminzer, Tetragonal field splitting of levels in MgO:Cr^{3+}, *Phys. Rev. B* **7**, 500–510 (1973).
180. S. B. Piepho, J. R. Dickinson, J. A. Spencer, and P. N. Schatz, High resolution absorption and magnetic circular dichroism spectra of Cs_2ZrCl_6:Os^{4+}, *Mol. Phys.* **24**, 609–630 (1972); P. Dorain, H. Patterson, and P. Jordan, Optical spectra of Os^{4+} in single cubic crystals at 4.2°K, *J. Chem. Phys.* **49**, 3845–3857 (1968).
181. P. George and D. S. McClure, The effect of inner orbital splitting on the thermodynamic properties of transition metal compounds and coordination complexes, in *Progress in Inorganic Chemistry* (F. A. Cotton, ed.), Vol. I, pp. 381–463, Interscience Publishers, New York (1959).
182. J. D. Dunitz and L. E. Orgel, Electronic properties of transition metal oxides. I. Distortions from cubic symmetry, *J. Phys. Chem. Solids* **3**, 20–29 (1957); II. Cation distribution amongst octahedral and tetrahedral sites, *J. Phys. Chem. Solids* **3**, 318–323 (1957).
183. A. Navrotsky and O. J. Kleppa, The thermodynamics of cation distribution in simple spinels, *J. Inorg. Nucl. Chem.* **29**, 2701–2714 (1967).
184. G. H. Dieke, H. M. Crosswhite, and B. Dunn, Emission spectra of the doubly and triply ionized rare earths, *J. Opt. Soc. Am.* **51**, 820–827 (1961); G. H. Dieke and H. M. Crosswhite, The spectra of the doubly and triply ionized rare earths, *Appl. Opt.* **2**, 675–686 (1963).
185. L. J. Nugent, R. D. Baybarz, and J. L. Burnett, Electron transfer and $f \rightarrow d$ absorption bands of some lanthanide and actinide complexes and the standard (III–IV) oxidation potentials for each member of the lathanide and actinide series, *J. Inorg. Nucl. Chem.* **33**, 2503–2530 (1971).
186. L. J. Nugent, R. D. Baybarz, J. L. Burnett, and J. L. Ryan, Electron-transfer and *f–d* absorption bands of some lanthanide and actinide complexes and the standard (II–III) oxidation potential for each member of the lanthanide and actinide series, *J. Phys. Chem.* **77**, 1528–1539 (1973).
187. C. K. Jørgensen, Electron transfer spectra, in *Progress in Inorganic Chemistry*, (S. J. Lippard, ed.), Vol. 12, pp. 101–158, Interscience Publishers, New York (1970).
188. C. K. Jørgensen, Differences between the four halide ligands, and discussion remarks on trigonal-bipyramidal complexes, on oxidation states and on diagonal elements of one electron energy, *Coord. Chem. Rev.* **1**, 164–177 (1966).
189. H. H. Tippins, Charge transfer spectra of transition-metal ions in corundum, *Phys. Rev. B* **1**, 126–135 (1970).
190. J. M. Langer, Comments on the theory of photoionization of transition metal impurities in semiconductors, *Phys. Stat. Sol.* **47**, 443–449 (1971).
191. J. Sabatini, Thesis, Princeton University, Department of Chemistry, 1973.
192. B. D. Bird, P. Day, and E. A. Grant, Low temperature charge transfer spectra of hexahalogeno iridates (IV) and osmates (IV), *J. Chem. Soc.* **1970**, 100–109.

193. G. N. Henning, A. J. McCaffery, P. N. Schatz, and P. J. Stephens, Magnetic circular dichroism of charge-transfer transitions: Low-spin d^5 hexahalide complexes, *J. Chem. Phys.* **48**, 5656–5661 (1968).
194. A. J. McCaffery, P. N. Schatz, and T. E. Lester, Magnetic circular dichroism of $IrCl_6^{2-}$ in crystalline $(CH_3NH_3)_2SnCl_6$, *J. Chem. Phys.* **50**, 379–385 (1969).
195. J. R. Dickenson, S. B. Piepho, J. A. Spencer, and P. N. Schatz, High resolution magneto-circular dichroism and absorption spectra of Cs_2ZrBr_6: Ir^{4+}, *J. Chem. Phys.* **56**, 2668–2677 (1972).
196. S. B. Piepho, J. R. Dickenson, J. A. Spencer, and P. N. Schatz, High resolution absorption and magnetic circular dichroism spectra of Cs_2ZrCl_6: Os^{+4}, *Mol. Phys.* **24**, 609–630 (1972).
197. C. K. Jørgensen and W. Preetz, Interpretation of electron transfer spectra of iridium(IV) and osmium(IV) mixed chloro-bromo complexes, *Z. Naturforsch.* **22A**, 945–954 (1967).
198. J. C. Collingwood, R. W. Schwartz, P. N. Schatz, and H. H. Patterson, Magnetic circular dichroism and absorption spectra of Cs_2ZrCl_6: Mo^{4+}, to be published.
199. F. Seitz, Interpretation of the properties of alkali halide–thallium phosphors, *J. Chem. Phys.* **6**, 150–162 (1938).
200. P. H. Yuster and C. J. Delbecq, Some optical properties of potassium iodide–thallium phosphors, *J. Chem. Phys.* **21**, 892–898 (1953).
201. K. Cho, Theory of stress and magnetic circular dichroism in alkali halide phosphors, *J. Phys. Soc. Japan* **27**, 646–658 (1969).
202. A. Fukuda, Structure of the *C* absorption band of Tl^+-type centers in alkali halides due to the Jahn–Teller effect, *J. Phys. Soc. Japan* **27**, 96–109 (1969).
203. S. Masunaga and E. Matsuyama, Structure of absorption bands in alkali halide crystals containing Tl^+-like ions under high pressure, *J. Phys. Soc. Japan* **34**, 1234–1239 (1973).
204. H. Overhol and J. Treusch, Energy bands of thallous chloride and bromide, *Solid State Comm.* **9**, 53–56 (1971).
205. A. Fukuda, K. Çho, and H. J. Pares, Relaxed excited states of KI:Tl^+ phosphors, in *Luminescence of Crystals, Molecules and Solutions* (F. Williams, ed.), pp. 478–488, Plenum Press, New York (1973).
206. A. P. Marchetti and G. L. Bottger, Optical absorption spectrum of AgF, *Phys. Rev. B* **3**, 2604–2607 (1971).
207. C. A. Allen and W. J. Fredericks, Optical properties of mercury in potassium chloride single crystals, *Phys. Stat. Sol. a* **3**, 143–152 (1970).
208. A. Fukuda and P. Yuster, Unusually large change in radiative lifetime of the *A*-band emission in KI: In^+ and KI: Sn^{2+} induced by a magnetic field, *Phys. Rev. Lett.* **28**, 1032–1034 (1972).
209. J. Höner zu Siederdissen and F. Fischer, Optical absorption of Bi^{+3}, Hg^0 and Cd^0 centers, *Phys. Stat. Sol. (b)* **48**, 215–222 (1971).
210. K. Hiraoka and W. Hamill, Excitation of luminescence from KCl:Tl^+, KB:Tl^+ and KI:Tl^+ by slow electron impact, *J. Chem. Phys.* **58**, 3868–3891 (1973).
211. D. Shoemaker and J. L. Kolopus, Pb^{++} as a hole trap in KCl.ESR and optical absorption of Pb^{3+}, *Solid State, Commun.* **8**, 435–439 (1970).
212. S. Radhakrishna and A. M. Karguppikar, Defect centers in antimony doped NaCl crystals, *J. Phys. Chem. Solids* **34**, 1497–1505 (1973).

213. Yasuo Uehara, Electronic structure of luminescence centers of ZnS phosphors, activated with impurity ions of S^2 configurations (I) ZnS: Bi Phosphors, *J. Chem. Phys.* **50**, 961–977 (1969).
214. D. Curie, *Luminescence in Crystals*, Methuen, London, New York (1963).
215. G. A. Slack, F. S. Ham, and R. M. Chrenko, Optical absorption of tetrahedral Fe^{2+} ($3d^6$) in cubic ZnS, CdTe and $MgAl_2O_4$, *Phys. Rev.* **152**, 376–402 (1966); E. M. Wray and J. W. Allen, Crystal field spectra of $3d^n$ impurities in zinc selenide, *J. Phys. C* **4**, 512–516 (1971).
216. M. R. Brown, A. F. J. Cox, W. A. Shand, and J. M. Williams, Sensitization of rare earth photoluminescence in II–VI compounds: II. Site sensitive excitation mechanisms in zinc selenide, *J. Phys. C* **5**, 502–510 (1972).
217. R. E. Dietz, H. Kamimura, M. D. Sturge, and A. Yariv, Electronic structure of copper impurities in ZnO, *Phys. Rev.* **132**, 1559–1569 (1963).
218. T. Yamaguchi and H. Kamimura, Dynamical Jahn–Teller interaction in optical and paramagnetic resonance spectra—application to a Cu deep impurity in II–VI semiconductors, *J. Phys. Soc. Japan* **33**, 953–966 (1972).
219. S. A. Kazanskii and A. I. Ryskin, Electron structure of ion–hole complexes attached to Ni ions in ZnS crystals, *Fiz. Tverd. Tela* **14**, 1818–1820 (1972) [English transl., *Soviet Phys.—Solid State* **14**, 1571–1572 (1972)]; S. A. Kazanskii and A. I. Ryskin, Phototransfer of holes from Ni^{2+} to Cu^{2+} ions in a Zns–Ni, Cu crystal, *Opt. Spectrosc.* **33**, 146–152 (1972).
220. S. A. Kazanskii, G. I. Khelko, A. L. Natadze, and A. J. Ryskin, Luminescence of II–VI compounds doped with transition metal ions, in *Luminescence of Crystals, Molecules and Solutions* (F. Williams, ed.), pp. 553–563, Plenum Press, New York (1973).
221. K. Era, S. Shionoya, and Y. Washijawa, Mechanism of broad-band luminescences in ZnS phosphors. I. Spectrum shift during decay and with excitation intensity, *J. Phys. Chem. Solids* **29**, 1827–1841 (1968).
222. B. Segall, Band structure, in *Physics and Chemistry of II–VI Compounds* (Aven and Prener, eds.), pp. 1–72, North-Holland Publishing Co., Amsterdam (1967).
223. T. Koda and S. Shinoya, Nature of the self-activated blue luminescence center in cubic CuS: Cl single crystals, *Phys. Rev. A* **136**, 541–555 (1964).
224. W. C. Holton, J. Schneider, and T. L. Estle, Electron paramagnetic resonance of photosensitive iron transition group impurities in ZnS and ZnO, *Phys. Rev. A* **133**, 1638–1641 (1964).
225. W. C. Holton, M. De Wit, R. K. Watts, T. L. Estle, and J. Schneider, Paramagnetic copper centers in ZnS, *J. Phys. Chem. Solids* **30**, 963–977 (1969).
226. J. L. Merz, H. Kukimoto, K. Nassau, and J. W. Shiever, Optical properties of substitutional donors in ZnSe, *Phys. Rev. B* **6**, 545–556 (1972).
227. C. H. Henry, K. Nassau, and J. W. Shiever, Optical studies of shallow acceptors in CdS and CdSe, *Phys. Rev. B* **4**, 2453–2463 (1971).
228. H. Mahr, Ultraviolet absorption of the mixed system KCl–KBr, *Phys. Rev.* **122**, 1464–1468 (1963).
229. Y. Nakai, T. Murata, and K. Nakamura, Exciton absorption in mixed alkali chlorides, KCl–RbCl and KCl–NaCl, *J. Phys. Soc. Japan* **18**, 1481–1485 (1963).

230. Y. Onodera and Y. Toyozawa, Persistence and amalgamation types in the electronic structure of mixed crystals, *J. Phys. Soc. Japan* **24**, 341–355 (1968).
231. P. N. Sen. Coherent potential approximation in a two-band model: Electronic and transport properties of semiconducting binary alloys, *Phys. Rev. B* **8**, 5613–5623 (1973); Excitons in mixed crystals of alkali halides and of cuprous halides: Coherent potential approximation, *Phys. Rev. B* **8**, 5624–5631 (1973).
232. J. P. Van der Ziel, Optical spectrum of Cr^{+3} pairs in $LaAlO_3$, *Phys. Rev. B* **4**, 2888–2905 (1971).
233. P. Kisliuk, N. C. Chang, P. L. Scott, and M. H. L. Pryce, Energy levels of chromium ion pairs in ruby, *Phys. Rev.* **184**, 367–374 (1969).
234. J. Owen and E. A. Harris, Pair spectra and exchange interactions, in *Electron Paramagnetic Resonance* (S. Geschwind, ed.), pp. 427–492, Plenum Press, New York (1972).
235. A. T. Casey, B. S. Morris, E. Sinn, and J. R. Thackeray, Magnetic behavior and structure of polymeric oxovanadium complexes, *Aust. J. Chem.* **25**, 1195–1205 (1972).
236. P. W. Anderson, Theory of magnetic exchange interactions: Exchange in insulators and semiconductors, in *Solid State Physics* (F. Seitz and D. Turnbull, eds.), Vol. 14, pp. 99–214, Academic Press, New York (1963).
237. N. L. Huang, Exchange interaction and energy transfer for $3d^3$ Ions: $^4A_2 \rightarrow {}^2E$ Transition for $KMgF_3$: V^{2+}, *Phys. Rev. B* **1**, 945–949 (1970).
238. S. Freeman, Molecular-orbital theory of the excited-state exchange interaction, *Phys. Rev. B* **7**, 3960–3986 (1973).
239. P. M. Levy, Anisotropic exchange, in *Magnetic Oxides* (D. J. Craik, ed.), Wiley—Interscience, New York, to be published.
240. J. W. Allen, R. M. Macfarlane, and R. L. White, Magnetic Davydov splitting in the optical absorption spectrum of Cr_2O_3, *Phys. Rev.* **179**, 523–541 (1969).
241. T. Fujiwara, Two-exciton transitions in $RbMnF_3$. I. Jahn–Teller effects in a pair of excited states, *J. Phys. Soc. Japan* **34**, 36–43 (1973); II. Optical absorptions, *J. Phys. Soc. Japan* **34**, 1180–1191 (1973).
242. J. Ferguson and H. J. Guggenheim, Electron transfer states of pairs of unlike transition metal ions in perovskite fluorides, *Phys. Rev. B* **1**, 4223–4228 (1970).
243. M. B. Robin and P. Day, Mixed valence chemistry—a survey and classification, in *Advances in Inorganic Chemistry and Radiochemistry* (Emeleus and Sharp, eds.), Vol. 10, pp. 247–422, Academic Press, New York (1968).
244. G. C. Allen and N. S. Hush, Intervalence transfer absorption. Part I, in *Progress in Inorganic Chemistry* (F. A. Cotton, ed.), Vol. 8, pp. 357–389, Interscience, New York (1967).
245. N. S. Hush, Intervalence transfer absorption. Part II, in *Progress in Inorganic Chemistry* (F. A. Cotton, ed.), Vol. 8, pp. 391–444, Interscience, New York (1967).
246. J. B. Goodenough, Metallic oxides, in *Progress in Solid State Chemistry* (H. Reiss, ed.), Vol. 5, pp. 145–399, Pergamon Press, Oxford (1971).
247. J. S. Prener and F. E. Williams, Activator systems in zinc sulfide phosphors, *J. Electrochem. Soc.* **103**, 342–346 (1956).

248. F. Williams, Radiative recombination at donor–acceptor pairs and at higher associates, *J. Luminescence* **7**, 35–50 (1973).
249. J. J. Hopfield, D. G. Thomas, and M. Gershenzon, Pair spectra in GaP, *Phys. Rev. Lett.* **10**, 162–164 (1963).
250. D. G. Thomas, M. Gershenzon, and F. A. Trumbore, Pair spectra and edge emission in gallium phosphide, *Phys. Rev.* **133A**, 269–279 (1964).
251. P. J. Dean and J. L. Merz, Pair spectra and "edge emission" in zinc selenide, *Phys Rev.* **178**, 1310–1318 (1969).
252. D. C. Reynolds and T. C. Collins, Donor–acceptor pair recombination spectra in cadmium sulfide crystals, *Phys. Rev.* **188**, 1267–1271 (1969).
253. P. Dean, Inter impurity recombinations in semiconductors, in *Progress in Solid State Chemistry* (J. O. McCaldin and G. Somorjai, eds.), Vol. 8, pp. 1–126, Pergamon Press, New York (1973).
254. N. Fontell, V. Hovi, and L. Hyvönen, *Ann. Acad. Sci. Fenn. AI* **1949**, No. 65.
255. W. T. Barrett and W. E. Wallace, *J. Am. Chem. Soc.* **76**, 366 (1954).
256. M. W. Lister and N. F. Meyers, *J. Phys. Chem.* **62**, 145 (1958).
257. M. A. Fineman and W. E. Wallace, *J. Am. Chem. Soc.* **70**, 4165–4169 (1948).
258. N. Fontell, *Soc. Sci. Fenn. Comm. Phys.–Math.* **1939**, No. 6.
259. V. Hovi, *Ann. Acad. Sci. Fenn. AI* **1948**, No. 55.
260. N. Fontell, V. Hovi, and A. Mikkola, *Ann. Acad. Sci. Fenn. AI*, **1949**, No. 54.
261. W. Bollman, P. Gorlich, P. Hank, and H. Mothes, *Phys. Stat. Sol.* **2**, 157 (1970).
262. Z. Kiss, Energy levels of Dy^{2+} in the cubic hosts CaF_2, SrF_2, and BaF_2, *Phys. Rev.* **137A**, 1749–1760 (1965).
263. W. Kaiser, S. Sugano, and D. L. Wood, Splitting of the emission lines of ruby by an external electric field, *Phys. Rev. Letters* **6**, 605–607 (1961).
264. D. S. McClure, Optical spectra of transition metal ions in corundum, *J. Chem. Phys.* **36**, 2757–2779 (1962).
265. J. Manthey, Crystal field and site symmetry of trivalent cerium ions in CaF_2: The C_{4v} and C_{3v} centers with interstitial fluoride charge compensation, *Phys. Rev. B* **8**, 4086–4098 (1973).
266. M. B. Robin, The color and electronic configurations of Prussian Blue, *Inorg. Chem.* **1**, 337–342 (1962).
267. H. Eyring, J. Walter, and G. Kimball, *Quantum Chemistry*, Wiley (1944).
268. D. S. McClure, in *Advances in the Chemistry of the Coordination Compounds* (S. Kirschner, ed.), pp. 498–508, Macmillan, New York (1961).
269. J. Ferguson, *Austral. J. Chem.* **21**, 307 (1968).
270. P. J. Dean and L. Patrick, *Phys. Rev. B* **2**, 1888 (1970).

2

The Imperfect Solid—
Color Centers in Ionic Crystals

W. Beall Fowler
Department of Physics
Lehigh University
Bethlehem, Pennsylvania

1. Introduction

1.1. General

The study of defects in ionic crystals through their interaction with electromagnetic radiation has been an important area of research for several decades. Such research has not yet directly resulted in the development of many important devices, although such developments are by no means precluded. In fact, much of the best work from a scientific point of view has been done on alkali halide crystals, which have not proved very useful technologically.

The importance of alkali halide color-center research for the solid-state chemist would seem to be that in these simple model solids so much can be learned: Sophisticated experimental tools such as spin resonance and magnetooptics have been applied, and detailed theoretical models have been developed and tested. This is truly a mature area in comparison with research in more complex insulators, and by virtue of its maturity one would hope to learn from it a good deal which could be transferred to studies of more complex systems. In particular, one might hope to avoid many of the pitfalls in both theory and experimental procedure which are a part of the development of a field of science.

1.2. Scope

This chapter is intended to emphasize those properties of color centers which are directly related to their interaction with electromagnetic radiation. It is designed to be tutorial, rather than a comprehensive review; the latter would be impossible within the space limitations, and would not appear to be in the spirit of this treatise. Alkali halide color centers will be used in many cases as examples, but centers in other ionic crystals will be discussed where appropriate.

Several comprehensive books, dealing mainly with alkali halide color centers, exist.[1–4] In addition, the compilations of abstracts submitted to the triennial color center conferences, the most recent of which is cited,[5] are quite useful.

1.3. Units and Definitions

Although optical spectra are commonly measured as a linear function of wavelength, analysis of the spectra requires a scale linear in frequency (or photon energy). One such common unit is cm^{-1}, but a more convenient unit, which we shall use, is the electron volt (eV). The conversion factors are the following[6]:

$$1\ \mathrm{cm}^{-1} = 1.2398541 \times 10^{-4}\ \mathrm{eV}$$

These lead to the approximate rules of thumb: To obtain eV, divide the number of cm^{-1} by 8065; or divide 12,400 by the wavelength in Å.

Optical absorption is often measured by means of a transmission technique, in which light impinges upon a surface and the amounts incident and transmitted are measured. If reflections are negligible, simple theory suggests that I_0, the incident light intensity, and I, the transmitted light intensity, should be related at a given photon energy E by

$$I(E) = I_0(E)e^{-\mu(E)d} \tag{1}$$

where d is the thickness of the crystal and $\mu(E)$ is called the absorption coefficient. The absorption coefficient is a characteristic of the material, and spectra are often plotted as $\mu(E)$ versus E. In some cases it is more convenient to plot or analyze a related quantity, the optical density (OD). The optical density is defined as

$$(\mathrm{OD}) = \log_{10}(I_0/I) \tag{2}$$

(OD) is clearly related to $\mu(E)$; however, to obtain the fundamental quantity μ from (OD), one must know d. A brief discussion of some of these points is given by Brown.[7] Theoretical expressions for $\mu(E)$ are given in Section 3.3.3.

2. Optical Properties—Perfect Crystals

2.1. The Spectrum

Perfect ionic crystals are typically transparent over a wide spectral range. In an alkali halide this range might extend from $\sim$0.05 eV to $\sim$5 eV (or in wavelength from $\sim$250,000 to $\sim$2500 Å). This transparent "window" facilitates the study of the optical properties of defects, many of which absorb light within this range.

2.2. The Infrared Region

Optical absorption in the infrared (IR) region is due to excitation of lattice vibrations by the incident light. Although analyses of infrared spectra and calculations of lattice vibrations were carried out early in the histroy of solid-state physics (see, e.g., Born and Huang[8]), developments in recent years of neutron diffraction techniques and computer codes have led to a detailed knowledge of the vibrations in a number of crystals.[9] Figure 1 illustrates the type of vibrational modes found in many alkali halide crystals.[10] A compilation of data for a number of alkali halides has been made by Knox and Teegarden.[10]

Interpretation of the infrared spectra follows from an understanding of the vibrational modes. While details of the interpretation are not always trivial, the main features of the IR spectra can usually be directly understood.[11] For example, the incident light can couple strongly with the transverse optical (TO) modes near $k = 0$, leading to a strong one-phonon process (photon absorbed, vibrational quantum created). The longitudinal optical (LO) mode near $k = 0$ may be excited by light of nonnormal incidence in a one-phonon process. Processes involving the creation of two or more phonons (not necessarily at $k = 0$) may also occur. A detailed analysis of both reflectivity and absorption is best provided through a calculation of the dielectric function; see, e.g., the discussion by Kittel.[11] An example of such an analysis for LiF and MgO is given by Jasperse *et al.*[12]

Crystals with more complicated structures have more modes at $k = 0$, and consequently their infrared spectra are more complex. Many of these more complex crystals have been studied, and references

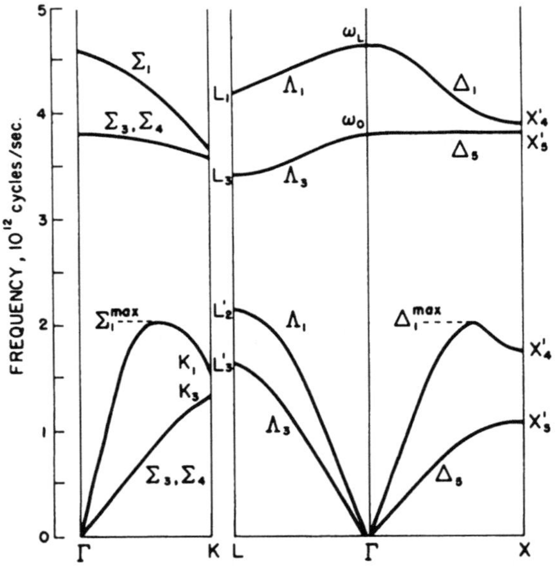

Fig. 1. Qualitative vibrational modes of an alkali halide such as KBr. Symmetry types and quantitative data are given in Ref. 10, which is also the source of this figure.

are given to work on calcium fluoride,[13] magnesium fluoride,[14] and alpha quartz.[15]

2.3. The Ultraviolet Region

Optical absorption in the ultraviolet (UV) region and beyond involves the excitation of electrons from valence or core bands in the solid (see Chapter 2 of Volume 1 for a discussion of energy bands). In some cases this excitation results in dissociation of the electron and hole, leaving the electron in a conduction band and the hole in a valence or core band; this is referred to as a band-to-band transition. In other cases a bound hole–electron pair, or exciton,[16] is formed. In most spectra both features exist and are sometimes difficult to distinguish from one another.

Until recently experimental difficulties limited high-precision optical absorption studies to the near-UV region. With the advent of electron accelerators as light sources, however, considerable far-UV and X-ray data have been and are being gathered on many solids by a number of groups. Detailed discussion of the latter is beyond the scope of this chapter, but several papers are cited which illustrate the techniques and results on alkali and magnesium halides.[17,18]

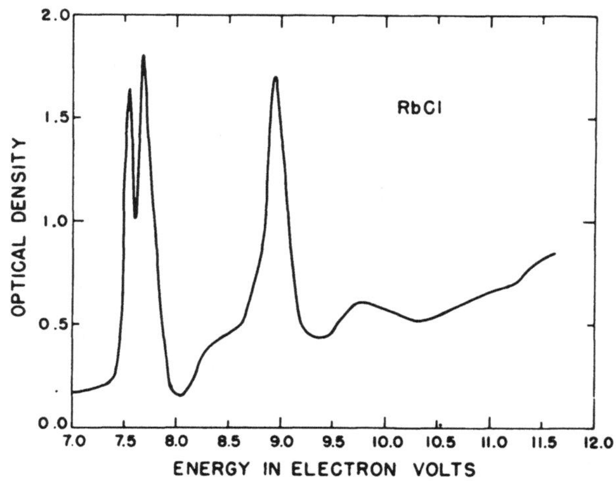

Fig. 2. The optical absorption spectrum of RbCl.[19]

Figure 2 illustrates, as an example, the absorption spectrum of RbCl from 7.0 to 12.0 eV.[19] The two peaks between 7.5 and 8.0 eV are due to excitons in which the hole is mainly on the halogens; the doublet reflects the spin–orbit splitting in the chlorine $3p$ shell. The shoulder at $\sim$8.2 eV is thought to result from band-to-band transitions, while the peak at $\sim$9 eV may well arise from an exciton degenerate with the conduction states.

Although detailed band structure calculations[20] support the idea that in most alkali halides the lowest-energy exciton involves a hole on the halogen sublattice, in certain other crystals such an assignment cannot always be made with confidence. Figure 3 illustrates why. One may make a zeroth-order estimate of the one-electron energies in a diatomic ionic solid by shifting the free-ion energies by the Madelung field in which the ions are located. The effect is to shift upward the states derived from positive ions, and downward those from negative ions. Although in the alkali halides the highest valence states are still largely halogenlike, this may not always be the case, as noted for thallium, silver, and copper halides. Here the zeroth-order states are close together in energy, and quantum mechanical admixture may create complex valence bands.[21] In some cases there is evidence that the low-energy excitons occur primarily on the positive ions,[22] as, for example, in TlBr and PbI_2.

Other optical absorption techniques, including photoconductivity,[23] photoemission,[24] and two-photon spectroscopy,[25] have

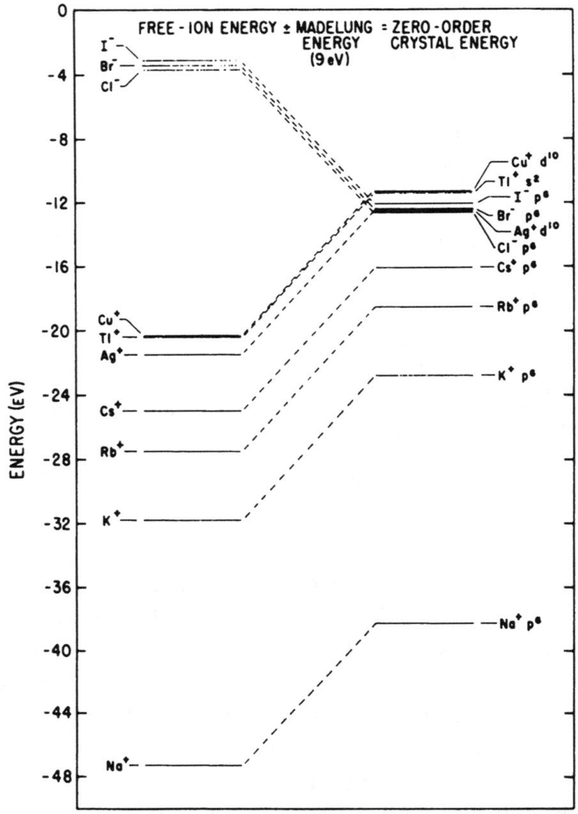

Fig. 3. The energies of outer electrons on various closed-shell ions, with and without Madelung energy corrections.

been used to study the electronic states in ionic crystals. Photoconductivity, discussed further in Section 3.5.1, can give information about the valence–conduction band gap; photoemission yields some knowledge of valence bandwidths and state densities; and two-photon spectroscopy gives information about transitions forbidden in one-photon spectroscopy.

In a number of ionic crystals luminescence is observed to occur following UV or X-irradiation. While this luminescence may involve permanent defects, in many cases it is intrinsic. However, this phenomenon involves recombination of an electron with a self-trapped hole; one has, in effect, a transient defect center, and its analysis properly belongs in Section 3.4 of this chapter.

3. Color Centers

3.1. Types of Centers

Optical absorption and luminescence in the visible region, or more generally within the infrared–ultraviolet "window" of the perfect crystal, occur as a result of the interaction of light with color centers.[1–5] In general, many types of color centers may exist. The simplest color centers may be categorized as being either impurity-related or defect-related. A common impurity center, for example, is Tl^+, which substitutes for alkalis in alkali halides. The simplest defect center is the *F* center, which in alkali halides consists of an electron trapped at a halogen vacancy. The V_K center, which is a self-trapped hole in an alkali halide, does not involve a lattice defect, but rather an electronic defect. Short-lived or transient[26] color centers may also exist. The recombination luminescence in an alkali halide involves an electron trapped at a V_K center; the center $(V_K + e)$ exists only until luminescence takes place, at most $\sim 10^{-6}$ sec.

Many centers exist which overlap the above categories, as, for example, an *F* center next to an impurity. Rather than provide an encyclopedic listing of center models, we present as examples some of the better-known centers in various materials. Figure 4 illustrates some of the most common color centers in alkali halides. The F_A center is an *F* center next to an alkali impurity, and the *M* center consists of two neighboring *F* centers. The *F′* center is an *F* center with an additional electron. The V_K center is a self-trapped hole.

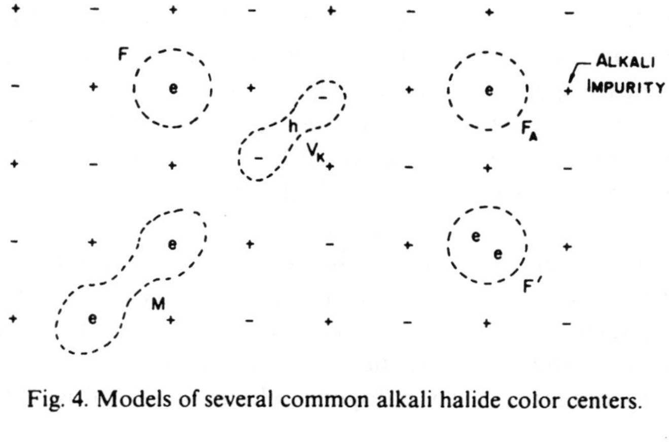

Fig. 4. Models of several common alkali halide color centers.

TABLE 1

Some Color Center Models in Various Crystals

Crystal	Designation	Model
Alkaline earth oxides (e.g., CaO)	F	Two electrons at O^{2-} vacancy
	F'	Three electrons at O^{2-} vacancy
	F^+	One electron at O^{2-} vacancy
Alkaline earth fluorides (e.g., CaF_2)	F	One electron at F^- vacancy
	M	Two F centers in (100) orientation
	F_3	Three F centers in (100) orientation
Alpha quartz (SiO_2)	E'	One electron at O^{2-} vacancy

Models for a few centers in other crystals are listed in Table 1. In some cases the assignments are tentative. A number of other centers have been studied in these crystals, but not all of these have been given names.

3.2. Color Center Creation

Color center creation[1,4] involves crystal chemistry and radiation.[27] Impurity centers include ions which may occur in natural specimens (e.g., aluminum in quartz, sodium in KCl) or which are introduced in one of several ways. The most common method of introducing chemical impurities is by doping the molten substance and then growing the crystal.[1] Alternative methods include ion implantation and additive coloration. Additive coloration involves heating the crystal in the presence of a vapor of some other material.[1] Depending upon the circumstances, one obtains one of several results. For example, heating an alkali halide crystal with halogen vapor at high pressure[1] ($\gtrsim 50$ Torr) leads to the formation of V bands, thus indicating that halogen diffuses into the crystal. If one heats sodium iodide in the presence of copper,[28] it is found that copper ions are readily incorporated into the crystal.

The technique of additive coloration may in some cases be used to create defect centers[1]; for example, it is one of the more commonly used methods for generating F centers. When an alkali halide is heated in alkali vapor the atoms in the vapor are deposited on the surface, where they combine with halogens which have diffused from the interior. The net result is F centers within the crystal and new planes of crystal at the surface.

Electrolytic methods may also be used to produce color centers.[1] For example, by applying ~ 100 V/cm across an alkali halide, using one electrode which is flat and the other which is pointed, one may create either F centers or V centers. The former are created if the pointed electrode is the cathode, the latter if it is the anode.

A more general technique of defect creation is through high-energy irradiation. It is clear that energetic particles can displace atoms and create defects, but of particular interest is the fact that in some cases X-irradiation also leads to defect formation. Since X-ray photons have essentially zero momentum, the primary process during X-irradiation is ionization, and the secondary processes leading to defect center formation are of considerable interest.

These effects have been widely studied in the alkali halides in connection with F-center production. Crawford's review article[27] traces the logic of the various investigations. It is significant that F centers can also be created through absorption of light in the exciton region,[27] in which the photon energy is ~ 5–10 eV. Furthermore, in the temperature range in which the creation efficiency is high the recombination luminescence efficiency of the $(V_K + e)$ center is low. These facts, and others, led to the suggestion[29] that the energy involved in creating the defect is that given up to the lattice when the $(V_K + e)$ center decays nonradiatively. Thus the excitation mechanism is irrelevant, provided that the system ends up as $(V_K + e)$. In the case of X-ray excitation $(V_K + e)$ is formed through the stages of ionization of the halogen ion, formation of V_K, and trapping of an electron into $(V_K + e)$, while with UV excitation $(V_K + e)$ can be formed directly.

The question of how this energy can be used to produce a defect is of considerable interest[27,29] and has received theoretical study. It turns out that there is a reasonable probability that the energy will be unequally shared by the two halogen ions, and that a sequence of collisions along the (110) direction can eventually lead to the ejection of a halogen from its normal position into an interstitial position.

Although there is theoretical reason[27] to believe that this vacancy creation process does not take place in alkaline earth oxides, its possibility in other ionic crystals should not be overlooked.

It should be pointed out that ease of coloration differs considerably among the various alkali halides; for example, it is very difficult to form F centers in NaI by standard techniques. Since the rate of creation ultimately depends upon energy balances within the crystal or at the surface, the existence of different rates in different crystals should not be surprising.

A second example of X-ray coloration involves the E' center in α-quartz. This center is thought to involve an oxygen vacancy. It is not formed under X-irradiation unless the sample contains germanium, in which case the defect which is formed is associated with a germanium ion.[30]

One can only speculate at this point as to the creation mechanism. It might be noted, however, that were the mechanism to involve vibrational energy apportioned between an oxygen and a silicon (or germanium), the large mass of the germanium would allow more of the kinetic energy to be imparted to the oxygen, thereby increasing the probability that the oxygen could escape.

It should finally be noted that many crystals, such as natural minerals, color when exposed to X-irradiation.[1] In most cases this involves transformation of a preexisting impurity into a different ionization state, one which then absorbs visible light. For example, SiO_2 containing Al may or may not be colored, depending upon whether the Al has trapped a hole. Such processes, strictly speaking, involve transformations rather than creation, and are discussed in Section 3.6.

3.3. Theory of Electronic Structure and Optical Properties

3.3.1. Examples

The F center in alkali halides has been widely studied, and a few of its optical properties (in KBr) are summarized here. Figure 5 illustrates the absorption and emission bands of the KBr F center at various temperatures.[31,32] The absorption peak is at 2.06 eV at 4°K, with a full-width at half-maximum (half-width) of 0.16 eV. At higher temperatures the half-width increases, while the area under the curve remains nearly constant. The emission peak is at 0.92 eV at 4°K and the half-width is 0.22 eV. The shift in energy, or Stokes shift, between absorption and emission peaks is quite large, 1.14 eV.

The absorption band is strong, with an oscillator strength of order one; the radiative lifetime is typically $\sim 10^{-6}$ sec, two orders of magnitude larger than observed for allowed transitions in atoms.

Figure 6(a) gives the absorption and emission bands[33] of the R' center in LiF at 4°K (the R' center consists of three adjacent F centers plus an extra electron). This spectrum is significantly different from

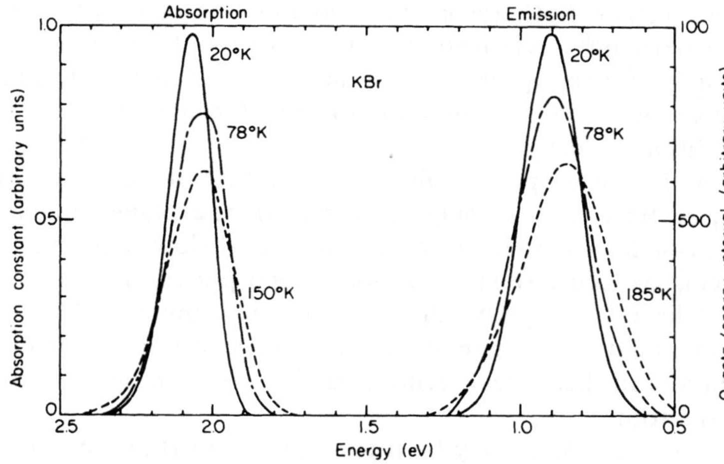

Fig. 5. *F*-center absorption and emission bands in KBr.[31]

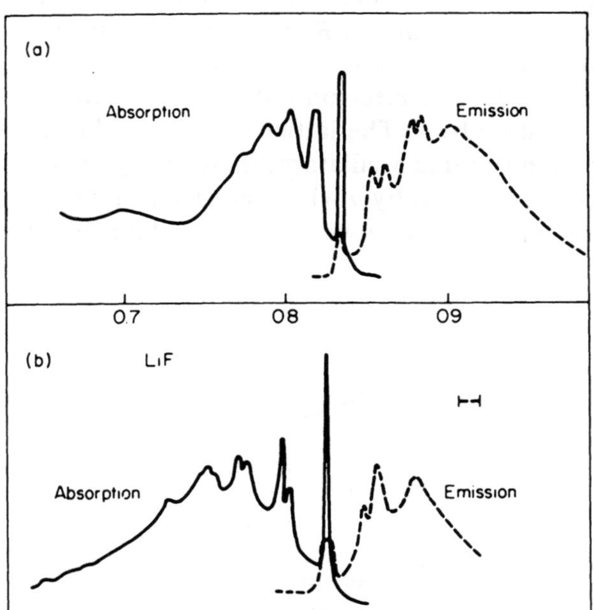

Fig. 6. Optical spectra of (a) *R′* center and (b) *M′* center in LiF at 4°K.[33]

that of the F center. Both absorption and emission bands are structured, the Stokes shift is relatively small (~ 0.2 eV), and absorption and emission spectra overlap each other and share a common peak. Similar behavior is seen in Figure 6(b) for the M' center, an M center plus an additional electron.

Both of these examples involve large numbers of identical, noninteracting centers, and their major features can be explained in terms of the electron–lattice interaction, as discussed below. However, in some systems, particularly when the host is amorphous, one may have "identical" defects in slightly different static environments. This can lead to wide, relatively temperature-independent absorption bands. Experimental and theoretical techniques[34] have been developed to treat this situation.

When all the defects may be treated as identical and noninteracting, the width of absorption and emission bands arises from thermal fluctuations, or lattice vibrations. Some quantitative aspects of the theory are given in Section 3.3.2. Qualitatively, the effect of lattice vibrations may be understood by considering Figure 7, a configuration coordinate diagram.[32,35,36] This is a plot of energy E versus some convenient vibrational coordinate R, analogous to the familiar plots of energy versus interatomic separation for H_2.

In Figure 7 the ground electronic state is denoted by m and an excited electronic state by k. The lattice will have vibrational states similar to those of harmonic oscillators; these are denoted by η when the electronic state is m and by γ when the electronic state is k. The optical transition probability between a pair of states (m, η) and (k, γ)

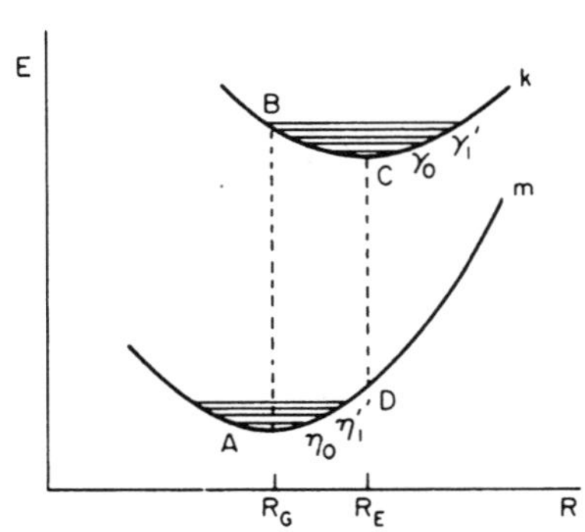

Fig. 7. Schematic configuration-coordinate diagram. See text for details.[38]

will depend upon energy conservation, electronic matrix elements and selection rules, and the overlap of vibrational wave functions η and γ. Two extreme cases are of interest. The case corresponding to the F center is one in which the ground-state minimum A and the excited-state minimum C are displaced a large distance in R. It turns out then that significant vibrational wave function overlap occurs only for states nearly vertical with respect to each other, e.g., A and B or C and D in Figure 6. The optical absorption peak is then $E_B - E_A$; after absorption the vibrational system is excited, and the lattice relaxes and phonons are emitted. The emission energy is $E_C - E_D$, after which another lattice relaxation takes place. One can see, then, that the Stokes shift represents energy given to the lattice. These are called "multiphonon" transitions.

The large, temperature-dependent widths arise from thermal fluctuations, which lead to dynamic changes in excitation energy. A detailed analysis also shows that the electronic wave functions change as R changes; this can lead to different transition probabilities[37–39] for the transitions $A \rightarrow B$ and $C \rightarrow D$.

Spectra such as that of the R' or M' centers[33] (Figure 6) may also be understood by referring to Figure 7. Here, however, the ground-state minimum A and the excited-state minimum C are rather close in R. There is significant overlap between low-lying vibrational wave functions in electronic states m and k, and under suitable conditions one may observe the discrete transitions $m, \eta \rightarrow k, \gamma$. Of particular interest is the "zero-phonon" absorption transition $\eta_0 \rightarrow \gamma_0$ and its inverse in emission, $\gamma_0 \rightarrow \eta_0$.

These examples, and many others, underscore the importance of understanding the electron–phonon interaction in interpreting color center spectra. The next section summarizes some aspects of that theory.

3.3.2. The Electron–Lattice Interaction

Theoretical treatments of the electron–lattice interaction and its effect on the optical properties of defects and impurities are many[32,33,36,40–42] and there are many aspects of the theory which will not be discussed here. Rebane's book[43] provides a concise treatment.

The Schrödinger equation for the electron–lattice system may be written

$$[T_e + T_N + U(\mathbf{r}, \mathbf{R})]\Psi(\mathbf{r}, \mathbf{R}) = E\Psi(\mathbf{r}, \mathbf{R}) \tag{3}$$

where $\mathbf{r}$ stands for all electronic coordinates, $\mathbf{r}_i$; $\mathbf{R}$ is for all nuclear coordinates $\mathbf{R}_I$; T_e is the electronic kinetic energy; T_N is the nuclear kinetic energy, and $U(\mathbf{r}, \mathbf{R})$ contains all of the potential energy terms.

For many situations this coupled equation may be written as two equations in the Born–Oppenheimer approximation.[36,41] The system wave function Ψ is approximated as a product

$$\Psi_{n\eta}(\mathbf{r}, \mathbf{R}) = \Phi_{n,\mathbf{R}}(\mathbf{r})\chi_{n\eta}(\mathbf{R}) \tag{4}$$

where Φ and χ satisfy

$$[T_e + U(\mathbf{r}, \mathbf{R})]\Phi_{n,\mathbf{R}}(\mathbf{r}) = E_n(\mathbf{R})\Phi_{n,\mathbf{R}}(\mathbf{r}) \tag{5}$$

and

$$[T_N + E_n(\mathbf{R})]\chi_{n\eta}(\mathbf{R}) = E_{n\eta}\chi_{n\eta}(\mathbf{R}) \tag{6}$$

Equation (5) yields electronic wave functions and energies, assumed to depend on the positions of the ions in an adiabatic fashion. The $E_n(\mathbf{R})$ would correspond to the smooth curves of Figure 7. Equation (6) yields vibrational wave functions and energies. The electronic energy functions $E_n(\mathbf{R})$ serve as potential energies for the vibrations, hence the vibrational levels are quantized as in Figure 7.

The qualitative discussion of configuration coordinate curves in Section 3.3.1 was based upon the equations of the Born–Oppenheimer approximation. This approximation is valid if small changes in the nuclear positions $\mathbf{R}$ produce small, continuous changes in the electronic energies $E_n(\mathbf{R})$ and wave functions $\Phi_{n,\mathbf{R}}(\mathbf{r})$. For nondegenerate electronic states this statement is equivalent to saying that because the nuclei are so much more massive than the electrons, the electrons respond rapidly to nuclear motion, but not vice versa. However, if electronic states are degenerate or nearly so, small nuclear motions can significantly affect admixtures and the Born–Oppenheimer approximation is no longer valid. The article by Henry and Slichter[42] discusses observables which are independent of this approximation, and a number of authors, most recently Ham,[44] have discussed models for which this approximation need not be made.

Further approximations must be made in applying Eqs. (5) and (6) to particular problems. In Section 3.3.4 we discuss aspects of the solution of Eq. (5), the electronic states for given nuclear positions. One formal aspect, which leads to a change in notation, is this: The vibrations may be classified group-theoretically into *normal modes* about the defect; for example, one mode in a cubic lattice with a substitutional point defect will be the totally symmetric Γ_1, in which ions in a given shell all vibrate in and out, in phase. When such a classification is made one considers not the explicit lattice positions **R**, but the normal coordinates **q**.

A simple, model analysis of the system is given by Fitchen,[33] illustrated in Figures 7 and 8. It is assumed that the function $E(\mathbf{q})$ for the ground electronic state m is

$$E_m = \tfrac{1}{2}M\omega^2 q^2 \tag{7}$$

where ω is the frequency of mode q, and M is an effective mass. For the excited state a linear coupling approximation is made:

$$E_k = \tfrac{1}{2}M\omega^2 q^2 + E_{mk} - A\hbar\omega(M\omega/\hbar)^{1/2} q \tag{8}$$

In Equation (8) A is a dimensionless constant which measures the strength of the linear interaction which displaces the equilibrium position of the mode without changing its frequency ω. Here E_{mk} is the separation between excited- and ground-state curves for $q = 0$ (the value of q corresponding to the minimum of the ground-state curve), and would be the transition energy for fixed $q = 0$.

Equation (6) can be solved for both E_m and E_k, yielding

$$E_{m\eta} = (\eta + \tfrac{1}{2})\hbar\omega \tag{9}$$

$$E_{k\gamma} = (\gamma + \tfrac{1}{2})\hbar\omega + E_{mk} - \tfrac{1}{2}A^2\hbar\omega \tag{10}$$

where η and γ are integer quantum numbers. The corresponding wave functions $\chi_{m\eta}(q)$ and $\chi_{k\gamma}(q)$ are also obtained.

The transition probability between states $m\eta$, $k\gamma$ is proportional to the quantity

$$|\mathbf{r}|^2_{m\eta,k\gamma} \equiv |\langle\Phi_m\chi_{m\eta}|\mathbf{r}|\Phi_k\chi_{k\gamma}\rangle|^2 \tag{11}$$

and if the q dependence of Φ is neglected, Eq. (11) is just

$$|\mathbf{r}|^2_{m\eta,k\gamma} = |\langle\Phi_m|\mathbf{r}|\Phi_k\rangle|^2|\langle\chi_{m\eta}|\chi_{k\gamma}\rangle|^2 \tag{12}$$

or a constant times the square of a vibrational overlap integral. This vibrational overlap integral may be calculated for various pairs η, γ, and the result is given in Ref. 33. For the case in which the $\eta = 0$ level is involved,

$$|\langle\chi_{m0}|\chi_{k\gamma}\rangle|^2 = (S^\gamma/\gamma!)e^{-S} \tag{13}$$

where

$$S = A^2/2 \tag{14}$$

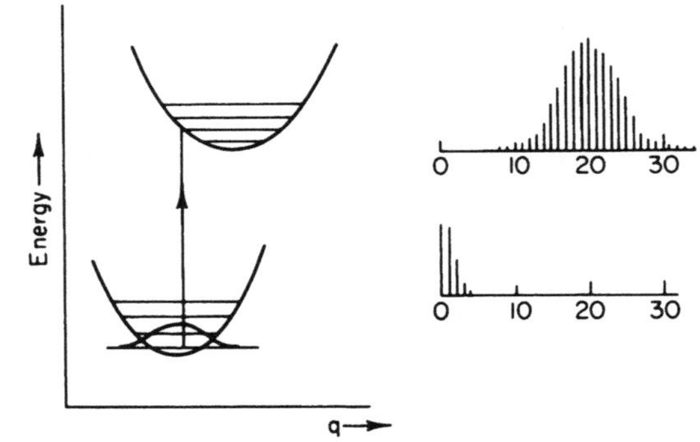

Fig. 8. Schematic configuration-coordinate diagram for linear coupling to one vibrational mode. Also shown are line shapes of weak ($S = 1$) and strong ($S = 20$) coupling.[33]

Equation (13) would give the $T = 0°\text{K}$ transition probability to the γth vibrational level. Note that the transition probability from $\eta = 0$ to $\gamma = 0$ is e^{-S}. This is the zero-phonon transition.

One thus predicts the $T = 0°\text{K}$ absorption spectrum to consist of a series of lines at

$$E_\gamma = (E_{mk} - S\hbar\omega) + \gamma\hbar\omega \tag{15}$$

and in emission

$$E_\eta = (E_{mk} - S\hbar\omega) - \eta\hbar\omega \tag{16}$$

a mirror image of the absorption. Note that the transition probabilities depend on both γ (or η) and S; Figure 8 illustrates the spectra for both strong ($S = 20$, upper curve) and weak ($S = 1$) coupling.

At finite temperatures a thermal average over ground vibrational states is made, leading to a more complex spectrum. One familiar result is the expression for the full-width at half-maximum of the envelope of transitions in the strong coupling limit:

$$H(T) \simeq 2.36[S(\hbar\omega)^2 \coth(\hbar\omega/2kT)]^{1/2} \tag{17}$$

Further complications arise in real systems from the contributions of more than one mode and the breakdown of the linear coupling approximation. These detailed difficulties should not, however, obscure the essential understanding which one has of the situation in which the Born–Oppenheimer approximation is valid.

When an electronic degeneracy or near-degeneracy occurs the Born–Oppenheimer approximation breaks down. According to the

Jahn–Teller theorem,[45] there will be interactions between the electrons and the vibrations which will tend to remove this degeneracy. If the interactions are strong, a static distortion occurs, the electronic degeneracy is removed, and one may apply the Born–Oppenheimer approximation to the distorted nondegenerate system. If the interaction is somewhat weaker, one has the more interesting (and difficult) situation in which dynamic distortions involving coupled electronic and vibrational states occur. The resulting "vibronic" wave functions are no longer simple products of electronic and vibrational parts.[44,45]

Since most optical transitions involve degeneracies in ground or excited states, vibronic effects are quite common. Since vibronic effects may be of the same magnitude as electronic perturbations such as spin–orbit or stress splittings, detailed analysis may become quite complex. It has been shown[42,46] that by measuring moments and changes in moments of optical bands, one could extract information about electronic perturbations independent of complex vibronic effects. This work has been of extreme importance in the analysis of broadband spectra. Particular results are discussed in Section 3.4.1.

3.3.3. Optical Transition Parameters

The basic theory of optical transitions in quantum mechanical systems is derived in most quantum mechanics texts, and its application to color centers appears in several places.[32,36] In this section we summarize some of the more familiar results.

If φ_k, φ_m are solutions of a time-independent Schrödinger equation with energies E_k, E_m, and if a monochromatic beam of N photons, polarized in the $\boldsymbol{\eta}$ direction, of energy $\hbar\omega$, is incident upon the system, then the transition probability per unit time W_{mk} is given by

$$W_{mk} = (4\pi^2 e^2 |E_k - E_m|/\hbar) N |\boldsymbol{\eta} \cdot \mathbf{r}_{km}|^2 \, \delta(|E_k - E_m| - \hbar\omega) \qquad (18)$$

in the electric dipole approximation. The dipole matrix element $\mathbf{r}_{km}$ is defined by

$$\mathbf{r}_{km} = \sum_q \int \phi_k{}^* \mathbf{r}_q \varphi_m d\tau \qquad (19)$$

a sum over all electrons. If the light is unpolarized, $|\boldsymbol{\eta} \cdot \mathbf{r}_{km}|^2$ is replaced by

$$\tfrac{1}{3}|\mathbf{r}_{km}|^2 \equiv \tfrac{1}{3}[(x_{km})^2 + (y_{km})^2 + (z_{km})^2] \qquad (20)$$

A number of related quantities may be defined. The oscillator strength f_{mk} is given by

$$f_{mk} = (2m/3\hbar)\omega_{mk}|\mathbf{r}_{km}|^2 \tag{21}$$

where $\omega_{mk} = |E_m - E_k|/\hbar$. The absorption coefficient μ is

$$\mu_{mk}(E) = (4\pi^2 e^2 N_A \hbar\omega_{mk}/3\hbar c)|\mathbf{r}_{km}|^2\, \delta(|E_k - E_m| - \hbar\omega) \tag{22}$$

where N_A is the number of atoms per unit volume. The lifetime of an excited state k against a spontaneous transition to a lower state m is

$$1/\tau_{km} = [4e^2(E_k - E_m)^3/3c^3\hbar^4]|\mathbf{r}_{km}|^2 \tag{23}$$

A number of modifications to these expressions occur for a center in a solid. Factors of order one must be included in some cases to account for the dielectric properties of the medium.(47) When multiphonon transitions occur sums over vibrational states lead to expressions for the transition rates in which k, m now refer only to the electronic states (e.g., Figure 7) evaluated for static nuclear positions $\mathbf{R}_0$ corresponding to the minima of the initial-state potential curves.

One commonly cited result is Smakula's equation:

$$N_A f_{mk} = (mc/2\pi^2 e^2 \hbar) n [E_0/E_{\text{eff}}]^2 \int \mu_{mk}(E)\, dE \tag{24}$$

where n is the index of refraction and $[E_0/E_{\text{eff}}]^2$ is a dielectric correction. If the Lorentz local field is used and the absorption band is Gaussian, one obtains

$$N_A f_{mk} = 0.87 \times 10^{17}\ \text{cm}^{-3} \times [n/(n^2 + 2)^2]\mu_{mk}(\text{max}) H_{mk} \tag{25}$$

where μ is in cm^{-1} and H_{mk}, the full-width at half-maximum, is in eV.

For the "generalized" versions of other quantities of interest one should see Refs. 32 or 36. It should be noted that uncertainties in dielectric corrections can lead to uncertainties of about a factor two in predicting absolute values of transition rates.(47)

3.3.4. Electronic Structure

As the preceding sections indicate, many aspects of the understanding of optical transitions in color centers involve knowledge of the electronic structure, that is, electronic wave functions and energy levels of the center with the nuclei assumed to be fixed. One thus has a many-electron Schrödinger equation [Eq. (5)] to solve, and since exact solutions are not possible, many approximate techniques have been developed to treat particular problems. We shall outline the ideas behind some of these approximations, but shall not treat them in detail. A more complete discussion of these approaches may be found in several review articles.[48]

Whether expressed formally in terms of group theory or in more intuitive terms, symmetry plays an important role in analyzing electronic spectra in solids. A free ion, for example, will have various energy levels as a result of electrostatic and spin–orbit interactions within the ion. In a crystal some of these levels will split and transition probabilities will be changed. One can attempt to calculate such changes, but even in the absence of detailed calculations much can be learned by analyzing the experimental results in terms of symmetries of the states involved.

One common method for treating the electronic Schrödinger equation is to assume a wave function in the form of an antisymmetrized product of one-electron functions; this leads to the coupled one-electron Hartree–Fock equations. Standard techniques exist[49] for solving the Hartree–Fock equations and many atoms and molecules[50] have been so treated. A few color center problems have been treated by Hartree–Fock theory,[51] although in most cases variations of this approach are used.

There appear to be two reasons why detailed Hartree–Fock solutions have not been attempted in many color center problems. One is complexity; a color center in a solid is, in effect, a very large molecule, and even if one can limit the number of neighboring ions which are treated in detail, computational difficulties exist. The second reason is physical. In Hartree–Fock theory each electron is treated as moving in the average Coulomb and exchange field of all the other electrons. But often the electrons have quite different binding energies from one another or, in classical terms, move at different speeds. A situation akin to the electron–lattice problem then exists, and one refers to "slow" and "fast" electrons. If the valence electrons of the center are "slow," i.e., weakly bound, while those of surrounding ions are "fast," then it is better to regard the "fast" electrons as responding to the instantaneous, rather than average, positions of the "slow" electrons, and to treat this response in terms of dielectric constants.[52] An extreme case of this situation is the shallow donor in a semiconductor,[53] where the interaction with the host ions is well approximated in part by including a dielectric constant in the Coulomb interaction between the weakly bound donor electron and the ion core.

In cases where Hartree–Fock theory is appropriate it is often approximated in one of various ways. For example, one might treat the problem of an impurity ion in a solid by using the free-electron, or Slater, exchange approximation[54] for the electronic interaction within the ion, and treat the neighboring ions separately, as discussed below. Hartree–Fock–Slater calculations for ions are facilitated by the availability of the Herman–Skillman computer program.[55]

The effects of interactions with neighboring ions may be treated in a number of ways. In a proper Hartree–Fock calculation one would include Coulomb and exchange interactions between the impurity ion and the neighbors, and the proper impurity wave functions would be orthogonal to functions on the neighboring ions. One sometimes invokes the cancellation theorem of pseudopotential theory[56] to argue that the orthogonality requirement may be lifted if the short-range Coulomb and exchange interactions are neglected. This is the theoretical basis for the so-called "point ion" approximation,[57] in which the surrounding crystal is treated as a collection of point ions whose charge is the net ionic charge (e.g., $+e$ for Na, $-e$ for Cl). A more general version of this is "ligand field" theory[58]; the field due to the neighbors may not be restricted to that of point ions, but it is assumed to exist independent of the state of the impurity, and the wave functions of the neighbors are not explicitly included in the calculation.

In a number of cases (e.g., Tl^+ in KCl) the electronic states of the impurity are nearly degenerate with those of the ligands. One then expects considerable quantum mechanical admixture, or covalency, and the states on neighboring ions must be treated equally with those of the impurity. One has, in effect, a large molecule imbedded in the solid. Some approximate techniques have been used to treat this problem, and covalency effects have been clearly demonstrated.[59,60]

In Section 3.4 some of the ideas outlined here are developed in terms of particular color centers.

3.4. Optical Properties—Results and Interpretation

In this section we shall discuss a number of particular centers. It is convenient here to establish three categories in terms of the relative importance of interactions within the defect or impurity versus interactions with the environment. These are the following: (1) Interactions within the defect are small; (2) interactions with the environment are small; (3) both interactions are important.

3.4.1. Electrons Trapped at Anion Vacancies

The limiting case of category 1 is that in which there is no central ion, and one or more electrons are trapped at one or more adjacent anion vacancies. (There is only limited experimental evidence[61] for the antimorph, holes trapped at cation vacancies, and we will not discuss this.) All of the interactions are then with the environment.

The simplest of these centers is the *F* center in alkali halides, consisting of one electron trapped at a halogen vacancy and illustrated in Figure 4. In this case electron–lattice coupling is strong and one observes broad absorption and emission bands. Some of the properties of the KBr *F* center are given in Section 3.3.1. We will discuss here some aspects of the theory of the electronic states of the *F* center, and will also discuss some more detailed optical experiments.

A vast number of articles and model calculations have appeared on alkali halide *F* centers, a very few of which are cited.* The peak positions of *F*-center absorption and the large oscillator strength can be predicted reasonably well by "vacancy-centered" models. Such models involve trial wave functions centered at the vacancy, often of simple form (i.e., hydrogenic). They also involve potential energy functions whose most significant feature is a well whose depth is of the order of the Madelung energy and whose radius is of the order of the nearest-neighbor distance. This well arises from the spherically symmetric term of a spherical harmonic expansion of the potential of a collection of point ions.[57]

Many variations of these models exist. For example, in the semicontinuum model[62,39] one treats the region of the crystal outside the vacancy according to effective-mass theory, while in the pseudopotential approach[63,64] attempts are made to include effects of the occupied orbitals on neighboring ions. More exact Hartree–Fock calculations have also been performed.[51,65] Some of these approaches lead to better quantitative results on absorption energies than others, but the absorption energy is not very sensitive to these modifications. The fact that the experimental absorption energies in various crystals scale with lattice constant (the Mollwo[66]–Ivey[67] law) is strong evidence that the point-ion potential well is of major significance in determining the absorption energies.

Other features of *F*-center absorption are more model-dependent. Several crystals exhibit a "*K* band" at an energy slightly above the *F* band. There appears to be good evidence[68] that this band is an envelope of the remainder of the Rydberg series $1s \rightarrow np$, of which the

* See Refs. 1, 2, 31, 32, 36–40, 42, 44, 46, 51, 52, 57.

F band corresponds to $n = 2$. If this is so, these higher p states can apparently be described rather well by effective-mass theory.

Another feature of F-center absorption is the spin–orbit splitting of the excited state. Because there is no central ion, the spin–orbit splitting must arise from interactions of the trapped electron with neighboring ions. Furthermore, because the electron–lattice interaction is large and the spin–orbit splitting in most crystals turns out to be small, techniques of observation(69,70) and analysis(46) are nontrivial. For example, the experiments of Mort *et al.*(69) involved a measurement of low-temperature Faraday rotation in the F-band region, using phase-sensitive detection methods. Their analysis of their data, on the basis of an "alkali-atom" model of the F-center states, showed that the spin–orbit splitting is opposite in sign to that obtained for an alkali atom. The magnitude is also rather small, i.e., 5×10^{-3} eV in NaCl, 3×10^{-2} eV in KI.

Smith(71) showed that both magnitude and sign could be computed if one used an excited-state wave function which was properly orthogonalized to the core orbitals; a simple smooth envelope function would not suffice, since it would not have the correct amplitude at the neighboring ions where the spin–orbit interaction is large.

Henry *et al.*(42,46) pointed out that the alkali-atom model could not be valid in a system where the electron–lattice interaction, which dynamically removes the degeneracy of the excited states, is much larger than the spin–orbit interaction. They proceeded to show how, by analyzing changes in various moments of the bands, one could understand the results of experiments such as those of Mort *et al.*(69) in terms of parameters calculated for the perfect lattice. This very powerful moment analysis technique has been used, along with phase-sensitive detection experiments, to determine the effects of other small applied fields(46,72) on the F center.

As discussed in Sections 3.3.1 and 3.3.2, the electronic states referred to in F-center absorption are those computed for a lattice configuration appropriate to the ground-state wave function. After absorption the lattice relaxes and the system is in a relaxed excited state (RES), whose nature may be quite different from that measured in absorption. Because the lifetime of the RES is only $\sim 10^{-6}$ sec, experiments on it are not easy. At one point in time it appeared that the long lifetime(73) and relatively small thermal activation energy (~ 0.1 eV) favored a description in terms of a pure p-like state whose wave function was very diffuse.(39) However, electric field experiments[74,75] indicated a strong admixture of *s*-like state, a result consistent with further calculations.[51,76] The most recent combined optical and magnetic resonance experiments[77] (discussed more fully in Section 3.5) suggest that, at least in KI, the RES is *p*-like and diffuse.

The complications that have arisen in the study of the "simplest" center are not shared by all of the other electron vacancy centers, although this may be because detailed experiments and analyses have not been performed on these systems. We will briefly summarize the situation regarding other electron–vacancy centers in the alkali halides before turning to other crystals.

The F' center (two electrons at a halogen-ion vacancy) is important in color center chemistry, as discussed in Section 3.6. Its intrinsic properties are somewhat difficult to study, partly because its absorption tends to overlap that of other centers. For example, the F' absorption band in KBr extends from ~0.5 eV to ~2.5 eV, with a tail at even higher energies.[78] The F' center tends to be thermally unstable at around room temperature, and it is optically unstable at arbitrarily low temperatures.

Several calculations have been made of the electronic states of the F' center in alkali halides. Because there are two trapped electrons, correlation effects must be included, and because the center has an extra charge, polarization and distortion effects may be important. Strozier and Dick[79] performed detailed calculations and concluded that only one stable bound state (a singlet) exists. They computed the optical absorption in KCl, assuming transitions to a continuum, and obtained reasonable results.

The properties of F-aggregate centers are discussed in detail in Ref. 80. These centers, consisting of two or more F centers, may be neutral or ionized, and they tend to have absorption bands which overlap one another as well as the F band. This overlap has caused considerable difficulty in understanding the chemistry of these defects. However, the basic models for many of the centers are well established and considerable knowledge has been obtained regarding their properties.

The M center, consisting of two adjacent F centers, is illustrated in Figure 4. Because this center is not cubic, dichroic effects may be observed in the optical spectrum. Studies on KCl[81] and on KBr[82] showed that in addition to the prominent M_1 band at ~1.5 eV there are strong transitions, called M_2 and M_2', underlying the F band, and weaker transitions in the same region. The M_1 band corresponds to transitions in which the dipole moment is along the axis of the center (Σ-polarized), while M_2 and M_2' correspond to perpendicular or Π-polarized transitions.

These transitions occur between the ground state and excited states, all of which are singlets. A system of triplet states also exists, and the lowest-lying triplet has a lifetime in KCl of ~1 min at low temperatures. This long lifetime allows spin resonance(83) and optical spectroscopy(84) to be carried out. One measures an optical absorption somewhat similar to that observed in the singlet system.

Herman *et al.*(85) treated the F_2 center as a hydrogen molecule embedded in a dielectric, and this technique yields reasonable transition energies for KCl. This approximate model does not account for many details, such as the M_2–M_2' splitting, but it gives a useful overall picture of the system. Detailed calculations(86,87) yielded more quantitative results, although some of the finer details of the spectra were not explained.

The most interesting properties of the R center, three adjacent F centers in an equilateral triangle, arise from the weak coupling to the lattice of the electronic states involved in the R_2 transition. This leads to zero- and few-phonon transitions, which have been thoroughly studied and are reviewed by Fitchen.(33) The basic theory of weak coupling is discussed in Section 3.3.2. The R center is particularly interesting, as pointed out by Silsbee,(88) because the ground state is degenerate and exhibits a dynamic Jahn–Teller effect.

Several theoretical papers on the electronic states of the R center have appeared.(88) It appears that a level scheme analogous to that of H_3 exists.

Ionized R and M centers, as well as more complex defects, have been studied.(89) Some of these exhibit few-phonon structure, as shown in Figure 6 for the R' center in LiF. Many of their features are discussed in the review articles cited above.

A few examples of trapped-electron centers in other materials will now be considered. Probably the most widely studied systems besides alkali halides have been the alkaline earth oxides.(90) Since these solids are divalent, the neutral centers will contain twice as many electrons as in the alkali halides, i.e., the F center consists of two electrons trapped at an oxygen vacancy, the F^+ center one electron. Because of the extra valence, both types of centers have bound excited states.

Many of the same experimental techniques as used in the alkali halides have been applied in the oxides. For example, magnetooptic studies have been performed on the F^+ center in MgO by Kemp *et al.*[91] They used a spin-optical double resonance detection method to study the absorption, centered at 4.95 eV. Their results indicated that the F^+ and F bands nearly coincide. Calculations by Bennett[92] on the F center yield transition energies in reasonably good agreement with experiment.

The situation is somewhat more interesting in CaO. Here the F^+ center absorption centered at 3.65 eV has zero- and few-phonon lines.[93] Extensive study has determined a spin–orbit parameter in the first excited state of about -7×10^{-3} eV; analyses of both the broad band and zero-phonon line[94] have yielded information about the Jahn–Teller effect in the excited state.

Luminescence studies have also been made on the F^+ and F centers in CaO.[95] The F^+ luminescence, like the corresponding absorption, exhibits zero- and few-phonon struture along with a broad band. The lifetime of the emission is short, $\sim 6 \times 10^{-8}$ sec, consistent with a highly allowed transition. Because the F center contains two electrons, it will have a more complex level scheme than the F^+ center, and in particular one may have both singlet and triplet excited states ($^1P_1$, $^3P_{0,1,2}$). It is thought that an emission band at ~ 2.5 eV arises from the transition $^1P \rightarrow {}^1S$, while a band at ~ 2 eV involves $^3P \rightarrow {}^1S$. The latter band exhibits phonon structure. Excited-state spin resonance studies[96] on this band have confirmed the triplet nature of the initial state; for example, the variation of the polarization of the emitted light with magnetic field is consistent with a level crossing within the triplet manifold.

Recent research[97] on BaO has led to the suggestion that the thermal ionization energy for the F center is only ~ 0.1 eV. This would lead to free-carrier absorption in the infrared. It is also argued that these F centers are the previously unidentified shallow donors in oxide-coated cathodes.

Electron-excess centers in the alkaline earth fluorides have been studied by various techniques.[98,99] An experimental complication here is that additive coloration tends to form both F and aggregate centers, all of which have transitions in the F-band region. However, X-irradiation under suitable conditions produces F centers. The optical absorption peak in CaF_2 at 4°K is at 3.3 eV, and the spin–orbit coupling constant in the first excited state is about -3.1×10^{-3} eV. These results are in reasonable agreement with theory.

The M center in CaF_2 has a $\langle 100 \rangle$ orientation.[99] The M band lies at ~ 2.4 eV, and other transitions overlap the F band. Luminescence from the M center has a lifetime of 3.6×10^{-8} sec. A linear R center, three F centers along a $\langle 100 \rangle$ direction, also exists.[100]

A final example of a trapped-electron center involves oxygen vacancies in quartz. A class of color centers called E' may be formed by particle bombardment of pure SiO_2 or by X-irradiation of Ge-doped SiO_2. There is considerable evidence that these centers involve an oxygen vacancy, which in Ge-doped SiO_2 is adjacent to a Ge site. Studies by electron paramagnetic resonance[30,101] are consistent with an unpaired electron primarily localized on one of the two adjacent cation sites in an sp^3-type orbital. Such might result, for example, if an O^- were removed from its normal site.

3.4.2. Rare Earth Defects

In certain cases the optical properties are relatively insensitive to the environment. Generally, these cases involve ions with unfilled d or f shells which are shielded from neighboring ions, and one observes zero- or few-phonon spectra. A few examples will be cited here, but the discussion will be brief. This topic is treated more thoroughly by McClure in Chapter 1 of this volume.

Rare earth impurities in rare earth crystals may exhibit essentially atomic spectra. For example, Varsanyi and Dieke[102] have studied Pr^{3+} ($4f^2$) in $LaCl_3$. They observed an interesting phenomenon, namely an energy transfer among impurity ions due to an electronic energy resonance. Although the impurity ions are not strongly coupled to the host, the weak coupling among them can lead to an energy transfer.

Extensive studies of the optical properties of rare earth ions in alkali and alkaline earth halides have been made by Bron and coworkers; Ref. 103 gives a number of earlier references. In some cases[104] the impurity-ion spectrum is very weakly coupled to the host; for example, the Sm^{2+} emission consists of sharp lines originating from electronic transitions within the $4f^6$ configuration. In other cases, such as Sm^{2+} absorption, the coupling is sufficiently strong to produce zero-phonon and multiphonon lines and broad bands, but sufficiently weak that these zero- and few-phonon lines are prominent. Because of the detailed spectral information which is thereby available, studies of these transitions have proven very useful in determining vibration spectra of the host solids. For example, in Ref. 103 the optical spectra of Sm^{2+} in CaF_2, $SrCl_2$, and EuF_2 are used to obtain vibrational modes for these solids, as well as to learn about the electron-phonon coupling.

3.4.3. Impurities Strongly Coupled to the Host

Most color centers lie between the extremes of trapped-electron centers, which have no internal structure but interact solely with the environment, and rare earth ions, in which the internal interactions are dominant. These are centers involving impurity ions whose electrons are not shielded from the host, or involving trapped holes. Only a few examples will be treated here, but these will illustrate many of the interesting phenomena which take place in systems in which both internal and external interactions are important. We will consider Tl^+, Ag^+, and self-trapped excitons in alkali halides.

Substitutional Tl^+ in alkali halides has been widely studied for many years, partly because of its usefulness as a radiation detector.[105,106] Other ions with outer s^2 electronic configurations, such as In^+ and Sn^{2+}, have also received extensive investigation.[107,108] The optical spectra of these s^2 ions have many features in common, although the presence of charge-compensating vacancies near the divalent ions[108] tends to complicate their spectra.

It was first pointed out by Seitz[109] that the Tl^+ absorption spectrum could be understood qualitatively by simple shifts in free-ion energy levels. The upward shift of the ground state is as indicated in Figure 3, while the excited states are expected to shift by a lesser amount since the more diffuse wave functions are not as strongly affected by the Madelung potential as the more compact ground-state wave functions. Figure 9 shows the correspondence of the levels. The notation for the levels is in terms of *LS* coupling, although in fact the coupling is intermediate; in the solid the point group notation for the cubic site is used.

One typically observes four absorption bands, denoted *A*, *B*, *C*, *D*; the 77°K transition energies in KBr are 4.79, 5.58, 5.93, and 6.50 eV, respectively.[107] As indicated in Figure 9, strong transitions to $^3\Gamma_1{}^0$ are not expected,[109] and the *D* band is usually attributed[32] to transitions in which the excitation involves the neighboring halogens.

Many calculations of the Tl^+ states have appeared, some of which are reviewed in Ref. 32. Early work by Williams and co-workers[110] provided a semiquantitative basis for understanding the system, and more recent work[59] using molecular orbital techniques has illustrated the crucial role which the partial covalency of the thallium–halogen

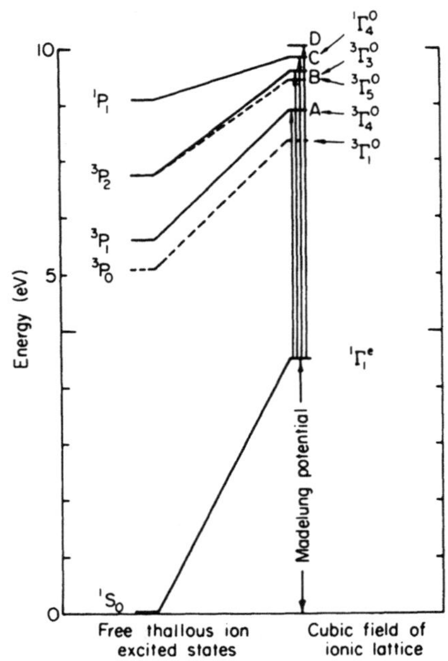

Fig. 9. The Seitz model of Tl^+ energy levels, both free and in a cubic field.[32]

interaction plays in determining the properties of the electronic states. This point was suggested some time ago[111,112] when it was pointed out that even in the absence of detailed calculations valuable information could be obtained from the spectra. The multiple peaks reflect the exchange and spin–orbit interactions within the ionic (*sp*) configuration, and a simple analysis[112] of relative absorption strengths and peak positions can yield numerical values of exchange and spin–orbit parameters. It turns out, for example, that in KCl:Tl the spin–orbit parameter is 77% of the free-ion value, while the exchange parameter is 24% of that observed for free Tl^+. The strong reduction in the exchange parameter is consistent with a large configuration interaction between the free Tl^+ states and states of the host. Bramanti *et al.*[59] calculated, for example, an effective charge on Tl of +0.85, and estimated an exchange parameter ~50% of the free-ion value.

Numerical values of the electronic parameters are listed in several places.[107,113,114] The analysis referred to above predicts that the ratio of absorption strengths R between $^3\Gamma_4{}^0$ and $^1\Gamma_4{}^0$ is related to a dimensionless function of energy differences, χ. Although experimental values of R versus χ tend to follow the universal curve, Hagston[115] has pointed out that the discrepancies which exist reinforce the importance of configuration interaction, probably with the states responsible for the D band.

One other interesting feature of the absorption spectrum is the existence of structure in certain of the bands. It appears generally agreed that this arises from dynamic Jahn–Teller effects[116] in the excited states, although in the divalent ions nearby vacancies can lead to splittings in the static lattice. These splittings in the simpler ions may be enhanced by applied uniaxial stresses.[117]

Luminescence in the $(s)^2$ ions is much more complex and less well understood than the absorption. Various bands appear as a function of temperature and of wavelength of exciting light. Space does not permit a detailed discussion of these phenomena, although several of the references[32,118–120] treat this topic.

Of considerable interest is a low-energy luminescence band that appears in several cases under appropriate conditions. In KI:In, for example, excitation in the A band at liquid helium temperature produces an emission band at 2.76 eV. Fukuda and Yuster[119] have found the luminescence lifetime to be $\sim 10^{-3}$ sec, and to decrease rapidly with applied magnetic field. They suggest that the lowest state is $^3\Gamma_1{}^0$, which is presumed to be only $\sim 10^{-3}$ eV below a Jahn–Teller-split $^3\Gamma_4{}^0$. Magnetic field admixture makes the $^3\Gamma_1{}^0$ to $^1\Gamma_1{}^0$ emission partially allowed and accounts for the strong field dependence of the lifetime. At higher temperatures thermal excitation promotes the system to $^3\Gamma_4{}^0$, from which emission occurs.

The corresponding emission in KCl:Tl is at 2.6 eV. Ranfagni *et al.*,[120] on the basis of low-temperature measurements, suggest that the luminescence here occurs directly from $^3\Gamma_1{}^0$, rather than by promotion to $^3\Gamma_4{}^0$. In this case the low-temperature lifetime is $\sim 0.2 \times 10^{-3}$ sec.

It is clear that the Tl-type ions represent a case in which the internal structure of the ions, the electronic states of the host, and the lattice interact in such a way as to yield complex but interesting phenomena.

When either an Ag^+ or a Cu^+ ion is substituted for an alkali ion in an alkali halide crystal the ionic ground state has a closed-shell d^{10} electronic configuration. Despite this similarity, the optical properties of Ag^+ and Cu^+ exhibit many differences, with Cu^+ the more difficult to understand.

Substitutional Ag^+ generally exhibits several absorption bands in the ~5-eV range.[32,121,122] These bands are weak and their intensity increases with increasing temperature. In KI, for example, bands denoted by A, B, C, D occur in the same spectral range as the free-ion transitions from $^1S_0$ to $^3D_3$, $^3D_2$, $^3D_1$ (~5 to 5.5 eV). The transition energies increase monotonically as one goes to crystals with smaller lattice constants; for example, the A band lies ~1 eV higher in NaCl than in KI.

It is generally accepted that these properties arise from optical transitions which are parity-forbidden (S to D).[121–123] This is consistent with the overall spectral location, the multiplicity, and the weakness and the temperature dependence of the absorption bands. That the bands lie above, rather than below, the free-ion transitions is, however, inconsistent with the argument used to rationalize the location of the Tl^+ bands. In a few cases the equilibrium position of the Ag^+ ion may be slightly displaced from the cubic position.

A detailed analysis of the d^9s multiplet in a cubic field was performed by Knox.[123] A fairly successful attempt was made to fit the positions and relative strengths of the Ag^+ bands in NaCl, by varying parameters such as exchange and spin–orbit coupling. Some modifications to this approach were made by Fussgaenger *et al.*,[121] and recently Hagston[124] has pointed out that in a strong field formalism more free parameters exist, and a better fit might be obtained.

The theory of phonon-assisted transitions[125] was applied to Ag^+ by Conway *et al.*,[126] whose estimate of an oscillator strength $\sim 10^{-3}$ agrees rather well with experiment. The simple theory, in which odd- and even-parity electron states are mixed by one mode ω_{eff}, predicts a temperature-dependent oscillator strength $f(T)$ to vary as

$$f(T) = f(0)\coth(\hbar\omega_{\mathrm{eff}}/2kT) \tag{26}$$

Refinements to the theory,[127] in which all the modes of vibration of the imperfect crystal are allowed to interact, predict a variation of f with T which varies at most ~2% from Eq. (26), and which predicts a value of ω_{eff} in good agreement with experiment for NaCl:Ag^+.

The location of the Ag^+ bands at higher energies than the free-ion values (as opposed to Tl^+) is somewhat difficult to understand. More dramatically, the transitions associated with Cu^+ in a number of crystals[28,128] lie at ~5 eV, while the free-ion $3d^{10} \rightarrow 3d^94s$ multiplet is in the ~3-eV region.[129] It has been suggested[60] that this shift is associated with a tendency of d-electron energies to be lower than those of s or p electrons in a crystal field; the small overlap of their wave functions with those of the neighbors may allow the *d* electrons to penetrate the ion cores, where the potential energy is lower.

The Cu^+ center has a number of other peculiar properties, some of which are discussed in Refs. 32 and 60. For example, in some instances the optical absorption is strongly allowed, while the luminescence appears to be dipole-forbidden.[130,131] Some of these effects may be explained if the ion is off-center in absorption but not in emission. It has recently been suggested that in fact the Cu^+ may be interstitial.[124]

The final topic to be treated in this section is the self-trapped exciton in alkali halide crystals. This is an unusual defect in that it is both intrinsic and transient. Its existence is based upon the existence of the V_K center,[132] a self-trapped hole[133] (see Figure 4) localized mainly on two adjacent halogen ions. The V_K center, stable at low temperatures, is characterized by the covalent bonding in the $(\text{halogen})_2{}^-$ complex.

When free electrons exist in the region of V_K centers, a significant fraction are trapped into an excited state. One then has trapped, bound electron–hole pairs, or excitons. These self-trapped excitons[134] may also be created by absorption in the exciton bands of the crystal.

The self-trapped excitons may then luminesce, with lifetimes[135] which are typically $\sim 10^{-9}$–10^{-6} sec. Although some studies have been made of optical absorption[26] associated with self-trapped excitons, their short lifetimes have made luminescence measurements the most fruitful. The luminescence measurements, coupled with theoretical analysis, offer strong evidence that the lowest states are triplet in character, with relative populations which are not in thermal equilibrium at low temperature.

These conclusions are discussed in several papers.[136–139] The variations of the radiative lifetimes in KI as a function of temperature were fit very well to a two-excited-state model in which the lower state had zero decay probability, and the relative populations were described by a nonequilibrium distribution; these two states are separated by 0.69×10^{-3} eV. This behavior is consistent with the triplet model,[137,138] which predicts that the lowest state should be of ${}^3A_u$ character, with zero luminescent transition probability in the static lattice. Slightly higher in energy lie ${}^3B_{2u}$ and ${}^3B_{3u}$, which may be nearly degenerate and which contain some singlet character through spin–orbit mixing. The observed ${}^3A_u$–${}^3B_u$ splitting is consistent with theory.[138]

A wealth of information has been obtained by studying the luminescence in an applied magnetic field[137,139] and most of it can be understood.[138] For example, in CsI sudden changes in lifetimes and polarizations are observed at an applied field of ~45 kG in crystals whose (100) axis is parallel to the field and to the direction of observation. Theoretical calculations of the energy levels of the center, using parameters deduced from other experiments, indicate at this field a crossing of the ^3A_u level and a ^3B_u level in the center oriented parallel to (100), and if thermal equilibrium is not present, a large change in lifetimes and polarizations will result. If the applied field is parallel to (111), the level crossing will not occur, and indeed smoothly varying lifetimes and polarizations are observed in this case.

The physics of the self-trapped exciton is thus in many ways the physics of triplet molecules, where the possible orientations of the "molecules" are few and are fixed. The variety of experimental data, including analysis of V_K-center electron spin resonance,[132,140] enables one to determine parameters in a detailed theoretical treatment.[138]

3.5. Other Experimental Probes

Several other experimental techniques have yielded important information on color center properties.

3.5.1. Photoconductivity

Electrons which are excited into the conduction band can carry a current in an applied electric field. The measurement of current versus photon energy thus gives information about the energy separation between the initial state and the conduction band, and may yield some information about state densities within the band. For example, photoconductivity studies[68] on the F center in RbCl supported the idea that transitions into the conduction band occur in the high-energy portion of the K band.

Similar studies may be used to gain information about the band gap. Emkey and Van Sciver[23] studied NaI in this way, and DiStefano and Eastman[141] have used photoconductivity measurements to establish a band gap of 9.0 eV in amorphous SiO_2. Some of the difficulties inherent in the technique and in interpretation of the results are discussed in Ref. 23. For example, in the band-to-band case most of the transitions occur near the surface, where the electronic states and the transport characteristics may vary from the bulk.

3.5.2. Electron Spin Resonance

Magnetic resonance techniques are very useful in both chemistry and physics. On the simplest level spin resonance signals are used to identify the presence of atomic species or defect centers, and on more sophisticated levels they are used to probe the phonon spectra, and are used in conjunction with other resonances to probe wave function densities.

Detailed discussions of the techniques, and complicated results, are beyond the scope of this chapter. Several good general books[142–144] on spin resonance exist, and the review article by Seidel and Wolf[61] treats applications to color center problems.

Spin resonance techniques are based on the fact that spin degeneracies are lifted in an applied magnetic field, with the splittings approximately proportional to the field. In contrast with optical spectroscopy, where the level separations are largely determined by internal forces, in magnetic resonance spectroscopy one creates his own level spectrum, within which transitions can be induced by applied ac fields. In fact, in electron spin resonance the transition energies are typically in the microwave region and it is convenient to operate at fixed microwave frequencies and to tune the static applied fields until resonance is obtained.

The most important feature of spin resonance spectra comes from the hyperfine structure. If the ions in the solid possess nuclear magnetic moments, these moments generate small internal magnetic fields which interact with the spin system under observation. Thus the net magnetic field interacting with a given spin will be the sum of the applied field and whatever internal magnetic fields exist at that site.

The internal fields in turn will vary in discrete amounts, for two reasons. First, the nuclear spins which generate the fields will exist in quantized states. Second, the nuclear spins will in general be situated on discrete lattice sites.

In a favorable situation, therefore, the application of a fixed-frequency microwave field and variation of the static applied field will lead to a spectrum of absorption versus field consisting of a number of discrete lines. By measuring the size of the splitting one obtains information about the electron density at the nucleus,[61] the so-called Fermi contact interaction. The number and relative intensity of resolved hyperfine lines yield information about the nuclear spin and the number of equivalent nuclei. By measuring the spectrum for various crystal orientations, one can learn about the symmetry of the center.

In some cases the hyperfine splittings are not resolved. In these cases it may be possible to extract some of the above information by a technique called electron-nuclear double resonance (ENDOR), discussed below.

Clearly, the most severe limitation of spin resonance spectroscopy is its inapplicability to spinless systems. And even in cases where an electron spin exists, magnetic nuclei which would generate a hyperfine field may be absent.

The V_K center in alkali halides is a good example of a simple, well-resolved hyperfine splitting.[132] The observed lines are in all cases consistent with the assumption that the unpaired spin interacts equally with two halogen nuclei. In LiF, then, one should observe a three-line spectrum; each F^- nucleus has spin $\frac{1}{2}$, so two of them have spin 1, and the allowed m values are then 1, 0, -1. This is indeed observed. In the chlorides and bromides the situation is complicated by isotope effects. For example, Cl^{35} and Cl^{37} have different abundances and slightly different nuclear moments, although both have spin $\frac{3}{2}$. Figure 10 illustrates in a schematic way the analysis of the KCl spectrum in terms of isotope effects. Spin resonance experiments[132] and their analysis were instrumental in determining the model of the V_K center as a self-trapped hole "shared" by two halogen ions.

The F center presents a more complicated situation.[61] In most cases the spectrum consists of a single broad, structureless line, of width ~ 100 G. This broad line is in fact the envelope of a multiple unresolved hyperfine spectrum, arising from the interaction of the F-center electron with the many nuclei in its environment. The ENDOR technique has been used to measure the hyperfine interaction in a number of cases, as discussed below.

To perform an ENDOR experiment[145] one must first have a saturated electron spin resonance transition. By saturated one means that the upper electron spin state has a sufficiently long lifetime against decay to the lower state that the upper-state population has reached its maximum value at some applied microwave field; further increases in the field intensity induce downward transitions at the same rate as upward transitions, so the net power absorbed does not change. Now, simultaneous application of a radiofrequency (rf) field at a frequency which induces nuclear spin transitions in the upper state will partially desaturate the electron resonance, and a change in microwave absorption will result. Thus one performs nuclear magnetic resonance,

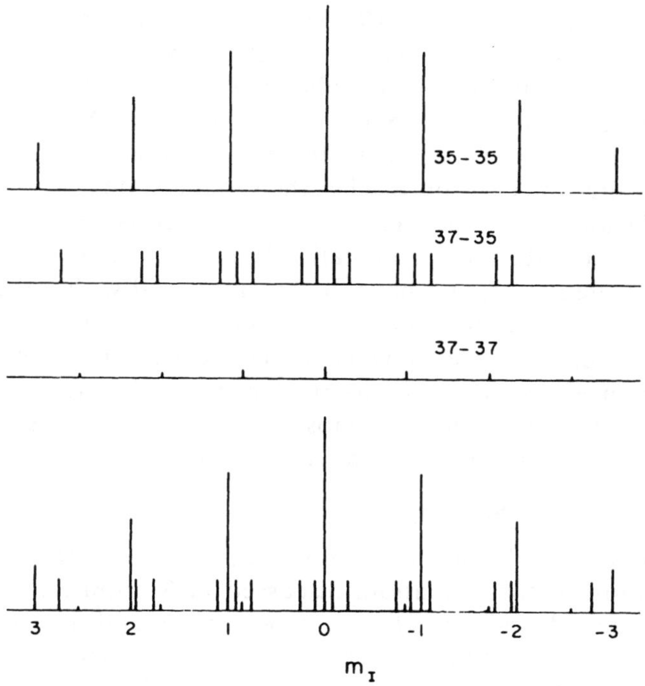

Fig. 10. Schematic illustration of isotope effects in the hyperfine spectrum of the KCl V_K center.[132] The bottom curve is the superposition of the three components above it.

and by using the electron spins as a detector, one determines the hyperfine interactions.

Considerable attention has been paid to the *F* center in KCl, where the hyperfine constants have been measured for many shells of neighboring ions.[146] These measurements have been interpreted in terms of an apparently anomalous electron density in the ground state, although these anomalies can apparently be resolved through an extremely careful theoretical analysis.[147]

Recently triple resonance techniques have been developed and applied to the *F*-center relaxed excited state in KI.[77] The triple resonance technique is based on the large spin memory in the *F* center optical pumping cycle; that is, if an *F*-center electron has a given spin orientation, there is a high probability that that spin will be preserved through a cycle of optical absorption, luminescence, and the accompanying lattice relaxation. This property is coupled with the fact that the optical circular dichroism[42,69] in an applied magnetic field is dependent upon which spin state the electron is in.

Now consider an ENDOR experiment on the relaxed excited state. Because of its short lifetime, direct detection of the change in electron spin resonance will be difficult. However, since the optical dichroism depends on the spin orientation, if an ENDOR process occurs in the relaxed excited state one will observe a change in dichroism after the electron reaches the ground state.

Mollenauer and Baldacchini[77] have measured, using this optical detection technique, a detailed ENDOR spectrum for the KI *F*-center excited state, and have fit this quite well with a theoretical model for the state in which the electron is in a diffuse *p*-like orbital.

Numerous references have been made earlier in this chapter to spin-resonance experiments on particular centers; in addition to those cited, it is a safe assumption that most centers which are paramagnetic have been studied by spin-resonance methods.

3.6. Color Center Chemistry

Study of the chemistry of color centers follows naturally from consideration of their creation, discussed in Section 3.2. There we briefly discussed primary creation processes by additive coloration, electrolysis, and irradiation; here we consider secondary processes, or transformations, induced in a variety of ways. This is an important topic, since some centers are created in large numbers only by secondary processes, and since in some potential applications transformations lead to degradation of the device.

A simple example involves a crystal initially containing only *F* centers. Optical absorption in the *F*-band region at sufficiently high temperature[148] ($\sim 175°K$ in KCl) leads to thermal ionization of the *F* center from the relaxed excited state. The freed electrons may be trapped at halogen vacancies, reforming the *F* centers; at lower temperatures the system could efficiently return to the ground state by emitting a photon, but in the present case it is more likely to be reionized. The electrons may also be trapped at *F* centers, forming F' centers, which are thermally stable below a certain temperature, and if the exciting light does not ionize the F' centers, a large steady-state concentration of F' centers and halogen vacancies (alpha centers) may be achieved. Subsequent irradiation in the F' band can then reform the *F* centers. Further aspects of this type of process are discussed in Section 4.

The F–F'–F cycle is also a good example of degradation effects. It is found that continued *F*-band irradiation at room temperature in KCl leads to the aggregation[149] of *F* centers and the eventual formation of M, R, and other multiple-vacancy centers. This means that the vacancy $-F'$ Coulomb interaction is such as to lead to migration and eventual aggregation. Furthermore, if alkali impurities are present, there are certain conditions[150] under which the F–F' process is accompanied by the creation of F_A centers. This occurs, for example, if after the F–F' conversion the crystal is held at the conversion temperature for some time. The rate of F_A-center formation then depends upon the number of F'–alpha pairs, and other experiments indicate that the F_A formation is associated with vacancy migration rather than F'-center migration.

Some interesting experiments on the $F \rightarrow F'$ conversion in an external magnetic field have been reported.[151] One can magnetically align the F-center spins at low temperature ($\sim 1.7°$K in KCl), and when the system is optically excited the F-center luminescence efficiency is increased by up to a factor of four times the value in zero field. This increase is due to the blocking of other decay channels and is consistent with the notion that the F' center does not have a bound triplet state. Conduction electrons resulting from F-center ionization apparently preserve their spins to a large degree at this temperature, and thus could only be trapped readily by F centers if the F' had a bound triplet. A second decay channel not discussed previously in this chapter involves tunneling of the excited F electron to a neighboring F center, forming a metastable F' and alpha, followed by tunneling back to the ground state; but this channel is also blocked by the nonexistence of a bound F' triplet.

Another simple example of transformations involves the impurity thallium, normally Tl^+. Ionizing radiation at low temperature leads to the creation of V_K centers, with the released electrons being trapped at thallium sites[152] to form Tl^0. Subsequent warming of the crystal to a temperature at which the V_K centers are mobile ($\sim 200°$K in KCl) leads to "glow peaks," luminescent bands occurring at particular wavelengths. These glow peaks arise from electron–hole recombination occurring when the V_K approaches the Tl^0. Some of the V_K centers also diffuse near a Tl^+, and the hole is trapped by the thallium forming Tl^{2+}. A second glow peak at 300°K results from thermal ionization of Tl^0 and recombination with Tl^{2+}.

Schulman and Compton[1] detail a number of other examples of temperature- and radiation-induced transformations. As suggested in Section 3.2, many systems have significant impurity concentrations, and ionizing radiation or heat treatment can lead to the release and trapping of charge, thereby changing the valence states.

An example of a system exhibiting such complicated chemical effects involves X-irradiated quartz containing Al, Ge, Na, Li, and H impurities.[153] A commonly observed pair of centers is a hole trapped near an aluminum impurity and an electron trapped near a germanium impurity. Many related centers have in the past been detected, involving the above in association with one or more hydrogen or alkali impurities. Detailed EPR experiments in connection with controlled specimen treatment (X-irradiation or heating) were carried out to measure various conversion processes among these centers. These data were then analyzed by solving coupled rate equations for several models of the conversion process. This analysis leads to a fairly clear choice for a particular end product, the β centers, formed by consecutive electron trapping at a germanium with three neighboring alkalis. This result seems to demonstrate the usefulness of detailed rate analysis in understanding such processes.

Some further examples of color center transformations are discussed in Section 4.

4. Applications: Information Storage

The phrase "information storage" is used here in a fairly general way, to convey the idea of externally induced changes in the crystal which can subsequently be detected to yield information about the source of the changes.

Various types of ionizing-radiation dosimeters are discussed in Ref. 1. One class, the thermoluminescent dosimeter, is of use in measuring relatively small doses of radiation. The principle of operation of this dosimeter is fairly simple: Some particular defect center is ionized by incident radiation, with the electron being trapped at some other site. Under suitable conditions subsequent heating of the crystal can cause thermal ionization of the trapped electron, which then returns to the ionized defect and recombines, emitting a photon in the process. Practical problems with this type of device often include the presence of shallow electron traps, which are thermally unstable at room temperature; this leads to a loss of dose indication with time.

Ginther and Kirk[154] were able to produce CaF_2:Mn without the troublesome shallow traps, and with one deep trap whose glow peak is at ~260°C. This device stores its information for about one month, and responds linearly to doses from $\sim 10^{-3}$ to $\sim 10^5$ R.[1]

Another system which has become widely used in dosimetry is LiF. Extensive studies[155] have indicated that the trapping centers involve holes, vacancies, and Mg^{2+} ions, while the luminescent center is the *F* center. Further studies[(156)] have indicated the existence of as many as ten glow peaks between 77 and 500°K.

At a more sophisticated level, one might wish to be able to store a high density of information, read it nondestructively, and erase it. This could be useful in computer memories, where the reading would be accomplished electrooptically, or in visual displays such as television screens.

"Photochromic" materials have been developed which have some attractive characteristics in this respect.[(157)] The F–F' system discussed in Section 3.6 is a good example. Suppose that only *F* centers are present. "Writing" might involve intense excitation into the *F* band, at ~ 2 eV. This converts *F* centers to F' centers, and reading might involve measuring absorption (using a weak source) in either *F* or F' regions. Erasing would involve intense F' excitation, say at ~ 1 eV, which would re-form the *F*'s.

As mentioned in Section 3.6, the F–F' system tends to degrade, and so is not a practical system. Some other systems have more desirable characteristics; several of these are discussed in various abstracts in Ref. 5, others are mentioned in Ref. 157. For example, CaF_2:Ce has normal absorption peaks at 3.1 and 1.8 eV; irradiation with light of energy greater than 2.8 eV leads to a new spectrum with absorption at ~ 1 eV and higher. The original spectrum is due to Ce^{3+}, the induced spectrum to Ce^{2+}. Irradiation at ~ 1 to 1.8 eV leads to re-formation of the Ce^{3+}.

Among other photochromic systems which have been investigated[(157)] are CaF_2:Sm^{2+}, Eu^{3+}; KCl:F_A; CaF_2:La; $SrTiO_3$: transition metals; halophosphates; glasses[(158)]; alkali halide sodalites.[(159)] Photochromic sunglasses are commercially available.

Photodichroic materials may also become useful in information storage. These materials involve centers of low symmetry which can be oriented by incident light and which, when oriented, give rise to an absorption whose intensity depends on the polarization of the incident light.

One system whose information storage capabilities have been studied is the KCl:Na F_A center.[(160)] The F_A center has an F_{A1} transition, corresponding to light polarized parallel to the vacancy–Na axis, and an F_{A2} transition with a perpendicular transition moment. Excitation into F_{A2} above 140°K leads to reorientation (at temperatures too high it can lead to ionization). By using polarized light, the F_A's can thus be aligned. Reading is accomplished by measuring F_{A2}

absorption at lower temperatures, ~55°K, where the reorientation probability is small. One thus must cycle the temperature in a write–read sequence.

Other potentially useful systems involve M or M_A centers in alkali halides.[161,162] The M center has an M absorption band with transition moment along the axis, and several M_F bands with moments perpendicular to the axis. Exciting into the M_F region leads to reorientation, and so by using polarized light, one can obtain aligned M centers. The reading is accomplished with M light, since this does not lead to reorientation. Several characteristics of the M center in NaF include[162] an estimated storage capacity of $\sim 10^8/cm^2$, a dark stability of at least three months, and $\sim 10^{10}$ reads without degradation. The energy required to write is 10–100 mJ/cm^2.

One disadvantage of the M-center system is the apparent need to maintain the system at low temperatures, at least in those cases thus far investigated, where the reorientation occurs without substantial degradation. It appears that reorientation is actually a several-step process; the first photon ionizes the M, forming M^+, and the second excites M^+, after which it reorients. The electron then returns from, e.g., an F' center and is trapped at M^+. It has been recently pointed out[163] that this property may be utilized in a two-beam device to produce three-dimensional storage. The second beam would be in the near-infrared and would inhibit M^+ reorientation by ionizing F' centers and thereby returning electrons to the M^+ centers before reorientation could take place.

Acknowledgments

I wish to thank those individuals whose granted permission for use of their published figures. Mr. T. F. Connolly, of the Research Materials Center at Oak Ridge National Laboratory, supplied a number of references and abstracts. This work was partially supported by the Office of Naval Research.

References

1. J. H. Schulman and W. D. Compton, *Color Centers in Solids*, Pergamon, New York (1962).
2. J. J. Markham, *F-Centers in Alkali Halides*, Academic, New York (1966).
3. W. B. Fowler (ed.), *Physics of Color Centers*, Academic, New York (1968).
4. J. H. Crawford, Jr. and L. M. Slifkin (eds.), *Point Defects in Solids, Vol. 1. General and Ionic Crystals*, Plenum, New York (1972).

5. *1971 Int. Conf. on Colour Centers in Ionic Crystals*, Reading, England (1971).
6. B. N. Taylor, W. H. Parker, and D. N. Langenberg, Determination of e/h, using macroscopic quantum phase coherence in superconductors: Implications for quantum electrodynamics and the fundamental physical constants, *Rev. Mod. Phys.* **41**, 375–496 (1969).
7. F. C. Brown, *The Physics of Solids*, Benjamin, New York (1967).
8. Max Born and Kun Huang, *Dynamical Theory of Crystal Lattices*, Oxford Univ. Press, London (1954).
9. A. D. B. Woods, B. N. Brockhouse, and R. A. Cowley, Lattice dynamics of alkali halide crystals. II. Experimental studies of KBr and NaI, *Phys. Rev.* **131**, 1025–1029 (1963).
10. R. S. Knox and K. J. Teegarden, in *Physics of Color Centers* (W. B. Fowler, ed.), pp. 1–51, Academic, New York (1968).
11. C. Kittel, *Introduction to Solid State Physics*, 4th ed., Wiley, New York (1971).
12. J. R. Jasperse, A. Kahan, J. N. Plendl, and S. S. Mitra, Temperature dependence of infrared dispersion in ionic crystals LiF and MgO, *Phys. Rev.* **146**, 526–542 (1966).
13. M. M. Elcombe and A. W. Pryor, The lattice dynamics of calcium fluoride, *J. Phys. C* **3**, 492–499 (1970).
14. R. S. Katiyar, Dynamics of the rutile structure. II. Lattice dynamics of magnesium fluoride, *J. Phys. C* **3**, 1693–1702 (1970).
15. D. A. Kleinman and W. G. Spitzer, Optical properties of quartz in the infrared, *Phys. Rev.* **125**, 16–30 (1962).
16. R. S. Knox, *Theory of Excitons*, Academic, New York (1963).
17. M. Cardona, R. Haensel, D. W. Lynch, and B. Sonntag, Optical properties of the rubidium and cesium halides in the extreme ultraviolet, *Phys. Rev. B* **2**, 1117–1131 (1970); F. C. Brown, C. Gähwiller, H. Fujita, A. B. Kunz, W. Scheifley, and N. Carrera, Extreme-ultraviolet spectra of ionic crystals, *Phys. Rev. B* **2**, 2126–2138 (1970).
18. P. Rabe, B. Sonntag, T. Sagawa, and R. Haensel, The optical absorption of MgF_2, $MgCl_2$, and $MgBr_2$ in the vicinity of the Mg L-shell transitions, *Phys. Stat. Sol. (b)* **50**, 559–569 (1972).
19. K. J. Teegarden and G. Baldini, Optical absorption spectra of the alkali halides at 10°K, *Phys. Rev.* **155**, 896–907 (1967).
20. A. B. Kunz, Combined plane-wave tight-binding method for energy-band calculations with application to sodium iodide and lithium iodide, *Phys. Rev.* **180**, 934–941 (1969).
21. W. B. Fowler, Dependence of silver halide energy bands on lattice constant and halogen, *Phys. Stat. Sol. (b)* **52**, 591–599 (1972).
22. G. Harbeke and E. Tosatti, Band-edge excitons in PbI_2: A puzzle?, *Phys. Rev. Letters* **28**, 1567–1570 (1972); H. Overhof and J. Treusch, The energy bands of thallous chloride and bromide, *Solid State Commun.* **9**, 53–56 (1971).
23. W. L. Emkey and W. J. Van Sciver, Photoconductivity due to excitonic-energy transfer to crystalline defects, *Phys. Rev. B* **5**, 610–617 (1972).
24. S. Kono, T. Ishii, T. Sagawa, and T. Kobayasi, X-ray photoelectron study of the valence bands in cuprous halides, *Phys. Rev. Letters* **28**, 1385–1387 (1972).
25. J. J. Hopfield and J. M. Worlock, Two-quantum absorption spectrum of KI and CsI, *Phys. Rev.* **137**, A1455–A1464 (1965).

26. R. G. Fuller, R. T. Williams, and M. N. Kabler, Transient optical absorption by self-trapped excitons in alkali halide crystals, *Phys. Rev. Letters* **25**, 446–449 (1970).
27. J. H. Crawford, Jr., Radiolysis of alkali halides, *Adv. Phys.* **17**, 93–151 (1968).
28. R. Bateman and W. J. Van Sciver, Optical absorption by the Cu^+ center in NaI, *Phys. Stat. Sol.* (*b*) **46**, 779–784 (1971).
29. H. N. Hersh, Proposed excitonic mechanism of color-center formation in alkali halides, *Phys. Rev.* **148**, 928–932 (1966); D. Pooley, *F*-Centre production in alkali halides by electron–hole recombination and a subsequent [110] replacement sequence: A discussion of the electron–hole recombination, *Proc. Phys. Soc.* (*London*) **87**, 245–256 (1966). D. Pooley, [110] Anion replacement sequences in alkali halides and their relation to *F*-centre production by electron–hole recombination, *Proc. Phys. Soc.* (*London*) **87**, 257–262 (1966).
30. F. J. Feigl and J. H. Anderson, Defects in crystalline quartz: Electron paramagnetic resonance of *E'* vacancy centers associated with germanium impurities, *J. Phys. Chem. Solids* **31**, 575–596 (1970).
31. W. Gebhardt and H. Kuhnert, Temperature dependence of *F*-centre absorption and emission, *Phys. Lett.* **11**, 15–16 (1964).
32. W. Beall Fowler, in *Physics in Color Centers* (W. B. Fowler, ed.), Academic, New York (1968).
33. D. B. Fitchen, in *Physics of Color Centers* (W. B. Fowler, ed.), Academic, New York (1968).
34. J. H. Thomas and F. J. Feigl, Electron trapping levels in silicon dioxide thermally grown on silicon, *J. Phys. Chem. Solids* **33**, 2197–2216 (1972).
35. A. von Hippel, Einige prinzipielle Gesichtspunkte zur Spektroskopie der Ionenkristalle und ihre Anwendung auf die Alkalihalogenide, *Z. Phys.* **101**, 680–720 (1936).
36. D. L. Dexter, in *Solid State Physics* (F. Seitz and D. Turnbull, eds.), Vol. 6, Academic, New York (1958).
37. W. Beall Fowler and D. L. Dexter, Relation between absorption and emission probabilities in luminescent centers in ionic solids, *Phys. Rev.* **128**, 2154–2165 (1962).
38. W. Beall Fowler and D. L. Dexter, Detailed balance in optical transitions in molecules and solids, *J. Chem. Phys.* **43**, 1768–1772 (1965).
39. W. Beall Fowler, Relaxation of the excited *F*-center, *Phys. Rev.* **135**, A1725–A1732 (1964).
40. S. I. Pekar, *Usp. Fiz. Nauk* **50**, 197 (1953).
41. M. Lax, The Franck–Condon principle and its application to crystals, *J. Chem. Phys.* **20**, 1752–1760 (1952).
42. C. H. Henry and C. P. Slichter, in *Physics of Color Centers* (W. B. Fowler, ed.), Academic, New York (1968).
43. K. K. Rebane, *Impurity Spectra of Solids: Elementary Theory of Vibrational Structure* (Transl. by John S. Shier), Plenum, New York (1970).
44. F. S. Ham, Solution of a vibronic model for the relaxed excited state of the *F*-center, *Phys. Rev. Letters* **28**, 1048–1050 (1972).
45. H. A. Jahn and E. Teller, Stability of polyatomic molecules in degenerate electronic states. I—Orbital degeneracy. *Proc. Roy. Soc.* (*London*) **A161**, 220–235 (1937); H. C. Longuet-Higgins, U. Öpik, M. H. L. Pryce, and R. A. Sack, Studies of the

Jahn–Teller effect. II. The dynamical problem, *Proc. Roy. Soc.* (*London*) **A244**, 1–16 (1958).
46. C. H. Henry, S. E. Schnatterly, and C. P. Slichter, Effect of applied fields on the optical properties of color centers, *Phys. Rev.* **137**, A583–A602 (1965).
47. D. Y. Smith and D. L. Dexter, in *Progress in Optics* (E. Wolf, ed.), Vol. 10, North-Holland, Amsterdam (1972).
48. M. I. Petrashen, I. V. Abarenkov, A. A. Berezin, and R. A. Evarestov, Calculations of the electronic structure of colour centres in ionic crystals (I), *Phys. Stat. Sol.* **40**, 9–29 (1970); (II), *Phys. Stat. Sol.* **40**, 433–460 (1970); A. M. Stoneham, Electronic structure of colour centres. Theoretical developments, *Phys. Stat. Sol.* (*b*) **52**, 9–24 (1972).
49. D. R. Hartree, *The Calculation of Atomic Structures*, Wiley, New York (1957).
50. P. O'D. Offenhartz, *Atomic and Molecular Orbital Theory*, McGraw-Hill, New York (1970); R. G. Parr, *Quantum Theory of Molecular Electronic Structure*, Benjamin, New York (1963).
51. U. Öpik and R. F. Wood, Point defects in ionic crystals. I. Methods of calculating the electronic structure, *Phys. Rev.* **179**, 772–782 (1969); R. F. Wood and U. Öpik, Point defects in ionic crystals. II. The *F* center in KCl, KBr, KI, and NaCl, *Phys. Rev.* **179**, 783–796 (1969).
52. W. B. Fowler, Influence of electronic polarization on the optical properties of insulators, *Phys. Rev.* **151**, 657–667 (1966).
53. W. Kohn, in *Solid State Physics* (F. Seitz and D. Turnbull, eds.), Vol. 5, Academic, New York (1957).
54. J. C. Slater, A simplification of the Hartree–Fock method, *Phys. Rev.* **81**, 385–390 (1951).
55. F. Herman and S. Skillman, *Atomic Structure Calculations*, Prentice-Hall, Englewood Cliffs, New Jersey (1963).
56. B. J. Austin, V. Heine, and L. J. Sham, General theory of pseudopotentials, *Phys. Rev.* **127**, 276–282 (1962).
57. B. S. Gourary and F. J. Adrian, in *Solid State Physics* (F. Seitz and D. Turnbull, eds.), Vol. 10, Academic, New York (1960).
58. C. J. Ballhausen, *Introduction to Ligand Field Theory*, McGraw-Hill, New York (1962).
59. D. Bramanti, M. Mancini, and A. Ranfagni, Molecular-orbital model for KCl:Tl, *Phys. Rev. B* **3**, 3670–3676 (1971).
60. K. L. Yip and W. Beall Fowler, The electronic structure of NaCl:Cu, *Phys. Stat. Sol.* (*b*) **53**, 137–146 (1972).
61. H. Seidel and H. C. Wolf, in *Physics of Color Centers* (W. B. Fowler, ed.), Academic, New York (1968).
62. J. H. Simpson, Charge distribution and energy levels of trapped electrons in ionic solids, *Proc. Roy. Soc.* (*London*) **A197**, 269–281 (1949).
63. J. K. Kübler and R. J. Friauf, Perturbation calculations of *F*-center wave functions with point ion and pseudopotentials, *Phys. Rev.* **140**, A1742–A1753 (1963).
64. R. H. Bartram, A. M. Stoneham, and P. Gash, Ion-size effects in color centers, *Phys. Rev.* **176**, 1014–1024 (1968).
65. F. Martino, Localized impurity states in the Hartree–Fock, LCAO approximation. I, *Int. J. Quantum Chem.* **2**, 217–233 (1968); II. The *F* center in KCl, *Int. J. Quantum Chem.* **2**, 233–263 (1968).

66. E. Mollwo, *Nachr. Ges. Wiss. Gött. II Math–Phys. Kl.* **1931**, 97.
67. H. F. Ivey, Spectral location of the absorption due to color centers in alkali halide crystals, *Phys. Rev.* **72**, 341–343 (1947).
68. G. Spinolo and D. Y. Smith, Optical absorption and photoconductivity in the *K* band of some colored alkali halides, *Phys. Rev.* **140**, A2117–A2120 (1965); D. Y. Smith and G. Spinolo, *K* band in colored alkali halide crystals, *Phys. Rev.* **140**, A2121–A2129 (1965).
69. J. Mort, F. Lüty, and F. C. Brown, Faraday rotation and spin–orbit splitting of the *F* center in alkali halides, *Phys. Rev.* **137**, A566–A573 (1965).
70. N. V. Karlov, J. Margerie, and Y. Merle d'Aubigne, Pompage optique des centres *F* dans KBr, *J. Phys. Rad.* **24**, 717–723 (1963).
71. D. Y. Smith, Theory of spin–orbit effects in the *F* band in alkali halides, *Phys. Rev.* **137**, A574–A582 (1965).
72. G. Chiarotti, U. M. Grassano, and R. Rosei, Stark effect of the *F* center in KCl, *Phys. Rev. Letters* **17**, 1043–1044 (1966).
73. R. K. Swank and F. C. Brown, Lifetime of the excited *F* center, *Phys. Rev.* **130**, 34–41 (1963).
74. L. D. Bogan and D. B. Fitchen, Stark effects in *F*-center emission, *Phys. Rev. B* **1**, 4122–4135 (1970).
75. H. Kühnert, Polarization of *F*-center luminescence in KCl due to electric fields, *Phys. Stat. Sol.* **21**, K171–K173 (1967).
76. W. B. Fowler, E. Calabrese, and D. Y. Smith, Excited states of the *F* center in alkali halides, *Solid State Commun.* **5**, 569–571 (1967).
77. L. F. Mollenauer and G. Baldacchini, Empirically determined wave function for the relaxed-excited state of the *F* center in KI, *Phys. Rev. Letters* **29**, 465–468 (1972).
78. R. S. Crandall and M. Mikkor, Photoconductivity of KBr containing *F* centers, *Phys. Rev.* **138**, A1247–A1249 (1965).
79. J. A. Strozier and B. G. Dick, A calculation of the electronic properties of the *F* center in KCl, *Phys. Stat. Sol.* **31**, 203–215 (1969).
80. W. Dale Compton and H. Rabin, in *Solid State Physics* (F. Seitz and D. Turnbull, eds.), Vol. 16, Academic, New York (1964).
81. F. Okamoto, Optical absorption of *M* centers in potassium chloride crystals, *Phys. Rev.* **124**, 1090–1097 (1961).
82. T. Neubert and S. Susman, *M* centers in potassium bromide. I, *J. Chem. Phys.* **43**, 2819–2829 (1965).
83. H. Seidel, Paramagnetic resonance of the metastable triplet state of the *M*-centre in KCl crystals, *Phys. Lett.* **7**, 27–29 (1963).
84. I. Schneider and M. E. Caspari, Photochemical study of *F* and *M* centers in additively colored alkali halides, *Phys. Rev.* **133**, A1193–A1201 (1964).
85. R. Herman, M. C. Wallis, and R. F. Wallis, Intensities of the R_1 and R_2 bands in KCl crystals, *Phys. Rev.* **103**, 87–93 (1956).
86. R. A. Evarestov, Origin of *F*(*M*) bands in the absorption spectrum of alkali halide crystals, *Opt. Spectry.* **16**, 198–199 (1964).
87. A. Meyer and R. F. Wood, Electronic structure of the *M* center in LiCl and LiF, *Phys. Rev.* **133**, A1436–A1442 (1964).
88. R. H. Silsbee, *R* center in KCl; stress effects in optical absorption, *Phys. Rev.* **138**, A180–A197 (1965); S. F. Wang and C. Chu, Calculation of low-lying quantum

states of the F_3 color center, *Phys. Rev.* **147**, 527–531 (1966); R. A. Evarestov, The electronic structure of the *R* center in alkali halide crystals, *Opt. Spectry. Suppl.* **4**, 13–16 (1970); R. C. Kern and R. H. Bartram, Wave functions for the *R* Center in KCl, Paper C42, *1971 Int. Conf. on Colour Centers in Ionic Crystals*, Reading, England (1971).

89. R. A. Evarestov, Theoretical studies on the *M'*-centre in alkali halide crystals, *Phys. Stat. Sol.* **31**, 401–406 (1969): M. A. Aegerter and F. Lüty, The $F_2{}^+$ center in KCl crystals. Part II: Optical properties and analogy with the $H_2{}^+$ molecule, *Phys. Stat. Sol.* (*b*) **43**, 245–261 (1971).
90. B. Henderson and J. E. Wertz, Defects in the alkaline earth oxides, *Adv. Phys.* **17**, 749–855 (1968).
91. J. C. Kemp, J. C. Cheng, E. H. Izen, and F. A. Modine, Magneto-optical structure of the MgO *F* band, *Phys. Rev.* **179**, 818–822 (1969).
92. H. S. Bennett, Two-electron *F'* centers in the alkaline-earth oxides and in the alkali halides, *Phys. Rev.* **B1**, 1702–1715 (1970).
93. J. C. Kemp, W. M. Ziniker, J. A. Glaze, and J. C. Cheng, Faraday rotation of the CaO *F* band and 3557 Å zero-phonon peak, *Phys. Rev.* **171**, 1024–1036 (1968).
94. Y. Merle d'Aubigné and A. Roussel, Magneto-optic study of the Jahn–Teller effect in the excited level of the F^+ centers in CaO, *Phys. Rev.* **B3**, 1421–1427 (1971).
95. B. Henderson, S. E. Stokowski, and T. C. Ensign, Luminescence from *F* centers in calcium oxide, *Phys. Rev.* **183**, 826–831 (1969).
96. P. Edel, C. Hennies, Y. Merle d'Aubigné, R. Romestain, and Y. Twarowski, Optical detection of paramagnetic resonance in the excited state of *F* centers in CaO, *Phys. Rev. Letters* **28**, 1268–1271 (1972).
97. B. H. Rose and E. B. Hensley, *F* and F^+ centers and free-carrier absorption in BaO, *Phys. Rev. Letters* **29**, 861–864 (1972).
98. W. Hayes and J. P. Stott, Electron nuclear double resonance of the *F* centre in CaF_2, *Proc. Roy. Soc.* (*London*) **A301**, 313–326 (1967); B. C. Cavenett, W. Hayes, I. C. Hunter, and A. M. Stoneham, Magnetooptical properties of *F* centres in alkaline earth fluorides, *Proc. Roy. Soc.* (*London*) **A309**, 53–68 (1969).
99. J. H. Beaumont and W. Hayes, *M* centres in alkaline earth fluorides, *Proc. Roy. Soc.* (*London*) **A309**, 41–52 (1969).
100. W. Hayes, Abstract H162, *1971 Int. Conf. on Colour Centers in Ionic Crystals*, Reading, England (1971).
101. R. A. Weeks, Paramagnetic spectra of E_2' centers in crystalline quartz, *Phys. Rev.* **130**, 570–576 (1963); R. H. Silsbee, Electron spin resonance in neutron-irradiated quartz, *J. Appl. Phys.* **32**, 1459–1462 (1961).
102. F. Varsanyi and G. H. Dieke, Ion-pair resonance mechanism of energy transfer in rare earth crystal fluorescence, *Phys. Rev. Letters* **7**, 442–443 (1961).
103. D. H. Kühner, H. V. Lauer, and W. E. Bron, Lattice dynamics and electron–phonon coupling field and vibronic spectra, *Phys. Rev.* *B* **5**, 4112–4126 (1972).
104. W. E. Bron and W. R. Heller, Rare-earth ions in the alkali halides. I. Emission spectra of Sm^{2+}-vacancy complex, *Phys. Rev.* **136**, A1433–A1444 (1964).
105. Herbert N. Hersh, Connections between color center phenomena and luminescence phenomena in general, *J. Electrochem. Soc.* **118**, 144C–149C (1971).

106. W. J. Van Sciver, Role of the tallium dimer in the NaI (Tl) scintillation process, *Phys. Lett.* **9**, 97–98 (1964).
107. A. Fukuda, Alkali halide phosphors containing impurity ions with $(s)^2$ configuration, *Sci. Light* **13**, 64–114 (1964).
108. S. G. Zazubovich, N. E. Lushchik, and Ch. B. Lushchik, Optical structure of luminescence centers in ionic crystals activated by mercury-like ions, *Opt. Spectry.* **15**, 203–207 (1963).
109. F. Seitz, Interpretation of the properties of alkali-halide-thallium phosphors, *J. Chem. Phys.* **6**, 150–162 (1938).
110. F. E. Williams, An absolute theory of solid-state luminescence, *J. Chem. Phys.* **19**, 457–466 (1951); F. E. Williams, Theory of the luminescence of impurity-activated ionic crystals, *J. Phys. Chem.* **57**, 780–784 (1953).
111. R. S. Knox, Configuration interaction in alkali halide phosphors, *Phys. Rev.* **115**, 1095–1106 (1959).
112. R. S. Knox and D. L. Dexter, Solid state luminescence theory and oscillator strengths in KCl:Tl, *Phys. Rev.* **104**, 1245–1252 (1956).
113. A. Honma, Energy parameters of In^+ and Tl^+ ions in KCl, *Sci. Light* **17**, 34–44 (1968).
114. R. Laiho, Energy parameters of Tl^+ in KBr, KI, and RbI, *Ann. Acad. Sci. Fennicae Sec. A, VI. Physica* **362**, 1–21 (1971).
115. W. E. Hagston, Theoretical formulation of the effects of charge transfer states on optical transitions in s^2 ions, *J. Phys. C* **5**, 680–690 (1972).
116. Y. Toyozawa and M. Inoue, Dynamical Jahn–Teller effect in alkali halide phosphors containing heavy metal ions, *J. Phys. Soc. Japan* **21**, 1663–1679 (1966); A. M. Lemos, M. C. Stauber, and J. F. Marion, Structure of the *A*, *B*, and *C* absorption bands in KCl:Tl, *Phys. Rev. B* **2**, 4161–4168 (1970).
117. W. Dultz, Uniaxial stress effects of the UV-absorption bands of In^+ and Tl^+ doped alkali halide crystals in a molecular treatment, *Phys. Stat. Sol.* (*b*) **48**, 571–580 (1971).
118. A. Fukuda, Jahn–Teller effect on the structure of the emission produced by excitation in the *A* band of KI:Tl–type phosphors. Two kinds of minima on the Γ_4^- $(^3T_{1u})$ adiabatic potential-energy surface, *Phys. Rev. B* **1**, 4161–4178 (1970).
119. A. Fukuda and P. Yuster, Unusually large change in radiative lifetime of the *A*-band emission in KI:In^+ and KI:Sn^{2+} induced by a magnetic field, *Phys. Rev. Letters* **28**, 1032–1034 (1972).
120. A. Ranfagni, G. P. Pazzi, F. Fabeni, G. Viliani, and M. P. Fontana, Origin of the 4750-Å emission in KCl phosphors, *Phys. Rev. Letters* **28**, 1035–1037 (1972).
121. K. Fussgaenger, W. Martienssen, and H. Bilz, UV absorption of Ag^+ doped alkali halide crystals, *Phys. Stat. Sol.* **12**, 383–397 (1965).
122. K. Fussgaenger, On the UV absorption of heavy metal ions in alkali halide crystals. I. Experimental results, *Phys. Stat. Sol.* **34**, 157–169 (1969); II. Interpretation of the temperature dependences of the UV dipole strength, *Phys. Stat. Sol.* **36**, 645–655 (1969).
123. R. S. Knox, The fine structure of absorption in NaCl:Ag, *J. Phys. Soc. Japan* **18**(Suppl. II), 268–274 (1963).
124. W. E. Hagston, Theory of the fine structure of the UV absorption of heavy metal ions in the alkali halides, *J. Phys. C* **5**, 691–701 (1972).

125. R. Kubo and Y. Toyozawa, Application of the method of generating function to radiative and non-radiative transitions of a trapped electron in a crystal, *Progr. Theoret. Phys.* (*Kyoto*) **13**, 160–182 (1955); A. D. Liehr and C. S. Ballhausen, Intensities in inorganic complexes, *Phys. Rev.* **106**, 1161–1163 (1957).
126. J. M. Conway, D. A. Greenwood, J. A. Krumhansl, and W. Martienssen, Oscillator strength for optical absorption by silver ions in alkali halide crystals, *J. Phys. Chem. Solids* **24**, 239–245 (1963).
127. T. P. Martin and W. Beall Fowler, Temperature dependence of phonon-induced electronic transitions in insulating solids, *Phys. Rev. B* **2**, 4221–4225 (1970).
128. R. Oggioni and P. Scaramelli, Optical absorption and photoluminescence of Cu-doped alkali halide crystals, *Phys. Stat. Sol.* **9**, 411–421 (1965); E. Krätzig, T. Timusk, and W. Martienssen, Optical absorption of Cu-centers in alkali halides, *Phys. Stat. Sol.* **10**, 709–723 (1965).
129. C. E. Moore, Atomic energy levels, Nat. Bur. Std. (U.S.) Circ. 467 (1949–1952).
130. G. Baldini, A. Jean, and G. Spinolo, Lifetime of excited Cu^+ in alkali halides, *Phys. Stat. Sol.* **25**, 557–561 (1968).
131. S. A. Mack and W. J. Van Sciver, Lifetimes of parity forbidden transitions of the Cu^+ center in the alkali iodides, *Phys. Stat. Sol.* (*b*) **46**, 193–201 (1971); M. Piccirilli and G. Spinolo, "Off-center" effect in the relaxed excited state of Cu^+ substitutional in alkali halides, *Phys. Rev. B* **4**, 1339–1343 (1971).
132. T. G. Castner and W. Känzig, The electronic structure of *V*-centers, *J. Phys. Chem. Solids* **3**, 178–195 (1957).
133. M. N. Kabler, in *Point Defects in Solids, Vol. 1. General and Ionic Crystals* (J. H. Crawford, Jr. and L. M. Slifkin, eds.), Plenum, New York (1972).
134. R. F. Edgerton, Ph.D. thesis, University of Rochester, 1962 (unpublished); M. N. Kabler, Low-temperature recombination luminescence in alkali halide crystals, *Phys. Rev.* **136**, A1296–A1302 (1964).
135. M. N. Kabler and D. A. Patterson, Evidence for a triplet state of the self-trapped exciton in alkali-halide crystals, *Phys. Rev. Letters* **19**, 652–655 (1967).
136. J. U. Fischbach, D. Fröhlich, and M. N. Kabler, Recombination luminescence lifetimes and the self-trapped exciton in alkali halides, *J. Luminescence* **6**, 29–43 (1973).
137. M. J. Marrone and M. N. Kabler, Magnetic circular polarization of luminescence from self-trapped excitons in alkali halides, *Phys. Rev. Letters* **27**, 1283–1285 (1971).
138. W. Beall Fowler, M. J. Marrone, and M. N. Kabler, Theory of self-trapped exciton luminescence in alkali halide crystals (to be published).
139. M. N. Kabler, M. J. Marrone, and W. B. Fowler, Magneto-optic effects in recombination luminescence from self-trapped excitons, in *Luminescence of Crystals, Molecules, and Solutions* (F. Williams, ed.), Plenum, New York (1973).
140. D. Schoemaker, The *g* and hyperfine components of V_k centers, *Phys. Rev. B* **7**, 786–801 (1973).
141. T. H. DiStefano and D. E. Eastman, The band edge of amorphous SiO_2 by photoinjection and photoconductivity measurements, *Solid State Commum.* **9**, 2259–2261 (1971).
142. C. P. Slichter, *Principles of Magnetic Resonance*, Harper and Row, New York (1963).
143. A. Abragam and B. Bleaney, *Electron Paramagnetic Resonance of Transition Ions*, Clarendon Press, Oxford (1970).

144. G. K. Wertheim, A. Hausmann, and W. Sander, *The Electronic Structure of Point Defects*, North-Holland, Amsterdam (1971).
145. G. Feher, Electron spin resonance experiments on donors in silicon. I. Electronic structure of donors by the electron nuclear double resonance technique, *Phys. Rev.* **114**, 1219–1244 (1959); G. Feher and E. A. Gere, Electron spin resonance experiments on donors in silicon. II. Electron spin relaxation effects, *Phys. Rev.* **114**, 1245–1256 (1959).
146. R. Kersten, Hyperfine and quadrupole interactions of the *F*-centre in KCl as measured by ENDOR, *Phys. Stat. Sol.* **29**, 575–586 (1968).
147. R. F. Wood, Hyperfine interaction of *F* centres in alkali-halide crystals, *Phys. Stat. Sol.* **42**, 849–854 (1970).
148. H. Fedders, M. Hunger, and F. Lüty, A comprehensive study of a simple photoconductive phosphor, *J. Phys. Chem. Solids* **22**, 299–307 (1961).
149. M. Schwoerer and H. C. Wolf, Elektronenspinresonanz-Untersuchungen zur Aggregation von *F*-Zentren in KCl, *Z. Physik* **175**, 457–479 (1963); C. J. Delbecq, A study of *M* center formation in additively colored KCl, *Z. Physik* **171**, 560–581 (1963).
150. F. Lüty, in *Physics of Color Centers* (W. B. Fowler, ed.), pp. 181–242, Academic, New York (1968).
151. F. Porret and F. Lüty, Luminescence and *F* formation involving spin-polarized *F* centers in KCl, *Phys. Rev. Letters* **26**, 843–846 (1971); C. Jaccard, Y. Ruedin, M. Aegerter, and P. A. Schnegg, Weak magnetic field enhancement of the luminescence from *F* centre pairs in alkali halides, *Phys. Stat. Sol.* (*b*) **50**, 187–198 (1972).
152. C. J. Delbecq, A. K. Ghosh, and P. H. Yuster, Trapping and annihilation of electrons and positive holes in KCl–TlCl, *Phys. Rev.* **151**, 599–609 (1966).
153. R. V. Lorenze, Jr. and F. J. Feigl, Generation and bleaching of paramagnetic centers in quartz by ionizing radiation, *Bull. Am. Phys. Soc.* **17**, 264 (1972).
154. R. J. Ginther and R. D. Kirk, The thermoluminescence of CaF_2:Mn, *J. Electrochem. Soc.* **104**, 365–369 (1957).
155. C. C. Klick, E. W. Claffy, S. G. Gorbics, F. H. Attix, J. H. Schulman, and J. G. Allard, Thermoluminescence and color centers in LiF:Mg, *J. Appl. Phys.* **38**, 3867–3874 (1967).
156. E. B. Podgoršak, P. R. Moran, and J. R. Cameron, Thermoluminescent behavior of LiF (TLD-100) from 77° to 500°K, *J. Appl. Phys.* **42**, 2761–2767 (1971).
157. Z. J. Kiss, Photochromics, *Phys. Today* **23**(1), 42–49 (1970).
158. G. K. Megla, Optical properties and applications of photochromic glass, *Appl. Optics* **5**, 945–960 (1966).
159. P. M. Heyman, I. Gorog, B. W. Faughman, and I. Shidlovsky, The application of colour centers in sodalite to storage and display, *1971 Int. Conf. on Colour Centers in Ionic Crystals*, Reading, England (1971).
160. H. Blume, T. Bader, and F. Lüty, Holographic information storage based on the dichroism of F_A centres in KCl:Na, *1971 Int. Conf. on Colour Centers in Ionic Crystals*, Reading, England (1971).
161. Irwin Schneider, Information storage using the anisotropy of color centers in alkali halide crystals, *Appl. Optics* **6**, 2197–2198 (1967).

162. I. Schneider, M. Marrone, and M. N. Kabler, Dichroic absorption of *M* centers as a basis for optical information storage, *Appl. Optics* **9**, 1163–1166 (1970).
163. I. Schneider, Three-dimensional optical storage element using *M* centers in a KCl crystal, *Appl. Optics* **10**, 980–981 (1971).

3

The Imperfect Solid—Dielectric Properties

G. E. Peterson
Bell Laboratories
Murray Hill, New Jersey

1. Introduction

This chapter is concerned with the influence of lattice defects on dielectric and optical properties of solids. It is an extension of the ideas presented in Volume 1, Chapter 4, which dealt with structure and composition in relation to properties. To a large measure it is restricted to a discussion of topics of current research interest to the solid-state chemist. In this regard heavy emphasis is placed upon nonlinear optical materials. Since ferroelectric crystals show enhanced nonlinear effects and are the most useful for present-day applications, we shall discuss them almost exclusively.

In Section 2 we review a few of the basic aspects of ferroelectric theory. This includes the polarization catastrophe and the concept of a soft mode. The general approach is along the lines of Kittel,[1] whose book is strongly recommended for additional reading at an elementary level. As the major concern of this chapter is with defects, our discussion is not extensive and has the prime purpose of giving the reader only enough background material to understand the recently developed ferroelectric theory of Lines.[2]

This theory, rather than being couched in terms of lattice dynamics with its enormous complexities, deals with the normal modes of a single unit cell. As such it is extremely useful for a qualitative and semiquantitative understanding of how defects influence ferroelectric properties. It is particularly well adapted to the needs of the solid-state chemist since it can be formulated in terms of familiar concepts such as bond strengths and atomic mass.

In Section 3 we consider ferroelectric pyroelectrics and with a few selected examples show how their properties can be advantageously altered. For example, a technique to eliminate accidental depoling due to a reverse bias field is considered. The choices made for discussion are such as to give the reader further insight rather than an encyclopedic coverage of the field.

In Section 4 optical properties of some nonlinear materials are examined with the same philosophy as in Section 3.

2. Introduction to Ferroelectrics

Of the 32 point groups (crystal classes) 21 lack a center of symmetry.[3,4] Of these, 20 develop a polarization when stressed, and are thus piezoelectric. The lone exception is point group 432 and even though this cubic group lacks an inversion center, other symmetry elements combine so as to exclude the effect. Out of the 20, ten have a unique polar axis and as a consequence have a "spontaneous" polarization. These ten classes are: 1, 2, 3, 4, 6, *m*, *mm*2, 3*m*, 4*m*, and 6*mm*. In general, the spontaneous polarization is a function of temperature, and the change in its magnitude when the sample is heated or cooled is the pyroelectric effect. Ferroelectric[5,6] crystals also belong to one of the above ten classes. They may be defined as pyroelectric crystals whose polarization may be reversed by the application of an electric field. Rochelle salt (sodium potassium tartrate tetrahydrate) was the first material known to have this remarkable property. A very brief look at its history is interesting.*

The first synthesis of this material appears to have been by Elie Seignette in about 1655, presumably for medicinal purposes. In 1880 Pierre and Jacques Curie studied its piezoelectric properties and in 1894 Pockels discovered anomalous dielectric behavior along the polar axis. Valasek in 1921 recognized the analogy between this dielectric phenomenon and ferromagnetism and called the temperature at which the effect disappeared the Curie temperature.

During World War I this material, because of its large piezoelectric coefficients, was investigated for possible underwater signalling applications. It was not until 1940 that Mueller, in his famous "interaction theory," showed that the anomalous piezoelectric and

* Also see van Klooster[7]; Cady,[4] pp. 513–515; and Känzig,[6] pp. 8–12.

electrooptic effects were a consequence of the peculiar dielectric behavior. In 1935 Busch and Scherrer discovered the second ferroelectric potassium di-hydrogen phosphate. Since then numerous ferroelectrics have been discovered and studied. Matthias at Bell Laboratories has been particularly active in leading the search for new materials.(8)

2.1. Outline of Ferroelectric Theory

The ferroelectric effect in solids may be described from two points of view: in terms of (1) polarization catastrophe, and (2) a transverse optical phonon of very low frequency. At first sight these seem to be quite unrelated descriptions of the phenomenon. However, this is not so, and as we shall show they can be connected by the famous Lyddane–Sachs–Teller(9) equation. We shall examine both viewpoints and to avoid complications with unrewarding algebra, we shall for simplicity consider a cubic crystal lattice containing a single diatomic molecule per unit cell.

Suppose an electric field E (the Maxwell field) is applied to a crystal. The internal field E_{loc}, that is, the field an atom in the lattice feels, can be expressed by

$$E_{\text{loc}} = E + \gamma P \tag{1}$$

where P is the polarization and γ is the Lorentz-field tensor. In the case of a cubic lattice $\gamma = 4\pi/3$; thus

$$E_{\text{loc}} = E + (4\pi/3)P \tag{2}$$

Generally, the Lorentz-field tensor will vary from site to site, but the average over all sites will still be roughly $4\pi/3$.

The dielectric constant of a cubic crystal is given by

$$\varepsilon = 1 + 4\pi(P/E) \tag{3}$$

and the atomic polarizability by

$$\alpha = P/E_{\text{loc}} \tag{4}$$

It is now easy to show that

$$\varepsilon = \frac{1 + N(\alpha^+ + \alpha^-)(8\pi/3)}{1 - N(\alpha^+ + \alpha^-)(4\pi/3)} \tag{5}$$

where N is the number of molecules per unit volume and α^+ and α^- are the polarizabilities of the two atoms in the diatomic molecule.

The + and − superscripts specify the positive and negative atoms, respectively. The above equation is the well-known Clausius–Mossotti relation (see, e.g., Ref. 10).

In the case of a ferroelectric crystal P will have a finite value even in the absence of an applied electric field. From Eq. (1) this implies that E_{loc} will also be nonzero. The Clausius–Mossotti relation shows that as $N(\alpha^+ + \alpha^-)(4\pi/3) \to 1$, $\varepsilon \to \infty$. This corresponds to a finite polarization even when the Maxwell field E is zero and this is just what we need for a ferroelectric. This is the so-called polarization catastrophe. We may view this catastrophe as a sort of "feedback effect." The local field causes a displacement which results in an increased polarization which in turn increases the local field. The polarization increases more rapidly than the elastic restoring force and finally higher-order restoring forces come into play and the displacement is limited.

Let us assume that the term $1 - N(\alpha^+ + \alpha^-)(4\pi/3)$ varies as $T - T_c$, where T_c is the critical temperature. In essence we are requiring that $1 - N(\alpha^+ + \alpha^-)(4\pi/3)$ goes to zero linearly with temperature. This leads to the following relationship:

$$\varepsilon \sim 1/(T - T_c) \tag{6}$$

This expression agrees rather well with what is at times observed experimentally. (See Fig. 1.[11])

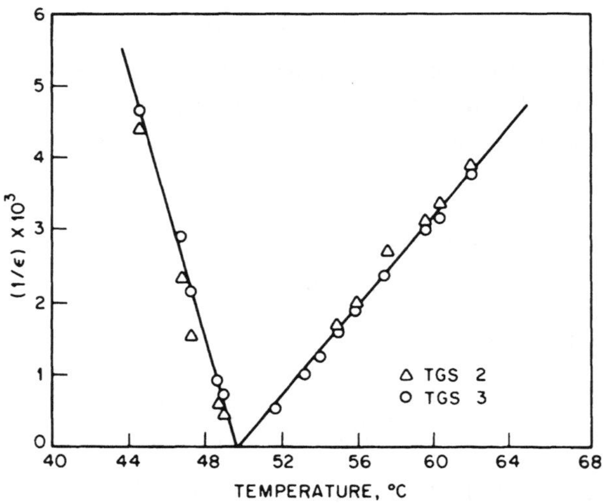

Fig. 1. Curie–Weiss law for triglycine sulfate. From Ref. 11.

The notion of a polarization catastrophe gives one a certain qualitative picture of what is happining in a ferroelectric as the Curie temperature is approached. It is very difficult to see, however, how these ideas could be used to predict, say, Curie temperatures as a function of stoichiometry.

A transverse optical phonon of very low frequency is a second way of looking at ferroelectric phenomena. Consider again the cubic lattice with a single diatomic molecule per cell and let us examine the ionic motion for a long-wavelength ($K = 2\pi/\lambda \rightarrow 0$) transverse optical mode. The atoms can be assumed to have equal but opposite charges and to vibrate against each other with a fixed center of mass. Such a vibration is illustrated in Fig. 2 and the direction of propagation of the wave is from the bottom to the top of the page. As the atoms are charged, the displacement from their equilibrium positions results

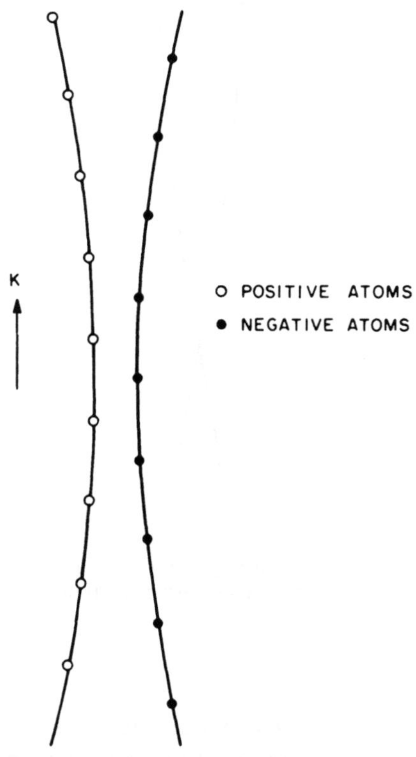

Fig. 2. A long-wavelength optical mode. The plane of positive atoms vibrates against the plane of negative atoms.

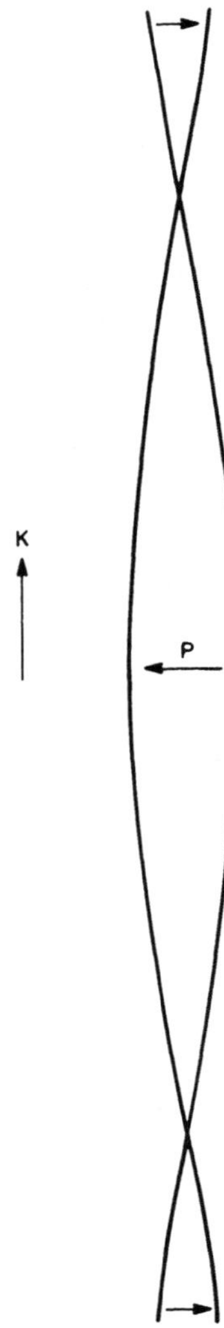

Fig. 3. A more or less uniformly polarized region of the lattice due to a long-wavelength optical mode.

in a polarization of the lattice. Since the modes under consideration are of relatively long wavelength, the polarization will be constant in magnitude and direction for large regions of the lattice. Of course it must go through zero at the nodes and reverse sign. (See Fig. 3.)

A uniformly polarized region of the sort illustrated in Fig. 3 will have a local field due to this polarization given by

$$E_{\text{loc}} = (4\pi/3)P \tag{7}$$

This field will have two very pronounced effects on the transverse optical mode. They are (a) accentuation of the lattice vibration amplitude, and (b) lowering of the optical mode frequency. If u represents the displacement of the positive lattice toward the negative lattice, the equation of motion can be written as

$$\mu\, d^2u/dt^2 = -Cu + qE_{\text{loc}} \tag{8}$$

$$= -Cu + q(4\pi/3)P \tag{9}$$

where μ is the reduced mass and C is the short-range restoring force. The term qE_{loc} represents the force applied to the atom due to the local field. The polarization P can be written in terms of the displacement u and the electronic contribution to the polarization β^+ and β^- for the positive and negative ions, respectively. Thus

$$P = Nqu + N(\beta^+ + \beta^-)E_{\text{loc}} \tag{10}$$

$$= Nqu + N(\beta^+ + \beta^-)(4\pi/3)P \tag{11}$$

or

$$P = Nqu/[1 - N(\beta^+ + \beta^-)(4\pi/3)] \tag{12}$$

and we find

$$\mu\frac{d^2u}{dt^2} = \left[-C + \frac{Nq^2(4\pi/3)}{1 - N(\beta^+ + \beta^-)(4\pi/3)}\right]u \tag{13}$$

The static ionic polarizability is defined by

$$\delta = q^2/C \tag{14}$$

Thus we can interpret Eq. (13) as a harmonic oscillator with an effective force constant given by

$$C' = C\left[1 - \frac{N\delta(4\pi/3)}{1 - N(\beta^+ + \beta^-)(4\pi/3)}\right] \tag{15}$$

We note that the force constant C' is smaller in magnitude than the short-range force constant C. This occurs because the long-range interactions due to the polarization of the lattice cancel the short-range forces. Again as in the polarization catastrophe model, a sort of feedback mechanism is operative. The local field reduces the

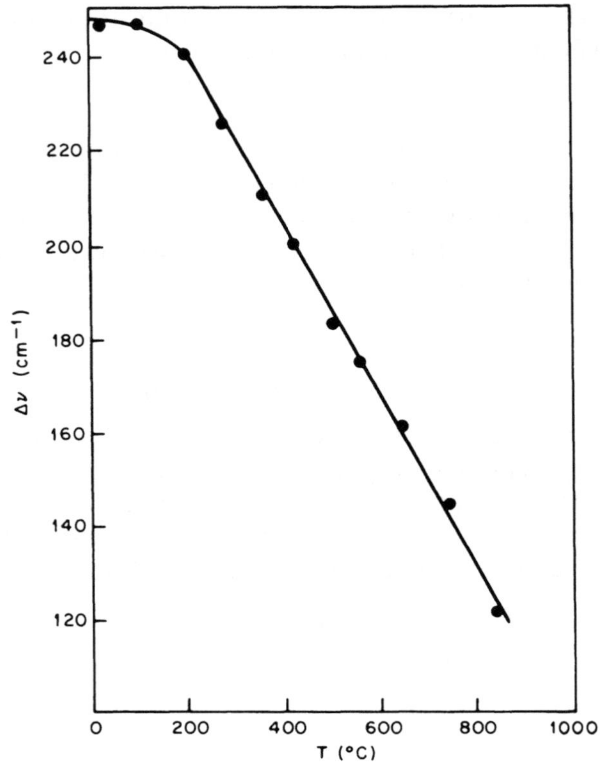

Fig. 4. Soft-mode frequency as a function of temperature for lithium niobate. From Ref. 12.

effective force constant, which causes an increased displacement. This in turn increases the polarization, which increases the local field. The frequency of the transverse optical mode associated with the force constant of Eq. (15) can be written as

$$\omega_T{}^2 = \frac{C}{\mu}\left[1 - \frac{N\delta(4\pi/3)}{1 - N(\beta^+ + \beta^-)(4\pi/3)}\right] \tag{16}$$

The reduction in frequency, or "mode softening" as it is often called plays a key role in current theories of ferroelectricity. Figure 4 shows the soft mode frequency as a function of temperature for lithium niobate.[12] The Curie temperature for this compound is 1210°C and one notes that the frequency is rapidly falling as this temperature is approached.

We shall now tie together the two ways of considering ferroelectricity: the polarization catastrophe and the soft mode. In a manner similar to that used for the transverse mode, we can determine a frequency for the longitudinal mode ω_L. Thus

$$\omega_L{}^2 = \frac{C}{\mu}\left[1 + \frac{N\delta(8\pi/3)}{1 + N(\beta^+ + \beta^-)(8\pi/3)}\right] \tag{17}$$

It is interesting to note that the effective force constant and frequency are not reduced by the polarization. This is made clear by noting that in the case of the transverse mode the term involving the polarization is substracted from the short-range force constant C, while in the case of the longitudinal mode it is added. By making use of the Clausius–Mossotti relation we can relate the two modes. This is the Lyddane–Sachs–Teller equation and it is given by

$$\omega_T{}^2/\omega_L{}^2 = \varepsilon(\infty)/\varepsilon(0) \tag{18}$$

where $\varepsilon(\infty)$ is the dielectric constant at a frequency high enough that ionic polarizability does not contribute and $\varepsilon(0)$ is the low-frequency dielectric constant. We note that as $\varepsilon(0) \to \infty$, $\omega_T \to 0$. In other words, as a finite polarization develops, the transverse optical mode tends toward zero and our two points of view are now related. When a material exhibits a large dielectric constant, say 1000, a low-frequency transverse mode ω_T is expected.

2.2. Statistical Theory of Lines

An elegant theory developed by Lines[2] is much better adapted to the understanding of the influence of defects than the lattice theory just discussed. The usual lattice-dynamic theories[13] are couched in terms of wave-vector-dependent parameters and unfortunately a detailed knowledge of the wave-vector dispersion relationships of all modes interacting with the soft mode or modes is necessary. Even worse, the few bulk dielectric properties that can be measured experimentally are related to a very large number of microscopic unknowns. For example, the detailed forms of the interionic force constants are required. To the solid-state chemist, who is accustomed to thinking in terms of molecules and bonds, such theories are difficult to interpret. It is fair to say, however, that the lattice-dynamic theories have reached a sophisticated level. They are particularly useful for neutron diffraction studies[14] where detailed microscopic motion can be measured directly.

In the approach of Lines, in direct contrast to the lattice theories, one isolates a single unit cell and determines the normal modes of vibration of this unit only. In essence one treats this cell exactly. Now this is a problem that is not at all unfamiliar to the solid-state chemist. The masses that go into the equation of motion are the atomic masses. The force constants are the familiar ones that represent ion–ion stretch and can be related to bond strength.

The Hamiltonian for the one cell under consideration can be written as

$$H = (1/v)\sum_{b}(P_b{}^2/2m_b) + V(q_b \ldots q_{b'}) \tag{19}$$

where the summation is over all atoms and v is the volume. This can be transformed to normal-mode coordinates using the familiar procedures found in standard texts[15] on lattice vibrations. For systems with a centrosymmetric nonpolar phase only even-order anharmonic terms are required. The result is

$$H = (1/v)\sum_{p}(\tfrac{1}{2}\pi_p{}^2 + \tfrac{1}{2}\Omega_p{}^2\xi_p{}^2 + A_p\xi_p{}^4 + B_p\xi_p{}^6) \tag{20}$$

where π_p and ξ_p are the conjugate momentum and displacement of the pth mode and terms beyond the sixth power have been neglected.

In order to incorporate the effects of other cells, we assume that the atoms are influenced by a local field. This varies from site to site and its source is the Maxwell field and a field due to the polarization P. We shall, as a first approximation, assume initially no correlations and express E_{loc} in terms of a thermally averaged polarization $\langle P\rangle$. Thus

$$E^b_{\text{loc}} = E + \gamma_b\langle P\rangle \tag{21}$$

If we let q_b represent the displacement of an atom of charge e_b from its equilibrium position, the interaction energy can be written as

$$V' = -(1/v)\sum_{b} e_b q_b(E + \gamma_b\langle P\rangle) \tag{22}$$

where the sum is again over the cell.

Clearly it is unreasonable to assume that there is absolutely no correlation between the motions of the atoms of one cell and those of another. An important aspect of ferroelectric phenomena is the concept of atoms moving together. This can be incorporated by adding an additional term to the expression for the local field. This becomes

$$E^b_{\text{loc}} = E + \gamma^b\langle P\rangle + \gamma^b_{\text{corr}}P \tag{23}$$

where γ^b_{corr} is the Lorentz parameter for the correlated motion of the atoms. To reiterate, the motions of a single unit cell are considered exactly. A field is applied to the cell consisting of three terms: (1) the Maxwell field, (2) that due to the thermally averaged polarization, and (3) that due to a correlated polarization.

Near a phase transition the ionic motion becomes dominated by a single mode (the soft mode) with excitation energy (or frequency) much smaller than the host of other optical modes. It is thus a good approximation to neglect all modes except that reflecting the symmetry of the soft mode. The Hamiltonian is now complete and transforming to the normal-mode coordinates π and ξ (where π and ξ refer to the soft mode only) yields

$$vH = \tfrac{1}{2}(\pi^2 + \omega_0{}^2\xi^2) + A\xi^4 + B\xi^6 - \eta S\xi[E + (\gamma S/v)\langle\xi\rangle] \quad (24)$$

where $\omega_0{}^2 = \Omega^2 - (\eta\gamma_{\text{corr}}S^2/v)$ and S is an effective charge parameter. η is a parameter describing the electronic contribution to the polarization. If electronic terms are absent, $\eta = 1$. The γ is now a "mode Lorentz parameter."

An examination of the transformed Hamiltonian shows that the dipolar contributions (i.e., those due to the local field) occur in two ways. First, the random phase component yields terms linear in the displacement variables, and second, the fully correlated part yields quadratic terms. Lines has shown that by using this Hamiltonian in conjunction with the conventional techniques of classical statistical mechanics, one readily obtains expressions for the temperature dependence of the bulk ferroelectric properties. With remarkable accuracy such properties as spontaneous polarization, soft-mode frequency, electric susceptibilities, and specific heat anomaly can be calculated.

For our purpose, which is primarily to qualitatively understand the influence of defects, we need not concern ourselves with complete solutions. It will be more than sufficient just to examine the potential energy terms in the Hamiltonian. Let us consider these individually.

(1) $\Omega^2\xi^2$ are the short-range interaction terms. ξ is a displacement amplitude parameter describing the motion of unit cell ions with the symmetry of the soft mode. This symmetry can be thought of as directly related to that ionic motion which results on going down through T_c into the ferroelectric phase. Ω^2 can be viewed as a sort of k/m factor, where k is a force constant and m is a reduced mass. Thus replacing one atom with a second that forms stronger bonds will increase this contribution. Likewise, substituting a heavier atom for a lighter one will decrease it. Figure 5 shows the general form of this function.

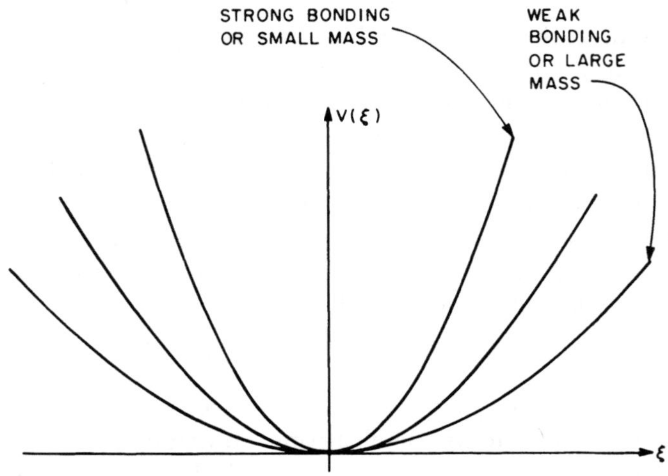

Fig. 5. The effects of bonding and mass on the function $\Omega^2\xi^2$.

(2) $A\xi^4$ and $B\xi^6$ result from the lack of harmonicity for small displacements. Their major effect is to stabilize the lattice in the polar phase. They are likely to be rather temperature independent.

(3) $\xi\langle\xi\rangle$ are intercell dipolar terms and are due the random phase component of the local field. Their major effect is to determine the spontaneous polarization, which varies as $\langle\xi\rangle$. As the Curie temperature is approached, the straight line representing this function rotates to the horizontal axis. This is because $\langle\xi\rangle$ is temperature dependent and goes to zero at T_c. These results are depicted in Fig. 6.

(4) $-\gamma_{\text{corr}}\xi^2$ are the intercell dipolar contributions that result from the completely correlated motion. Their effect increases as the Curie temperature is approached.

There is little question that complicated functions can be generated by linear combinations of 1–4. Potential energy curves appropriate to both displacement and order–disorder ferroelectrics can readily be synthesized and Lines has given some excellent examples of these. The original papers are very strongly recommended for further study.(2) We shall consider a fairly simple case that is useful for displacement ferroelectrics below the Curie temperature.

By combining terms 1–4, we get the function shown in Fig. 7. This can be characterized by two parameters. They are the depth of the well Δ and the displacement of the well Γ. It is easy to show that

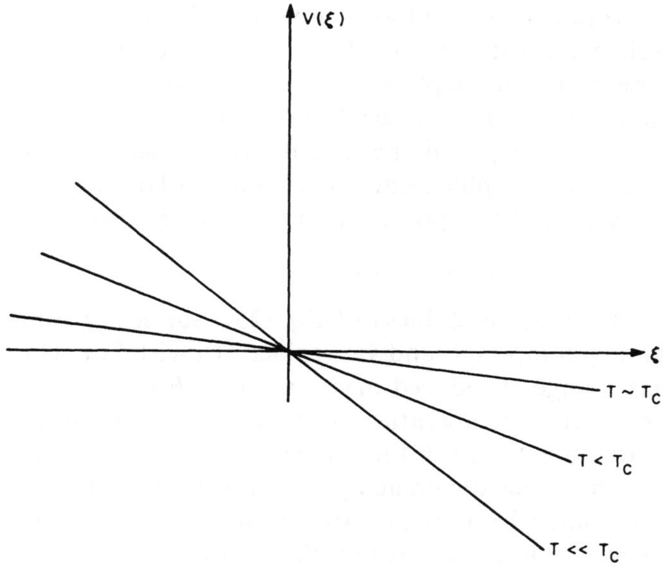

Fig. 6. The function $\xi\langle\xi\rangle$ and its variation with temperature.

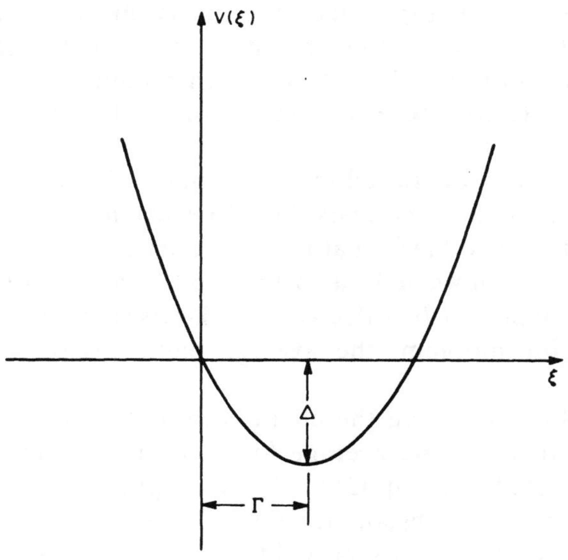

Fig. 7. Potential energy function illustrating depth of well Δ and shift of minimum Γ.

Δ is inversely proportional to Ω and consequently $\Delta \sim m/k$. So, to deepen the well, we must increase the mass or decrease the bond strength. Furthermore, the depth of the well is proportional to the Curie temperature T_C. As long as we do not change the gross crystal structure, both γ and γ_{corr} will be essentially unchanged and the contributions for random phase and correlated dipolar terms remain constant. This leads to the important and very useful result:

$$T_C \sim m \quad \text{or} \quad 1/k \tag{25}$$

Let us illustrate the usefulness of Eq. (25). Suppose we have a lithium-containing compound and we replace ^{7}Li with ^{6}Li. This is a *large* fractional change in reduced mass and from Eq. (25) we would predict a drop in Curie temperature. Indeed such a mass effect has been found by Bergman[16] for lithium niobate. We might further ask what would be the effect of replacing niobium by tantalum in the same compound. Since Ta–O bonds are stronger than Nb–O bonds, consequently k will be larger in the tantalate.* As a result, we predict that T_C must drop. One finds experimentally that the Curie temperature[18] decreases from roughly 1200°C to 700°C in agreement with our prediction. Of course there is a change in reduced mass but this is fractionally small and its effect would be minimal in comparison to the increase in bond strength.

The parameter Γ in Fig. 7 (i.e., the displacement of the well minimum from the y axis) can be shown to be proportional to the spontaneous polarization as $T \to 0$. Arguments similar to that just given for the Curie temperature are useful in rationalizing the magnitude of this quantity.

Usually we will be interested in how crystals will be altered for rather small variations in composition. The extension to partial atomic substitution or to the formation of vacancies is fairly straightforward. For example, if we make a partial substitution of ^{6}Li for ^{7}Li, the average lithium mass will be decreased. Likewise if we incorporate some tantalum for niobium, the average bond strength will be increased.

In general for estimating the changes in a "mildly" defective system we should determine average force constants and average masses. It is expected that Eq. (25) will often apply to the average quantities and will be of substantial value in "predicting" what will happen. It is hoped that the reader will try to rationalize for himself

* This can be inferred from electronegativity differences.[17]

what is occurring in the crystals to be discussed in the following sections, preferably by drawing likely potential energy functions.

3. Pyroelectricity

3.1. Introduction

Pyroelectricity is a very old phenomenon. It can be described as the change in magnitude of the electric charges on some faces of certain crystals with temperature. According to Cady,* the attractive power of tourmaline crystals was recognized in Ceylon and India from time immemorial. Ancient people had observed that when this crystal was heated in a fire, the ashes were first attracted and then repelled. The phenomenon of pyroelectricity was fascinating enough to hold the attention of such notables as Lord Kelvin, A. C. Becquerel, and Pierre and Jacques Curie. Lord Kelvin was responsible for adopting the point of view that a pyroelectric crystal behaved as if it had a spontaneous polarization whose magnitude varied with temperature. In more modern times, in particular during the last ten years, the search for infrared detectors and photomixers[(19)] has rekindled a great interest in pyroelectrics. To a very great measure, work has largely been directed toward ferroelectric pyroelectrics, the reason being that the effect is so much larger in these systems. The pyroelectric effect in polar crystals is also useful for the study of spontaneous polarization,[(20a,20b)] ferroelectric domain reversal, phase transitions, optical absorption, and optical relaxation.[(21)]

One can distinguish between two types of pyroelectricity, primary and secondary. In the first case the crystal is prohibited from expanding with temperature. In other words, it is clamped. If the crystal is allowed to expand freely as heated, the expansion causes an additional polarization because of the piezoelectric effect. This is the secondary contribution. This polarization is added to the first and the resultant is observed. As mentioned in Section 2, there are ten crystal classes that theoretically can show pyroelectricity under uniform heating or cooling. They are 1, 2, 3, 4, 6, *m*, *mm*2, 3*m*, 4*mm*, and 6*mm* (Ref. 3, pp. 78–80).

For applications of pyroelectrics it is desirable to be able to alter the material in various ways. One searches for methods to increase or decrease the pyroelectric coefficients, stabilize the crystal

* See Ref. 4, pp. 1–8, 699–712.

against the effects of depoling, enhance the sensitivity and induce new phenomena by doping, and produce pyroelectric effects in more or less amorphous materials such as plastics.

In this section we shall examine how the above can be accomplished, by discussing in detail certain selected examples. The methods of altering the materials will include stoichiometry variations, impurity additions, and mechanical treatments.

3.2. Pyroelectric Measurements

We shall define the pyroelectric coefficient p by the following equation:

$$\Delta P = p\,\Delta T \tag{26}$$

where ΔP is the change in polarization and ΔT is the change in temperature. For an infinitesimal change

$$p = dP/dt \tag{27}$$

There are two excellent ways for making a quantitative study of the pyroelectric effect. They are (1) the dynamic technique of Chynoweth[20a] and (2) the continuous charge integration method of Glass.[20b]

In the first, the dynamic technique of Cynoweth, the crystal is pulse heated with a modulated light beam. The pyroelectric signals are square waves and the current is given by

$$i \propto (dP/dT)(dT/dt) \tag{28}$$

If the modulating frequency is high enough, the rate of heating will depend only on the specific heat C and the current will be

$$i \propto (1/C)(dP/dT) \tag{29}$$

The dynamic technique is well suited for the study of the variation of dP/dT with temperature, since it has a very high resolution. Absolute measurements are only possible by quantitative measurement of the flux absorbed, and this usually renders the technique less accurate than charge integration. The concept of a pyroelectric detector for infrared radiation follows naturally from the dynamic technique. Figure 8 shows such a detector utilizing pyroelectric polyvinylidene fluoride films (see Section 3.6) as the active element. Since the rate of heating is a constant,

$$V \propto R\,dP/dt \tag{30}$$

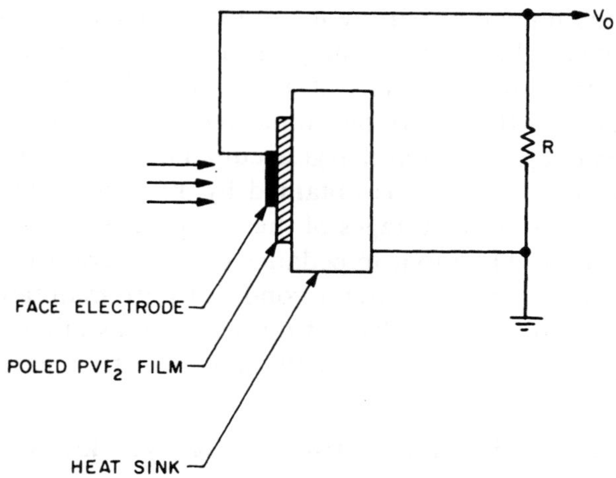

Fig. 8. PVF_2 pyroelectric detector. From Ref. 32.

In the second method, the charge developed by heating the crystal is continuously integrated by means of an operational amplifier; the experimental setup is shown in Fig. 9. The output voltage is proportional to the change in polarization between the two temperatures T_1 and T_2 over which the crystal is heated. This may be expressed by the following relation:

$$V_{\text{out}} = [\Delta P]_{T_1}^{T_2}/C \tag{31}$$

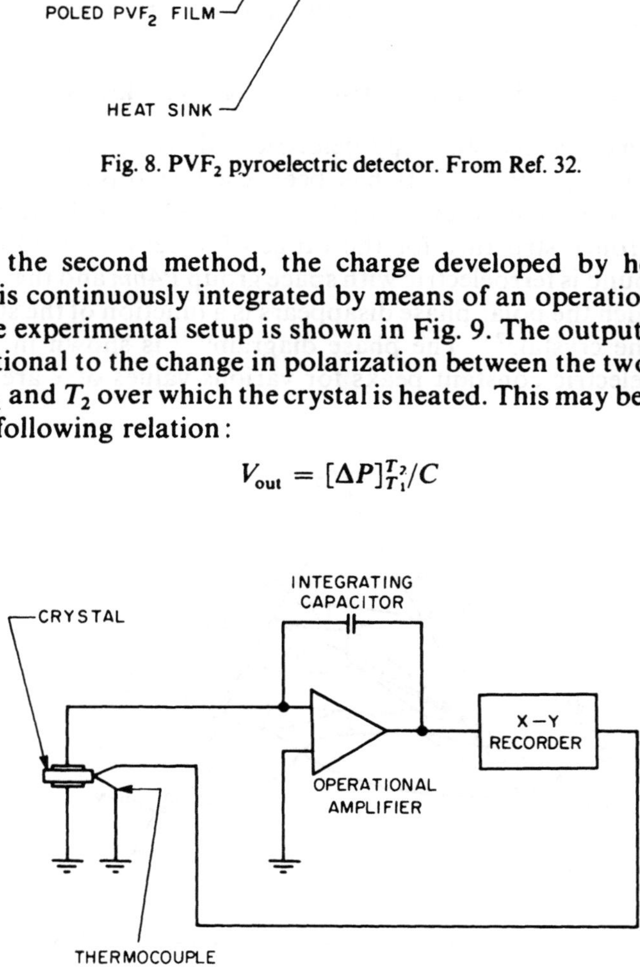

Fig. 9. Continuous charge integration technique for pyroelectric measurements. From Ref. 20b.

where C is the magnitude of the integrating capacitor of the operational amplifier. The output voltage from the operational amplifier can be connected to the y axis of an xy recorder while the x axis can be driven by a thermocouple connected to the crystal. This gives a very convenient plot of total charge accumulated versus temperature. The pyroelectric coefficient is obtained by differentiating the curve. There are two major advantages of the charge integration technique just described: (1) The input impedance of the operational amplifier is essentially zero, so short-circuit conditions are maintained across the crystal; (2) polarization discontinuities, such as might occur at a first order phase transition, are easily and correctly measured.

3.3. Effect of Stoichiometry in Strontium Barium Niobate

A very successful pyroelectric detector material is $Sr_{1-x}Ba_xNb_2O_6$ (SBN).[22] It has been used at room temperature with a response time of less than 1 nsec.[23] SBN has the tetragonal tungsten bronze structure for the range of x between 0.25 to 0.75. The compound is ferroelectric with space group *P4bm* and the temperature at which the polar phase disappears is a function of the stoichiometry of the crystal.[24] The phase diagram[25] is shown in Fig. 10 and the dielectric constant peaks for various values of x are shown

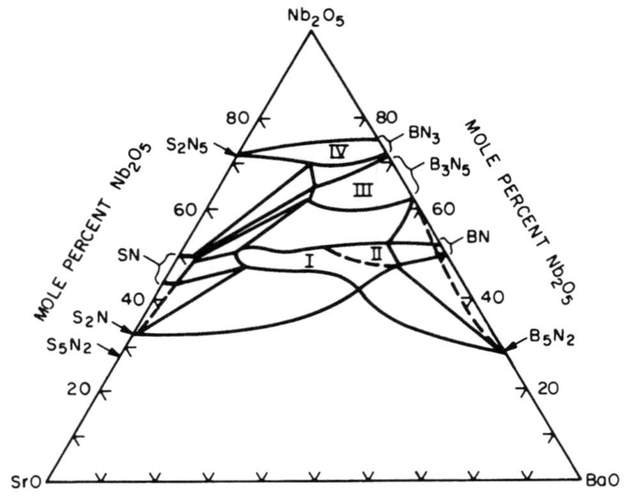

Fig. 10. Ternary phase relationships in the room-temperature isotherm of the system $SrO–BaO–Nb_2O_5$. From Ref. 25.

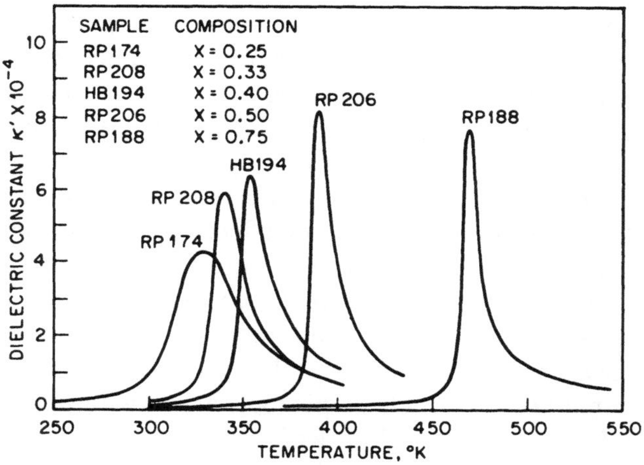

Fig. 11. Dielectric constant as a function of temperature for various compositions of SBN. From Ref. 20b.

in Fig. 11. This peak is an approximate indication of the polar to nonpolar transition temperature. We observe that the transition temperature rises with barium content. This we easily understand from Eq. (25).

By making use of the charge integration technique described in Section 3.2, we can study the change in ΔP with temperature. These results are plotted in Fig. 12 for various values of x. We note

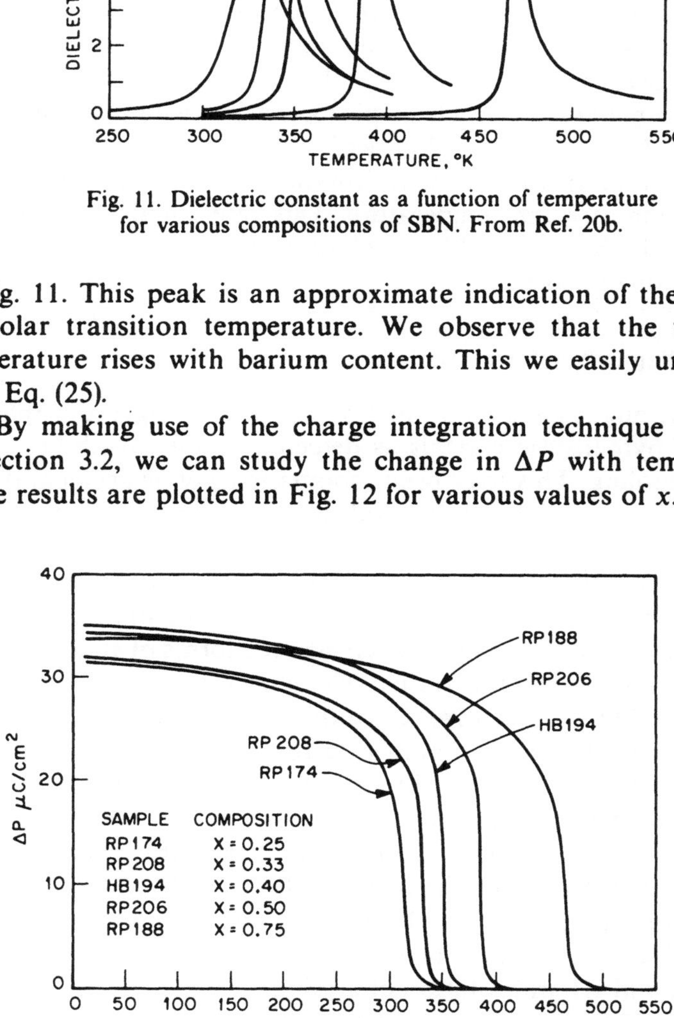

Fig. 12. The change in polarization of various SBN compositions as a function of temperature. From Ref. 20b.

that in general the curves are all of similar shape and tend very rapidly to zero near the transition temperature. This shape seems to be characteristic of many pyroelectrics. One aspect of a good pyroelectric detector is a large value of dP/dT at the temperature of operation, which is usually room temperature. Clearly, then, to maximize dP/dT, one chooses an SBN crystal with x near 0.25. As a general rule, if varying stoichiometry affects the transition temperature, to maximize dP/dt, move the transition temperature as close as possible to the ambient temperature. Of course, a large value of dP/dt may not necessarily be the only requirement for the optimum detector. Other properties such as electrical conductance and dielectric constant or phenomena such as piezoelectric oscillations may also play a role (Ref. 19, pp. 251–280; Ref. 26).

An additional problem with low-Ba-content SBN crystals is a slow depoling at room temperature. This rate could be as large as 7% per day. In the case of SBN depoling can be prevented by the application of a small bias field.

3.4. Depoling in Triglycine Sulfate

As one might guess from the last section, the biggest problem in the use of pyroelectric crystals as detectors is the possibility of the crystals depoling. This may occur spontaneously or accidentally. In the latter case this might be the result of overheating or the accidental application of a reverse bias field. Studies have shown that it is possible to inhibit this in triglycine sulfate (TGS) by the addition of alanine to the crystal.[27] Figure 13 shows the structure of the glycine molecule and both the right- and left-handed forms of alanine.

Below the Curie temperature, which occurs at 49.4°C, triglycine sulfate crystallizes[28] into space group $P2_1$. Above the phase transition the symmetry is $P2_1/m$. The ferroelectric crystal structure is shown in Fig. 14. There are three types of glycine molecules in the unit cell and those designated I and III are completely planar if the hydrogen atoms are ignored. Type II is only partially planar since the nitrogen is displaced from the plane of the carbons and the oxygens. The structure contains a complicated hydrogen-bonding scheme and the ferroelectric switching involves exchange of protons between glycines II and III and rotation of glycine I.

When TGS crystals are grown containing about 0.1% alanine (these are called ATGS) the hysteresis loops are very strongly affected. Figure 15 shows typical curves of pure and doped crystals. One notes

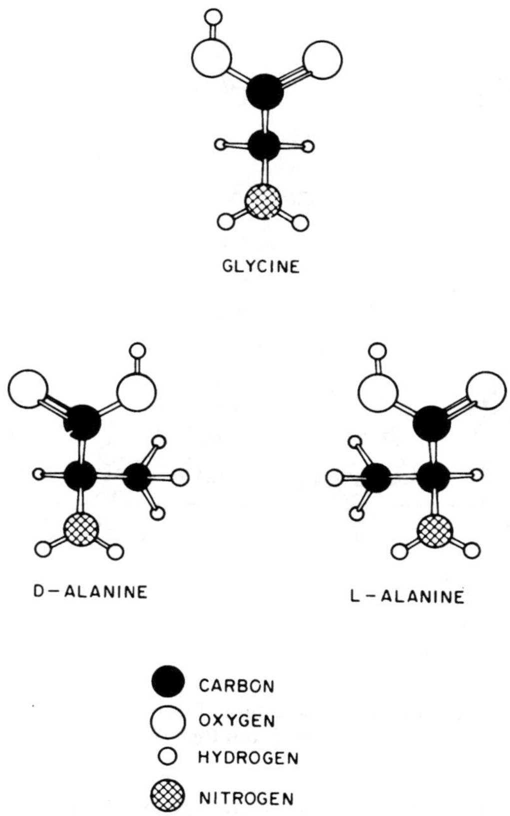

Fig. 13. Structure of glycine and the right- and left-handed forms of alanine.

that the doped loops are strongly biased. This means that at zero applied field the crystal is in a state of maximum polarization. Crystals containing alanine will retain this self-bias even after prolonged heating or after the application of a large reverse bias electric field. In fact, when crystals were held for several hours at 60°C (which is above the Curie temperature) they were found to repole themselves upon cooling.

The mechanism for the stabilization is a most interesting one.[29] The principal reversible dipole in the ferroelectric structure is glycine I. This molecule rotates through a large angle when the reversal of the polarity occurs, and this is illustrated in Fig. 16. Since the symmetry of the molecule is high, the rotation changes the molecule into its mirror image.

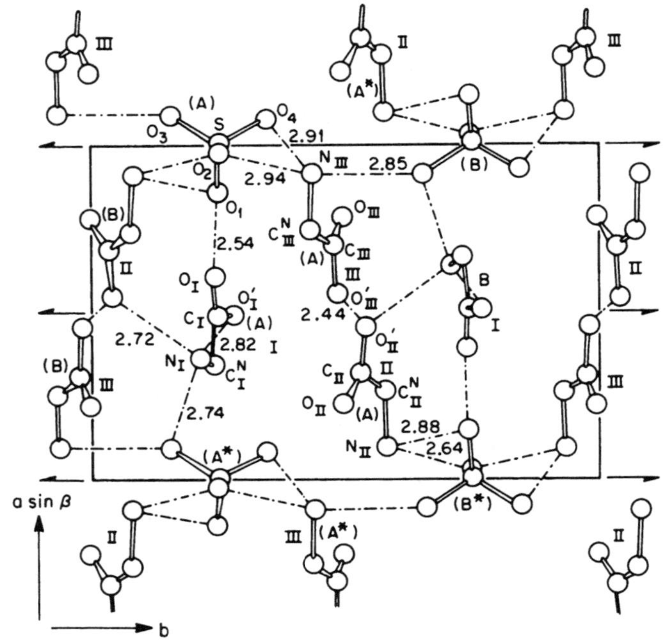

Fig. 14. Projection of the structure of triglycine sulfate along the *c* axis. From Ref. 28.

Suppose now that an L-alanine molecule replaces one of the glycine molecules. Due to the lack of equivalence of the CH_3 group and the hydrogen, there is only one way that the molecule can fit into the lattice and satisfy the hydrogen bonding. It is also clear that if a number of L-alanine molecules are incorporated, they all must have the same orientation, for the same reason. When the alanine sub-

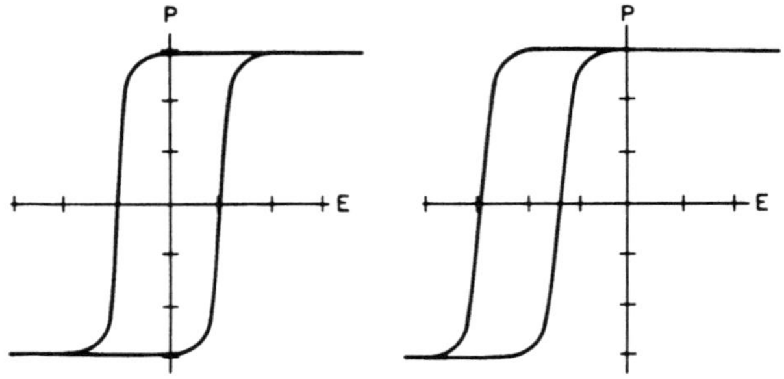

Fig. 15. Hysteresis loops of normal TGS and ATGS. From Ref. 27.

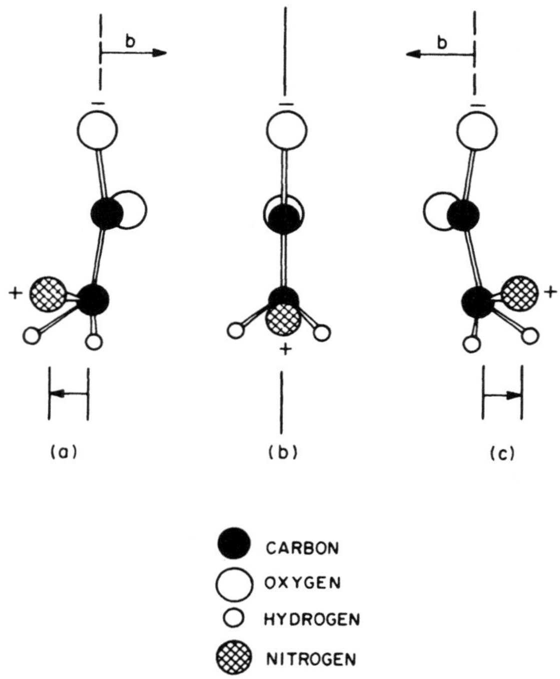

Fig. 16. Glycine I molecule shown in (a) one polar state, (b) nonpolar state, and (c) state with polarity opposite to (a). From Ref. 29.

stitutes for a glycine there must be some lattice strain due to the larger CH_3 group. Again, when a number of alanines are substituted, a correlated distortion must occur in the lattice and a polar strain field is generated. It is convenient to look upon this as a dipolar field with a long range of influence. Since the alanine molecule cannot rotate into its mirror image, the dipolar field is irreversible and permanent. Thus we have a bias on the polarization and a large degree of stability.

3.5. Transition Metal Doping of Lithium Tantalate

The doping of pyroelectric crystals with transition metal impurities leads to several new and interesting phenomena. Considerable work has been done on lithium tantalate containing copper and chromium impurities below the Curie temperature.[21] Lithium tantalate has the same space group as lithium niobate, *R3C*. The most obvious effect of the impurity ion is to increase the absorption coefficient for the incoming radiation. If the ion is excited by an optical pulse $P(t)$, after relaxation of the ion the lattice will be heated by an amount ΔT. The resulting polarization is

$$P = p\,\Delta T \tag{32}$$

Figure 17 shows an approximate energy level diagram for chromium in lithium niobate. While the true symmetry of the ion is C_3, the

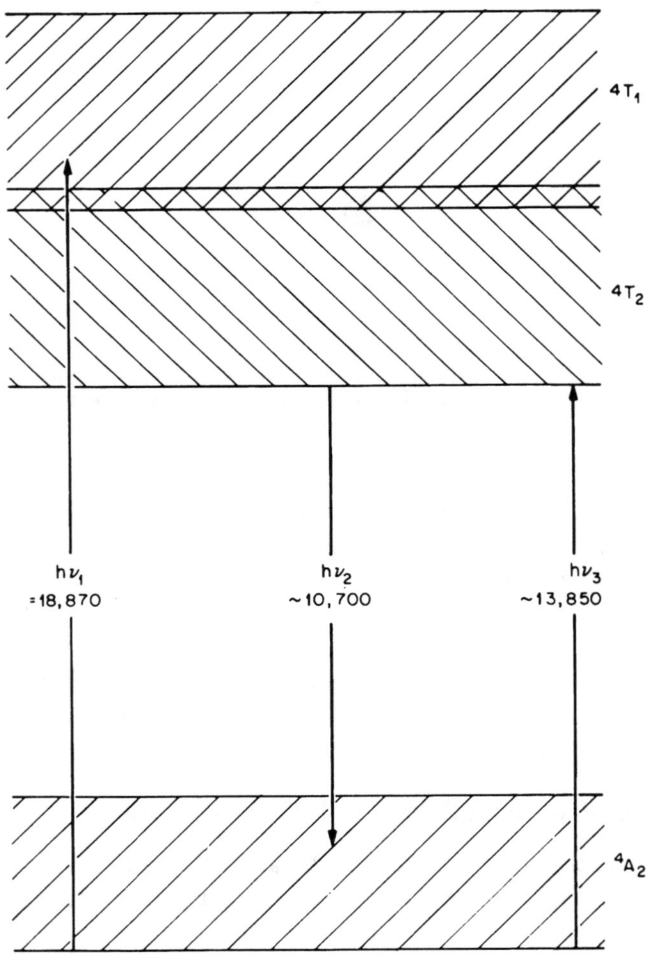

Fig. 17. Simplified energy level diagram of Cr^{3+} in $LiNbO_3$ and $LiTaO_3$. From Ref. 21.

state designations for O_h are used. The optical absorption is the $^4A_2$ to $^4T_1$ transition. Nonradiative relaxation to the $^4T_2$ occurs and the lattice is heated. Emission of the Stokes-shifted radiation at 10,700 cm^{-1} or nonradiative decay to the ground state follows. If the decay is nonradiative, further heating occurs.

The current response per unit area can be expressed as

$$i(t) = epP(t)/c \tag{33}$$

where c is the heat capacity and e the fraction of the incident radiation absorbed. The absorbed radiation can be written as:

$$e = R[1 - \exp(-\alpha d)](1 - \eta v_2/v_1) \tag{34}$$

where R is a constant that accounts for reflection loss, α is the absorption coefficient, d is the thickness, and η is the quantum efficiency.

The pyroelectric response time for this mechanism will be limited by the relaxation time of the excited state. For Cr^{3+} this will be of the order of microseconds. It is obvious that a great deal can be learned about relaxation processes in transition-metal-doped pyroelectric crystals by studying the polarization produced after a pulse.

A second effect can also occur and this is probably the most interesting. Since the host crystal is acentric, the transition metal ion may have a dipole moment in both its ground and excited states. In fact, upon excitation, not only will there be a redistribution of charge at the impurity site, but also at the ligands and the surrounding lattice due to the dipole on the impurity. Hence

$$P = N[\mu(\text{excited}) - \mu(\text{ground})] \tag{35}$$

where N is the density of impurity ions and μ specifies the resulting moment of the total charge redistribution near the center. The optical excitation of the impurity gives rise to an instantaneous macroscopic polarization, while that due to the pyroelectric effect develops during nonradiative decay. Thus this effect may be used at frequencies above the reciprocal relaxation time of the excited state, while the pyroelectric mechanism may be used below this frequency. In this regard it is useful for optical mixing and rectification.[30] Figure 18 illustrates a remarkable example of optical rectification. A sample of $LiTaO_3:Cu^{2+}$ is bonded to a larger crystal of undoped $LiTaO_3$. The Cu^{2+} ions are excited by ~ 16 psec pulses at 1.06 μm. The "transducer" (i.e., the Cu-doped $LiTaO_3$) generates a voltage pulse by optical rectification. This pulse propagates through the undoped crystal and is made "visible" by means of the electrooptical effect and

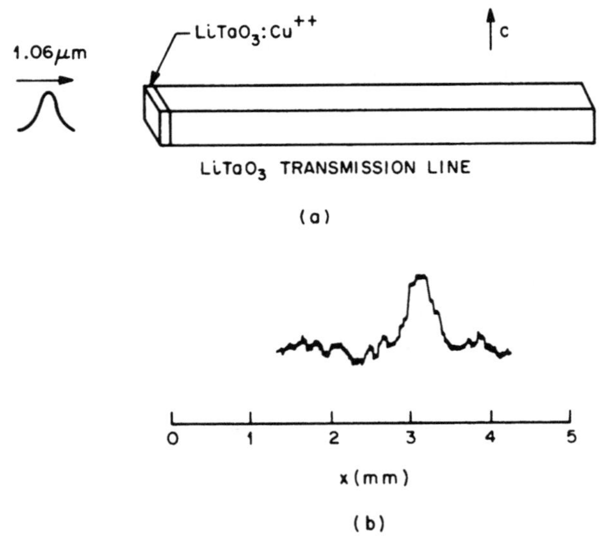

Fig. 18. (a) Electrooptic transmission line with $LiTaO_3$:Cu^{2+} transducer bonded to one end. (b) Induced birefringence in the transmission line due to the propagating electrical pulse. From Ref. 30.

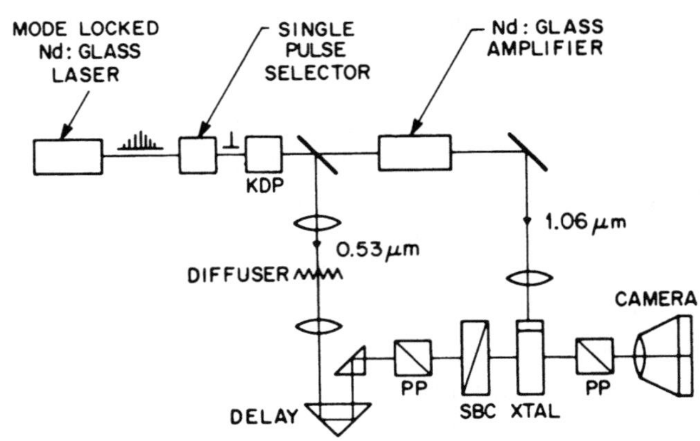

Fig. 19. Schematic of experiment used to generate and measure the electric pulses. SBC is a Soleil Babinet compensator. PP is a polarizing prism. The KDP crystal was used for second harmonic generation. From Ref. 30.

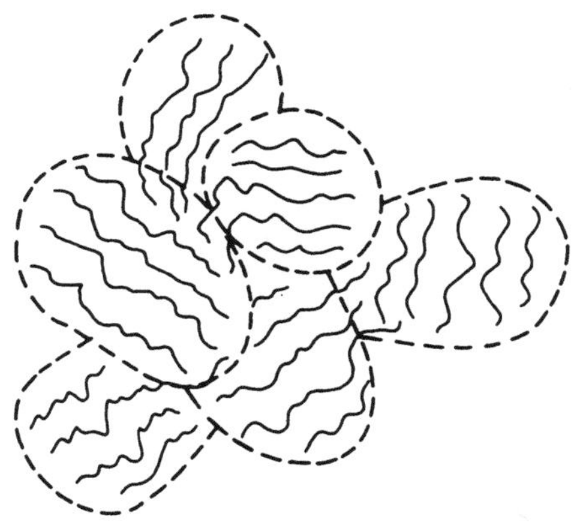

Fig. 20. A largely crystalline polymer showing crystallites. See also Ref. 31.

a delayed illumination by a 0.53-μm pulse. The experimental setup is shown in Fig. 19.

3.6. Pyroelectric Polyvinylidene Fluoride Polymer

Polymers are frequently used in industry as structural materials, for example, as insulation in high-frequency coaxial cables. X-ray diffraction shows some polymers to have a considerable degree of order. There are large crystalline regions and between them amorphous zones.[31] Figure 20 shows a schematic diagram of crystallites in a largely crystalline polymer. The temperature T_g below which this property sets in is called the glass temperature. There is also a higher temperature T_m at which the crystallites melt. If a largely crystalline

Fig. 21. An oriented crystalline polymer produced by drawing in the horizontal direction. See also Ref. 31.

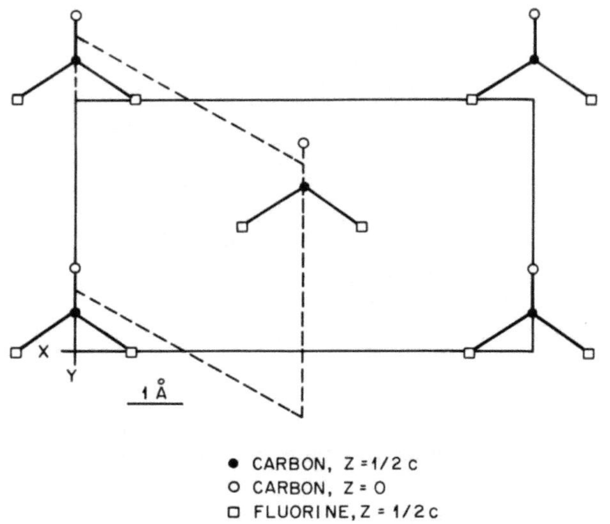

Fig. 22. Structure of PVF_2 drawn at 50°C. The projection is onto the *ab* plane. From Ref. 33.

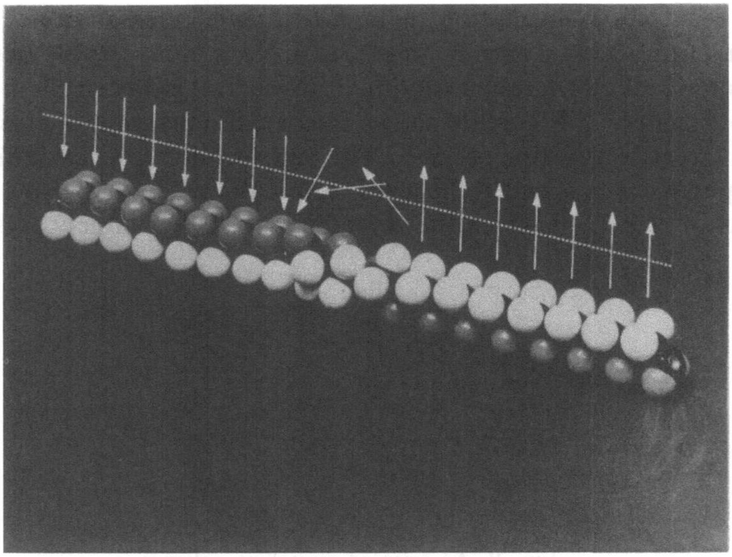

Fig. 23. PVF_2 molecular model illustrating a 180° rotation of molecular dipoles about the polymer chain axis. From Ref. 32.

polymer is subjected to a stress and heated above T_g so that some flow can occur, the material will elongate. The crystallites will line up along the direction of the stress and this will produce an oriented crystalline polymer. Figure 21 shows such a polymer.

It is possible to induce pyroelectric effects in polyvinylidene fluoride (PVF_2) films.[32] If a sheet of this material is stretched to several times its length at a temperature of about 100°C, the material becomes an oriented polycrystalline film. The space group for the crystals is *Cm*2*m* with two polymer units in the cell.[33] The repeat unit is CH_2CF_2 and the cell constants are $a = 8.47$ Å, $b = 4.90$ Å, and c (along the chain axis) = 2.56 Å. The space group, *Cm*2*m*, is noncentrosymmetric with point symmetry *mm*2. Thus a polarization is possible. Figure 22 shows the structure of polyvinylidene fluoride projected onto the *ab* plane. The net moment of a unit cell can be most easily visualized as the sum of the CH and CF dipoles. Clearly, the net moment is perpendicular to the polymer chain.

There is no reason to believe that the moments of each unit cell in the oriented polymer necessarily point in the same direction. In fact, it is very likely that the randomness is such as to give a zero net moment. It is also possible that the moments along a chain may point in different directions. Figure 23 shows a PVF_2 molecular model illustrating a 180° rotation about the polymer axis.[34]

Since there will be in general a zero net moment, poling is necessary to make the film pyroelectric. The procedure is to evaporate aluminum contacts onto both sides of the film. An electric field of

TABLE 1

Birefringence and Pyroelectric Coefficient of Poled PVF_2 Films as a Function of the Method of Film Preparation*

Description	$n_y - n_z$	$n_x - n_z$	Pyroelectric coefficient, $nC/cm^2\,°C$
14 μm (biaxially stretched)	0.018	0.011	2.4 ± 0.7
19 μm (biaxially stretched)	0.019	0.011	2.4 ± 0.7
50 μm (uniaxially stretched in the *x* direction)	0.000	0.013	0.7 ± 0.2
125 μm	0.000	0.000	0.3 ± 0.1

* From Ref. 32. Indices measured at 6328 Å with an Abbé refractometer; $n_{av} = 1.425$; n_z corresponds to the film normal or polar axis while n_x and n_y refer to the orthogonal directions in the plane of the film having the smallest and largest indices, respectively. All films were obtained from Kureha Chemical Co., Tokyo, Japan.

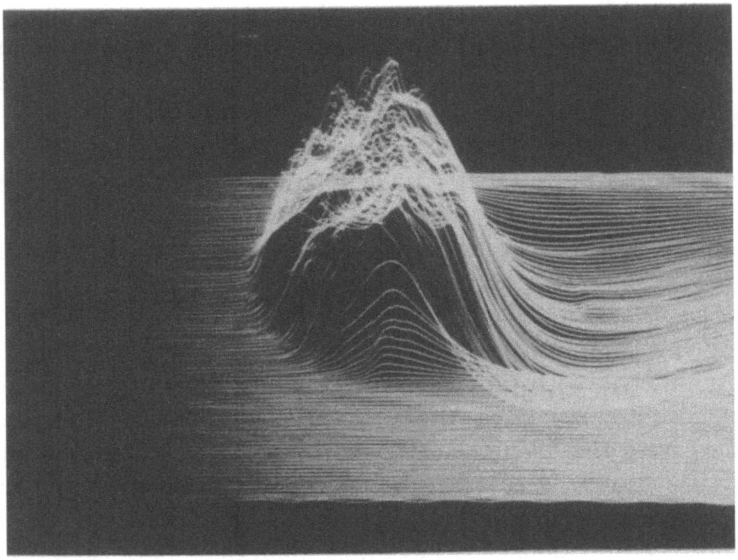

Fig. 24. Pyroelectric scanning displays of a PVF_2 film having a poled central region. From Ref. 34.

about 10^6 V/cm is applied while the material is maintained at about 120°C. The film is then cooled with the field on. The induced moment is found to be permanent. Some typical values for the birefringence and pyroelectric coefficient are indicated in Table 1. The values for the pyroelectric coefficient are comparable to those found for lithium niobate.

An excellent way to study the polarization characteristics of PVF_2 films is by means of pyroelectric scanning. Figure 24 shows the response of a pyroelectric detector in which only the central 2-mm-diameter region was poled. The detector was scanned with a He–Ne laser beam. It is evident that the response is quite nonuniform and also that there is no response outside the poled region, suggesting that there is no domain structure.

4. Nonlinear Optical Materials

4.1. Introduction

The interaction of light with electric and magnetic fields as it passes through crystals, glasses, and liquids has been known for close to 100 years.* Around 1845 Michael Faraday showed that the plane of polarization of a light wave can be rotated when it propagates through a material in a magnetic field. In about 1875 John Kerr found a similar effect for electric fields. In this case the refractive index of the material was slightly altered when electric fields of sufficient strength were applied. These phenomena can effectively be described as the interaction of a light wave with a "dc" field. By this we mean that the optical properties of the material are being altered rather slowly in comparison to the optical frequency. These effects, however, are suggestive of more exotic phenomena in which two light beams interact or a single beam interacts with itself. Here we have the high-frequency "ac" electric or magnetic field changing the material at an optical rate.

It was not until the advent of lasers that sufficient field intensity could be achieved to allow the production and observation of these more interesting effects. In 1961 Franken and co-workers[38] at the University of Michigan produced the first optical harmonic generation. An intense beam of ruby laser light at a wavelength of 6943 Å was focused onto a quartz crystal. A second harmonic beam in the ultraviolet was generated.

The method by which this conversion takes place is not difficult to understand. When low-intensity light propagates through a crystal the electric field of the beam polarizes the material. The magnitude of the oscillating polarization is proportional to the electric field. Thus

$$P = \alpha E \tag{36}$$

and if

$$E = E_0 \cos \omega t \tag{37}$$

then

$$P = \alpha E_0 \cos \omega t \tag{38}$$

The oscillating charges radiate and their only effect is to slow the propagation of the light. On the other hand, when the crystal is exposed to very high-intensity laser light the effect is quite different. Since the electric fields are large ($> 10^7$ V/cm), the polarization is no longer directly proportional to the electric field. In a piezoelectric crystal like quartz it can be expressed as

$$P = \alpha E + \beta E^2 \tag{39}$$

* References 35–37 are interesting review articles.

Again letting $E = E_0 \cos \omega t$, we get

$$P = \alpha E_0 \cos \omega t + \beta (E_0{}^2/2)(1 + \cos 2\omega t) \tag{40}$$

Thus we see that the polarization has a component oscillating at *twice* the optical frequency of the light. These oscillating charges radiate at both ω and 2ω and thus ultraviolet radiation is generated from the red beam. This sort of harmonic conversion is very common at both radio and audio frequencies and is sometimes called second harmonic distortion.

Because of optical dispersion, the fundamental beam and the second harmonic will usually travel at different speeds in a crystal. To convert as much of the fundamental beam to the second harmonic as possible, it is necessary that both travel at the same speed. This is called phase matching and is an important key to the design of efficient nonlinear[39] devices. In other words, phase matching requires that the index of refraction of the fundamental and harmonic beams be the same. This may be accomplished by finding a direction in the crystal such that the refractive index of the ordinary wave at ν_1 is equal to the refractive index of the extraordinary wave at ν_2. This is often possible but unfortunately the two beams tend to separate because the wavefronts will not usually be parallel. This leads to a rather short interaction length and less than optimum conversion.

Sometimes one is able to pick a path through the crystal so that the two beams remain collinear and the velocities are the same as well. In a uniaxial crystal, which is the most useful for many nonlinear applications, this is at 90° to the symmetry axis.[40] Then, by varying the temperature, it may be possible to adjust the two refractive indices such that the two beams travel at the same speeds and for appreciable distances (cm).

Of the 32 crystal classes, 11 are centrosymmetric and are not useful for nonlinear applications.[41–43] However, a centrosymmetric crystal may be made acentric by the application of a stress and this is the basis of the acoustooptic effect in which acoustic and optical waves interact (Ref. 3, pp. 241–258; Ref. 44). Of interest to us in this chapter will only be nonlinearities which occur in unstrained crystals. Of the remaining 21 noncentrosymmetric classes, cubic class 432 has "so much symmetry" that it has no nonlinear effects. Thus there are 20 potentially useful classes. Cubic crystals in general cannot be phase matched and so they are not candidates for harmonic generator and similar applications. There are 13 uniaxial crystals that can be used for second harmonic generation. They are: $\bar{6}2m$, $6mm$, 622, $\bar{6}$, 6,

3*m*, 32, 3, $\bar{4}2m$, 4*mm*, 422, $\bar{4}$, and 4. Two of these, 622 and 422, obey a special symmetry called Kleinman's symmetry[45] and almost always show small nonlinear effects.

Although not suitable for harmonic generator applications, the cubic nonlinear crystals, since they have zero birefringence, are excellent for modulators. By a modulator we mean a nonlinear device that adds radio or microwave frequency sidebands to the optical carrier. Phase matching is not ordinarily required and the zero birefringence makes adjustments very noncritical.

It is obvious from Eq. (39) that two optical beams of different frequencies can be "mixed" (heterodyned) to form the sum and difference frequencies. This is the principle behind the optical parametric oscillator and amplifier. Another phenomenon is parametric fluorescence. This is the spontaneous decomposition of a beam at frequency ν_3 into two beams such that $\nu_1 + \nu_2 = \nu_3$. We shall have occasion to mention this in Section 4.3.

Because of the high optical fields employed in all nonlinear devices, optical damage frequently occurs.[46] One aspect of this is an induced refractive index change that may happen at even fairly low power densities ($< kW/cm^2$). Until recently the solution to this problem seemed insurmountable. It now appears to be understood in at least some materials and can probably be controlled.

It should be clear from the previous discussion that the solid-state chemist can well be faced with a number of problems in the preparation of nonlinear materials. Some of these are improvement of optical homogeneity, adjustment of phase match temperature, and reduction of optical damage.

We shall illustrate how the above can be accomplished by reference to lithium niobate. This material is of such technological importance that it has been studied very extensively. It is fortunate indeed that so many facets of crystal chemistry can be illustrated by a single material.

4.2. Techniques for Optical Homogeneity Determination

There are three techniques that are useful for studying homogeneity of nonlinear optical crystals. The first is to visually map the birefringence as a function of crystal position.[47] In the case of a uniaxial crystal, which shall be of major concern to us, the simplest procedure is to examine it between crossed polarizers. An optical beam polarized at 45° to the optical axis will excite both ordinary and extraordinary waves and the phase shift between them will be determined by the birefringence and path length. After passing through the crystal the two beams combine to form an elliptically polarized beam. When it passes through a second polarizer, intensity variations are produced that are a function of the birefringence.

To simplify the analysis of the resulting interference pattern, it is advantageous to polish the crystal in the form of a slight wedge. This causes a series of dark bands to appear in the fringe pattern and the distortion of the bands helps locate the regions of poor homogeneity. This method of testing a crystal is quite effective in quickly determining its overall quality. It is also useful in making a crude assessment of the effects of melt composition.

The second procedure[48] is to quantitatively measure the birefrigence as a function of crystal position by means of an optical compensator. Assume again the polarized incident beam is at 45° to the optical axis and the second polarizer is crossed with respect to the first when no crystal is present. Now the optical compensator can be set to give either a maximum or minimum in the transmission. If Γ is the net phase retardation induced by the compensator, then the birefringence δB is given by

$$\delta B = -B(\delta h/h) - (\delta \Gamma/h) \tag{41}$$

where h is the thickness. Figure 25 shows a typical mapping for a crystal of lithium niobate.

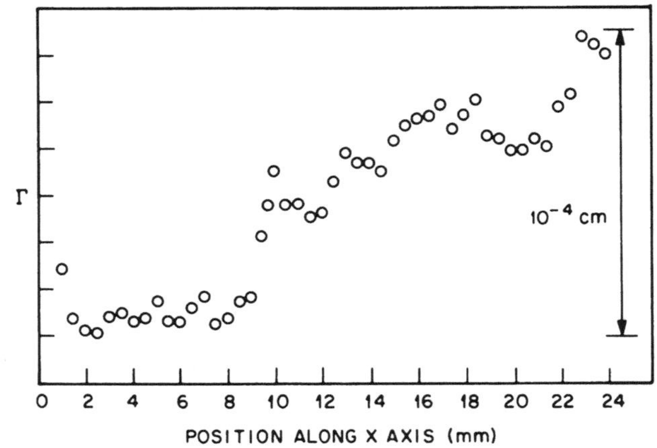

Fig. 25. Birefringence mapping at 0.63 μm along the X axis for a lithium niobate crystal grown from a stoichiometric melt. From Ref. 48.

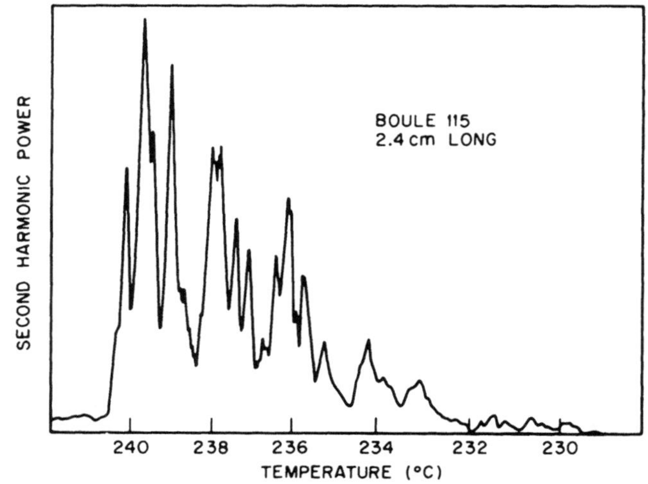

Fig. 26. Phase-match temperatures for a rather poor lithium niobate crystal. The fundamental beam was at 1.15 μm. From Ref. 49.

The third way (perhaps the most elegant but not necessarily the most informative) is to study optical inhomogeneities by means of second harmonic generation.[48] As mentioned in Section 4.1, the best way to make a harmonic generator is to have both the fundamental and harmonic beams travel normal to the optical axis. Phase matching is achieved by adjusting the temperature. Now if the crystal is inhomogeneous, then the phase match temperature will vary throughout. Thus if the path length is fairly long, there will be a very complicated dependence of second harmonic power on temperature. Figure 26 shows this sort of effect. Instead of the second harmonic power peaking at one temperature, there are a large number of peaks. It is obvious that this method is a good qualitative appraisal of crystal perfection. Also, by looking at thin sections, the phase match temperature throughout the crystal can be studied.

4.3. The Method of Parametric Fluorescence

When laser radiation impinges on a nonlinear optical crystal the incoming photon ν_3 can decompose into two beams such that $\nu_1 + \nu_2 = \nu_3$. This sort of spontaneous decomposition is called parametric fluorescence.[49] This emission can be obtained without the complications of optical resonators and at power levels much

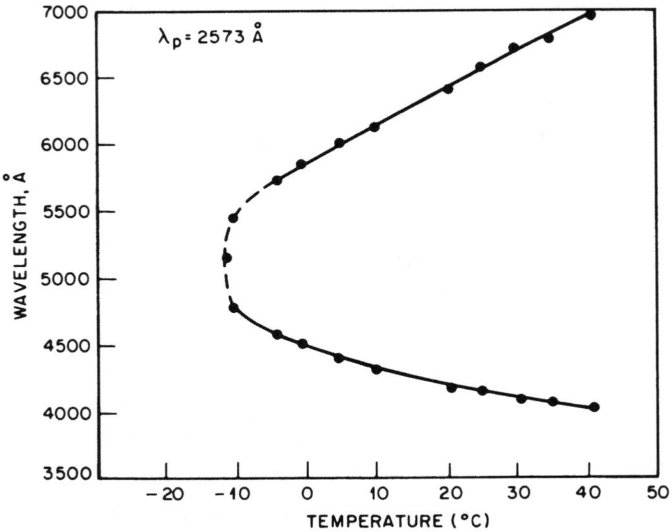

Fig. 27. Temperature tuning of ADP using the parametric emission technique. From Ref. 49.

below those needed for parametric oscillation. It is a very useful tool for determining optical nonlinearities and also for obtaining temperature and angular tuning curves of nonlinear materials. It has been a rather neglected technique that the solid-state chemist may well wish to have in his "bag of tricks." Figure 27 shows a temperature tuning curve for ADP (ammonium dihydrogen phosphate).* Note that the pump frequency ν_3 is 2573 Å and the fluorescence ν_1 ranges from ~7000 Å to ~5100 Å while the fluorescence ν_2 ranges from ~4000 Å to ~5100 Å. When $\nu_1 = \nu_2$ the system is called degenerate. If we define the degenerate frequency to be ν_0, then

$$\nu_1 = \nu_0(1 - \gamma) \tag{42}$$

$$\nu_2 = \nu_0(1 + \gamma) \tag{43}$$

$$\nu_3 = 2\nu_0 \tag{44}$$

where γ measures the departure from degeneracy. Combining Eqs. (42) and (43), we can show that

$$\gamma^2 = \left[\frac{(\nu_1/\nu_2) - 1}{(\nu_1/\nu_2) + 1}\right]^2 \tag{45}$$

* M. M. Dowley. See Ref. 49.

Now Boyd and Ashkin[50] have shown that

$$\gamma \sim C(T - T_0)^{1/2} \tag{46}$$

where C and T_0 are constants. We have

$$\gamma^2 = \left[\frac{(\nu_1/\nu_2) - 1}{(\nu_1/\nu_2) + 1}\right]^2 = C^2(T - T_0) \tag{47}$$

Thus a plot of γ^2 versus T is linear. We shall make use of this result in Section 4.6.

4.4. Index Damage Measurements

There are two techniques that can be used for a quantitative analysis of laser-induced index damage. In the first[51] a rather intense beam (the damaging beam) of either green or blue laser light is focused onto the crystal. This causes a change Δn_e in the extraordinary index of refraction, which can be probed by means of a second laser beam. The wavelength of this beam is usually in the red to minimize the production of further damage.

The red beam is polarized at 45° to the optical axis. In the absence of the green or blue damaging beam the red light after passing through the crystal is elliptically polarized. The exact character of the polarization depends upon the natural birefringence of the crystal. A compensator is used to convert the elliptically polarized beam back to a linearly polarized beam and a second polarizer (analyzer) is set to null the light.

When the green or blue beam is turned on, the resulting Δn_e upsets the null. The transmission through the analyzer becomes

$$T = \sin^2(\pi/\lambda \cdot \Delta n_e) \tag{48}$$

Thus from measuring the increase in transmission an estimate can be made of Δn_e. The sensitivity of this method is such as to allow a change in Δn_e of the order of 10^{-5} to be detected. This method of study has the advantage that high powers can be employed in the damaging beam. Consequently it gives a direct appraisal of the usefulness of a crystal for nonlinear applications. There may be certain difficulties in interpretation because the thermally induced birefringence change may exceed the laser damage.

A much more accurate method[52,53] for the quantitative measurement of index damage is the holographic technique. In the previous method the change in birefringence was measured, while in the holographic technique the index change is determined. The

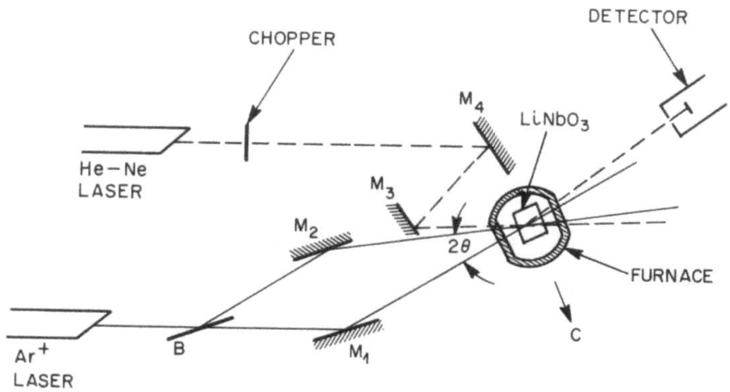

Fig. 28. The holographic method of laser damage measurement. From Ref. 66.

experimental setup is shown in Fig. 28. The two beams from the 5145-Å argon laser intersect in the crystal and since the beams are coherent, the interference pattern gives the following intensity distribution:

$$I \propto \cos^2[(\pi/L)x] \tag{49}$$

where $2L \sin \theta = \lambda_0$ and 2θ is the angle between the beams. A refractive index variation proportional to the intensity is formed in the crystal. The diffraction efficiency of this grating (i.e., hologram) can be expressed by

$$\eta = \sin^2[\pi(\Delta n)d/\lambda \cos \theta] \tag{50}$$

where Δn is the modulation amplitude of the index change and d is the crystal thickness. During measurement the 5145-Å writing beam is σ polarized and the reading beam at 6328 Å is π polarized. The C axis of the crystal is normal to the grating planes. With the holographic method a Δn of about 10^{-7} can be measured. Thermal effects are unimportant but vibrations may cause some difficulty. The greater apparent sensitivity may also be offset by the lower power densities.

4.5. Optical Homogeneity

Perhaps the most difficult problem that the solid-state chemist must deal with in the preparation of nonlinear optical materials is homogeneity. Since these materials are so easily polarized, the electric field associated with defects causes substantial changes in refractive indices. Extensive studies have been made of the homogeneity problem in lithium niobate and the solution is now well in hand. We shall examine this important material in some detail.

The crystal structure of lithium niobate has been determined with substantial accuracy by Abrahams and co-workers.[54] The compound crystallizes into the noncentrosymmetric space group *R*3*c* and the point symmetry is 3*m*. There are six molecules per hexagonal (two per rhombohedral) unit cell. The lithium and niobium atoms occupy sites with symmetry three, while the oxygen atoms lie in sites with symmetry one. Some of the physical properties of $LiNbO_3$ and the isostructural compound $LiTaO_3$ are listed in Table 2. Figure 29 shows the structure. At the Curie temperature, which occurs at about 1200°C, the center of symmetry is lost and the space group is $R\bar{3}$. Unlike many ferroelectrics, the displacements of the atoms from the positions in the centrosymmetric phase are quite large. For Nb it is 0.897 Å and for Li it is 0.714 Å. The movement of these atoms with respect to the oxygen planes is also shown in Fig. 29 along with the direction of the dipole moment of the cell.

Large crystals of lithium niobate can readily be grown by the Czochralski technique[55] and poled to be single domain.[56] Until recently, the procedure was to sinter or melt equal molar amounts of Li_2CO_3 and Nb_2O_5. A large, clear single crystal could then be pulled whose dimensions were of the order of 2 cm across and 10 cm long. Unfortunately, the homogeneity was in general rather poor. Figure

TABLE 2

Properties of Stoichiometric $LiNbO_3$ and $LiTaO_3$*

Property	$LiNbO_3$	$LiTaO_3$
Melting point, °C	1250[59]	1650[5]
Curie temperature, °C	1190[58]	680[72]
Birefringence	−0.105[58]	−0.0020[72]
Net charge†		
Nb or Ta	+1.59[73]	+1.21[73]
O	−0.86[73]	−0.74[73]
Electronegativity difference M − O (M is Nb or Ta)	2.27[17]	2.17[17]
Debye temperatures,ᵃ† °K		
Heat capacity (4–10°K)	583[74]	487[74]
Elastic constants	571‡	504‡
X ray (Debye–Waller factor)	503[77]	452[78]

* From Ref. 79.
† These crystals were niobium or tantalum rich.
‡ Computed from the elastic constant date of Ref. 75 using Anderson's method.[76]

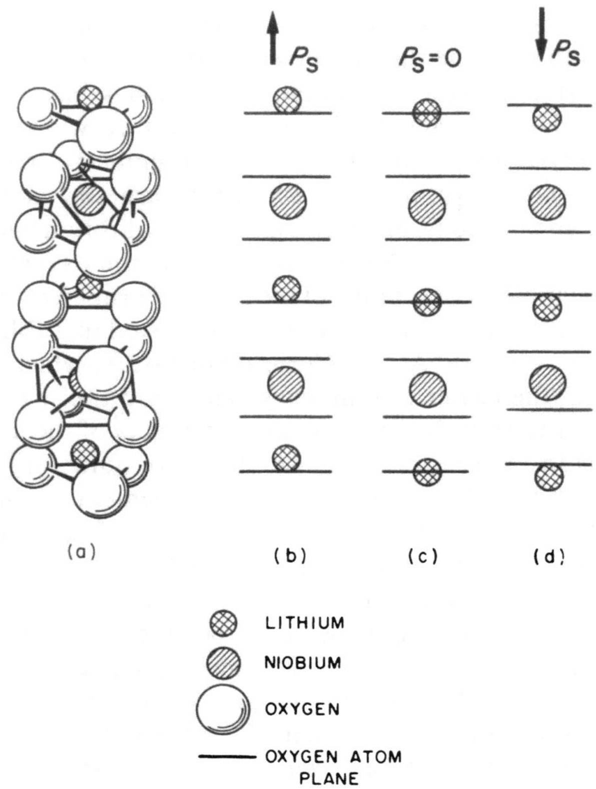

Fig. 29. Crystal structure of lithium niobate, illustrating both the polar and nonpolar phases. See Ref. 54.

26 shows typical second harmonic power versus temperature for an a-axis-grown crystal. The laser light is 90° phase matched and also propagates along a. Thus we are examining the overall homogeneity. It is clear that crystals of this sort leave a great deal to be desired.

Niobium-95 NMR studies[57] indicated that the electric field gradients of crystals pulled from such melts were quite random. Because of the large quadrupole moment of the niobium nucleus (0.2 b), defects drastically broaden the NMR lines. Studies of this sort are extremely valuable in probing defects at an atomic level. If a highly polarizable material such as lithium niobate contains distribution of defects, the electric fields that result can alter the birefringence through the electrooptic effect.

The NMR work not only showed that the crystals contained many defects, but that the defect concentration varied rather smoothly

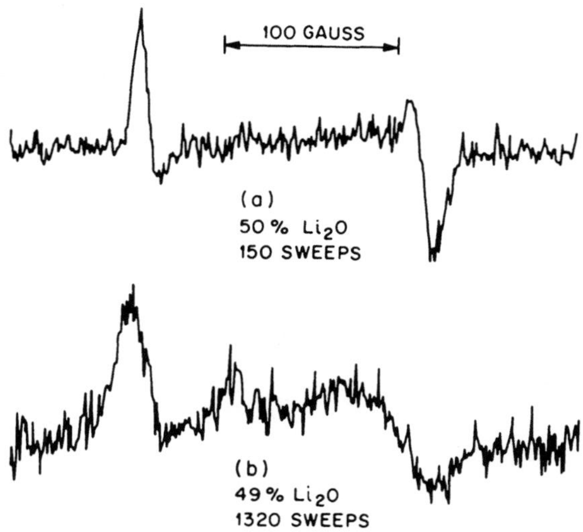

Fig. 30. Niobium-93 NMR for (a) stoichiometric and (b) nonstoichiometric lithium niobate. From Ref. 57.

along the length of the ingot. The last part to be pulled from the melt contained fewer defects than the first. An NMR examination of 1:1 molar ceramic specimens showed very much sharper lines than any single-crystal sample. This prompted a further examination of sintered powders. The results of such a study are shown in Figs. 30 and 31. It is obvious that the broadening of the lines can be correlated

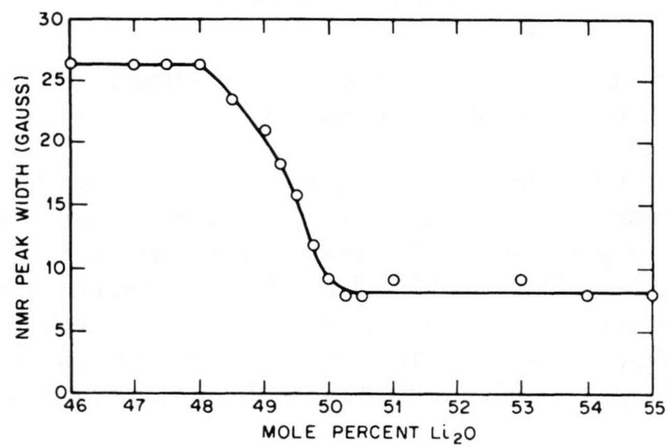

Fig. 31. Half-width of low-field derivative peak as a function of composition for lithium niobate. From Ref. 57.

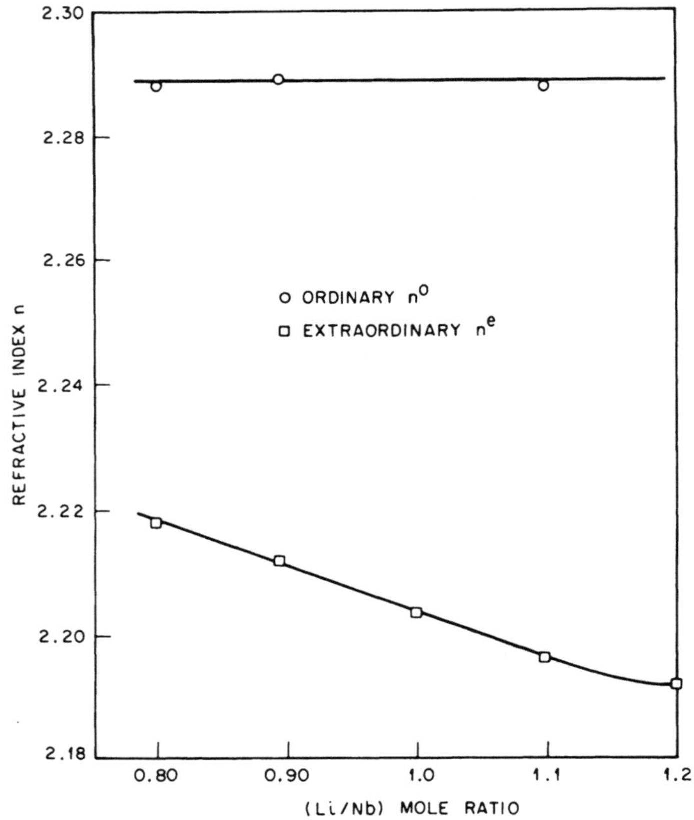

Fig. 32. Refractive index at 6328 Å versus melt stoichiometry for lithium niobate. From Ref. 58.

with the stoichiometry of the sintered specimens. This strongly suggested that the broadening in the single crystals was also due to stoichiometry.

Indeed, Bergman and co-workers[58] had shown that the extraordinary index of refraction, the Curie temperature, and the phase matching temperature were functions of melt stoichiometry. Figure 32 shows the refractive index data of these workers. The phase equilibrium work of Lerner and co-workers[59] strongly suggested that the compound did not melt congruently and that the maximum in the melting was shifted toward the niobium-rich side of the phase diagram.

It is obvious that if the solidus and liquidus curves are not touching at the 1:1 melt composition, two phenomena will occur:

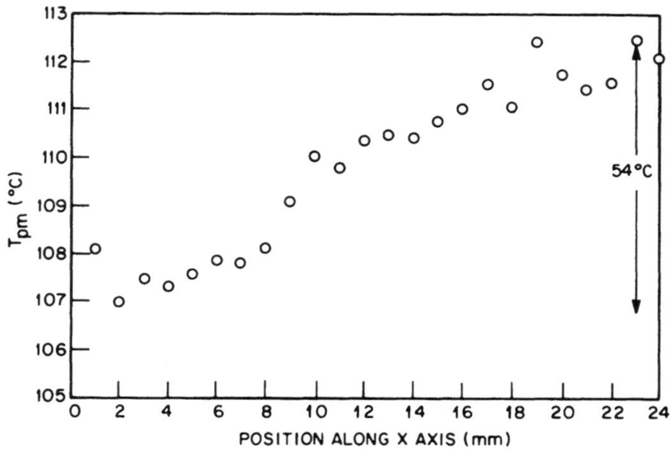

Fig. 33. Phase-match temperature along the X axis for a lithium niobate crystal grown from a stoichiometric melt. From Ref. 48.

(1) The crystal will not be stoichiometric, and (2) the stoichiometry will vary along the crystal. This is vividly illustrated by Fig. 33 taken from the work of Nash,[(48)] which shows the 90° phase-match temperature as a function of position along the ingot. A very nice discussion of crystal growth and stoichiometry can be found in the book by Hannay,[(60)] which is recommended for further study.

The homogeneity problem resolves itself into finding a melt composition such that the crystal composition always matches it. This is of course the congruent composition and must occur at the maximum in the melting curve.

To find this "special" composition, which from the work of Lerner must be in the range from 48–49 mole % Li_2O, two lines of attack were followed. The first[(61)] was to methodically melt and then freeze ceramic specimens of various compositions and then use NMR to decide which composition was unaffected by melting. The second[(62)] was to determine an accurate phase diagram. Both of these studies converged to the composition 48.6 mole % Li_2O and 51.4 mole % Nb_2O_5. To determine the phase diagram with the required accuracy, the following unique combination of techniques was employed: NMR,* Curie temperature measurements, and differential thermal analysis. The NMR was most valuable near stoichiometry, while the Curie temperature technique was employed for samples further removed. Figure 34 shows the phase diagram.

* The NMR technique is quite unusual, but powerful. See Ref. 57.

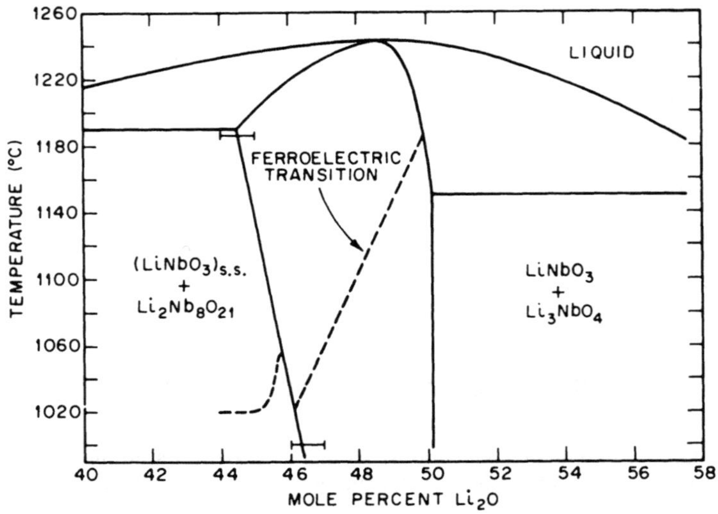

Fig. 34. Phase diagram for lithium niobate in the Li_2O–Nb_2O_5 system. From Ref. 62.

Knowledge of an accurate phase diagram and an accurate congruent composition allowed workers at Bell Laboratories and elsewhere to grow highly homogenous crystals. Figure 35 shows the second harmonic generation in a crystal grown at Stanford University using the congruent composition.* Note that there is only one strong peak.

4.6. Changing Phase-Match Temperature

Correcting one problem may cause others. Shifting from a stoichiometric melt to a congruent melt for lithium niobate causes an enormous improvement in optical homogeneity but also a reduction in the 90° phase-match temperature. Table 3 illustrates this decrease. at first sight this may seem inconsequential and possibly even advantageous. This unfortunately is not the case. In general the lower the operating temperature, the greater the possibility of laser damage. For laser light of 5145 Å, complete elimination of all damage requires a temperature of 250°C, while substantial removal occurs at roughly 170°C. One therefore seeks ways to raise the phase-match temperature while still retaining the high homogeneity of the congruent melt.

* Information concerning congruent melt composition was communicated by J. R. Carruthers.

TABLE 3

SHG Parameters for 90° Phase Matching in $LiNbO_3$*

Pump wavelength, μm	T_0, °C (Stoichiometric)	T_0, °C (Congruent)
1.064 Nd:YAlG	47	−8
1.084 He–Ne	92	42
1.152 He–Ne	210	172

* From Ref. 47.

It is not particularly difficult to alter the Curie temperature and phase-match temperature by melt additions, but usually the crystal composition will vary along the ingot in undesirable ways. An ingenious way to modify the phase-match temperature in lithium niobate and still keep the crystal homogeneous was developed by Bridenbaugh and co-workers.[63]

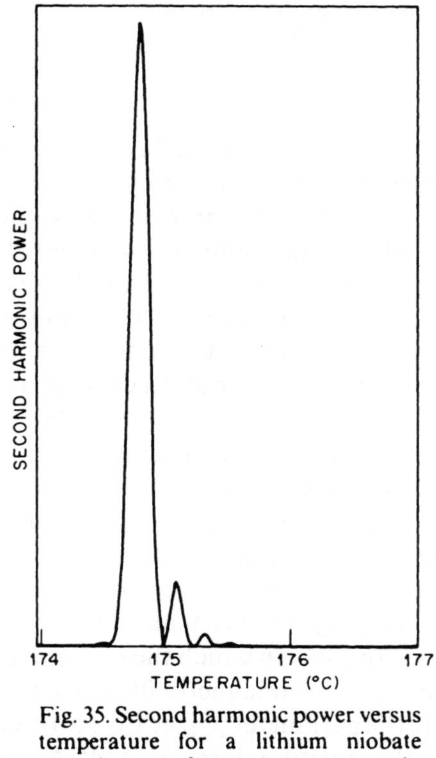

Fig. 35. Second harmonic power versus temperature for a lithium niobate crystal grown from a congruent melt. From Ref. 47.

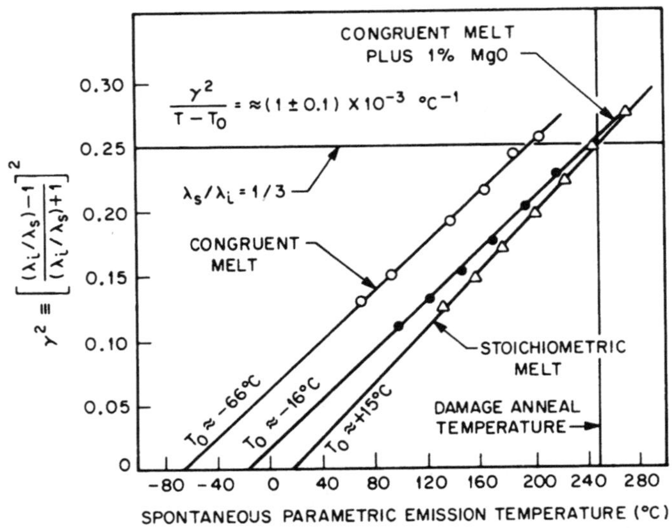

Fig. 36. Wavelength departure from degeneracy in optical parametric fluorescence versus temperature. The pump wavelength is 5145 Å. From Ref. 63.

It was known that the addition of MgO to the melt would raise the Curie temperature and also presumably the phase-match temperature. The distribution coefficient for MgO was found to be 1.3 by Nassau (Ref. 56, p. 264) and an examination of the ternary oxide phase diagram showed that MgO additions resulted in excess Li_2O in the melt. The distribution coefficient for excess lithium was established to be 0.30 and it also increased the Curie temperature. Thus we have a case where one oxide has a distribution coefficient greater than one while the second has a distribution coefficient less than one. Consequently the concentration of excess lithium will be increasing along the length of the crystal while the concentration of Mg will be decreasing. Both of these oxides increase the Curie temperature and thus compensate each other as the crystal is grown. The net result is as follows: (1) The Curie and phase-match temperatures are increased and (2) high optical homogeneity is achieved.

We can illustrate the way in which the material has been modified by a study of the optical paramagnetic fluorescence versus temperature. Figure 36 plots the parameter γ^2 [see Eq. (46)] versus temperature. The frequency of the exciting laser is 5145 Å and the following definitions allow us to relate the notation of Fig. 36 to Eq. (46). We assume that the interchange of frequency and wavelength will be obvious to the reader.

$$\lambda_p = 5145\ \text{Å pump wavelength} \equiv \lambda_3$$
$$\lambda_s = \text{signal wavelength} \equiv \lambda_1$$
$$\lambda_i = \text{idler wavelength} \equiv \lambda_2$$

The desired region of operation is above 250°C to ensure minimum laser damage and at $\lambda_s/\lambda_i = 1/3$ for good resonator design. From the figure we see that such operation is not possible with the highly homogeneous congruent melt crystals. It is in principle possible with the highly inhomogeneous stoichiometric melt crystals. However, optical losses are in general too great. It is possible with the quite homogeneous MgO doped crystal.

This method of using one impurity to compensate another appears to be very powerful. If a crude ternary phase diagram is known, likely ions for double doping can be specified. Even if a phase diagram is not known, a good guess may suffice. The effect upon the Curie temperature and probably change in phase-match temperature can be inferred from the theory of Lines. [See Sec. 2.2.]

4.7. Minimizing Optical Index Damage

Optical index damage has been observed in a number of very important materials.[64] These include $LiNbO_3$, $LiTaO_3$, KTN, $BaTiO_3$, $Sr_{1-x}Ba_xNb_2O_6$, and CdS. This seems to be a very general sort of phenomenon in nonlinear materials. The extensive work done on lithium niobate has lead to a very clear understanding of the nature and solution of this most vexing problem. Even in the first device[65] (a 21-MHz modulator) built with this material in 1964, unusual optical phenomena which can now be ascribed to index damage were observed. Almost everyone who works with this material "rediscovers" damage.

The extensive studies that followed lead to a number of unexplainable observations. Some of these are as follows.

(1) No optical luminescence could be associated with damage.
(2) No laser-induced 10-GHz EPR signals were found.
(3) All crystals annealed at 1000°C in oxygen damaged less than crystals annealed in oxygen partial pressures of less than 100 mm Hg.
(4) Reduction of the crystal definitely increased the damage.

(5) Electric field annealing at ~600°C definitely reduced the damage.
(6) Crystals from the same source, the same stoichiometry, and identical heat treatments varied widely in their damage susceptibility.
(7) Changing stoichiometry had little if any effect.
(8) Chemical analysis by commercial laboratories showed several impurities, none of which was found to have any effect.
(9) Any photoconductivity associated with the damage process was masked by the pyroelectric effect.

Though this information to some degree characterized index damage, until recently there was very little progress toward its elimination, enhancement, or understanding. It was felt by workers at Bell Laboratories[66] that damage was somehow associated with an impurity and for unknown reasons had not been detected. Based on some theories of damage developed by Chen and Johnston,[67] it was speculated that the damage center might possibly be a transition metal. If this were so, EPR would be an excellent way to detect it. However, previous EPR work on undoped lithium niobate at 10 GHz had been unsuccessful. To circumvent this difficulty, a spectrometer was designed and built for 0.5 GHz. It was hoped that if the lines were "g" broadened, dropping to this lower frequency might sharpen them and allow more optimum detection for a given amount of magnetic field modulation. Apparently this was the case, for signals were detected and identified as iron. Figure 37 shows the signals for an undoped powder specimen and Fig. 38 some reference spectra.

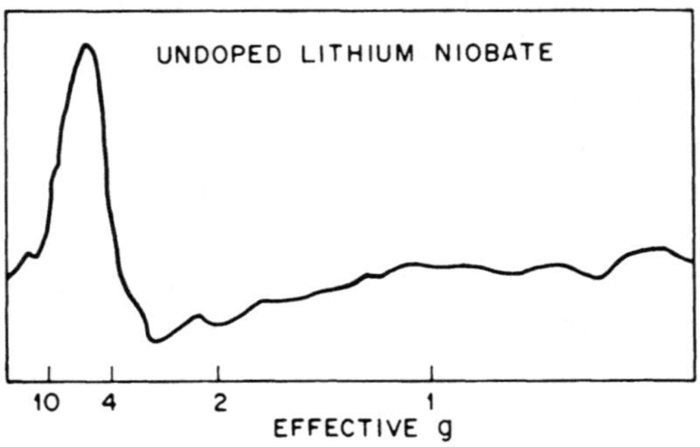

Fig. 37. Low-frequency EPR signals from an undoped sample of lithium niobate.

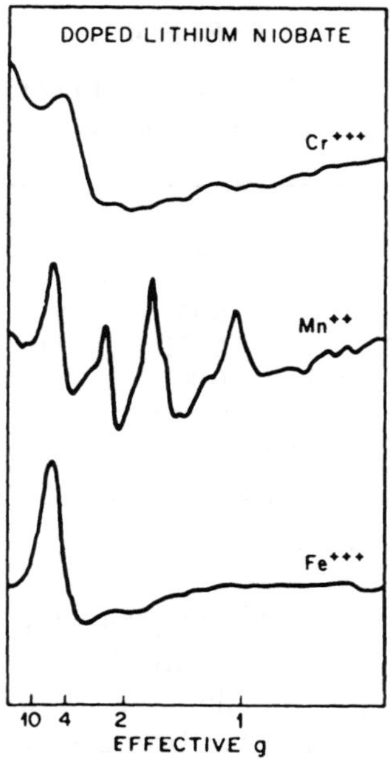

Fig. 38. Reference EPR spectra.

Every crystal examined showed iron impurity, usually in the range 10–100 ppm.

Once it was shown that crystals contained iron it was a simple matter to increase its content and show that the damage also increased. By coupling damage measurements with EPR studies, it was found that Fe^{2+} is the active center.

The mechanism of the index damage is via the reaction

$$Fe^{2+} + h\nu = Fe^{3+} + e\,(\text{conduction})$$

Once the electrons are in the conduction band they drift or diffuse out of the illuminated region and are trapped, possibly by Fe^{3+} ions by the reverse process of the above equation. Thus there is a polarization change due to a redistribution of electrons. This causes a change in the local field and alters the birefringence through the electrooptic effect. It is now easy to understand most of the observations about damage previously listed. The only obscure observation is number 5, which concerns electric field annealing. It is believed that the electric field sweeps some of the iron out of the crystal and thus improves it. Table 4 shows the diffraction efficiencies for various crystals and treatments. The low-iron crystals contain about 1 ppm of this impurity.

TABLE 4

Diffraction Efficiencies (%) of 0.2-cm-Thick $LiNbO_3$ Crystals after Various Heat Treatments*

Treatment	Undoped	0.05% Fe_2O_3	Low Fe #1	Low Fe #2
Poled from 1150°C and quenched	0.20	95†	0.019	0.010
Annealed for 75 hr at 600°C	0.021	1.0	0.0039	Not detected
Field annealed at 600°C for 15 min	0.007	‡	0.0004	Not detected

* After Peterson *et al.*[66] The low-Fe crystals are the highest-purity crystals obtained by careful crystal growth.
† This was the maximum diffraction efficiency observed. An efficiency of 50% was still obtained with this crystal thinned down to 0.02 cm.
‡ This crystal had very uneven Fe^{2+} distribution.

Prior to the realization of the great importance of iron it was thought that certain vacant sites in the lithium niobate structure might be responsible for the damage. A number of compounds related to the tetragonal tungsten bronzes, which were said to have a more "filled up" structure, were investigated.[68] Examples of these are $Ba_2NaNb_5O_{15}$, $K_3Li_2Nb_5O_{15}$, and $(Ba, Sr)_5Nb_{10}O_{30}$. These compounds were found to be somewhat more resistant to room-temperature laser damage than the early (iron-contaminated) lithium niobate. However, they all damage badly at reduced temperatures. EPR examinations reveal iron contamination. Thus this damage could probably be reduced or eliminated. The better behavior at room temperature is associated with the different kinetics of the $Fe^{3+} \rightarrow Fe^{2+}$ conversion. There is little reason to believe that the "filling of sites" plays any role in reducing damage.

It is probable that other multivalent ion impurities such as Mn^{2+}–Mn^{3+} or Cu^{+}–Cu^{2+} can give rise to index damage.[69] Unfortunately iron contamination of crystals is so universal and its influence upon damage so large it is very difficult to ascertain the true effects of other ions at low concentrations. Another problem is that at higher concentrations the added ion may be influencing the valence state of the low-level Fe impurity. Preliminary EPR studies of lithium niobate simultaneously doped with Cu and Fe show a preference for configuration Cu^{2+}–Fe^{2+}.

The procedure for growing crystals with reduced iron has been discussed by Bridenbaugh.[70] One starts with high-purity Li_2CO_3 and Nb_2O_5. The particular batch of Nb_2O_5 is checked before use by EPR. Care is exercised to keep contact with ceramics and ceramic dust to a

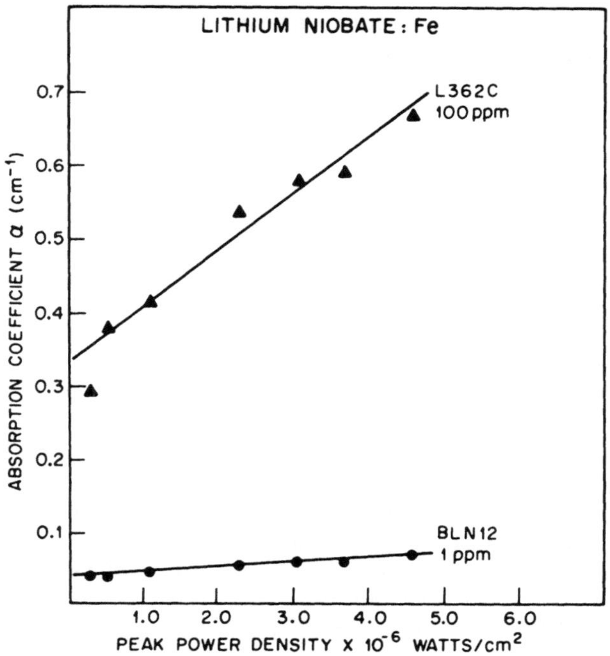

Fig. 39. Absorption coefficient in the green versus peak power in megawatts. From Ref. 71.

bare minimum. Instead of using beads to support the crucible, a sapphire pedestal is employed and the oxide and carbonate are reacted in the rf coil rather than firing in a furnace. Finally, the conversion of the crystal to a single ferroelectric domain (i.e., poling) utilizes a low-Fe environment and low-Fe electrodes. Crystals grown using the above procedure are reasonably low in iron ($\sim$3 ppm) and have a reduced susceptibility to damage.

Another phenomenon that appears to be associated with Fe in crystals is filamentary optical damage. This occurs for peak powers in the megawatt range and has been discussed by Schinke and Smith[71] for barium sodium niobate and lithium niobate. Figure 39, which was provided by these two workers, shows the absorption coefficient in the green versus peak power density for lithium niobate containing different Fe levels. One notes that the absorption coefficient increases much more rapidly for the crystal with higher iron concentration. This nonlinear absorption process is detrimental to high-power applications. Heat is poured into the crystal near the beam and causes a refractive index variation. The beam self-focuses, resulting in further heating and focusing. Finally crystal damage occurs.

References

1. C. Kittel, *Introduction to Solid State Physics*, Wiley, 3rd ed., New York (1968).
2. M. E. Lines, *Phys. Rev.* **177**, 797 (1969); **2B**, 690 (1970); **2B**, 698 (1970); **5B**, 3690 (1972).
3. J. F. Nye, *Physical Properties of Crystals*, Oxford University Press (1960).
4. W. G. Cady, *Piezoelectricity*, Dover, New York (1964).
5. F. Jona and G. Shirane, *Ferroelectric Crystals*, Macmillan, New York (1962).
6. W. Känzig, *Solid State Phys.* **4**, 1 (1957).
7. H. S. van Klooster, *J. Chem. Ed.* **36**, 314 (1959).
8. B. T. Matthias, in *Ferroelectricity* (E. Weller, ed.), p. 176, Elsevier, New York (1967).
9. R. H. Lyddane, R. G. Sachs, and E. Teller, *Phys. Rev.* **59**, 673 (1941); A. S. Barker, Jr., *Phys. Rev.* **136A**, 1290 (1964); W. Cochran, *Z. Krist.* **112**, 465 (1959).
10. W. K. H. Panofsky and M. Phillips, *Classical Electricity and Magnetism*, pp. 29–35, Addison-Wesley, Reading, Mass. (1955).
11. S. Triebwasser, *IBM J. Res. and Dev.* **2**, 212 (1958).
12. W. D. Johnston, Jr., and I. P. Kaminow, *Phys. Rev.* **168**, 1045 (1968).
13. N. Boccara and G. Sarma, *Physics* **1**, 219 (1965); E. Pytte and J. Feder, *Phys. Rev.* **187**, 1077 (1969).
14. J. Feder and E. Pytte, *Phys. Rev.* **1B**, 4803 (1970).
15. M. Born and K. Huang, *Dynamical Theory of Crystal Lattices*, Clarendon Press, Oxford (1954).
16. J. G. Bergman, Jr., Private communication.
17. F. A. Cotton and G. Wilkinson, *Advanced Inorganic Chemistry*, p. 103, Interscience, New York (1966). G. E. Peterson and P. M. Bridenbaugh, *J. Chem. Phys.* **48**, 3402 (1968).
18. G. E. Peterson, J. R. Carruthers, and A. Carnevale, *J. Chem. Phys.* **53**, 2436 (1970).
19. J. Fox, *Submillimeter Waves*, pp. 251–294, Polytechnic Press, New York (1970).

20a. A. G. Chynoweth, *J. Appl. Phys.* **27**, 78 (1956).

20b. A. M. Glass, *J. Appl. Phys.* **40**, 4699 (1969).

21. A. M. Glass and D. H. Auston, *Optics Comm.* **5**, 45 (1972).
22. A. M. Glass, *Appl. Phys. Letters* **13**, 147 (1968).
23. D. H. Auston, A. M. Glass, and A. A. Ballman, *Phys. Rev. Letters* **28**, 897 (1972).
24. M. H. Francombe, *Acta Cryst.* **13**, 131 (1960); I. G. Ismailzade, *Kristallografiya* **5**, 268 (1960); P. B. Jamieson, S. C. Abrahams, and J. L. Bernstein, *J. Chem. Phys.* **48**, 5048 (1968).
25. J. R. Carruthers and M. Grasso, *J. Electro. Chem. Soc.* **117**, 1426 (1970).
26. A. M. Glass and R. L. Abrams, *J. Appl. Phys.* **41**, 4455 (1970).
27. P. J. Lock, *Appl. Phys. Letters*, **19**, 390 (1971).
28. S. Hoshino, Y. Okaya, and R. Pepinsky, *Phys. Rev.* **115**, 323 (1959).
29. E. T. Keve, K. L. Bye, P. W. Whipps, and A. D. Annis, *Ferroelectrics* **3**, 39 (1971).

30. D. H. Auston and A. M. Glass, *Appl. Phys. Letters* **20**, 398 (1972).
31. J. D. Roberts and M. C. Caserio, *Basic Principles of Organic Chemistry*, pp. 1088–1092, Benjamin, New York (1965).
32. J. H. McFee, J. G. Bergman, Jr., and G. R. Crane, *Ferroelectrics* **3**, 305 (1972).
33. J. B. Lando, H. G. Olf, and A. Peterlin, *J. Polymer Sci.* **4**, 941 (1966) (Part A-1).
34. A. M. Glass, J. H. McFee, and J. G. Bergman, Jr., *J. Appl. Phys.* **42**, 5219 (1971).
35. J. A. Giordmaine, *Sci. Am.* **1964** (June).
36. R. W. Terhune, *International Sci. Tech.* **1964** (August).
37. J. R. Carruthers, in *Encyclopedia of Chemical Technology*, p. 623, Wiley, New York (1971).
38. P. A. Franken, A. E. Hill, C. W. Peters, and G. Weinreich, *Phys. Rev. Letters* **7**, 118 (1961).
39. J. A. Giordmaine, *Phys. Rev. Letters* **8**, 19 (1962); P. D. Maker, R. W. Terhune, M. Nisenoff, and C. M. Savage, *Phys. Rev. Letters* **8**, 21 (1962).
40. R. C. Miller, G. D. Boyd, and A. Savage, *Appl. Phys. Letters* **6**, 77 (1965).
41. J. A. Giordmaine, *Phys. Rev. Letters* **8**, 19 (1962).
42. J. E. Midwinter and J. Warner, *Brit. J. Appl. Phys.* **16**, 1135 (1965).
43. M. V. Hoben, *J. Appl. Phys.* **38**, 4365 (1967).
44. R. W. Dixon, *IEEE Trans. on Electron Devices* **3**, 229 (1970).
45. D. A. Kleinman, *Phys. Rev.* **126**, 1977 (1962).
46. A. Ashkin, G. D. Boyd, J. M. Dziedzic, R. G. Smith, A. A. Ballman, H. J. Levinstein, and K. Nassau, *Appl. Phys. Letters* **9**, 72 (1966).
47. R. L. Byer, J. F. Young, and R. S. Feigelson, *J. Appl. Phys.* **41**, 2320 (1970).
48. F. R. Nash, G. D. Boyd, M. Sargent, and P. M. Bridenbaugh, *J. Appl. Phys.* **41**, 2564 (1970).
49. S. E. Harris, *Proc. IEEE* **57**, 2096 (1969).
50. G. D. Boyd and A. Ashkin, *Phys. Rev.* **146**, 187 (1966).
51. A. M. Glass, G. E. Peterson, and T. J. Negran, pp. 15–26 in *Proceedings of the ASTM:NBS Symposium on Damage in Laser Materials* (1972).
52. R. L. Townsend and J. T. LaMacchia, *J. Appl. Phys.* **41**, 5188 (1970).
53. F. S. Chen, J. T. LaMacchia, and D. B. Fraser, *Appl. Phys. Letters* **13**, 223 (1968).
54. S. C. Abrahams, H. J. Levinstein, and J. M. Reddy, *J. Phys. Chem. Solids* **27**, 1019 (1966); S. C. Abrahams, J. M. Reddy, and J. L. Bernstein, *J. Phys. Chem. Solids* **27**, 997 (1966); S. C. Abrahams, W. C. Hamilton, and A. Sequeira, *J. Phys. Chem. Solids* **28**, 1693 (1967).
55. A. A. Ballman, *J. Am. Cer. Soc.* **48**, 112 (1965).
56. K. Nassau, in *Ferroelectricity*, p. 268, Elsevier, Amsterdam (1967).
57. G. E. Peterson and J. R. Carruthers, *J. Solid State Chem.* **1**, 98 (1969).
58. J. G. Bergman, A. Ashkin, A. A. Ballman, J. M. Dziedzic, H. J. Levinstein, and R. G. Smith, *Appl. Phys. Letters* **12**, 92 (1968).
59. P. Lerner, C. Legras, and J. P. Dumas, *J. Cryst. Growth* **3/4**, 231 (1968).
60. N. B. Hannay, *Solid State Chemistry*, pp. 202–207, Prentice-Hall, Englewood Cliffs, New Jersey (1967).
61. G. E. Peterson and P. M. Bridenbaugh, unpublished.
62. J. R. Carruthers, G. E. Peterson, M. Grasso, and P. M. Bridenbaugh, *J. Appl. Phys.* **42**, 1846 (1971).
63. P. M. Bridenbaugh, J. R. Carruthers, J. M. Dziedzic, and F. R. Nash, *Appl. Phys. Letters* **17**, 104 (1970).

64. A. Ashkin, G. D. Boyd, J. M. Dziedzic, R. G. Smith, A. A. Ballman, H. J. Levinstein, and K. Nassau, *Appl. Phys. Letters* **9**, 72 (1966).
65. G. E. Peterson, A. A. Ballman, P. V. Lenzo, and P. M. Bridenbaugh, *Appl. Phys. Letters* **5**, 62 (1964).
66. G. E. Peterson, A. M. Glass, and T. J. Negran, *Appl. Phys. Letters* **19**, 130 (1971).
67. F. S. Chen, *J. Appl. Phys.* **40**, 3389 (1969); W. D. Johnston, *J. Appl. Phys.* **41**, 3279 (1970).
68. W. A. Bonner, W. H. Grodkiewicz, and L. G. Van Uitert, *J. Cryst. Growth* **1**, 318 (1967).
69. J. J. Amodei, W. Phillips, and D. L. Staebler, *Appl. Optics* **11**, 390 (1972).
70. P. M. Bridenbaugh, submitted to *J. Crystal Growth*.
71. D. P. Schinke and R. G. Smith, submitted to *J. Appl. Phys.*
72. A. A. Ballman, H. J. Levinstein, C. D. Capio, and H. Brown, *J. Am. Ceram. Soc.* **50**, 657 (1967).
73. G. E. Peterson and P. M. Bridenbaugh, *J. Chem. Phys.* **48**, 3402 (1968).
74. J. G. Bergman, Jr. and J. P. Maita, Private communication.
75. A. W. Warner, M. Onoe, and G. A. Coquin, *J. Acoust. Soc. Am.* **42**, 1223 (1967).
76. O. L. Anderson, *J. Phys. Chem. Solids* **24**, 909 (1963).
77. S. C. Abrahams, J. M. Reddy, and H. J. Levinstein, *J. Phys. Chem. Solids* **27**, 997 (1966).
78. S. C. Abrahams and J. L. Bernstein, *J. Phys. Chem. Solids* **28**, 1685 (1967).
79. G. E. Peterson, J. R. Carruthers, and A. Carnevale, *J. Chem. Phys.* **53**, 2436 (1970).

4

The Imperfect Solid—Transport Properties

David Adler
Department of Electrical Engineering and
Center for Materials Science and Engineering
Massachusetts Institute of Technology
Cambridge, Massachusetts

1. Introduction

1.1. Types of Disorder

Any discussion of imperfect solids must necessarily define both what is meant by a perfect solid and the various types of imperfections that are commonly encountered. A complete discussion of chemical and structural defects in solids has already been given in Chapter 5 of Volume 1. However, since transport properties are so dependent on imperfections and since we wish to discuss amorphous solids in detail, a brief outline of order and disorder is presented here.

A perfect solid is one in which the environment of any atom, no matter how far away we look, is precisely the same as the environment of any other equivalent atom in the material. This is perfect short-range as well as long-range order, the range simply being an indication of just how far away from the atoms under consideration the environments are identical. It is obvious that perfect infinite-range order is impossible, since all solids are of finite size, and surface atoms do not have identical environments to atoms in the bulk. Internally, grain boundaries can exist in polycrystalline solids, and these can reduce the range of the order still further. For the purposes of a discussion of bulk transport properties, the range of the order need not be very long—1000 Å is more than sufficient. However, it is important to take precautions to eliminate the effects of grain boundaries, surfaces, and contact effects in any experiment; i.e., it must be verified that bulk properties of a single-crystalline phase are being measured. The fact that this has not always been the case with regard to the materials discussed in this chapter will become evident shortly.

Even within 1000-Å grains, perfect order cannot exist. Thermodynamically, entropy considerations lead to imperfections at finite temperatures, and finite cooling rates lead to some of these being frozen in at very low temperatures. In this section, a classification of the types of imperfections that can exist in any particular material will be presented. Any solid can be looked at as a collection of ion cores, whose equilibrium positions define the structure of the material, surrounded by an electron gas, which is characterized by a charge density that can vary in both space and time. The ion cores vibrate around their equilibrium positions, and the quantized normal modes can be described in a "particle" picture as phonons.[(1)] The separation of electronic from phonon motion is called the Born–Oppenheimer approximation, which can become invalid if the electron–phonon interaction is strong. In such cases, which occur regularly in some of the materials discussed in this chapter, changes in the electronic state lead to dynamic lattice imperfections, and thus we must be very careful in interpreting the experimental results. However, when the electron–phonon interaction is small, we can decouple the electronic and vibrational problems and ignore what might be called *electronic* disorder, brought about by excitations of the outer electrons from their ground states.

The remaining types of disorder are compositional, positional, and magnetic. It is useful to distinguish between *static* and *dynamic* imperfections. Static imperfections in *compositional* order include impurities and antistructure, among others; static imperfections in *positional* order include vacancies, interstitials, dislocations, etc.; for systems in which the ion cores have a nonvanishing magnetic moment, a static imperfection in *magnetic* order is a reversed spin. We shall be concerned also with dynamic imperfections such as phonons, which introduce positional disorder, and magnons (spin waves), which introduce magnetic disorder.

Perfect order implies that if we are given the position, spin, and type of any particular atom in the solid together with an orientation, we can predict the position, spin, and type of all the other atoms with 100% accuracy. The opposite situation, perfect disorder, implies the atoms are all at random with respect to the atom under consideration. Clearly, materials can exist for which the nearest-neighbor environment is known precisely, but beyond a distance l the atoms are essentially at random. Such a material has short-range order, but long-range disorder; l is a measure of the range of the order. When we refer to disordered materials we usually imply only long-range disorder, beyond 20 Å or so. Thus, a ferromagnetic material above its Curie temperature has no long-range magnetic order, an alloy above its ordering temperature has no long-range compositional order, and a liquid has no long-range positional order. But in all of these situations a great deal of short-range order usually exists. Amorphous solids are defined as solids with no long-range positional order. When they are made up of varying percentages of many different chemical types they can have much short-range disorder as well. We shall discuss amorphous as well as imperfect crystalline solids in this chapter.

1.2. Types of Bonding

Before entering into a general discussion of transport, it is essential to review the different types of bonding that occur in solids, since each category provides us with a sharply distinct conduction mechanism. We shall not consider pure van der Waals bonding, since the solids that result from such bonding melt at very low temperatures and hardly conduct electricity at all. The remaining classes of solids are metallic, covalent, and ionic. Of course, any real solid is a combination of all of these classes, that combination which minimizes its free energy at the appropriate temperature and pressure, but we usually can classify the solid by identifying the major contribution to its cohesive energy.

Metallic solids are characterized by a delocalization of the outer electrons, and it is this delocalization which is responsible for the entire cohesive energy. The reduction of energy of the solid compared to the gas results from a lower kinetic energy of these outer electrons; this is a quantum mechanical effect due to the reduction in momentum which can occur when we spread out a localized electron over many atomic distances. Typical cohesive energies of metallic solids are 1–10 eV.

Covalently bonded solids are very different from metallic solids, despite the fact that a covalent bond between two atoms exists for the same reason that metallic bonds exist, i.e., the reduction of kinetic energy arising from a delocalization of electrons. In a covalent bond, however, the electrons are spread over just two atoms, not the entire solid. The strength of the covalent bond arises from the fact that each atom contributes at least one electron to the bond; both electrons delocalize, and as long as they have opposite spins they can both occupy the orbital with the minimum kinetic energy without violating the exclusion principle. Covalently bonded solids typically have cohesive energies of 4–13 eV.

Ionic bonding is the simplest to understand. It can exist only if we have at least two atomic types present. If an electron is removed from an electropositive atom and placed on an electronegative atom a long distance away, the atomic energy levels often increase only very little. (In fact, for one material, CsF, the energy actually *decreases* if we form Cs^+ and F^- ions.) The actual binding then arises from the electrostatic attraction of an optimum array of alternating positive and negative ions. Ionic bonding produces solids with cohesive energies typically in the 6–18 eV range. The importance of the different bonding mechanisms to transport properties of solids will become evident in the next section.

2. Electronic Structure of Perfect and Imperfect Solids

2.1. Energy Band Theory of Solids

The energy band theory of solids is discussed in detail in Chapter 2 of Volume 1 of this Treatise. In this section a brief qualitative description is presented in order to distinguish among the different classes of solids and pave the way for the discussion of electrical conductivity. Ordinary energy band theory is based on several assumptions that are not universally valid. The first is the Born–Oppenheimer approximation, discussed in the last section, which fails when the electron–phonon interaction is strong. The second is the assumption that the ion cores at their equilibrium positions form a perfectly periodic lattice, which as we discussed previously is invalid when surfaces, imperfections, or other types of disorder exist. Furthermore, there is the one-electron approximation, in which it is assumed that each outer electron feels only the average electrostatic repulsion due to all the other electrons in the solid. This assumption would appear to be justified only when the electron–electron interactions are small, for otherwise two electrons should be able to correlate their motion to avoid each other, and thus sharply reduce the energy of the solid. However, a simple calculation can indicate just how large the intra-atomic electronic repulsion typically is. If we consider an atom about

2 Å in diameter with an atomic dielectric constant ε of approximately 2, then the intraatomic electronic repulsion U is greater than

$$U \gtrsim e^2/\varepsilon d \sim 3.6 \quad \text{eV} \tag{1}$$

Such energies cannot be neglected even in a *qualitative* description of a solid. The reason why the one-electron approximation is successful in most cases despite the neglect of electronic correlation is that there ordinarily are many mobile electrons present in a solid. Since these electrons are free to respond quickly to changes in charge density, the effects of excess charges in solids are rapidly neutralized by the electrons already present. This phenomenon, known as *screening*, will occur provided there are relatively mobile electrons in the material. If the outer electrons are localized for any reason, the one-electron approximation becomes suspect.

The one-electron approximation, together with several less essential simplifying assumptions discussed in Volume 1, Chapter 2, enable us to calculate the states available to the outer electrons in any perfectly periodic solid. The results of such calculations are most easily envisioned by building up the solid from what might be called the atomic limit, the limit in which the material is pulled apart uniformly while maintaining the appropriate periodic structure, until the separation between nearest-neighbor atoms a is very large. This is a simple limit to consider, since for large a we have a collection of isolated atoms whose energy levels are known.(2) Furthermore, in the limit of large a all these energy levels are the same for each atomic type present in the solid. The energy levels can be labeled by the electronic configuration of the atoms, using the familiar spectroscopic notation $1s$, $2s$, $2p$, . . . for single electrons and the appropriate many-electron notation,(3) if necessary. As we lower the interatomic separation from infinity to finite values, electrons on different atoms begin to interact with each other. This results in a spreading of the sharp atomic levels into nearly continuous regions of allowed electronic energies. These regions are called *bands*; the regions of energy separating the bands, in which no electronic states are possible, are called *gaps*. As a decreases, the widths of the bands increase. Sketches of the energy band structure of the three major classes of solids as functions of $1/a$ are shown in Figures 1–3. The atomic limit is on the left, where $1/a$ is small, and the bands have zero width.

The sketches are very different for the three classes of solids. Figure 1 shows the energy band structure for a typical metallic solid,

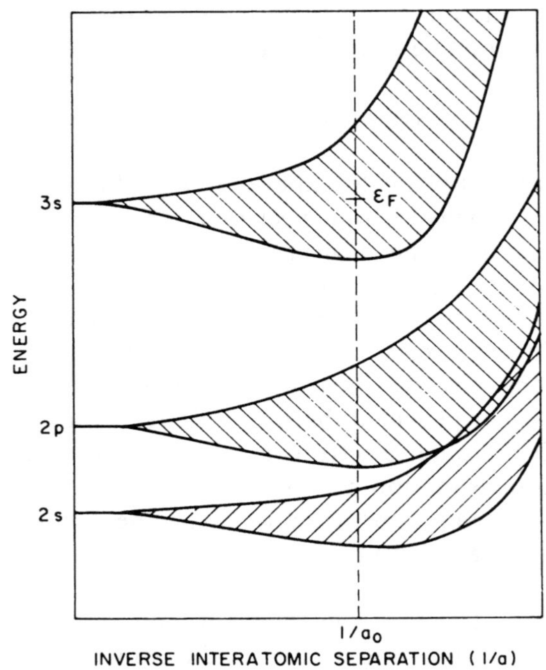

Fig. 1. Energy band structure of a metallic solid as a function of the reciprocal of the nearest-neighbor interatomic separation. The example shown is Na. The shaded regions indicate allowed electronic energy values. The equilibrium interatomic separation is labeled a_0 and the Fermi energy is indicated by ε_F.

such as Na. Each band initially spreads rather symmetrically around its atomic-limit value as the electronic degeneracies are split via interatomic interactions. However, since the electrons present in the material (e.g., Na has N 3s electrons, where N is the total number of atoms in the solid) are insufficient to fill all the bands in the diagram (since the 3s band has 2N states), the total energy of the material decreases as the atoms are brought together. This reflects the decrease in kinetic energy brought about by delocalization of the outer electrons, as discussed in the last section, and represents the cohesive energy of the solid. It is clear from Figure 1 that if all of the bands which contain electrons were filled, the cohesive energy of a metallic solid would be very small. Metallic solids are thus characterized by partially filled bands.

Figure 2 shows the corresponding sketch for a covalently bonded solid, such as Ge. The major feature in this case is the splitting of a single atomic level into two bands. The lower of these bands is usually called a bonding band and the upper an antibonding band, in analogy to the bonding and antibonding orbitals of a diatomic molecule. In fact, the center of the two bands basically represents the average energies of the bonding and antibonding orbitals of such a diatomic molecule on the interatomic separation; the bandwidth results from the presence of all the other atoms in the solid. A comparison of Figures 1 and 2 indicates why the cohesive energy of a covalent solid is generally greater than that of a metallic solid for equivalent outer bandwidths—the appearance of the energy gap adds to the stability of the covalent solid relative to the atomic limit.

The analogous sketch for an ionic solid, such as NaCl, is shown in Figure 3. The atomic limit shown in this case does not show the

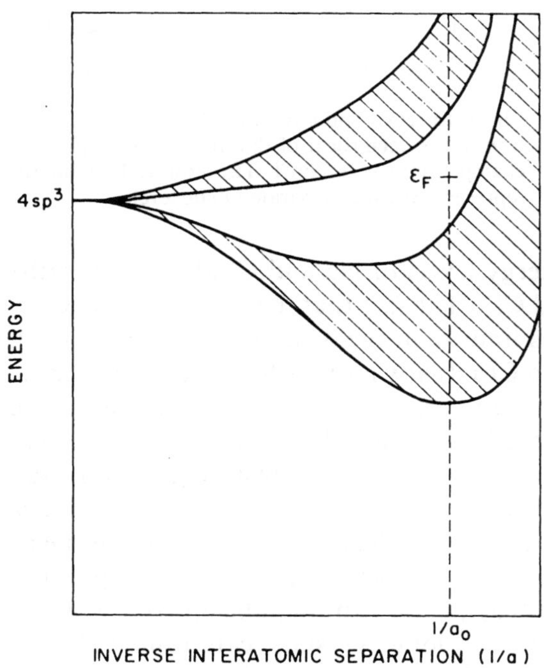

Fig. 2. Energy band structure of a covalent solid as a function of the reciprocal of the interatomic separation. The example shown is Ge. The notation is the same as in Fig. 1.

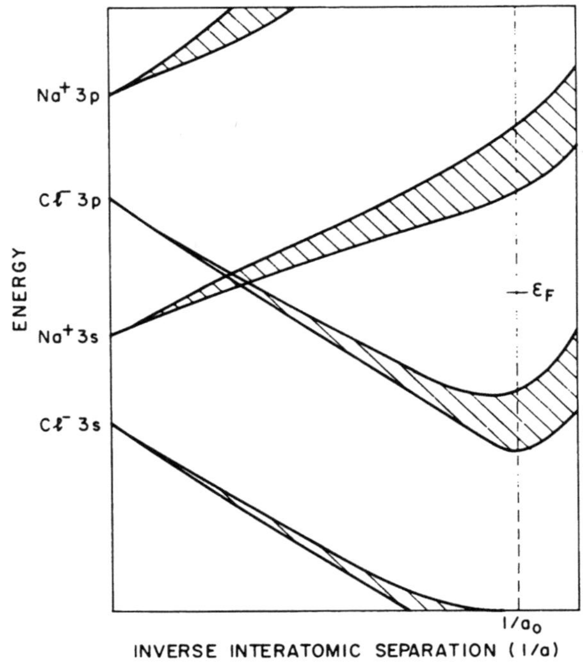

Fig. 3. Energy band structure of an ionic solid as a function of the reciprocal of the interatomic separation. The example shown is NaCl. The atomic limit in this case represents the well-separated *ionic* levels.

electronic energy levels of Na and Cl atoms, but rather those of the *ions* Na^+ and Cl^-. The fact that the $3p$ level of the Cl^- ion has higher energy in this limit than the $3s$ level of Na^+ implies that the neutral atoms Na and Cl actually give the lowest energy at infinite separation. However, as the interatomic separation decreases, the energy of an electron on a Cl^- ion sharply decreases since it begins to feel the electrostatic attraction of the nearest-neighboring Na^+ ions. On the other hand, the energy of an *electronic* state on a Na^+ ion sharply increases, since it is repelled by the surrounding negatively charged Cl^- ions. At an interatomic separation of approximately 10 Å the energy levels cross, before any significant bandwidth is attained. The band structure of ionic crystals can usually be estimated quite accurately purely from a consideration of the ionic levels and simple electrostatic effects.

For all classes of solids there exists an equilibrium interatomic spacing a_0 in which the total energy is a minimum. This occurs because below a sufficiently small value of a all electronic energy levels increase with decreasing interatomic distance. This effect, shown explicitly in Figures 1–3, is a direct consequence of the exclusion principle, which prevents electrons of the same spin from spatially approaching each other. When the atoms are sufficiently close that their electronic charges overlap, the exclusion principle forces some of them to have higher energy.

2.2. Effects of Imperfections in Nearly Perfect Crystals

The energy band theory of solids depends on Bloch's theorem, which can be proved for perfectly periodic potentials only. The introduction of defects or impurities into an otherwise periodic solid invalidates the applicability of Bloch's theorem, and thus the energy band structure derived from its use. However, in crystals containing only small densities of imperfections it can be assumed that the nonperiodic part of the potential is sufficiently small to be treated as a perturbation. In such cases the Bloch states are approximate eigenstates, and the effect of imperfections is to introduce a scattering between states with different values of the wave vector $\mathbf{k}$. We can define parameters such as the lifetime τ, the average time that an electron remains in a Bloch state, or alternatively, the mean free path Λ, the average distance traversed by an electron in a given state before it is scattered into another Bloch state. The exact eigenstates are no longer Bloch states, but are linear combinations of states of different $\mathbf{k}$.

A simple problem of the effects of a single imperfection in an otherwise perfectly periodic crystal was solved by Koster and Slater[4] using perturbation theory. Below a critical strength of the perturbing potential they found that the density of states remains unchanged. However, when the perturbation is sufficiently intense, two new states asymmetrically placed around the center of each band. If these states remain in the band, they resonate with the other states at the same energy. However, for sufficiently large perturbations some of these states may fall in the gap regions, in which case they represent localized bound states. In metallic crystals these states are not very important since the Fermi energy falls in the middle of a band and only states near the Fermi energy contribute to transport. However, in semiconductors or insulators the bound states can dominate the transport behavior.

A net attractive (negative) perturbing potential preferentially lowers the energies of the unperturbed states from which the impurity states arise, while a net repulsive (positive) perturbation increases the energy levels. Thus, for example, an isoelectronic impurity that is more electronegative than the atom for which it substitutes is likely to provide an excess localized attraction compared to the perfect crystal, and this could cause a state to split off from the bottom of a band into the gap. If this state came from above the Fermi energy in the perfect crystal but moves below it because of the perturbation, the impurity becomes negatively charged. In a metal this rearrangement of charge density leads to essentially complete screening of this excess charge. In a covalent semiconductor, however, the Fermi energy is in the gap. If the state was previously unoccupied and is lowered from the conduction band into the gap, the Fermi energy at least at low temperatures must fall between the impurity state and the top of the valence band. The impurity then remains neutral, but acts as an acceptor. If the state actually falls below the top of the valence band, it would appear that the impurity can ionize and leave a hole in the valence band, even at $T = 0$. However, the situation is now one in which the impurity is negatively charged and thus will attract the mobile hole. The problem must be redone with the perturbation being the negatively charged impurity. If the hole is retrapped, the state remains an acceptor level. The latter problem is usually handled in a hydrogenic approximation,[5] which is valid only if the ground-state orbit is large compared to the interatomic separation. If, on the other hand, the impurity is not isoelectronic with the atom for which it substitutes but contains an extra electron, the situation in a covalent semiconductor is very different. If the impurity state does not split off from the conduction band, then the extra electron could reach equilibrium in the bottom of the conduction band and contribute to transport. However, this implies that the impurity becomes positively charged, and once again a bound state could result from its electrostatic attraction with the electron. But if the state is below the bottom of the conduction band, it remains filled at $T = 0$, and the neutral impurity acts as an ordinary donor. These problems have rarely been treated adequately in a quantitative manner, but the importance of self-consistency is evident. Some of the possibilities for a single impurity level in a covalent semiconductor are sketched in Figure. 4.

The possibilities in ionic crystals are as varied as in covalent crystals. Here it is important to maintain charge neutrality for the solid as a whole, so that introduction of an atom of different valence than the one for which it substitutes must be compensated for either by a change in the charge state of one of the host ions or by a vacancy

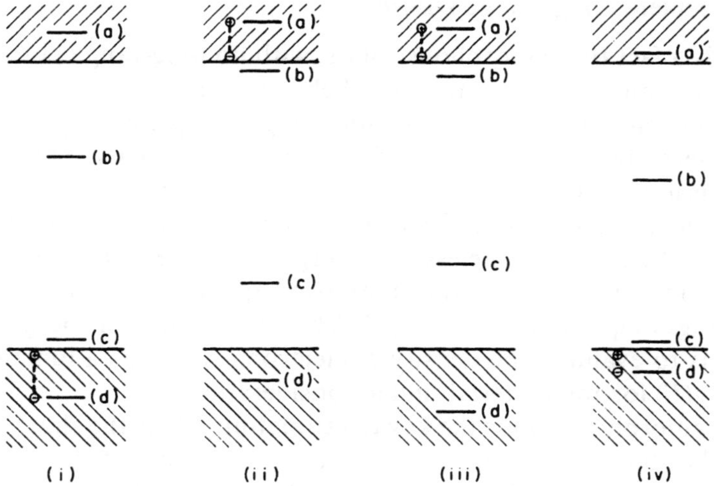

Fig. 4. Energy band structure of a nearly perfect covalent semiconductor containing a single impurity. (i) Isoelectronic impurity more electronegative than the atom for which it susbstitutes: (a) no effect of impurity upon transport; (b) deep acceptor level, little effect on transport; (c) shallow acceptor level, dominates transport at low temperatures; (d) ionized impurity, free hole even at $T = 0$. (ii) Electronegative impurity with an extra electron: (a) ionized impurity, free electron in conduction band; (b) shallow donor level; (c) deep donor level; (d) resonant state in valence band. (iii) Isoelectronic electropositive impurity: (a) ionized impurity, free conduction electron; (b) shallow donor level; (c) deep donor level; (d) resonant valence band state. (iv) Electropositive impurity with one fewer electron: (a) resonant conduction band state; (b) deep acceptor level; (c) shallow acceptor level; (d) ionized impurity, valence band hole.

that is capable of electrostatically trapping charge. The electronic structure can often be simply calculated in ionic solids,[6] since the perfect crystal usually contains only ions with filled electronic shells.[7]

The purpose of this qualitative discussion is to point out the many possibilities which can occur for even a single imperfection in an otherwise perfect crystal. It is clear, however, that (1) the effects depend sensitively on the type of solid, (2) imperfections can lead to localized states in the forbidden gaps of crystals, and (3) the bulk of the states in the bands are not greatly affected by the deviations from periodicity, provided the concentration of imperfections is low; they remain delocalized, and are linear combinations of Bloch states.

2.3. Complete Long-Range Disorder

It is a fact that amorphous solids, with no long-range positional order, exist and that their transport behavior is not generally qualitatively different from that in corresponding near-perfect crystals. Yet all of energy band theory is based on perfect periodicity. The reason for this is not naivety or deliberate deception, but simply that quantitative calculations have thus far proved to be impossible without perfect periodicity. Nevertheless, the surge of interest in amorphous solids in recent years has led to both some qualitative results and the beginnings of quantitative theory.[8] The first purpose of energy band theory is to evaluate the density of electronic states as a function of energy $g(E)$ for the material under consideration.

In amorphous solids the key to estimating $g(E)$ is the short-range order that must exist from chemical considerations if the state is to be metastable. From the viewpoint of the tight-binding approximation[8] of ordinary crystalline band theory the electronic energy levels of a solid depend primarily on the type of atom which contributes the electron under discussion and on the type and position of its nearest neighbors. Structural studies[9,10] indicate, however, that the nearest-neighbor environment is often essentially identical in corresponding crystalline and amorphous solids. The effects of slight

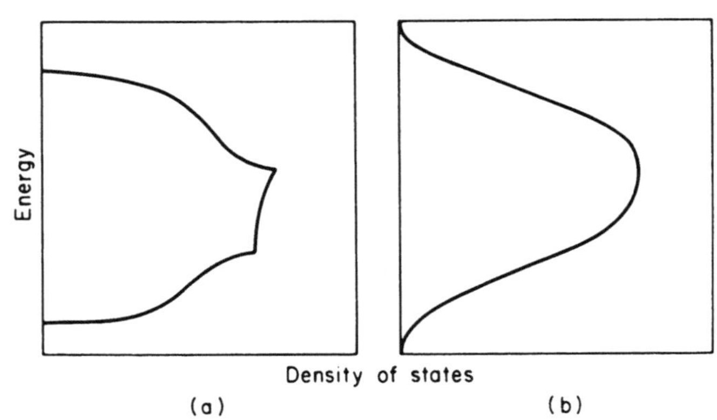

Fig. 5. (a) Density of electronic states as a function of energy for a single band of crystalline solid. The sharp behavior at the band edges and in the interior are the Van Hove singularities. (b) Density of electronic states as a function of energy for a single band of an amorphous solid. All van Hove singularities have disappeared.

variations in nearest-neighbor orientations and consequently increasingly larger positional deviations in farther neighbors in the amorphous as compared to the crystalline material can then be introduced as a perturbation on the calculable eigenstates of the crystal. Consider the simplest possible density of states for a single, nondegenerate band of a crystal, shown in Figure 5(a). The sharp structure, both at the band edges and internally, represents the van Hove singularities,[11] which are a necessary consequence of three-dimensional periodicity. If we introduce long-range structural disorder as a non-**k**-conserving perturbing potential, we should expect all the states to shift somewhat in energy, some up and some down. On the average the perturbing potential should be neutral, since the atoms are identical in the amorphous and crystalline solids; thus, the center of gravity of the band should remain unchanged by the perturbation. The extent of the shifts in energies of the states depends on the amount of short-range disorder present, but in any event the perturbation removes the van Hove singularities. In particular, the band edges should no longer be sharp, and some states should exist beyond the crystalline band extrema. The resulting density of states of a band in an amorphous solid is shown in Figure 5(b). The states beyond the sharp extrema of the crystalline bands are called *band tails.*

Recent quantitative attempts at estimating $g(E)$ for amorphous Ge and Si are reviewed by Weaire.[8] Unfortunately, most of the models studied to date are most inaccurate in the vicinity of the band tails, and as we shall see in the next section, the transport properties of amorphous semiconductors are dominated by states in the valence and conduction band tails.

3. Transport Properties of Solids

The transport properties of any material are obtained by imposing one or more thermodynamic forces, such as external electric or magnetic fields or temperature gradients, across a sample of the material and then measuring the resulting flows of electrical current and heat. The simplest effects are the direct ones: the electrical current arising from an electric field, which gives the electrical conductivity σ; and the heat flow due to a temperature gradient, which yields the thermal conductivity κ. Cross-effects include the thermoelectric power α and the Hall coefficient R_H. These transport coefficients are important because they can ordinarily be simply related to the microscopic energy band structure and can be used to either verify theoretical predictions or set values on important parameters which are useful in estimating some of the other properties of the solid.

The most significant transport coefficient for these purposes is the electrical conductivity. Microscopically, electrical current results only from the flow of charged particles in an applied electric field; thus, in a solid the conductivity must be either electronic or ionic. In any event, the conductivity can be expressed microscopically as

$$\sigma = nq\mu \tag{2}$$

where n is the number of particles of charge q per unit volume available to flow in the applied field and μ is the average velocity per unit field attained by these particles. If parallel conduction mechanisms exist, we must sum their contributions to σ. The importance of Eq. (2) is that it relates a macroscopically measurable quantity σ to a microscopic parameter μ which can often be calculated. The parameter μ is called the microscopic mobility.

3.1. Transport in Metals

The most important characteristic of a metallic solid, evident from Figure 1, is the presence of a partially filled energy band. This means that there are unoccupied states available at essentially the same energy as the highest occupied states for electrons to scatter into if a small field is applied. These states enable the electrons near the Fermi energy to move easily, even in very small fields. Thus, metals have high electrical conductivity. If the metal were perfect, the electrons in the partially filled band would initially accelerate in a direction opposite (since they are negatively charged) to that of an applied field, the acceleration increasing with the time the field is on. This implies that the electrical conductivity becomes infinite if the field is applied for a long time. Actually, this is not the case, because if any electron reaches the top of the band, it undergoes a Bragg reflection and makes a large negative contribution to the conductivity. In fact, in a perfect metal, if we wait sufficiently long, the conductivity would in fact vanish, since the positive and negative contributions exactly cancel as an electron passes through all the states in a band. However, all this is irrelevant, since perfect crystals do not exist, and any deviation from periodicity leads to scattering processes. Since the electric field produces a nonequilibrium distribution of electrons and scattering is always such as to tend to restore equilibrium, the scattering processes act to reduce the energy of the electrons. Thus, the electrons lose the energy they gain from the field to the ion cores; this is the mechanism for the I^2R Joule heating that always occurs when a current flows through a material. But, more important, the electrons with the highest energies are scattered back toward the bottom of the band, and Bragg scattering is highly unlikely. Thus, real metals are characterized by large but finite conductivities. The mobility is usually written

$$\mu = e\tau/m^* \tag{3}$$

where e is the magnitude of the electronic charge, m^* is the effective mass of carriers in the band (determined only from the energy band structure at equilibrium), and τ is the relaxation time, a quantity which depends only on the scattering processes. The greater the scattering, the smaller is the value of τ. Any deviation from perfect periodicity contributes to τ. In particular, surfaces, grain boundaries, stacking faults, dislocations, vacancies, interstitials, impurities, and positional, compositional, and magnetic disorder all can be significant, depending on the sample being investigated. One of the most important effects is dynamic positional disorder, or phonons, since they are always present at finite temperatures. This produces a temperature-dependent scattering and yields a relaxation time inversely proportional to temperature. At high temperatures phonon scattering usually predominates in high-quality metallic samples. At low temperatures the relatively temperature-independent scattering from imperfections eventually controls the relaxation time. Quantitative calculations of phonon scattering and scattering by static imperfections have been carried out.[12] The thermoelectric power of a metal can be expressed as

$$\alpha = \frac{\mu^2 k^2 T}{3e}\left[\frac{d \ln \sigma(E)}{dE}\right]_{\varepsilon_F} \tag{4}$$

where $\sigma(E)$ is the contribution to the conductivity associated with electrons in states at energy E.[13] Thus, α should be proportional to temperature and of the order of 1 μV/°K at room temperature. For a spherical band the Hall coefficient is

$$R_H = -1/nec \tag{5}$$

where n is the carrier concentration.[14] Thus, R_H should be temperature independent and relatively small for metals.

3.2. Transport in Covalent Semiconductors

Since semiconductors contain a band gap between the highest filled state and the lowest empty state, the electrical conductivity vanishes at $T = 0$. There can be no conduction in a filled band because the field can only redistribute the electrons over the same states, and thus no net charge can be transported. At sufficiently high values of the applied field electrons can tunnel from the valence band into the conduction band, thereby creating free carriers even at $T = 0$; however, such fields are rarely achieved in practice with typical semiconductor band gaps, of the order of 1 eV. In a perfect semiconductor the dominant conduction mechanism is thermal excitation of electrons into the conduction band. Since such excitation leaves a free hole in the valence band, ambipolar conduction is possible, and Eq. (2) must be generalized to

$$\sigma = ne\mu_e + pe\mu_h \tag{6}$$

where p is the free-hole concentration and μ_e and μ_h are the electron and hole mobilities, respectively. The carrier concentrations of electrons and holes are equal, and are determined from the density of states and Fermi–Dirac statistics. At temperatures sufficiently low that the energy gap E_g is large compared to kT, the Boltzmann limit is applicable and in the effective-mass approximation[12]

$$n = p = \tfrac{1}{4}(2m^*kT/\pi\hbar^2)^{3/2} \exp(-E_g/2kT) \tag{7}$$

where m^* is the geometric mean of the electron and hole effective masses.

As is the case for metals, the mobility of carriers in a semiconductor depends on the predominant scattering mechanism. The scattering can be quite different in the two bands, particularly if the two bandwidths are not the same order of magnitude. If scattering by acoustic phonons predominates,

$$\mu_A = AT^{-3/2} \tag{8}$$

where A is independent of temperature. If the scattering is predominantly via optical phonons, for temperatures less than $\hbar\omega_o/k$, where ω_o is the optical phonon frequency, then the mobility can be expressed by

$$\mu_P = (\hbar^3 k/2em^*\omega_o)^{1/2}(e^{\hbar\omega_o/kT} - 1) \tag{9}$$

On the other hand, for $kT > \hbar\omega_o$ the mobility decreases with temperature as

$$\mu_P = BT^{-1/2} \tag{10}$$

where B is a constant. For ionized-impurity scattering

$$\mu_I = CT^{3/2}/[\ln(DT^2) - 1] \tag{11}$$

where C and D are constants. The mobility resulting from scattering by neutral impurities is given by

$$\mu_N = e^3 m^*/20 N_N \hbar^3 \varepsilon_0 \tag{12}$$

where ε_0 is the static dielectric constant and N_N is the neutral-impurity concentration. Although the mobility given by Eq. (12) looks constant, in many situations N_N increases sharply as the temperature is lowered. When more than one scattering mechanism is important, the actual mobility is the harmonic sum of the mobilities given in Eqs. (8)–(12). But, in general, the bulk of the temperature dependence of the electrical conductivity of an intrinsic semiconductor comes from the free-carrier concentration, Eq. (7), rather than from the mobility.

When imperfections are present drastic modifications of the conductivity of covalent semiconductors can result. Nevertheless, the basic techniques are the same, given the actual density of states $g(E)$ including the contributions from imperfection levels. Fermi–Dirac statistics and $g(E)$ are used to find the Fermi energy as a function of temperature. Once $\varepsilon_F(T)$ is known, the free-electron and free-hole concentrations immediately follow. For example, if there is only one type of impurity present, e.g., a concentration N_d of shallow donor levels at an energy E_d below the conduction band minimum, then at very low temperatures ($kT \ll E_d$) the Fermi energy is halfway between the donor level and the bottom of the conduction band; then the free-electron concentration is given by

$$n = N_d^{1/2}(m^* kT/2\pi\hbar^2)^{3/4} \exp(-E_d/2kT) \tag{13}$$

At temperatures large compared to E_d/k all the donors are ionized, and the free-carrier concentration becomes

$$n = \tfrac{1}{2}N_d[1 + (1 + 4n_i^2/N_d^2)^{1/2}] \tag{14}$$

where n_i is the intrinsic carrier concentration, which follows from

$$n_i = \tfrac{1}{4}(2m^* kT/\pi\hbar^2)^{3/2} \exp(-E_g/2kT) \tag{15}$$

If a temperature region exists for which $kT \gg E_d$, but kT is still sufficiently small compared to E_g that $n_i \ll N_d$, then an exhaustion region is obtained in which

$$n = N_d \tag{16}$$

independent of temperature. At much higher temperatures n_i becomes much larger than N_d and conduction is intrinsic. Conduction in the exhaustion region is called *extrinsic*. The free-hole concentration always follows from the neutrality condition

$$p = n - N_d^+ \tag{17}$$

where N_d^+ is the concentration of ionized donors.

For ambipolar conduction the thermoelectric power is given by

$$\alpha = (\alpha_n\sigma_n + \alpha_p\sigma_p)/(\sigma_n + \sigma_p) \tag{18}$$

where σ_n and σ_p are the contributions to conduction from electrons and holes, respectively, and α_n and α_p are

$$\alpha_n = -(k/e)[(E_c - \varepsilon_F)/kT + a_n] \tag{19}$$

$$\alpha_p = (k/e)[(\varepsilon_F - E_v)/kT + a_p] \tag{20}$$

Here E_c and E_v are the conduction band minimum and valence band maximum, respectively, and a_n and a_p are constants of the order of 2–4, depending on the predominant scattering mechanism. These constants represent the transfer of kinetic energy through the sample due to the fact that the average carrier energy relative to the band edge is greater at the hot than at the cold end. In an intrinsic semiconductor the Fermi energy is essentially equidistant from the valence and conduction band edges. Thus, from Eqs. (19) and (20) α_n and α_p are nearly equal and opposite, and Eq. (18) shows that the thermoelectric power will be dominated by the carriers that have the larger mobility. Thus, the sign of α is usually a good guide to whether electrons or holes dominate the conduction.

The low-magnetic-field Hall coefficient for ambipolar conduction is given by a somewhat different averaging:

$$R_H = \frac{\sigma_p\mu_p - \sigma_n\mu_n}{(\sigma_p + \sigma_n)^2}\frac{r}{c} \tag{21}$$

where r is a constant of order unity that depends on the predominant scattering mechanism. Thus, the Hall coefficient in the intrinsic region will also be dominated by the carriers of higher mobility. However, note that the squares of the mobilities enter R_H, while the expression for α contains only the first powers of the mobilities. Thus, for ambipolar conduction there can be an extensive temperature region for which α and R_H have opposite signs.

3.3. Transport Properties of Ionic Solids

The energy band structure of an ionic solid may not differ very much from that of a covalent solid, except perhaps for the magnitude of the energy gap. In most ionic materials the component ions contain filled electronic shells. In such cases it always takes a great deal of energy to promote an electron from a filled shell to the lowest unfilled energy level. In the solid, as Figure 3 indicates, this means that a large separation in energy usually exists between the highest filled band and the lowest empty band. Thus, ionic solids are generally wide-gap insulators, and if the gap is very large, no measurable electronic conductivity occurs in perfect crystals at room temperature (e.g., for $E_g = 10$ eV and $T = 300°$K, $e^{-E_g/2kT} \simeq 10^{-87}$).

But in ionic solids another mechanism of electrical conduction becomes possible: the motion of the ions themselves in the applied electric field. This mechanism is called ionic conduction, and can exist even in a near-perfect crystal. However, ionic conduction is usually negligible compared to electronic conduction, for several reasons. Since ions are generally about 10^5 times as massive as electrons, the velocities that ions attain at any given energy are down by a factor of 10^2–10^3 compared to those of electrons; thus, ionic mobilities are correspondingly low, and rarely exceed 10^{-3} cm^2/V-sec. Furthermore, ions can move through the solid only by means of diffusion, and this generally requires the existence of vacancies or interstitials. If the solid is initially perfect, these imperfections must be created. It takes an energy of approximately 1–5 eV to create an interstitial–vacancy pair, and then an energy of equivalent magnitude to move the vacancy or interstitial through the material. The total activation energy for intrinsic ionic conduction is half the energy of creation of a defect plus the energy necessary to move it, i.e., of the order of 5 eV. Clearly, intrinsic ionic conduction is unobservable at all temperatures unless the electronic energy gap exceeds twice this value. Even then, because of the greater mobility of electrons, electronic conduction must dominate at sufficiently high temperatures.

In imperfect crystals interstitials and vacancies may already be present. In such cases the activation energy for ionic conduction is reduced to just the energy necessary to move the defect. This is called extrinsic ionic conduction, and it can dominate the conductivity at lower temperatures provided the defect does not produce electronic states in the gap with a lower activation energy than that required to move the vacancy or interstitial ion. For filled-shell electronic configurations the presence of such defects generally does not reduce the electronic activation energy very much, and extrinsic ionic conduction is often observed.

If impurities are present, particularly transition metal impurities, shallower electronic states can be produced in the gap. Since this leads to a low-activation-energy conduction process, it predominates at sufficiently low temperatures even if the impurity concentration is quite small.

3.4. Mott Insulators

The energy band theory of solids discussed in Section 2 is based on the one-electron approximation, which as indicated previously is not always valid. As discussed in Section 2.1, the outer electrons of metallic solids delocalize and spread into bands, and since the uppermost occupied band is only partially filled, the electronic energy of the solid is lower than that of the well-separated atoms. Thus, the cohesive energy of a metallic solid arises from a decrease in kinetic energy of the outer electrons via delocalization. However, we pay a price for this decrease in kinetic energy, namely an increase in potential energy. This comes about because, in the delocalized state, any two electrons with antiparallel spins, which never could have approached each other if they were localized on different atoms, must for a time be simultaneously present on the same atom. In that case their electrostatic repulsion acts to increase the potential energy of the solid. Usually, because of screening effects from the many free carriers in a metal, this intraatomic Coulomb repulsion is reduced from its free-atom value of 5–20 eV to the order of tenths of an eV. On the other hand, the kinetic energy reduction arising from delocalization is of the order of one-quarter of the bandwidth, as can be seen from Figure 1, and the width of the conduction band in metals is typically 10–20 eV.

Mott[15] was the first to point out that for materials in which the outer electrons form a *narrow* band the kinetic energy reduction upon delocalization need not exceed the potential energy increase, and the ground state of such a system is thus one in which these electrons remain localized on their atoms. If such a situation occurred for a simple metal like Na, there would be no cohesive energy, and the material would not condense at all. However, if other types of bonding were responsible for the cohesive energy, some outer electrons need not delocalize. There are two simple examples of this effect. Consider a rare earth solid, such as Ce. Cerium atoms contain four outer electrons, two $4f$ electrons, and two $6s$ electrons. When these atomic states spread and form bands, states from three atomic levels, the $4f$, $5d$, and $6s$ levels, all have energies near the Fermi energy. Since the bandwidth is approximately proportional to the overlap of the electronic wave functions from nearest-neighboring atoms and the spatial extent of the wave functions increases with increasing principal quantum number, the $6s$ band is quite wide, the $5d$ band is narrower, and the $4f$ band has essentially no width at all (less than 0.01 eV). Thus, the $6s$ electrons delocalize and begin to fill the overlapping $6s$–$5d$ bands, but the $4f$ electrons remain localized despite having energies near ε_F. Consequently, the $6s$ electrons are responsible for the cohesive energy of Ce; the $4f$ electrons make essentially no contribution. Since the $6s$–$5d$ band has room for 12 electrons per atom and only 2 are present, Ce is a metal despite the localization of the $4f$ electrons.

A more interesting example is CoO, a primarily ionic compound composed of Co^{2+} ions and O^{2-} ions; the latter have a closed-shell electronic configuration, but each Co^{2+} ion has seven outer $3d$ electrons, three fewer than necessary to fill an electronic shell. Because of the rather large separation between Co^{2+} ions and the rather small covalency effects with the O^{2-} ions, the $3d$ band of CoO is quite narrow.(16) But the only electrons near the Fermi energy of CoO are these $3d$ electrons; thus, the lack of delocalization of the $3d$ electrons implies that instead of a 70%-filled band, CoO has its filled bands separated from its empty bands by an energy gap. This has major implications for transport, since instead of being a metal, CoO is an insulator. In order to differentiate such insulators from ordinary band insulators, they can be called *Mott insulators*.

An immediate problem with regard to Mott insulators is accounting for their cohesive energy. Certainly, in CoO the $3d$ electrons make little or no contribution because they are localized. The answer is, of course, that these materials are bonded *ionically*—the transfer of charge from the cobalt atoms to the oxygen atoms results in an electrostatic attraction which stabilizes the solid. The additional metallic bonding possible via the $3d$ electrons does not occur because the intra-atomic Coulomb repulsion would actually lead to a decrease in the cohesive energy.

A more substantial problem is determining when electrons do not delocalize. It would be nice if this could be determined from the results of a band structure calculation, but the major force preventing delocalization is a correlation effect, which cannot be handled within the one-electron approximation.

The important parameters in determining whether Mott localization occurs are the bandwidth Δ and the intraatomic Coulomb repulsion U. The bandwidth is obtainable, in principle, from a careful calculation, but U is difficult to estimate. In the atomic limit U is just the difference between the ionization potential and the electron affinity of the free ion, but the screening in the solid both from electronic effects and ionic polarization is difficult to estimate. Once Δ and U are known it is clear that the condition $\Delta \gg U$ leads to ordinary band formation, whereas $\Delta \ll U$ results in Mott localization. A model that exhibits these limits in a simple manner was suggested by Hubbard.[17,18] In this model all *interatomic* Coulomb repulsion effects between electrons are neglected, an approximation equivalent to replacing the real (screened) Coulomb interaction by an extremely short-range repulsion that vanishes for distances greater than one atomic diameter. For a single, nondegenerate band Hubbard suggested the Hamiltonian

$$H = \sum_{\mathbf{k},\sigma} \varepsilon(\mathbf{k}) n_{\mathbf{k}\sigma} + U \sum_i n_{i\uparrow} n_{i\downarrow} \tag{22}$$

where the $\varepsilon(\mathbf{k})$ are the one-electron energies as a function of the wave vector $\mathbf{k}$, the $n_{\mathbf{k}\sigma}$ are the operators that give the number of electrons of spin σ in the one-electron state $\mathbf{k}$, and the $n_{i\sigma}$ are the operators that give the number of electrons of spin σ on the atom at $\mathbf{R}_i$. The Hamiltonian (22) consists of two terms, the first being just the one-electron Hamiltonian. The second term increases the energy of the system by U whenever a spin-up electron and a spin-down electron are simultaneously present on the same atom. Thus, this term explicitly adds intraatomic correlations to the one-electron Hamiltonian. The Hamiltonian (22) also has the virtue of being exact in the two opposite limits of $U \ll \Delta$ and $U \gg \Delta$. If $U \simeq 0$, the Hamiltonian clearly reduces to the ordinary band limit. On the other hand, if $\Delta \simeq 0$, $\varepsilon(\mathbf{k}) = E_a$, independent of $\mathbf{k}$, and the first term reduces to NE_a, where N is the number of atoms (assuming each atom has one outer electron). Then the Hamiltonian is diagonal in the i, σ representation, the eigenvalues being

$$E_n = NE_a + nU \tag{23}$$

where n is the number of atoms with two outer electrons simultaneously present. This is clearly the correct total energy in the atomic limit, in which the energy of each unoccupied atom (i.e., positive iron core) is taken as the zero of energy. Then each singly occupied atom (neutral atom) has energy E_a, the negative of the ionization potential of the atom, and each doubly occupied atom (negative ion) has energy $2E_a + U$, U being the difference between the ionization potential and the electron affinity of the atom. If n of the atoms are doubly occupied, then n must be unoccupied, leaving $(N - 2n)$ singly occupied atoms. Thus, the total energy of the system is $E_n = (N - 2n)E_a + n(2E_a + U) = NE_a + nU$, in agreement with (23). The ground state in the atomic limit is clearly

$$E_0 = NE_a \tag{24}$$

corresponding to the situation in which exactly one electron is on each atom at all times. The first excited state is

$$E_1 = NE_a + U \tag{25}$$

which is an energy U above the ground state. Thus, an energy gap equal to U exists, and the material is an insulator. Ordinary band theory, of course, could result only in a half-filled band for the case

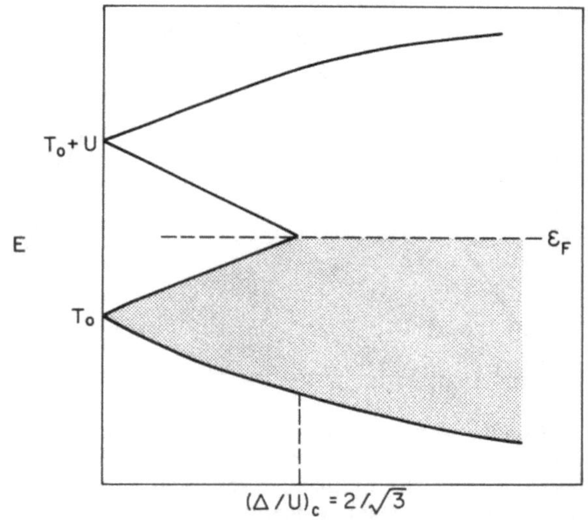

Fig. 6. Sketch of the quasiparticle energies as a function of the ratio of the bandwidth Δ to the intra-atomic Coulomb repulsion U for a material with a half-filled nondegenerate band. From Hubbard.[18]

of one electron per atom, since there are two states per atom (one of each spin) in a nondegenerate band.

Unfortunately, even the simplified Hamiltonian (22) cannot be solved with any degree of accuracy, and most attempts have been based on further approximations or on perturbation expansions around the atomic or band limits. Since the atomic limit is highly degenerate, degenerate perturbation theory must be used in this case, adding more difficulties. Nevertheless, several widely different analyses have indicated that the transition between an insulating ground state and a metallic ground state occurs rather near $\Delta \simeq U$.[18–20] The band structure found by Hubbard using a simple model for the one-electron density of states is shown in Figure 6, in which the transition occurs at $\Delta = 1.16U$.

The transport properties of Mott insulators have been investigated by means of the Hubbard model.[21–23] The results of Bari

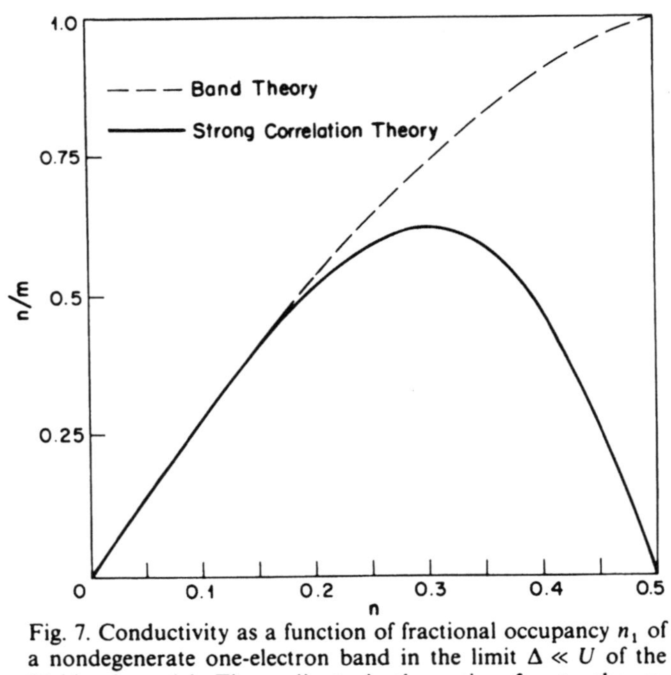

Fig. 7. Conductivity as a function of fractional occupancy n_1 of a nondegenerate one-electron band in the limit $\Delta \ll U$ of the Hubbard model. The ordinate is the ratio of n to the renormalized effective mass, a quantity proportional to the dc conductivity, expressed in units of $\Delta a^2/\pi$. The results of ordinary band theory are indicated by the dashed lines. From Bari *et al.*[21]

et al.[21] for the region in which the bandwidth is small compared to U are shown in Figure 7. The point $n = 0.5$ is the case of one electron per atom; here strong correlation theory gives semiconducting behavior instead of the metallic behavior predicted by ordinary band theory. The asymmetry of the solid curve between the top and bottom of the quasiparticle band illustrates that transport is somewhat different for an ordinary band semiconductor and a Mott insulator. The region just below $n = 0.5$ represents conduction in an imperfect Mott insulator. The presence of donors or acceptors leads to a temperature-dependent deviation of n from its perfect-crystal value of 0.5, depending on the number of ionized defects. The mobility must also be evaluated in order to obtain the full conductivity. This still depends on the mechanism dominating the scattering of carriers; however, spin-disorder scattering is intense in Mott insulators,[23,24] and can thus be expected to limit the mobility in nearly perfect crystals over a wide range of temperatures.

The Hall coefficient in the high-field limit has been calculated by Langreth.[25] It can be shown[26] that near the atomic limit the Hall coefficient is a factor of four smaller than that for an ordinary semiconductor, due to the strong correlations. When spin-disorder scattering dominates the mobility Brinkman and Rice[27] have shown that the Hall mobility $\mu_H \equiv c\sigma R_H$, which should be of the same order of magnitude as the microscopic mobility μ if only one type of carrier is present [see Eq. (21)], can become much smaller than the microscopic mobility at low temperatures. Since this is the region, however, where the spin-disorder scattering becomes quite small, other scattering processes might well mask this effect.

In the region in which the Hubbard model indicates metallic behavior it might be expected that the transport properties should be describable in terms of the ordinary expressions for metals given in Section 3.1. However, solutions to the model in this region have proved to be very difficult to obtain. Hubbard's own attempt,[18] based on an ad hoc Green's function decoupling scheme, does not possess a Fermi surface.[28] Brinkman and Rice[2] applied a simple variational approach due to Gutzwiller[30] and found an effective-mass enhancement which diverges near the point of transition from metal to insulator. Although the variational approach cannot be expected to be applicable in the region of the metal–insulator transition, this result does suggest variations from ordinary metallic transport in strongly correlated solids.

3.5. Polarons and Hopping Conduction

The other major approximation in addition to the one-electron approximation inherent in the band theory of solids discussed in Section 2 is the Born–Oppenheimer approximation. It is not difficult to see that this approximation can break down in ionic solids. For example, during a longitudinal optical lattice vibration positively and negatively charged ions move in opposite directions, the amplitude of this motion varying with position along the direction of oscillation. Thus, this type of vibration sets up alternating regions of positive and negative charge density in excess of the equilibrium values. Consider a delocalized (band) electron with most of its amplitude near a particular position in the primitive cell of an ionic crystal. The electron, being negatively charged, can lower its energy by inducing a lattice polarization which brings extra positive charge near the positions where its amplitude is large, and which pushes away some negative charge into regions where its amplitude is small. This means that there must be a strong coupling between the electrons and the longitudinal optical phonons, and such interactions are neglected in the Born–Oppenheimer approximation. This problem is handled by introducing a fictitious particle, the *polaron*, which is an electron that always moves around together with the associated lattice polarization that minimizes its energy. The electronic problem is replaced by the polaronic problem, and the remaining polaron–phonon interactions are assumed to be small. The polaron clearly has lower energy than the electron alone, but it has a larger effective mass, since it must carry its lattice distortion with it as it moves.

Polaron theory has been analyzed in various different limits, depending on the strength of the electron–phonon interaction, the extent of the lattice polarization, and the bandwidth of the electrons. In the limit in which the associated lattice deformation extends over a large number of lattice parameters, the lattice may be replaced by a continuum, much as in the Debye model for specific heat. It can be shown[31] that the extent of this deformation is given by

$$r_0 = (\hbar/2m^*\omega_o)^{1/2} \tag{26}$$

where m^* is the band effective mass and ω_o is the average frequency of the longitudinal optical phonons; r_0 is called the polaron radius. If r_0 is much larger than the lattice spacing, the polaron is called large. In large-polaron theory the parameter that determines the strength of the electron-phonon interaction is

$$\alpha \equiv \frac{1}{2}\left(\frac{1}{\varepsilon_\infty} - \frac{1}{\varepsilon_0}\right)\frac{e^2}{r_0}\frac{1}{\hbar\omega_o} \tag{27}$$

where ε_0 and ε_∞ are the static and high-frequency dielectric constants, respectively. For $\alpha < 4$ the coupling is called weak, and the electronic energy levels are modified only slightly. The energy of the electron is lowered by $\alpha\hbar\omega_o$, and its band effective mass is increased by a factor of $(1 + \alpha/6)$.[32] For $\alpha > 10$ strong coupling theory[33] applies, the decrease in energy becomes $0.1\alpha^2$, and the increase in effective mass is then a factor of $0.02\alpha^4$. When α is very large the polaron effective mass becomes so great that the polaron hardly moves in an applied electric field.

When r_0 is small compared to the interatomic spacing a the continuum approximation breaks down. Furthermore, if the energy E_p given by

$$E_p \equiv \frac{1}{\pi^2}\frac{r_0}{a}\alpha\hbar\omega_o \tag{28}$$

is large compared to the electronic bandwidth Δ, *small polaron* theory applies.[34,35] This situation comes about if the bandwidth is very small or if the coupling constant α is very large. The quantity E_p defined in Eq. (28) is the approximate reduction in energy due to small-polaron formation.

While the formation of large polarons, even in the strong coupling limit, does not lead to serious modifications of the transport properties of band electrons, quite the reverse is the case for small polarons. Holstein[35] found that for the perfect crystal small-polaron states overlap sufficiently to form a polaron band, in which ordinary band conduction can take place. This conduction should predominate at low temperatures. However, the width of this band can be shown to decrease exponentially with increasing temperature, and in the vicinity of half the Debye temperature the bandwidth becomes less than the uncertainty in energy due to the finite lifetime of polaron states. Above this temperature electrical transport in the polaron band is negligible, and the small polaron can be thought of as localized. This localization comes about because the electron can form a bound state with the local lattice deformation that it induces.[36] The deformation extends only to nearest neighbors in the small-polaron limit.

Once trapped, the only means by which the polaron can contribute to conduction is by "hopping" from one lattice position to an equivalent one. But for an equivalent position to exist, either the lattice around an unoccupied site must distort, the lattice around the polaron must "undistort," or some combination of the two must take place. All of these lattice deformations require energy in the form of longitudinal optical phonons. At low temperatures there are few such phonons present, and the hopping probability is small; but at high temperatures there are exponentially more phonons present, and the hopping probability is much larger. Hopping conduction can be considered simply as a diffusion of carriers through the lattice with the assistance of phonons. Since the diffusion constant is related to the mobility by an Einstein relation, it is clear that the mobility of a localized carrier conducting by hopping from site to site is of the form

$$\mu = \mu_0 e^{-W/kT} \tag{29}$$

where W is the minimum energy necessary to obtain two equivalent sites. The energy W can be easily evaluated in terms of the small-polaron binding energy E_p by noting that for small deviations around any equilibrium configuration the potential must vary quadratically with a suitably defined distortion parameter. Thus, the minimum energy to obtain an equivalent site must be the situation in which the region around the carrier "undistorts" half-way and the region around a neighboring site distorts half-way; this only requires a total of half the polaron binding energy, since the latter requires a full "undistortion" of the polaron site and the energy increases quadratically with distortion. Thus, Eq. (29) can be written

$$\mu = \mu_0 e^{-E_p/2kT} \tag{30}$$

The preexponential μ_0 can be somewhat temperature dependent. An algebraic factor of T^{-1} enters from the Einstein relation relating mobility to diffusion,

$$\mu = eD/kT \tag{31}$$

where D is the diffusion constant. For the small polaron at high temperatures Holstein[35] finds an additional factor of $T^{-1/2}$ in D, so that μ_0 varies as $T^{-3/2}$. Of course, the exponential itself dominates the temperature dependence of the conductivity.

Hopping conduction via localized states is a very different process from band conduction through delocalized states. But since both processes require temperature activation in semiconducting solids, they cannot be distinguished by electrical conductivity measurements alone. Equation (2) expresses the conductivity as a product of the carrier concentration n and the mobility μ. For both hopping conduction and semiconduction we expect

$$\sigma = \sigma_0 \, e^{-E_a/kT} \tag{32}$$

where E_a is the experimentally determinable activation energy. The unique feature of hopping conduction, however, is Eq. (30), an activation energy of the *mobility*. None of the scattering mechanisms discussed for free carriers, Eqs. (8)–(12), exhibits any such temperature dependence. Essentially all of the contribution to E_a in band semiconductors comes from the carrier concentration, and E_a can be associated with half the energy gap in the intrinsic region, Eq. (7). For hopping conduction there still can be a contribution to E_a from n, e.g., if the polarons first must be freed from the vicinity of a defect site to which they are electrostatically bound in order to participate in dc hopping conduction, but there also must be contribution to E_a from μ. Thus, for hopping we may write

$$E_a = E_b + E_p/2 \tag{33}$$

where E_b represents the energy necessary to free the carrier from a defect site.

In ordinary semiconduction, as Eqs. (18)–(21) indicate, thermoelectric power or Hall effect measurements can be used to separate mobility from carrier concentration. Analogous results can be derived for hopping conduction. The thermoelectric power becomes[37]

$$\alpha = \frac{k}{q}\left(\frac{E_b}{kT} + a\right) \tag{34}$$

where a could be very small if no vibrational energy is transferred by the hopping. The Hall effect arising from the hopping of small polarons has proved more difficult to evaluate. Initial work was contradictory,[38–40] but recent calculations[41–43] have confirmed the original result of Friedman and Holstein[38] that for a triangular lattice (e.g., face-centered or body-centered cubic) the Hall mobility for small-polaron hopping in the high-temperature limit is

$$\mu_H = \mu_0 \, e^{E_p/6kT} \tag{35}$$

where μ_0 varies as $T^{-1/2}$. Thus, the Hall mobility has only one-third the activation energy of the microscopic mobility μ and is generally larger. Furthermore, because the preexponential decreases with increasing temperature, the Hall mobility reaches a maximum when $kT = E_p/3$; at higher temperatures the Hall mobility decreases with increasing T. But perhaps the most intriguing result obtained is that the Hall coefficient for small-polaron *holes* is negative (n type), just as for electrons. This suggests that when the predominant conduction mechanism is hopping of small-polaron holes, the sign of the thermoelectric power should be p type but that of the Hall coefficient should be n type. For a square lattice[38] the Hall mobility has been estimated as being very close to the actual mobility, and the sign anomaly for holes does not occur. Thus, the local coordination of the atoms is extremely important in analyzing Hall-effect experiments.

It should be pointed out that small-polaron theory does not apply to metallic crystals, because of the large free-carrier concentrations that exist in these materials. The lattice cannot distort around one carrier per site, since the same atoms would have to move in opposite directions to lower the energy of two electrons in nearest-neighboring primitive cells. The theory begins to become invalid for imperfect semiconductors also, when the free-carrier concentration becomes too large. Even in the extreme small-polaron limit of nearest neighbor distortions only, an 8% donor or acceptor concentration is too large for full polaron stabilization in a face-centered cubic or hexagonal close-packed lattice, for which each atom has 12 nearest neighbors. This focuses on another important distinction between small and large polarons. In large-polaron theory the strong electron–phonon interaction is treated as a zeroth-order term in the Hamiltonian. Instead of looking for electron states, Schrödinger's equation is solved for polaron states; electrons without their associated phonon distribution never enter the problem. On the other hand, in small-polaron theory the ordinary electronic states are found. For semiconductors this gives the usual valence and conduction bands. Only after free carriers are created, by either intrinsic or extrinsic processes, does the lattice distort and the polarons form.

Hopping conduction is not associated only with small-polaron formation, and the same process can occur whenever the carriers are in localized states. At low temperatures in any imperfect semiconductor or insulator there are always carriers bound to donors or acceptors and thus in localized states. If there are both donors and acceptors present and some donor levels are above some of the acceptor levels, thermal equilibrium requires that equal concentrations of donors and acceptors are ionized even at $T = 0$. Let us assume a concentration N_d of shallow donors and N_a of shallow acceptors, with $N_d > N_a$. Then all N_a acceptors and N_a of the donors per unit volume will be ionized at very low temperatures, while a concentration $(N_d - N_a)$ of the donors will be localized. The random electric fields from the ionized donors and acceptors will spread the donor levels somewhat. Furthermore, the environments of an occupied and of an unoccupied donor are very likely quite different because of the presence or absence of the electron itself. In such a situation phonon-assisted hopping of the electrons through the localized donor levels becomes a possible conduction mechanism. The activation energy W is a measure of the spread in energy of the donor states, and should ordinarily be small compared to the ionization energy of the donor. Thus, at sufficiently low temperatures this mechanism of conduction, known as impurity conduction, should predominate. Detailed treatments of impurity conduction have been carried out by Miller and Abrahams[(44)] and by Mott and Davis.[(45)]

3.6. Amorphous Semiconductors

In Section 2.3 the expected one-electron density of states $g(E)$ of amorphous semiconductors was discussed. The most striking prediction of the present-day qualitative theory is the existence of disorder-induced band tails in place of the sharp band edges of crystalline solids. However, $g(E)$ by itself does not suffice in analysis of transport properties, since a knowledge of $g(E)$ only serves to set the Fermi energy and determine the equilibrium distribution of electrons at any temperature. In order to evaluate the electrical conductivity, the mobility of a carrier in any of the states must be known. In particular, we are interested in the mobility of electrons above ε_F and holes below ε_F. Given the ground-state configuration, we can place a carrier in any particular state and ask for the velocity it will obtain in an applied electric field. The possibility of quantum mechanical tunneling between any two states at the same energy implies that a single mobility can be ascribed to all states at that energy; thus we write $\mu(E)$. The electrical conductivity can then be written

$$\sigma = \int_{-\infty}^{\infty} dE g(E) f(E)[1 - f(E)] e\mu(E)/kT \tag{36}$$

where $f(E)$ is the Fermi–Dirac distribution function:

$$f(E) = (e^{(E-\mathcal{E}_F)/kT} + 1)^{-1} \tag{37}$$

An important point which is clear from the outset is that $\mu(E)$ must depend sensitively on $g(E)$. For example, consider a situation in which all of the states under discussion are localized. These states must then have zero mobility if they do not spatially overlap other states with the same energy. However, if a sufficient density of overlapping localized states exists at a particular energy, tunneling between them leads to a finite mobility. This situation reflects the fact that extended linear combinations of the degenerate localized states can be constructed, and these must also be eigenstates of the same Schrödinger equation. Since the tunneling probability falls off exponentially as the distance between the states increases, we can conclude that at energies where the density of states $g(E)$ is large, the mobility $\mu(E)$ can be orders of magnitude greater than at energies for which the density of states is small. This argument suggests that $\mu(E)$ falls off rapidly in the band-tail regions. Mott[46,47] proposed that at a critical value of energy in each band tail the mobility sharply decreases from a finite value to essentially zero. The suggested critical energies have come to be known as mobility edges, since they should play the same role in amorphous-semiconductor transport that the band edges play in crystalline-semiconductor transport.

There has been a great deal of recent theoretical work attempting to justify the existence of mobility edges and also to investigate their sharpness,[48–51] although the question remains an open one at present due to the general lack of rigor in the calculations. Most of the theoretical models investigated assume random potential wells on a periodic lattice, an approach more suited to the disordered-alloy problem than to the amorphous-semiconductor problem, for which long-range positional disorder is an essential feature. A quantitative model which does exhibit the effects of positional disorder is one based on a semiclassical, percolation-theoretic approach[49,52] This model assumes a disordering potential superimposed on the ordinary periodic potential, and uses the semiclassical condition that an electron is localized if its energy is less than all of the potential barriers within an enclosed region around its position. For a perfectly periodic crystal all electrons are localized if their energies are below that of the saddle point in the potential located between nearest-neighboring atoms and they are delocalized at all higher energies. Thus, a sharp conduction band edge exists at this critical energy. When a Gaussian disordering potential is superimposed on the periodic potential the problem becomes an application of classical percolation theory.[53] The percolation problem is essentially one of placing conducting spheres in random positions on the nodes of a particular lattice in such a way that they just touch if they are on nearest neighbors. The calculation usually performed is an evaluation of the fraction of conducting sites that lie on a completed path between the electrodes, $P(p)$, expressed as a function of the fraction of nodes that are conducting, p. This function can be shown to have a sharp threshold p_c below which the probability of any completed path through the system is essentially zero. It turns out[53] that $P(p)$ very sharply increases from zero to one in a relatively narrow region of p above p_c. If the mobility of a percolating carrier is taken as proportional to $P(p)$, a relatively sharp (exponential) mobility edge is obtained.[49] However, it is more reasonable that the carrier mobility should be taken as proportional to the conductivity of the percolating system rather than the percolation probability $P(p)$.[54] As is shown in Figure 8, the behavior of the conductivity $\sigma(p)$ in three dimensions is much more gradual than is $P(p)$,[55] and thus, semiclassically, the mobility edge is not sharp at all. Mott[51] has recently presented qualitative arguments that a more complete quantum mechanical analysis would yield a sharp edge.

If a sharp mobility edge does exist, then integration of Eq. (36) would always produce a temperature-activated conductivity, Eq.

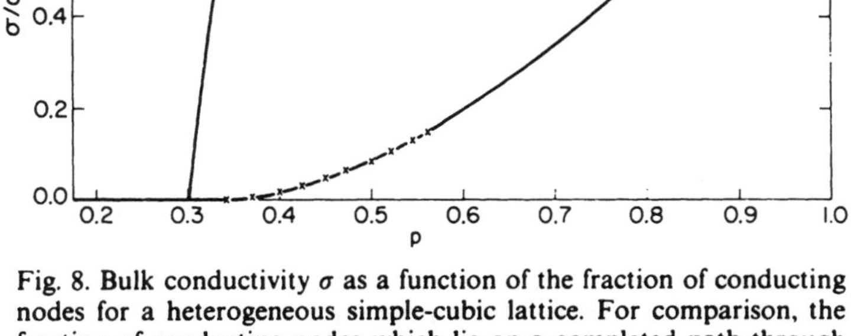

Fig. 8. Bulk conductivity σ as a function of the fraction of conducting nodes for a heterogeneous simple-cubic lattice. For comparison, the fraction of conducting nodes which lie on a completed path through the system, P, is also shown. From Adler *et al.*[55]

(32), as in ordinary semiconductors. The activation energy for holes and electrons would then be the difference between the Fermi energy and the valence and conduction band mobility edges, respectively. However, as opposed to the situation for ordinary crystalline semiconductors, even in the intrinsic region the Fermi energy need not be near the center of the gap.[10] At low temperatures ε_F should lie half-way between the actual band edges rather than the mobility edges, and if the band tails are extensive and asymmetric, ε_F could be quite temperature dependent. Furthermore, the free-electron and free-hole concentrations can be very different, since the Fermi energy can be much nearer one of the mobility edges than the other.

Cohen *et al.*[56] suggested the possibility that in extremely disordered systems the valence and conduction band tails could actually overlap in the gap, as is shown in Figure 9; this has come to be known as the CFO model. The proposed band structure would produce metallic behavior, due to the presence of a finite density of states at the Fermi energy, were it not for the fact that states in the band tail are localized. If a sharp mobility edge exists, ordinary semiconduction, as described by Eq. (32), still results. But there are still several unusual features of overlapping band tails. First, when the tail states are localized an important distinction between valence and conduction band states must be made. Since a covalent semiconductor

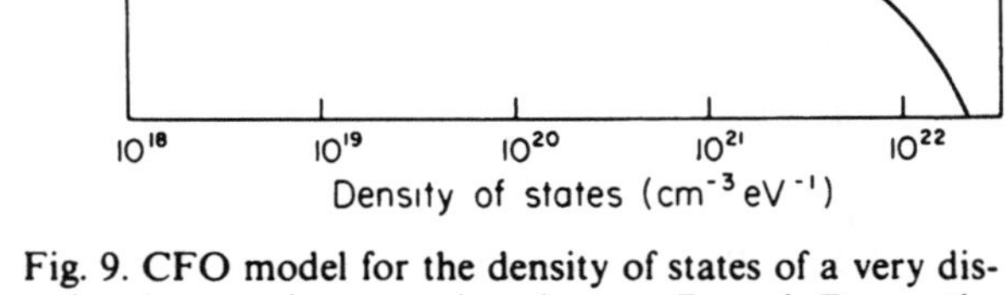

Fig. 9. CFO model for the density of states of a very disordered amorphous semiconductor. E_v and E_c are the valence and conduction band mobility edges, respectively. E_g is the mobility gap, and ε_F is the Fermi energy. The positions of the positively and negatively charged traps at zero temperature are schematically indicated.

ordinarily contains just enough electrons to fill the valence band while leaving the conduction band empty, localized valence band states represent neutral atoms only when they are filled; localized conduction band states represent neutral atoms only when they are empty. Thus, an empty state in the valence band tail must represent a positively charged ion, while a filled conduction band state corresponds to a negatively charged ion. Second, since the band tails overlap, there are states in the valence band, hence ordinarily filled, that have higher energy than ordinarily unfilled states in the conduction band. Therefore, in order to attain thermal equilibrium, a redistribution of electrons must take place, yielding the ground state shown in Figure 9. The filled states in the conduction band tail represent negatively charged ions, the empty states in the valence band represent positively charged ions. States deep in the valence and conduction band tails must be localized in different regions of space,(10) so that these charged traps are spatially separated. The entire model is based on the assumption that the energy band structure is independent of the occupation of the states, an assumption which is clearly invalid when the states involved are highly localized.(10,57) Nevertheless, since the CFO model is qualitative in nature, modifications due to taking electronic interactions into account can leave most of its essential features intact.

The CFO model assumes that ordinary donor and acceptor levels have insignificant concentrations in covalent amorphous semiconductors. This idea is based on a suggestion of Mott(46) that atoms in such solids almost always can find positions upon formation such that their valence requirements are locally completely fulfilled. However, donor and acceptor levels are easily introduced into the model by treating donor levels as states originating from the valence band (i.e., neutral when occupied) but located in the vicinity of the conduction band, and acceptor levels as states tailing from the conduction band to positions near the valence band.

When a finite density of states exists at the Fermi energy another conduction mechanism becomes possible, as discussed in Section 3.5. In fact, this mechanism, phonon-assisted hopping among states near ε_F, should predominate at sufficiently low temperatures, when the free-carrier concentration has become negligible. If W is the hopping activation energy and R is the average separation between states near ε_F, the hopping mobility is proportional to

$$\mu \propto e^{-\alpha R} e^{-W/kT} \tag{38}$$

where α gives the rate of fall-off of the localized wave functions in states near the Fermi energy. Mott[58,59] pointed out that as the temperature is decreased still further, because the term $e^{-W/kT}$ in (38) becomes extremely small, the carrier can increase its mobility by tunneling to farther than nearest neighbors, thereby finding a state whose energy is closer to its own. If the carrier hops an average distance of nR, it finds n^3 as many states available to hop into, and can on the average find a state only W/n^3 above its own energy (assuming the density of states is not a rapidly varying function of energy near ε_F). Thus the mobility should be a sum of terms of the form

$$\mu_n \propto [\exp(-\alpha nR)] \exp(-W/n^3kT) \tag{39}$$

Treating n as a continuous parameter, as is appropriate only if $n \gg 1$, the predominant term is the one that maximizes the exponent in (39); this occurs for

$$n = (3W/\alpha RkT)^{1/4} \tag{40}$$

Substitution of (40) in (39) shows that the mobility for what might be called "phonon-assisted tunneling" takes the form

$$\mu = A \exp(-T_0/T)^{1/4} \tag{41}$$

where $T_0 = 256\alpha^3R^3W/27k$. Such a mobility yields a conductivity whose logarithm is not a straight line as a function of T^{-1}, as is the case for ordinary semiconduction or phonon-assisted hopping. Instead, the logarithm of conductivity should be linear only as a function of $T^{-1/4}$ in the temperature region for which phonon-assisted tunneling predominates. Although crudely derived, it now appears that Eq. (41) is quite accurate, even in much more sophisticated models.[60,61] We should note, however, that at sufficiently high temperatures phonon-assisted hopping to nearest neighbors should eventually predominate, returning to the behavior of Eq. (32).

If a sharp mobility edge does not exist, as the percolation calculations suggest, then conductivity in delocalized states also will not give the simple activated behavior of Eq. (32), but a plot of the logarithm of σ as a function of T^{-1} will exhibit a continuously increasing slope with increasing T, qualitatively similar to the behavior of Eq. (41). In fact, for reasonable values for the extent of the band tail (determined by the variance of the Gaussian disorder introduced) a plot of $\log \sigma$ vs. $T^{-1/4}$ is linear over a fairly large temperature range (100–300°K).[55] This illustrates one of the problems inherent in interpreting the experimental data: When $\log \sigma$ is a linear function of T^{-1} conduction could be bandlike, or could be via phonon-assisted hopping either in one of the band tails or near ε_F; when $\log \sigma$ is a linear function of $T^{-1/4}$ conduction could be via phonon-assisted tunneling or could be bandlike just above a gradual mobility edge.

When there is a finite density of states at the Fermi energy these states can also dominate the thermoelectric power. In such cases the ordinary metallic expression for α, Eq. (4), should be applicable, and thermoelectric power should be proportional to the temperature. As noted in Section 3.5, when phonon-assisted hopping of localized holes predominates the Hall coefficient could be n type, depending on the local coordination.

If the strength of the electron–phonon interaction is large, polaron theory could be applicable to amorphous semiconductors. Particularly in highly localized states, such as should exist in band tails, the small-polaron limit can be expected to be appropriate. However, because of the absence of long-range order, the small-polaron band regime will be absent, and conduction will be only via hopping down to the lowest temperatures.

4. Insulator–Metal Transitions

As the discussion in Sections 2 and 3 indicates, both insulating and metallic states are possible for a wide class of materials, depending on the details of the density of states. It might be asked whether a given material can undergo a transition from one such state to another as a parameter such as temperature, pressure, or impurity concentration is varied. In fact, many such mechanisms have been proposed, and such transitions are quite commonly and spectacularly observed with accompanying conductivity discontinuities of factors of up to 10^{13}. In this section the suggested mechanisms are classified and discussed.

4.1. Band Overlap Transitions

A particularly simple type of transition can result from band effects entirely within the one-electron approximation, and without any necessary accompanying symmetry changes. Such transitions can be induced by either temperature or pressure and can lead to large discontinuities in conductivity. For example, if a material is a semiconductor at low temperatures and has a small energy gap that decreases with increasing temperature, then at a critical temperature the gap may pass through zero, bringing about a semimetallic state. Similarly, if the gap is a decreasing function of pressure, the gap can go through zero at a critical pressure, also bringing about a transition to a semimetal. But since a narrow-gap semiconductor has relatively high conductivity for an insulator and a semimetal with small overlap has relatively low conductivity for a metal, a simple band overlap transition should be rather gradual.

Most observed transitions are not gradual, but, in fact, are quite sharp. However, several modifications of band overlap transitions can be envisioned that sharpen the conductivity jump. The simplest is allowing for a thermodynamically first-order transition, in which the gap suddenly changes from positive to negative at the critical value of the parameter being varied. Since the primitive-cell volume depends on electronic structure, a first-order transition can be expected as a function of pressure. Furthermore, since total entropy is a sensitive function of the density of states, a first-order temperature-induced transition is also a possibility.

A second mechanism for attaining sharp band overlap transitions is via a change in crystal structure. Since the electronic density of states can vary qualitatively with crystal symmetry, it is not difficult to conceive of a band overlap semiconductor–metal transition accompanying a phonon-induced crystallographic phase transformation.

Still another possibility is that a different type of insulating phase forms at low temperatures—an excitonic insulator.[62] Such a state can arise if the one-electron band structure of a material indicates that it is a narrow-gap semiconductor or a semimetal with small band overlap. In any material with small carrier concentrations a free electron and a free hole attract each other with a Coulomb interaction approximately given by

$$V_C(e) = -e^2/\varepsilon_0 r \tag{42}$$

where r is the distance between the two carriers. The interaction (42) has bound-state solutions, indicating that the electron and hole move around together as a single quasiparticle called an exciton. If the binding energy of the exciton is larger than the band gap in a semiconducting state or the overlap in a semimetallic state, the material can lower its energy by forming bound excitons from electrons just below ε_F and holes just above ε_F. Since these relatively high-energy electrons and low-energy holes are depleted as they form excitons, the energy necessary to create the next electron–hole pair is an increasing function of exciton density. The spontaneous formation of excitons continues only as long as the energy for pair creation remains smaller than the exciton binding energy. Since the bound excitons have no net charge, they cannot contribute to conduction. If essentially all of the ordinarily available electrons and holes at any given temperature are tied up as excitons, conduction is significantly retarded, and the state is called an excitonic insulator.

The excitonic insulator can be considered an example of Bose condensation and, as such, is stable only below a critical temperature. The finite-temperature properties have been worked out only in the large-exciton limit, in which the Bohr radius of the exciton is large compared to the interatomic separation.[63] Formation of an excitonic insulator necessarily induces a lowering of the symmetry of the crystal. This symmetry reduction manifests itself either as a segregation of opposite spins on different sublattices (i.e., antiferromagnetic ordering) or a crystalline distortion. The latter occurs only if the electron–phonon coupling is sufficiently large. Above a critical temperature a second-order phase transition to a semiconducting or semimetallic state occurs. A first-order transition cannot be ruled out, especially when interexcitonic coupling is considered.[64] The second-order transition from excitonic insulator to semiconductor or semimetal shows no conductivity discontinuity.[65]

4.2. Electronically Induced Phase Transformations

This class of insulator–metal phase transitions makes use of the fact that in narrow-band materials a metallic state can be unstable at low temperatures toward a crystallographic distortion to lower symmetry.[66] This arises because the *electronic* energy of a metal always decreases upon distortion. For example, an infinitesimal pairing of atoms along a particular crystal axis splits all bands in two (by doubling the periodicity and hence halving all Brillouin zones). In general, the lower states of the band are reduced in energy and the upper states are increased in energy by the distortion. But in a metal the band is partially filled, so that at least some of the upper states are empty at $T = 0$. Thus, more filled states decrease than increase their energy as a result of the pairing, and the total electronic energy of the metal is lowered. The maximum reduction in ground-state electronic energy occurs for those particular distortions that decrease the energy of just that fraction of the band states that are filled and increase the energy of the fraction of empty states. For example, if a band is two-thirds filled and we can envision a distortion that lowers the energy of two-thirds of the states and raises the energy of the remaining one-third (as a trigonal distortion from cubic symmetry), the decrease in total ground-state electronic energy is clearly maximized. But this is just the distortion that tends to yield a semiconducting state. For very small distortions the band splittings are relatively small, and the probability of opening up a real energy gap in the three-dimensional density of states is not very large. However, if the band is quite narrow, even a small distortion can result in a real gap.

The crystal, however, does not spontaneously distort at $T = 0$ unless the *total* energy, electronic plus lattice, is reduced. In the harmonic approximation the lattice energy can be written

$$E_{\text{lattice}} = E_u + B\varepsilon^2 \tag{43}$$

where E_u is the undistorted lattice energy, ε is a dimensionless parameter that measures the fractional distortion, and B is a constant. No linear term in (43) appears since it is assumed that the undistorted lattice is the equilibrium configuration in the absence of electronic effects. Calculation of the ground-state electronic energy as a function of distortion is more difficult, but if the undistorted solid has a half-filled, isotropic band, then in the wide-band limit

$$E_{\text{electronic}} = N[E_0 - 0.15\Delta] - A\varepsilon^2 \tag{44}$$

where E_0 is the average band energy, Δ is the bandwidth, and A is a constant that depends on the strength of the electron–phonon coupling. On the other hand, for narrow bands

$$E_{\text{electronic}} = NE_0 - \tfrac{1}{3}NA\varepsilon - O(\Delta/\varepsilon) \tag{45}$$

Equation (44) shows that the decrease in electronic energy is proportional to ε^2 in the wide-band limit, and, since $A \ll B$, it never overcomes the increase in lattice energy, Eq. (43), for any value of ε. Thus, the distortion does not take place in wide bands, and wide-band metals are stable. On the other hand, as Eq. (45) indicates, in narrow bands the decrease in electronic energy is proportional to ε, and some finite value of ε minimizes the total energy. The energy at this minimum is approximately

$$E(\varepsilon) \simeq E_u + N(E_0 - 0.2E_g - 0.02\Delta^2/E_g) \tag{46}$$

where E_g is the energy gap introduced by the distortion. But Eqs. (43) and (44) show that the total energy for $\varepsilon = 0$ is

$$E(0) = E_u + N(E_0 - 0.15\Delta) \tag{47}$$

A comparison of Eqs. (46) and (47) indicates that the distortion takes place at $T = 0$ only if $\Delta < 2E_g$. Since energy gaps introduced by distortions of the order of a few percent are generally in the range of 0.2–1.0 eV, electronically induced crystalline distortions should characterize materials that possess an outer electronic band less than about 2 eV wide.

Once it is established that the distorted state is the actual ground state of the solid, it is important to investigate the finite-temperature behavior. The crucial point here is that the energy gap is due entirely to the distortion itself. For example, for a pairing distortion the lower band can be thought of as a bonding band (as in a covalent semiconductor), in which the electrons from the paired atoms are in bonding orbitals. The empty upper band is then an antibonding band in which the electronic density is concentrated away from the area between the nearest-neighboring atoms. For such a situation excitation of an electron across the gap actually leads to a decrease in the magnitude of the gap, since the excited electron no longer contributes to the chemical binding. It can be shown(66,67) that the gap decreases with free-carrier concentration as

$$E_g = E_g(0) - \beta n \tag{48}$$

where β can be determined from the fractional distortion. Since the gap decreases with carrier concentration, the total energy is a rapidly increasing function of temperature. In addition, the electronic entropy increases as the gap begins to decrease, the electronic entropy of the metallic state being the maximum possible value. These conditions lead to a semiconductor–metal transition at a critical temperature that depends on the initial energy gap and the fractional distortion. The transition can be of either first order(66) or second order(68) thermodynamically, depending on the model used for the electronic band structure. A typical result for the transition temperature for a 2% distortion is(69)

$$T_0 = E_g(0)/10k \tag{49}$$

Thus for low-temperature energy gaps of the order of 0.3 eV the semiconductor–metal transition should occur in the vicinity of room temperature.

Although the model was originally worked out for the case of a solid whose high-temperature state is metallic, it should apply equally well to a semimetallic state.[70] In this case the type of distortion that takes place below T_0 is one which stabilizes the energy splitting that is already present. No increase in the size of the primitive cell is necessary. If the stabilization is sufficiently large, a real energy gap can be introduced, and the low-temperature state is then semiconducting. In such a case free-energy considerations show that a first-order semiconductor–semimetal transition occurs at a critical temperature. The model can be generalized in other respects as well. For example, with any reduction of symmetry, even without a change in the size of the primitive cell, if the band under consideration is degenerate, additional crystalline-field splittings could also produce the energy gap necessary to obtain a semiconducting ground state.[71] Furthermore, even a lowering of symmetry itself is not an absolute necessity, particularly when the high-temperature state already has a complex structure. In this situation motions of the ions within a single primitive cell could, in principle, yield an energy gap by intensifying crystalline-field and structure-type splittings that already exist but are insufficient by themselves to produce a gap in the density of states. This type of transition is semiconductor-to-semimetal. It could manifest itself structurally via sharp changes in either the lattice parameter or the basis. Ultimately, once the electron–phonon interaction is introduced in detail the transition could be driven merely by the softening of a phonon mode.[72,73]

An analogous, although theoretically more complex, mechanism involves an antiferromagnetic ordering of spins.[74] Slater[75] was the first to point out that antiferromagnetism is equivalent to a pairing-type distortion as far as its effects on the electronic band structure is concerned. If any antiferromagnet has a half-filled band in the absence of magnetic ordering, the antiferromagnetic state could then be semiconducting. However, in calculating the finite-temperature behavior, subtle complications arise.[74,76] It is necessary to include the exchange energy term explicitly in the one-electron Hamiltonian. This term arises from electronic interactions, and without it no magnetic order is possible. Once the exchange energy is introduced an energy splitting results, just as in the crystalline-distortion model. This splitting could then produce an antiferromagnetic semiconducting state provided the nonmagnetic state is that of a metal with a half-filled band. But the energy gap arises solely from the sublattice magnetization, and, for self-consistency, it must decrease with decreasing magnetization.(74)

Since the thermal excitation of carriers across the gap is equivalent to reversing spins,(66) thus decreasing the magnetization, the energy gap must decrease with increasing carrier concentration. Therefore, an expression completely analogous to Eq. (48) is appropriate also for an energy gap due to antiferromagnetism. However, the finite-temperature properties in this case are further complicated by the fact that the exchange energy is a term arising from electronic interactions. In calculating and comparing the free energies of the antiferromagnetic and paramagnetic states, the total energy must be evaluated. But since an interaction term appears explicitly in the Hamiltonian, the total energy is not simply the sum of the occupied one-electron energies, since that sum counts the electronic interactions twice; the total interaction energy must be subtracted from the above sum to obtain the correct total energy.(74) When these effects are taken into account a catastrophic collapse of the energy gap occurs at the Néel temperature, bringing about a semiconductor–metal transition. For extremely narrow bands the transition is of second order, and takes place at the critical temperature,

$$T_N = E_g(0)/4k \tag{50}$$

For wider bands the Néel temperature and the order of the transition depend on the details of the band structure.

Recent work on electronically induced semiconductor–metal transitions has concentrated on the effects of explicitly introducing electron–phonon,(77–81) electron–electron,(82,83) and electron–spin(84) interactions into simple models. However, the qualitative nature of the theory as outlined here remains generally unaffected.

4.3. Mott Transitions

In Section 3.4 narrow-energy-band theory was discussed in detail, and it was pointed out that a transition from a Mott insulator to a metal should occur at a critical value of the ratio of the bandwidth Δ to the intraatomic Coulomb repulsion U. Such a transition is known as a Mott transition. Mott(15) has presented strong arguments that such a transition from insulator to metal should be a sharp one. The essence of his reasoning is that for small carrier concentrations in a Mott insulator a free electron and a free hole attract each other with a Coulomb interaction, Eq. (42), and form a nonconducting exciton; however, for large carrier concentrations the Coulomb interaction becomes screened at large distances,[85] and is more accurately written

$$V_s(r) = -(e^2/\varepsilon_0 r)\, e^{-\beta r} \tag{51}$$

where β is the appropriate screening parameter, which increases monotonically with the carrier concentration. Above a critical value of β, $V_s(r)$ has no bound-state solutions, and no excitons form. Thus, at a critical value of the carrier concentration there should be a sudden transition from no carriers participating in conduction to a large number, i.e., a sharp insulator–metal transition. Hubbard[18] found that the effective band gap introduced by electronic correlation in his simplified model decreases gradually to zero as Δ/U increases to its critical value, and that the overlap of the two bands increases continuously with Δ/U on the metallic side; therefore, this represents a gradual rather than a sharp transition. However, Mott's argument depends on electronic screening, which is an effect of *interatomic* correlations, and is thus beyond the scope of Hubbard's model, which considers only intraatomic effects.

It is clear that a Mott transition could occur as a function of pressure, provided the solid condenses just on the insulating side at atmospheric pressure. As high pressures are then applied the interatomic spacing decreases, thus increasing the bandwidth Δ. Consequently, the ratio Δ/U could increase through the critical value. It is more difficult to envision a temperature-induced transition. Frohlich[86] suggested that screening effects could bring about such a transition; i.e., thermal excitation of free carriers across the Hubbard gap could reduce the *effective* value of U via screening effects, thus increasing the ratio Δ/U from below to above its critical value. Few quantitative calculations have been performed in this model, but it can be shown[87] that the transition should occur at temperatures of the order of E_H/k, where E_H is the Hubbard gap. Thus, these transitions occur above the melting temperature of the solid unless the gap is very small, i.e., the solid is very near the critical value of Δ/U at $T = 0$. An alternative model[88] is that the Hubbard gap is reduced by free-carrier excitation, as in Eq. (48), thus inducing a first-order phase transition at much lower values of the temperature.[66] A mechanism for reduction of the Hubbard gap with n has been suggested by Doniach,[88] who noted that in the presence of a large free-carrier concentration many sites are unoccupied, thus presenting a reduced potential to a tunneling electron; the *average* potential then behaves like Eq. (48). However, it is not clear that this reduction in average potential should lead to a catastrophic collapse of the gap at a critical value of n as in the crystalline-distortion model.

A completely different approach to insulator–metal transitions has been introduced by Falicov and Kimball,[89] who proposed a model based on the possible simultaneous existence of localized and extended states in a particular solid. The model assumes that the material is a Mott insulator whose valence band is the lower of the Hubbard subbands of Figure 6; as in the Hubbard model, interatomic correlations are neglected. On the other hand, in the Falicov–Kimball model the intraatomic Coulomb repulsion U is taken to be so large that double occupancies essentially never occur. Furthermore, it is also assumed that an empty conduction band with only negligible intraatomic Coulomb repulsions exists not too far above the valence band. If the separation between the valence and conduction bands is taken to be E_g, creation of a single electron–hole pair costs not E_g but $E_g - V$, where V is the intraatomic *attraction* between the free electron and the localized hole. Furthermore, it takes an additional energy of only $E_g - 3V$ to excite the second electron–hole pair, since the two itinerant electrons *each* feel the attraction of both localized holes. Thus, the effective energy gap E_{eff}, the energy necessary to excite the next electron–hole pair, can be written

$$E_{eff} = E_g - (2n + 1)V \tag{52}$$

where n is the number of electron–hole pairs already present. But Eq. (52) is of the same structure as Eq. (48), and thus this model exhibits an insulator–metal transition as a function of temperature, provided $V < E_g$; otherwise the metallic state is stable at all temperatures. Falicov and Kimball found in the narrow-band limit that above a critical ratio $V/E_g = 0.63$ the insulator–metal transition is thermodynamically of first order.

There are several objections that can be raised with regard to the Falicov–Kimball model. First, the interatomic Coulomb interaction between neighboring localized holes in the valence band could be significant in the limit of large U upon which the model is based. This repulsion should add a positive term, which increases monotonically with n, to Eq. (52); such a term could even suppress the transition. Second, Falicov and Kimball assume that V is independent of n, an approximation valid only if the screening is small. If significant screening exists at carrier concentrations below the critical value. the transition temperature is actually well above the calculated T_0. Finally, a sharp transition in this model requires a very delicate balance between E_g and V, physically uncorrelated quantities. For example, a conductivity jump of a factor of 50 requires V to be between $0.9E_g$ and $1.0E_g$.

The Falicov–Kimball model has been extended to the case of finite bandwidths by Ramirez *et al.*[90] and to include the possibility of the formation of magnetically ordered states by several groups.[91–93] The molecular-field-theory nature of Eq. (52) has been recently improved through use of the coherent-potential approximation[94] to better handle the random arrangement of the localized holes, although the qualitative nature of the results is not seriously modified.[95,96]

5. Experimental Results

In Sections 1–4 the theory of transport in imperfect solids has been presented in some detail. It is tempting to conclude that interpretation of the experimental data is now merely mechanical, once the type of solid and the class of the predominant imperfections is identified. Unfortunately, the situation is, of course, never so straightforward in real materials, whose bonding is often a complex combination of the four major types and whose imperfections are varied and difficult to characterize. In this section we shall present the experimental data on several particular solids that can serve as prototypes for various groups of unconventional materials which are now attracting interest; in each case we shall attempt to interpret these results on the basis of the theory outlined in Sections 1–4. In each subsection an attempt will also be made to generalize the results that are presently available on the entire group of materials, although in most cases it is dangerous at best to conclude that a basic understanding has already been obtained.

5.1. Metal Oxides

The ordinary (non-transition) metal oxides include BeO, B_2O_3, Na_2O, MgO, Al_2O_3, SiO_2, K_2O, CaO, SrO, CdO, and SnO_2, among others. Of these, MgO has been studied quite extensively, and can serve as the prototype material. MgO has the (cubic) rocksalt structure, containing one molecule per primitive cell. It can be considered a purely ionic crystal[97,98] consisting of Mg^{2+} and O^{2-} ions. Both of these ions have identical electronic configurations, the closed-shell Ne structure. Thus, we do not expect the perfect crystal to exhibit much electronic conductivity. For such materials as a starting point in estimating the density of states $g(E)$ we can neglect all bandwidth effects and use the extreme tight-binding limit.[7] In pure, stoichiometric material only two possible ionic processes result in electronic conduction; they can be represented by the chemical equations

$$Mg^{2+} + O^{2-} \longrightarrow Mg^{+} + O^{-} \tag{53}$$

$$Mg^{2+} + Mg^{2+} \longrightarrow Mg^{+} + Mg^{3+} \tag{54}$$

The first process creates an electron–hole pair by placing an excess electron on a magnesium site, leaving a hole on an oxygen site. The second process creates an electron–hole pair on two magnesium sites. The closed-shell configurations of both Mg^{2+} and O^{2-} represent filled $2p$ shells and empty $3s$ shells. Thus, (53) represents an excitation from the $2p$ band of the O^{2-} ions to the $3s$ band of the Mg^{2+} ions, while (54) represents an excitation from the $2p$ band of the Mg^{2+} ions to the $3s$ band of the Mg^{2+} ions. We can ignore the $3s$ band of the O^{2-} ions since O^{3-} has enormous ground-state energy, and even the Madelung (electrostatic) potential of the highly ionic crystal does not stabilize it. In fact, even the O^{2-} ion is unstable in free space, having a *negative* ionization potential of 9 eV.[99] This makes process (53) very exothermic for spatially separated free ions, since a free O^{2-} ion spontaneously releases an electron, thus lowering its energy by 9 eV. In addition, since Mg^{2+} has an electron affinity of 15 eV, (53) occurs spontaneously in the limit of free ions, thus releasing 24 eV. However, in MgO crystals the Madelung potential resulting from the strong electrostatic attraction between Mg^{2+} and O^{2-} is 24 eV, and it stabilizes both Mg^{2+} and O^{2-} relative to Mg^{+} and O^{-}. Thus, (53) in a rigid-ion MgO crystal requires 24 eV in the zero-bandwidth limit. In the real material this value is reduced somewhat by screening of the electrostatic interactions due to the presence of electrons that can adjust their charge density to minimize the total energy of the crystal. Part of this screening is due to the core electrons on the ions themselves and part is due to polarization of the charge densities on the surrounding ions in the crystal and the effects of whatever covalent bonding exists. The former is an effect that occurs also in free ions, and can be estimated by noting that the experimental Slater–Condon parameters, which measure intraionic electrostatic interactions, are about 15% smaller than the calculated values.[100] The polarization and covalency effects can be estimated from the fact that the multiplet splittings of transition metal ions in MgO are approximately 20% less than in the free ions.[101] These effects together thus result in a reduction of the effective average energy of (53) in MgO crystals to approximately 16 eV.

In addition, we must consider the effects of finite bandwidth. The spreading of atomic energy levels into bands leads to a decrease in the minimum energy of excitations such as (53). In NiO, which has the same structure and almost exactly the same lattice separation as MgO, the $2p$ band of the O^{2-} ions is about 4 eV wide.[102] The conduction band is the $3s$ band of the Mg^{2+} ions, and a width of 12 eV is a reasonable estimate.[7] Assuming symmetric spreading of the bands, we can conclude that (53) can occur with a minimum energy of 8 eV, which should be approximately the band gap of MgO.

On the other hand, since the $2p$ electrons of Mg^{2+} ions are bound by about 80 eV in the free ion, the process (54) can be shown to require

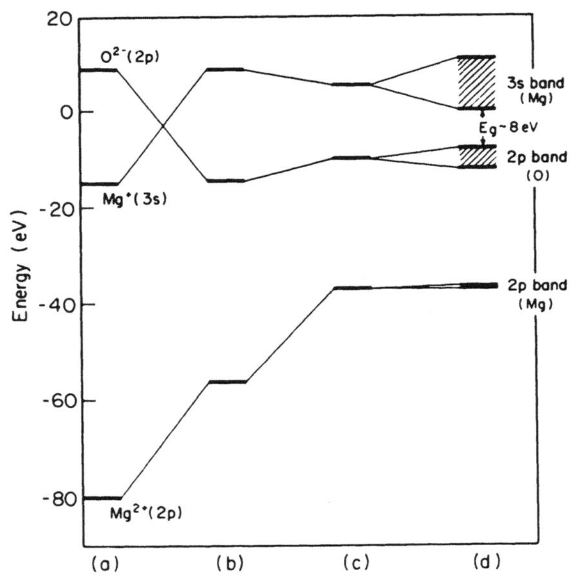

Fig. 10. Estimate of the energy band structure of pure, stoichiometric MgO: (a) free-ion energies; (b) including effects of the Madelung potential; (c) including screening and covalency effects; (d) including the effects of finite bandwidth.

an energy of at least 38 eV.[7] It is clear that no intrinsic conduction can result from such a high-energy process. A sketch of these estimates of the band structure of MgO is shown in Figure 10.

The band structure estimated in this way is in good agreement with optical[103] and X-ray[104] measurements. But in this chapter we are more concerned with transport properties of imperfect crystals. In fact, MgO is usually quite imperfect—e.g., every one of the 3*d* transition metal ions has been observed in carefully prepared single crystals of MgO.[105] These ions enter the lattice substitutionally for Mg^{2+}. In general, the ionization potentials of transition metal ions are much smaller than the 80 eV value for Mg^{2+}, so that it is much easier to excite a conduction electron in impure MgO than in the pure material. For example, the process

$$Ni^{2+} + Mg^{2+} \longrightarrow Ni^{3+} + Mg^{+} \tag{55}$$

which creates an electron in the 3*s* conduction band (but no free hole in the valence band, since the hole is localized on the Ni^{3+} impurity), takes 45 eV less energy than the intrinsic process (54). In fact, the minimum energy of (55) in an impure MgO crystal can be estimated to be only 7 eV, less than the intrinsic gap.[7] Still smaller energies result if Fe^{2+}, Ti^{2+}, or V^{2+} ions are present. In addition, transition metal ions can be incorporated into MgO in their trivalent state, e.g., as Fe^{3+} or Cr^{3+}; in such cases a process like

$$Fe^{3+} + O^{2-} \longrightarrow Fe^{2+} + O^{-} \tag{56}$$

creates a hole in the valence band. Process (56) can be estimated as requiring an energy of only 3 eV. These estimates are somewhat unreliable because of the polaron effects discussed in Section 3.5, which should reduce the energy necessary to create free carriers in (55) and (56), but these distortion energies are of the order of only tenths of electron volts.[16] But the major point is that the presence of transition metal ions in MgO leads to an increase in conductivity for the same reason that the presence of Ga increases the conductivity of Ge, namely by reducing the energy necessary to create a free carrier. However, whereas in Ga-doped Ge the energy needed to excite a hole in the valence band is reduced from 0.7 to 0.01 eV, in Fe^{3+}-doped MgO this reduction is from 8 to 3 eV. Thus, even in heavily doped MgO we do not expect much hole conduction at low temperatures; e.g., $e^{-3\,\mathrm{eV}/2kT} \sim 10^{-25}$ at 300°K. Even at 1200°K we should not expect conductivities exceeding 10^{-4} $(\Omega\text{-cm})^{-1}$ from process (56).

Electrical conductivity of MgO has been measured by several groups,[106–112] with chaotic results, primarily due to the lack of sample characterization. Early work was carried out only at high temperatures, 1000–1600°K, since the observed conductivity did not exceed the order of 10^{-10} (Ω-cm)$^{-1}$ even at 1000°K. The conductivity could generally be fit by the ordinary semiconductor relationship (32) with activation energies E_a of the order of 2.7 eV. Since this is much less than half the optical gap of 8 eV, it undoubtedly does not represent intrinsic electronic conduction, but rather either impurity-controlled electronic conduction or ionic conduction. For a particularly pure sample Mitoff[109] found an activation energy of 3.5 eV for temperatures greater than 1300°K, but only 0.9 eV from 100 to 1300°K. The higher value is not inconsistent with intrinsic electronic conductivity. In later work Mitoff[110] found that both electronic and ionic conduction were present between 1200 and 1800°K, but that electronic conduction was more likely to predominate at higher temperatures and in less pure crystals. The experiments of Davies[111] indicate that the 3.5-eV activation energy found at high temperatures in samples of high purity does not arise from intrinsic electronic conduction, but rather from intrinsic Mg^{2+} ionic conduction, a result consistent with a direct measurement of diffusion of radioactive Mg^{2+} in MgO.[113] Low-temperature data have been reported by Lewis and Wright,[112] who concluded that the conductivity of MgO could be characterized by a single activation energy of 2.8 eV from 650 to 2000°K. The conductivity at 650°K was 10^{-17} (Ω-cm)$^{-1}$, one of the lowest values ever reported for any material. The data of Lewis and Wright are compared with those of some earlier workers in Figure 11. The 2.8-eV activation energy does not appear to be extrinsic O^{2-} ionic conduction[114] as originally suggested, but it may well be extrinsic Mg^{2+} conduction via migration of Mg vacancies.

The important point is that since the intrinsic electronic activation energy in MgO is about 4 eV but the activation energy associated with creating and moving an Mg vacancy is only 3.5 eV, then in ultra pure MgO crystals ionic conduction predominates at low temperatures, but electronic conduction, because of the larger mobility, must eventually predominate at high temperatures. However, because of the large activation energies for both intrinsic processes, extrinsic conduction is usually observed in imperfect samples.

Similar considerations hold for the other alkaline earth oxides. When sufficiently pure, BeO has been shown to be a good ionic conductor.[105] Surplice[115] investigated CaO and SrO and found high-

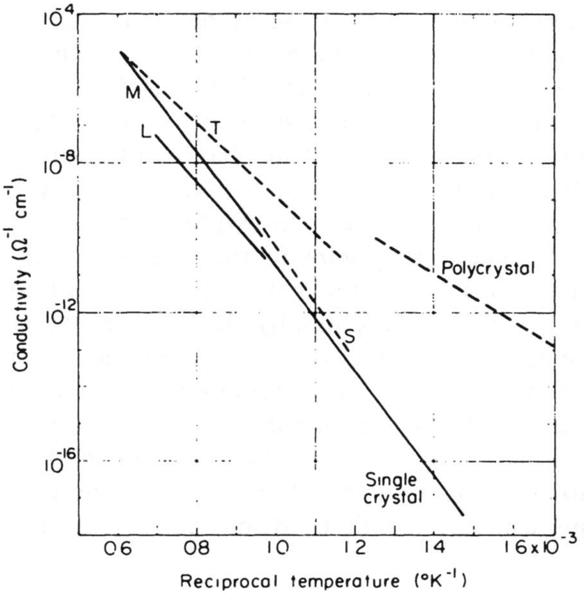

Fig. 11. Electrical conductivity of MgO as a function of temperature, as measured by several investigators. From Lewis and Wright.[112]

temperature activation energies of 3.5 and 2.0 eV, respectively. On the basis of the previous conclusions for MgO the activation energy for CaO can be assumed to be that for intrinsic Ca^{2+} migration. In SrO the activation energy has been determined to be due to formation of O^{2-} vacancy–interstitial pairs (Frenkel defects) and subsequent migration of the O^{2-} interstitials.[116] Good electronic conduction can be obtained in wide-gap semiconductors, however, as has been demonstrated vividly in SnO_2 crystals, in which Hall mobilities in excess of 13,000 cm^2/V-sec have been observed.[117]

5.2. Transition Metal Oxides

The transition metal oxides are an enormous group of materials whose diverse transport properties run the gamut from those characteristic of excellent metals to those characteristic of dielectric insulators. Two classes of these materials can immediately be distinguished—the saturated oxides, in which all of the *d* electrons of the transition metal atoms have been given off to the oxygen atoms, and the unsaturated oxides, in which there is an unfilled shell (i.e.,

1–9) of *d* electrons remaining on some of the transition metal ions *after* they reach their appropriate valence state. The former class, which includes Si_2O_3, TiO_2, V_2O_5, CrO_3, ZrO_2, Nb_2O_5, MoO_3, and WO_3, among others, are always semiconducting, and have a filled oxygen 2*p* band as their valence band and ordinarily an empty metallic *d* band as the conduction band. An energy band sketch, analogous to the one in Figure 10 for MgO, for rutile (TiO_2) is shown in Figure 12.[7] The energy gaps in these materials are sufficiently smaller than those in ordinary metal oxides so that conduction in even high-purity crystals is generally electronic rather than ionic. In addition, the usual presence of other transition metal impurities or of nonstoichiometry results in the predominance of *extrinsic* electronic conduction at all ordinary temperatures.[7] The interesting feature of the saturated oxides is that for either reduced material or *n*-type doped samples conduction is in the *d* band of the transition metal ions, a band which, as shall be discussed shortly, could be quite

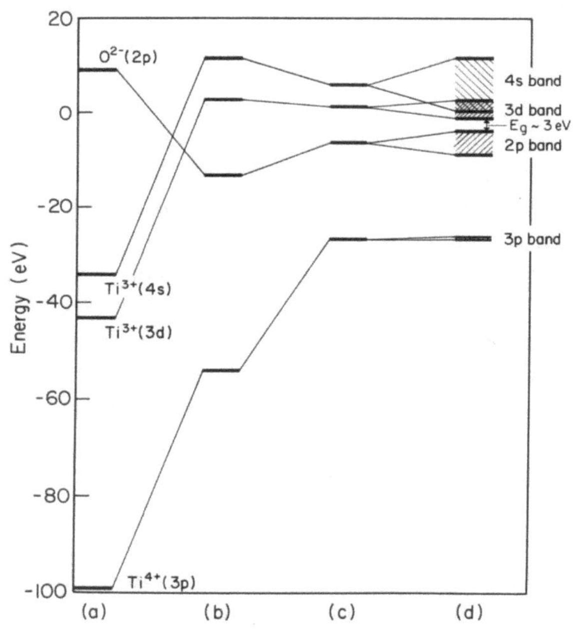

Fig. 12. Sketch of the energy band structure of pure, stoichiometric TiO_2: (a) free-ion energies; (b) including effects of the Madelung potential; (c) including screening and covalency; (d) including finite-bandwidth effects.

narrow. Under such conditions if the strength of the electron–phonon coupling is large, as should be the case in these primarily ionic crystals, small polarons should form, and hopping conduction should be observable. Whether it has been or not in, for example, TiO_2 is still a matter of debate, although the recent weight of evidence is for *large-polaron* formation.[7]

Although the saturated oxides are interesting for other reasons as well, the class of unsaturated oxides is so diverse in its transport behavior and so poorly understood in general at present that we shall concentrate on those materials in this section. The overall experimental situation with regard to transport can be summarized by a glance at Figure 13, which shows the electrical conductivity as a function of temperature for 20 unsaturated, binary transition metal oxides.[118] But the theoretical implications of this diversity of behavior needs further amplification. One of the basic problems is that for several of the most insulating of these materials (e.g., FeO, CoO) ordinary energy band theory predicts partially filled bands and thus metallic behavior.[118] The resolution of this type of situation has already been discussed in Section 3.4: These materials must be Mott insulators, the outer 3*d* bandwidths being small compared to the intraionic Coulomb repulsion *U*. The real difficulty, however, is in the *a priori* predictions of Mott insulation, since some excellent metals appear to have bandwidths of the same order of magnitude as the Mott insulators. Some qualitative trends have recently been established,[119,120] but no successful quantitative calculations have been attempted in real solids. We shall discuss each of the three classes of electrical transport behavior observed—metallic oxides, insulating oxides, and oxides which exhibit insulator–metal transitions. A short discussion of the ternary bronzes, some of which exhibit every conceivable type of transport behavior over a narrow composition range, will also be presented.

5.2.1. Metallic Oxides

A large class of transition metal oxides are metallic, including TiO, VO, NbO, CrO_2, ReO_2, RuO_2, OsO_2, MoO_2, RhO_2, WO_2, IrO_2, and ReO_3. The cubic monoxides in this group all form quite defective structures, even when stoichiometric; e.g., TiO, VO, and NbO contain, respectively, 15%, 16% and 25% vacancies of both type of ions when pure and stoichiometric. This occurs because the energy to create a vacancy in these materials is small and the presence

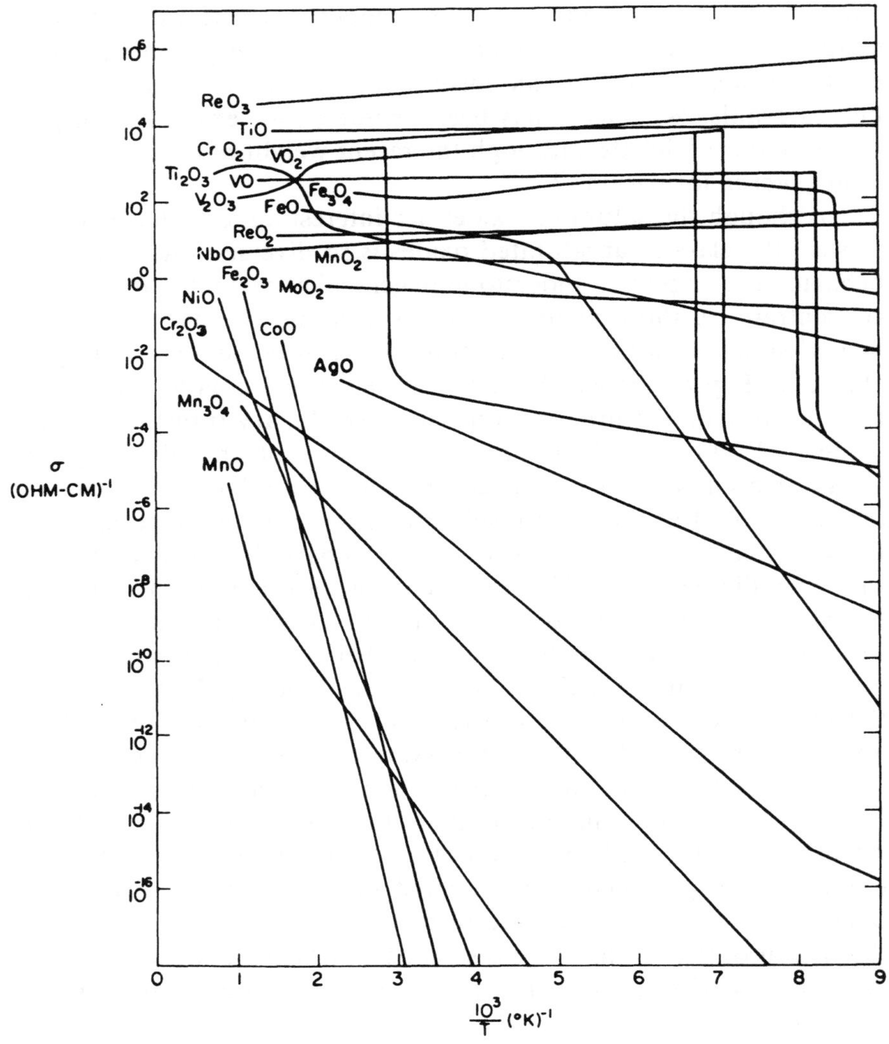

Fig. 13. Electrical conductivity as a function of temperature for several unsaturated, binary transition metal oxides. (The data for VO have since been shown to be incorrect.) From Adler.[118]

of vacancies lowers the effective density of the conduction electrons and thus their kinetic energy. We shall consider ReO_3 as the prototype material, since it has been the most extensively investigated.

ReO_3 has a cubic structure in which the Re^{6+} ions are at the cube corners and the O^{2-} ions are at the center of the cube edges. The

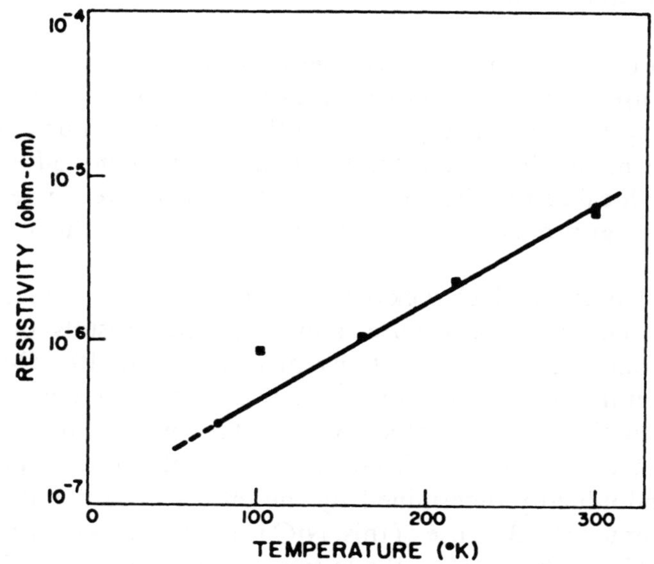

Fig. 14. Electrical resistivity as a function of temperature for a single crystal of ReO_3. From Ferretti *et al.*[121]

structure is closely related to that of perovskite, which has a second metallic ion at the center of the cube. The electrical resistivity as measured by Ferretti *et al.*[121] is shown in Figure 14. The remarkably high conductivity, essentially equal to that of pure copper at 77°K, is particularly amazing for an oxide. The thermoelectric power at 300 K is *n* type and only 0.4 μV/°K in magnitude, also indicative of good metallic conductivity.

A calculation of the band structure of ReO_3 has been carried out by Mattheiss,[122,123] who found a three-piece Fermi surface. There has since been a great deal of experimental evidence[124,125] confirming the predicted shape of the Fermi surface. Although the Fermi surface geometry is not a very sensitive function of the one-electron potential, the conclusion that can be reached from the agreement between theory and experiment is that as long as the solid is on the metallic side of the Mott transition ordinary energy band calculations appear to be applicable. However, it is still an unfortunate fact that the presence of Mott insulation cannot be predicted by the results of such calculations, no matter how sophisticated.

5.2.2. Insulating Oxides

Most of the common 3*d* (iron series) transition metal oxides are excellent insulators; this group includes MnO, FeO, CoO, NiO, CuO, Fe_2O_3, Cr_2O_3, and Mn_3O_4, among others. There is much evidence that these materials are all Mott insulators, with the 3*d* electrons remaining localized in the solid. By far the most investigated member of this group is NiO, which we shall use as the prototype material.

NiO has the (cubic) rocksalt structure, with one molecule per primitive cell. Nominally pure, stoichiometric crystals ordinarily have a room-temperature resistivity of greater than 10^{13} Ω-cm.[126] However, ultrapure but nonstoichiometric crystals have been grown with resistivities at 600°K of as low as 10 Ω-cm.[127] Since different samples have such vastly different resistivities, the observed conduction is undoubtedly determined by the effects of lattice defects or random impurities. Because of this, NiO is usually doped with known amounts of Li, which enters the lattice substiutionally for Ni. Since the second ionization potential of Li is much larger than the third ionization potential of Ni, the Li exists in NiO as Li^+; thus, to maintain charge neutrality, one Ni^{3+} should be formed for every Li^+ introduced. However, this need not always be the case. For example, one O^{2-} vacancy can compensate for two Li^+ ions. Since a Ni^{3+} ion is just a Ni^{2+} ion with an extra hole, Li^+ impurities act as acceptors, and the resulting extrinsic conduction is *p* type. In nominally pure NiO conduction is usually also *p* type, evidence that nickel vacancies are more prevalent than oxygen vacancies.

The conductivity almost always increases exponentially with increasing temperature, as is normal in semiconductors, but the activation energy depends on the sample and the temperature range. For undoped samples the activation energy is about 0.9 eV from 200 to 500°K,[128,129] but it decreases to approximately 0.6 eV in the 600–1000°K range.[129] Above 1000°K the activation energy usually increases to about 1.0 eV,[130,131] although values near 1.9 eV have also been measured.[132,133] At very low temperatures, 10–100°K, activation energies in the vicinity of 0.004 eV are observed.[134,135]

Doping with Li does not affect the activation energy at either very high or very low temperatures. However, in the intermediate temperature range moderate doping (~1% Li) induces a drop in activation energy from 0.9 to about 0.3 eV below 500°K and from 0.6 to about 0.2 eV above 500°K.[129] Heavy doping (5–10% Li) results in a further decrease of activation energy to 0.18 eV below 500°K and 0.14 eV above 500°K.[129,136]

The Hall mobility of both doped and undoped samples between 200 and 400°K can be represented[135,137,138] by

$$\mu_H = (0.01\ \text{cm}^2/\text{V-sec})e^{0.075\ \text{eV}/kT} \tag{57}$$

The Hall effect is p type in this range, but above 400°K the Hall mobility begins to decrease more rapidly than indicated in (57), and at about 600°K it reverses sign.[139]

Thermoelectric power above 300°K is proportional to the logarithm of the resistivity,[135,137,140] as Eq. (20) suggests for unipolar free-hole conduction. There is a maximum in α near 300°K in undoped material which decreases to about 150°K in Li-doped samples.[137,140] The sign of the thermoelectric power indicates p-type conduction predominates, even above 600°K, where the Hall effect is n type.

Analysis of the results on several well-characterized samples of Li-doped NiO[135,137,140] shows that a consistent picture can be obtained using large-polaron theory, provided it is assumed that the Li acceptors are partly (2–10%) compensated by oxygen vacancies.[16] Such an interpretation leads to the conclusions that the conduction observed in the 200–1000°K range in Li-doped NiO is predominantly via band transport of holelike large polarons with an effective mass of about six free-electron masses. The mobility in this temperature range obeys the relation

$$\mu = (0.3\ \text{cm}^2/\text{V-sec})e^{0.075\ \text{eV}/kT} \tag{58}$$

proportional to the Hall mobility, Eq. (57), but a factor of 16 greater. This implies that the constant r in Eq. (21), ordinarily of order unity, is 0.06 in NiO, an unusual result. Optical phonon scattering appears to dominate the mobility.[16] Direct drift mobility measurements indicate general agreement with Eq. (58), but suggest a somewhat lower optical phonon frequency and somewhat higher preexponential.[141]

For the *bound* carriers, on the other hand, there seems to be no doubt that small-polaron hopping dominates the transport.[16,142,143] These holelike small polarons have a very large effective mass, and the hopping time at 300°K is of the order of 10^{-10} sec.[144]

A theoretical analysis of the electronic structure of NiO is fraught with apparent paradoxes. Once it is concluded that NiO is a Mott insulator the band structure can be handled in terms of the Hubbard model described in Section 3.4. It is then clear that the intraionic Coulomb energy U must be much larger than the bandwidth for the $3d$ band of the Ni^{2+} ions. The $2p$ band of the oxygen ions and the $4s$ band of the nickel ions must also be considered explicitly; these bands should be well on the metallic side of the Mott transition, and can thus be handled by ordinary band theory. Band calculations for NiO have, in fact, been carried out,[145,146] and yield $2p$ and $4s$ bands, in agreement with optical experiments. The ground-state configuration of Ni^{2+} is $3d^8$, and analysis of the $3d^8$ band is more of a problem because of the strong correlations. However, there are several indirect guides to the width of the $3d^8$ band. Reinen[147] measured the absorption spectra of dilute solutions of Ni in MgO and found that the crystalline-field peaks observed were essentially identical in position and width to those of pure NiO. This correlation is shown in Figure 15, and indicates a rather narrow width for the $3d^8$ band.[16] In addition, no Mott transition as a function of pressure has been observed in NiO at pressures below 2 Mbar; recently a transition to a metallic state has been observed at about

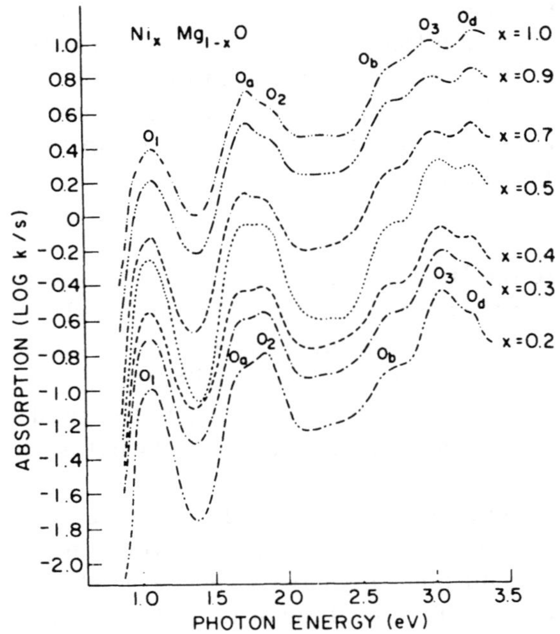

Fig. 15. Absorption spectra of $Ni_xMg_{(1-x)}O$ as a function of photon energy from 1.0 to 3.5 eV. From Reinen.[147]

2.5 Mbar after two other first-order transitions that could represent structural transformations.[148] This need not be a Mott transition, and, in fact, more likely represents an overlap of the $2p$ or the $4s$ band with the Fermi energy under the high-pressure conditions. In any event the extremely high pressures necessary to obtain metallic NiO also suggest a narrow $3d^8$ band. Finally, there has been some very recent photoemission work that indicate a $3d^8$ bandwidth of less than 0.3 eV.[149,150] There seems to be little doubt at present that the $3d^8$ band is extremely narrow. The other important parameter in the Hubbard Hamiltonian is U, which can be estimated as approximately 13 eV for NiO.[16] With this value of U an estimate of the band structure of pure NiO can be made; a sketch of how the relevant bands build up from the atomic limit is shown in Figure 16. This sketch is simplified in the sense that the $3d^7$ (Ni^{3+}) cores are all taken to be in their ground states at all times. (If all possible excitations of a strongly correlated d band are taken into account, over 10^6 pseudoparticle bands result.[26]) The optical spectrum of NiO predicted from this calculation[16] is in agreement with the experimental observations.[151]

In order to analyze the transport properties of Li-doped or nonstoichiometric NiO, the effects of these impurities or defects must be carefully analyzed. Once again, it is the strongly correlated $3d^8$ band which makes this problem formidable. Let us consider the simpler case of Li doping. The lowest-energy state of the doped material is one in which all uncompensated Li^+ ions have a Ni^{3+} ion for a nearest neighbor. This arises because a Li^+–Ni^{3+} pair on the nickel sublattice forms an effective electron–hole pair which binds electrostatically, just like an exciton; the binding energy can be estimated as about 0.4 eV.[142] These bound Li^+–Ni^{3+} complexes lead to low-energy excitations which do not exist in the perfect crystal. These new excitations arise because Ni^{3+} represents a $3d^7$ configuration, which is not present in pure, stoichiometric NiO. Furthermore, because of the strong correlations in the $3d$ bands, the energies of the extrinsic excitations cannot be obtained from the previously calculated density of states of perfect NiO. The two most important of these excitations can be expressed[152]

$$[Ni^{3+}] \longrightarrow Ni^{3+} \tag{59}$$

$$[Ni^{3+}] + O^{2-} \longrightarrow Ni^{2+} + O^{-} \tag{60}$$

where the brackets on the left indicate the hole on the Ni^{2+} is bound to the nearest-neighboring Li^+ impurity. Process (59) represents the freeing of the bound hole from the vicinity of the Li^+ site, while

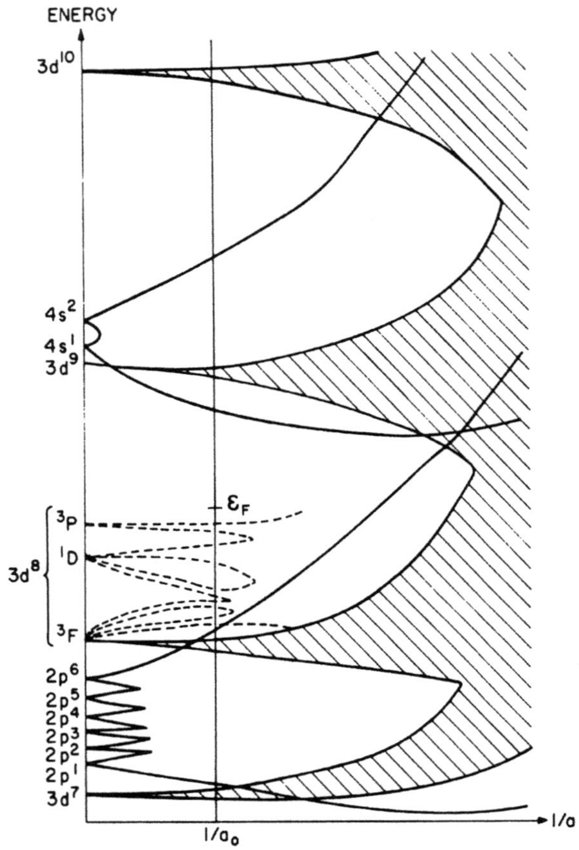

Fig. 16. Pseudoparticle band structure as a function of inverse lattice parameter for bands near the Fermi energy of pure NiO. The correlated ground-state $3d^n$ bands are shaded; excited crystal-line-field-split bands of the $3d^8$ configuration are dashed.

process (60) represents the creation of a hole in the oxygen $2p$ band. In either case, once freed, the hole can participate in conduction. If the $3d$ band is extremely narrow, as it appears to be, small-polaron formation would be expected. There is, in fact, much evidence that the bound holes form massive small polarons, as discussed previously. Thus, if process (59) represents the predominant conduction mechanism, small-polaron hopping should be observed. However, this is in conflict with the experimental results previously reviewed, which indicate bandlike conduction via large polarons predominates. On the other hand, if process (60) is responsible for the observed conduction, there is no difficulty in accounting for the bandlike behavior, since the $2p$ band is 4 eV wide,[149] and an estimate of the large-polaron mass of holes in this band is completely in accord with experiment.[16] The only apparent difficulty with this interpretation is that optical experiments indicate that the $2p$ band begins 1.5 eV below the $3d^8$ band.[149] But this difficulty is only illusory, and arises from the neglect of the strong correlations. The optical experiments measure only the separation between the $2p$ and $3d^8$ bands for the *pure* material, since few Ni^{3+} ions are present relative to the large density of Ni^{2+} ions. But since the intrinsic analog of process (60),

$$Ni^{2+} + O^{2-} \longrightarrow Ni^{+} + O^{-} \tag{61}$$

requires creation of a Ni^+ ($3d^84s$) state, which is energetically unfavorable, the intrinsic $2p$ band appears to be well below the $3d^8$ band.[16,152] On the other hand, process (60) does not require any Ni^+ ion to be created, and this lead to the top of the $2p$ band being very near the $3d^8$ band in Li-doped NiO.[16] The calculated electrical band structure of Li-doped NiO is shown in Figure 17. The calculations are certainly not reliable to an accuracy of better than 0.3 eV, so the relative positions of the top of the $2p$ and $3d^8$ bands could well be reversed, and, in fact, experimentally this appears to be the case.[16] An interesting feature here is that the $3d^8$ band can be below the $2p$ band without the material being metallic, because the $3d^8$ band is in the strong-correlation limit. Thus the major features of both the transport and the optical properties of NiO can be explained in a consistent manner.

Several recent experimental results[141,149,150,153,154] are consistent with the predictions of the strong-correlation model. One striking result[141] is that drift mobility experiments yield an extremely low efficiency of hole extraction, about 10^{-4} of the generation rate. This is easily explained if the vast majority of holes are generated in the high-density $3d^8$ band while only a small fraction are generated in $2p$ band, which has a density of states near the Fermi energy equal only to the uncompensated Li^+ or nickel-vacancy concentration.

Recently a one-electron calculation but assuming localized $3d$ states has been carried out for NiO,[155] using the $X\alpha$ cluster method.[156] Although the results can be interpreted in terms of the

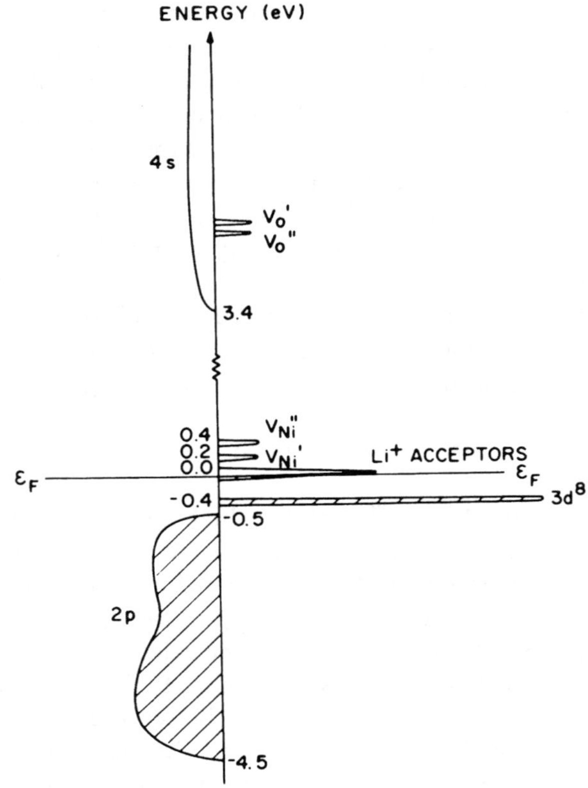

Fig. 17. Electrical density of states of Li-doped NiO. States in which carriers form large polarons are drawn to the left; states in which carriers form small polarons are drawn to the right. States that are occupied at $T = 0$ are shaded. Partial self-compensation by means of oxygen vacancies is assumed. V_x' and V_x'' refer to singly and doubly ionized x vacancy levels, respectively. From Adler.[152]

optical and photoemission experiments, it is not clear how these calculations can be used to describe the extrinsic transport properties.

Major unresolved problems in NiO are the anomalous sign of the Hall effect above 600°K and the fact that the Hall mobility appears to be a factor of 16 lower than the microscopic mobility.

Although other insulating transition metal oxides can be described in a manner analogous to NiO,[16] the proximity of the top of the oxygen $2p$ band and the $3d^8$ band appears to be more or less accidental in NiO. Thus, free-small-polaron hopping should be observable in some of the other oxides, and there is some evidence for the predominance of this mechanism in MnO and FeO. However, none of the other materials has been studied with anywhere near the varied number of techniques that have been used on NiO.

5.2.3. Oxides Which Exhibit Insulator–Metal Transitions

Several transition metal oxides are semiconducting at low temperatures but at a given temperature or over a small temperature range their conductivity suddenly jumps by a large factor, as great as 10^7, and at higher temperatures the materials are metallic. Since we have already mentioned many transition metal oxides that are insulating and many that are metallic, perhaps it is not too surprising to find that a few of the materials exhibit insulator–metal transitions. However, the number of materials which show such behavior is astounding. To date insulator–metal transitions have been observed in Ti_2O_3, Ti_2O_5, Ti_4O_7, Ti_5O_9, Ti_6O_{11}, V_2O_3, V_3O_5, V_4O_7, V_5O_9, V_6O_{11}, V_8O_{15}, VO_2, V_6O_{13}, NbO_2, and Fe_3O_4, among others. Since these transitions all occur at temperatures between 100 and 1000°K, it seems clear that all of them cannot be explained by a chance near-cancellation of energies of the order of 1 eV (12,000°K) or more.

If we look for a thread of consistency with regard to the observed transitions, then it must be the change in crystalline symmetry which accompanies all of the insulator–metal transitions listed previously, with the exception of Ti_2O_3. But even in Ti_2O_3 there is a rapid change of the lattice parameters and a 20% softening of a phonon mode through the transition. In the vast majority of cases the low-temperature structure is a distortion of the high-temperature structure, and an order parameter, the fractional distortion, can be introduced. Most of the transitions appear to be thermodynamically of first order, although several are most likely of second or higher order. Several, but not all, of the materials exhibit antiferromagnetic order. For just one of the materials, V_2O_3, the long-range antiferromagnetic ordering which exists at low temperatures vanishes at the same temperature that the insulator–metal transition and the accompanying crystalline phase transformation occur. Although V_2O_3 is thus unique rather than typical, it has certainly been studied more than any of the others, and will be the one material considered in detail in this section.

At high temperatures V_2O_3 has the (trigonal) corundum structure, but below about 150°K a monoclinic distortion occurs.[157]

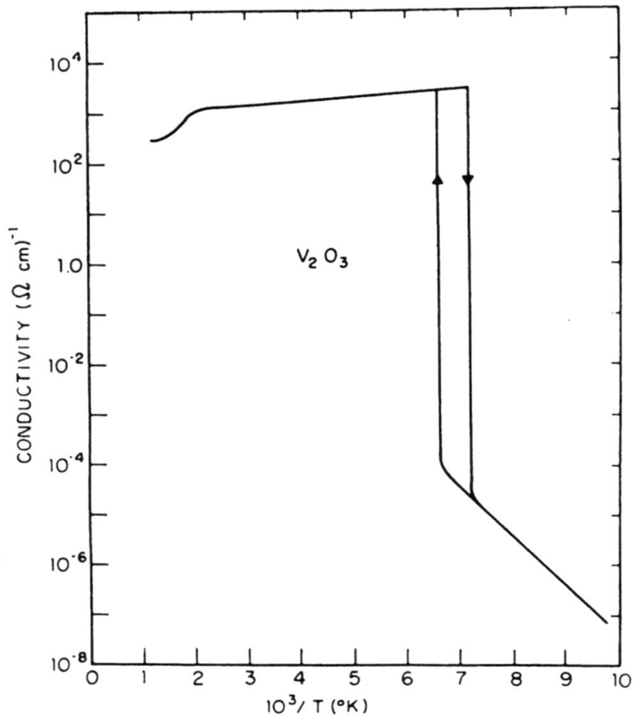

Fig. 18. Electrical conductivity as a function of inverse temperature for a single crystal of V_2O_3. From Feinleib and Paul.[158]

The transition is of first order, and a large latent heat has been observed.[158] Below the transition temperature V_2O_3 is antiferromagnetic, with a moment of about 1.2 μ_B per vanadium ion.[159] The low-temperature state is semiconducting, with an activation energy of approximately 0.15 eV; at high temperatures the material is metallic, with an anomaly in the vicinity of 600°K.[158] At the transition the conductivity sharply jumps by a factor of greater than 10^7 in good single crystals. A curve of conductivity as a function of temperature is shown in Figure 18. Optical experiments indicate that an energy gap of about 0.3 eV opens up in the low-temperature phase.[158,160,161] The Hall constant is independent of temperature above the transition and indicates the presence of approximately one free carrier per vanadium ion[162]; the Hall mobility is of the order of 0.1 cm^2/V-sec in the metallic region. The magnetic susceptibility, relatively independent of temperature below 150°K, increases by about a factor of two at the transition point, and very slowly decreases with increasing temperature in the metallic phase.[163]

V_2O_3 is all things to all men, and the transition has been attributed to every one of the models discussed in Section 4. But the major problem which is the center of controversy at present is whether the transition is primarily a Mott transition or primarily due to an electronically induced phase transformation. The latter point of view[164] is based on the fact that a 2% crystalline distortion occurs as the temperature is lowered through the transition temperature, and that the pressure and stress dependences of the transition temperature and the energy gap are in accordance with theoretical predictions.[66] This theory also is capable of explaining almost all of the other observed transitions in a consistent manner, as well as providing for an understanding of the frequency of such transitions in transition metal compounds.[69] Thus, it has a great deal of appeal. However, it is clear from the low mobility and very small mean free path that the high-temperature phase of V_2O_3 is strongly correlated.[164] Furthermore, there is evidence for strong electronic interactions from the resistivity behavior.[165] Thus, V_2O_3 is certainly not very far from being a Mott insulator, and it probably cannot be treated by ordinary one-electron theory. In addition, the electronically induced phase transition model does run into some difficulties with regard to V_2O_4, since the lack of a doubling of the primitive cell in the distorted region rules out a simple band splitting model. Nevertheless, the multiplicity of $3d$ bands present in V_2O_3 does enable the distortion to stabilize an insulating state,[119] and essentially all the experimental data on pure V_2O_3 can be explained by a straightforward modification of the model. However, all this is independent of whether or not the low-temperature state is that of a Mott insulator or a normal band semiconductor, and so the applicability of the distortion model does not rule out an accompanying Mott transition.

The opposite point of view, that the transition in V_2O_3 is basically a Mott transition, with the lattice distortion being magnetorestrictive in origin and of little consequence, has recently been forcefully presented.[166–174] The bulk of the evidence presented comes from a mapping out of the phase diagram of the $(V_{1-x}Cr)_2O_3$ system. In general, the insulating phases in this system exhibit relatively large separations between nearest-neighboring vanadium ions, and the susceptibility experiments particularly suggest the presence of localized spins.

However, several complications exist in the pseudobinary system, not the least of which is the disorder induced by the Cr ions. As discussed in Section 3.6, disorder itself can induce a localization of electronic states, and this additional localizing force can shift the phase diagram considerably. But for the $(Cr_xV_{1-x})_2O_3$ system the continuity of the phase diagram at small values of x implies that the low-temperature state of pure V_2O_3 retains its basic character out to $x = 0.2$, at which point the Cr_2O_3 ground-state configuration takes over.[163] Pure Cr_2O_3 is almost surely a Mott insulator,[26] so the entire phase diagram can be important, in the sense that if the low-temperature states of Cr_2O_3 and V_2O_3 are continuous, then the Mott-insulating nature of the ground state of pure V_2O_3 would be established. However, the phase boundary near $x = 0.2$ shows this is not the case.[163,175] A further complication is the apparent existence of a third type of antiferromagnetic order in the intermediate-x, intermediate-temperature region, not previously investigated.[175] The phase diagram is shown in Figure 19. Only the AFM I region is known to be Mott insulating.

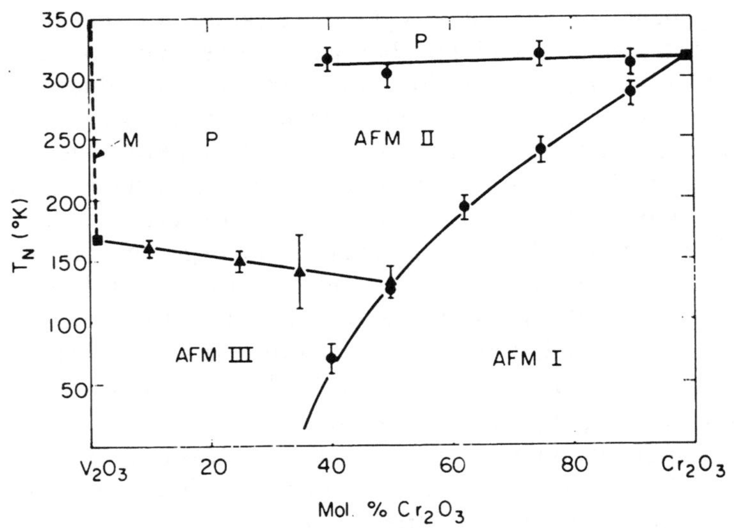

Fig. 19. Phase diagram for the $(Cr_xV_{1-x})_2O_3$ system determined by magnetic and crystallographic boundaries. The complete boundary between the AFM II and P states has not yet been determined. P: paramagnetic; AFM I: Cr_2O_3-type antiferromagnetic ordering corundum lattice; AFM II: different antiferromagnetic structure, corundum lattice; AFM III: V_2O_3-type antiferromagnetic ordering, monoclinic lattice; M: metallic state, corundum lattice. From Reid *et al.*[175]

The question of the nature of the AFM III region of Fig. 19 can be answered only by experimentally determining whether the state is one in which the $3d$ electrons are localized or itinerant. An itinerant antiferromagnetic state which tends to stabilize low-temperature semiconducting behavior has been discussed in Section 4.2, and this is perfectly consistent with the observed moment of 1.2 μ_B, a value significantly lower than the saturation moment of 2 μ_B expected for localized spins. A localized $3d^2$ configuration could experimentally manifest itself in several ways. The crystalline-field excitations (Frenkel excitons) of the localized electrons could be observable in the optical spectrum, as occurs in Mott insulators such as Cr_2O_3 and NiO; none have yet been found in V_2O_3.[160] The absorption edge may well mask all of these localized $3d^2$ excitations, but the question remains why the peaks are not shifted to lower energies by overlap and screening effects. The so-called "white line" X-ray spectrum characteristic of Mott insulators[176] has not been observed in V_2O_3, and, in fact, no change in the X-ray emission spectrum has been found between the low-temperature and the high-temperature states.[177] This is difficult to reconcile with the occurrence of a sharp Mott transition. In addition, absolutely no change in the positron lifetime between the two states has been measured in positron annihilation experiments,[178] and even the momentum distributions are very similar in the semiconducting and metallic states.[179] One might expect sharp changes in the momentum distribution upon delocalization. Finally, a large mobility change at the transition or at least a change to hopping conduction in the low-temperature state would be indicative of a localized–itinerant transition, but neither has been observed. In short, although there are some indications that the semiconducting state of V_2O_3 is that of a Mott insulator, there is no direct evidence as yet that this need be the case.

Whether or not the low-temperature phase of pure V_2O_3 is a Mott insulator, it appears naïve at present to dismiss the lattice distortion as irrelevant to the transition. Of the eight oxides of vanadium which exhibit insulator–metal transitions, only V_2O_3 exhibits antiferromagnetic order up to the transition temperature. Since the antiferromagnetism is necessary to stabilize the Mott-insulating state, the other seven transitions cannot be Mott transitions. On the other hand, in every case that has been investigated a lattice distortion does occur. But perhaps the most convincing piece of evidence that the crystalline transformation is vital for the occurrence of the insulator–metal transition comes from experiments on amorphous films,[180] prepared by sputtering the material onto a metallic substrate. When the substrate temperature was above that of the insulator–metal transition during deposition the amorphous films were metallic at all temperatures. On the other hand, when the substrate temperature was held below 150°K during deposition, the resulting films were semiconducting at all temperatures. In either case, the insulator–metal transition did not occur. Crystallized films, however, exhibited the transition at the usual temperature.

It is clear from our discussion of amorphous semiconductors in Section 2.3 that the short-range order is ordinarily virtually identical in corresponding amorphous and crystalline material. Thus, a simple explanation of the experiments on amorphous films is that when they are deposited on substrates held above 150°K the short-range order of the film is the same as that of trigonal V_2O_3 and the film is thus an amorphous metal. When cooled below 150°K the crystalline distortion does not occur because in the absence of long-range order the material cannot pick out a particular direction along which the V^{3+} ions can conveniently pair. However, when the substrate is held below 150°K the short-range order is that of monoclinic V_2O_3 and the material is therefore semiconducting. Once again, the amorphous nature of the film prevents a long-range crystallographic transformation at 150°K, and the transition is suppressed. Thus, it appears that if the lattice distortion can be suppressed, the insulator–metal transition does not occur. (An alternative explanation of these experiments is that the stoichiometry of the amorphous films depends critically on the substrate temperature, with excess oxygen occurring at low deposition temperatures, and suppressing the metallic state. However, it would appear that a range of compositions in which the transition does occur at ordinary temperatures must be produced, so that this explanation is rather unlikely.)

Another conclusion reached from an intensive investigation of the phase diagram of Figure 19 is that the high-temperature anomaly in the vicinity of 600°K (see Figure 18) in pure V_2O_3 represents a continuous (supercritical) metal–insulator Mott transition.[165–174] Indeed, the dashed line in Figure 19 has been shown to end in a critical point.[181] This suggests another problem, for once again the evidence is not yet clear. The actual behavior of the resistivity above the anomaly itself is in doubt—McWhan *et al.*[173] found that the conductivity above 600°K turned over and began increasing with further increase in temperature, although two other independent studies[158,162] observed a monotonic decrease of conductivity with increasing temperature up to 900°K. Furthermore, the Hall coefficient is constant from 150 through at least 700°K,[162,182] implying a constant carrier concentration of one carrier per vanadium atom throughout the transition. Furthermore, no anomalous temperature dependence between 400 and 600°K can be determined from the NMR Knight shift.[183] One of the main pieces of evidence in favor of the Mott transition interpretation is that the necessary higher entropy of the very high-temperature state can be easily explained by a localization of the electrons, thus leaving a set of randomly arranged spins.[171] The alternative picture of a band uncrossing would ordinarily decrease the entropy. However, Zeiger[184] has recently shown that a simple two-band model can account not only for the high-temperature anomaly of pure V_2O_3, but also for the dashed portion of the entire phase diagram of the $(V_{1-x}Cr_x)_2O_3$ system shown in Figure 19. In this model in the metallic region below 600°K the Fermi energy lies in a wide band, with a narrow empty band somewhat above. Because of the interband electronic interactions, the upper band moves down relative to the lower band with its thermal population, and the high-temperature anomaly merely represents the repopulation brought about by the Fermi energy crossing through the narrow band. This model explains the first-order nature of the transition in the $(V_{1-x}Cr_x)_2O_3$ system as well as the observed critical point, without the necessity of invoking a Mott transition. Clearly, the final words on V_2O_3 have not yet been written.

5.2.4. Bronzes

The term bronze has been given to an entire series of doped, saturated transition metal oxides which often have a metallic luster. Typically they consist of oxides such as WO_3, MoO_3, V_2O_5, or Nb_2O_5 arranged on the perovskite or perovskite-type structure with alkali or other metallic ions at the body centers. In particular, the sodium tungsten bronzes (Na_xWO_3), potassium molybdenum bronzes (K_xMoO_3), and sodium vanadium bronzes ($Na_xV_2O_5$) have been investigated in some detail. These are treated in a special subsection because specific bronzes fit into all three classes previously discussed, depending on the value of x.

The sodium tungsten bronzes are metallic for values of x greater than about 0.25, and some of the materials even become superconducting at low temperatures. The ordinary cubic pervoskite structure is stable for $0.30 < x < 1.00$. Two tetragonal forms are also stable, one for $0.10 < x < 0.45$, which is the superconducting phase, and an antiferroelectric phase, which leads to insulating behavior.[185] An interesting feature of the entire class of alkali tungsten bronzes is that the energy levels of the alkali atoms universally appear to be above the bottom of the conduction band of the saturated oxide. For the sodium tungsten bronzes the conduction band is the lowest $5d$ band of WO_3. Since the sodium energy levels are above the bottom of this $5d$ band, the sodium atoms ionize spontaneously and each contributes one electron to the conduction band, ordinarily resulting in metallic conductivity. Thus, the alkali atoms are always ionized in the bronzes and need not be further considered in a discussion of transport behavior. This determination has been made primarily via NMR measurements,[186] which exhibit no Knight shift in these materials; thus, the carriers are not in the s band of the alkali atoms and must therefore be in the $5d$ band of the tungsten ions.

The Hall coefficients are independent of temperature for the sodium tungsten bronzes, as is expected for metals[187]; the Hall mobility is of the order of 20 cm^2/V-sec at room temperature, also typical of good metals. Each sodium atom contributes one free carrier to the conduction band. The thermoelectric power is linear in temperature,[188] still further evidence of ordinary metallic behavior. The effective mass is of the order of the free-electron mass, and the plasma frequency scales with x. This latter fact is responsible for the changing metallic luster with composition that characterizes the system.

As far as we are concerned, a particularly interesting question is whether or not the change to semiconducting behavior below $x \simeq 0.25$ is an example of a Mott transition. At present there is little evidence that this is the case. An interesting material to compare to Na_xWO_3 is the nonstoichiometric crystal WO_{3-x},[189,190] which is electronically very similar to the sodium tungsten bronzes, but structurally very different. Three crystalline forms of stoichiometric WO_3 exist, a high-temperature monoclinic form, an intermediate-temperature triclinic form, and a low-temperature triclinic or monoclinic form.[191] Large conductivity discontinuities occur at the transformation temperatures, and it can be concluded that the transitions are of first order.[190] Hall-effect studies suggest the first transition, at 230°K, is due to a sharp increase (factor of 200) in carrier concentration, while the second transition, at 290°K, is primarily brought about by a relatively small mobility increase. In the high-temperature state the material appears to be metallic, with mobilities of the order of

10 cm^2/V-sec.[189] As x increases, the carrier concentration increases, the mobility decreases, and the transitions are slowly washed away. The mobility appears to be dominated by optical phonon scattering at very low x, but above $x = 0.001$ the mobility becomes thermally activated. Although this could indicate the onset of small-polaron hopping, the actual mobilities in this range are not very different than in the region of predominantly optical phonon scattering, and ionized-impurity scattering cannot be ruled out. Furthermore, there is evidence for an ordering of the oxygen vacancies in the range $x > 0.001$,[189] an effect which could provide additional band splittings. In none of the materials is there any firm evidence for a Mott-insulating ground state.

Sodium vanadium bronzes, $Na_xV_2O_5$, have also been studied somewhat. The structure is tetragonal for the range $0.22 < x < 0.40$.[192] In this form the material is semiconducting and the mobility is slightly thermally activated ($E_a \sim 0.05$ eV).[193] The room-temperature Hall mobility is of the order of 0.2 cm^2/V-sec. This material could well be an example of one in which small-polaron hopping predominates, but much more work is necessary.

Finally, blue potassium molybdenum bronze, $K_{0.30}MoO_3$, is an interesting material, which forms a monoclinic layered structure.[194] Although the closely related red potassium molybdenum bronze, $K_{0.33}MoO_3$, is a semiconductor at all temperatures,[195] the blue phase is metallic above 180°K but transforms to a semiconductor below that point.[195,196] Below 10°K the conductivity looks degenerate, although it is small in magnitude.[197] The Hall mobility in the semiconducting region decreases with increasing temperature but jumps by about one order of magnitude near the 180°K transition. The carrier concentration is constant in the metallic region. No change of symmetry has been noted at the transition temperature. The gradual nature of the transition indicates perhaps that a simple band overlap model is applicable, but once again more detailed work is essential.

5.3. Rare Earth Oxides and Chalcogenides

Just like the class of transition metal oxides, the rare earth chalcogenides (oxides, sulfides, selenides, and tellurides) exhibit the gamut of transport behavior, containing metals, insulators, and materials which undergo metal–insulator transitions. Initially this is disconcerting, since the $4f$ states of rare earth ions differ from the $3d$ states of transition metal ions in the sense that the former should always be well on the insulating side of the Mott transition. However, order is restored by the observation that the metallic conduction never appears in the $4f$ band, but rather in the $5d$ or $6s$ conduction band of the rare earth ions. But because the $4f$ states tend to pin the Fermi energy, these materials are generally n-type conductors. The material whose transport behavior has been studied in the most detail is EuO, which we shall use as the prototype for this group.

EuO crystallizes in the (cubic) rocksalt structure and is ferromagnetic below about 70°K.[198] The pure material can be made either oxygen rich or europium rich. Oxygen-rich samples exhibit semiconducting behavior at all temperatures, with an activation energy of 0.6 eV.[199] On the other hand, europium-rich crystals are only semiconducting above the Curie temperature, and the activation energy is only 0.3 eV. Below 70°K the conductivity increases by more than a factor of 10^{13}, and the material becomes metallic.[200] Electrical conductivity as a function of temperature is shown in Figure 20.[201] Because of the large effects on the resistivity in the vicinity of the Curie temperature, EuO and related materials are known as magnetic semiconductors.

The stoichiometric material consists of Eu^{2+} and O^{2-} ions. Eu^{2+} has a $4f^7$ configuration, outside of a closed-shell Xe core. Photoemission experiments[202] indicate that the $4f^7$ ground state is about 1.1 eV below the conduction band and approximately 3 eV

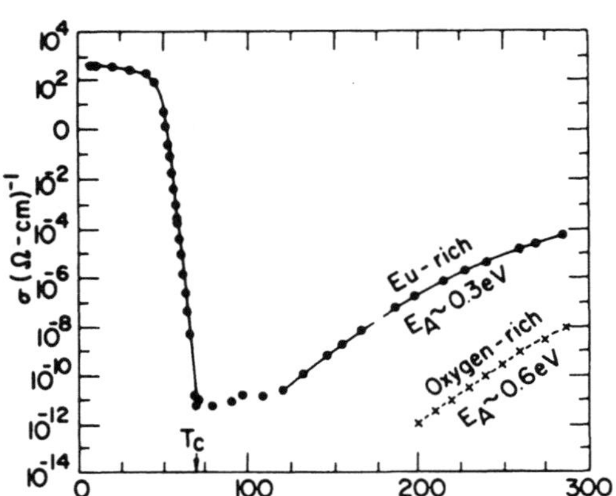

Fig. 20. Electrical conductivity as a function of temperature for Eu-rich and O-rich EuO. From Torrance *et al.*[201]

above the top of the oxygen $2p$ band. It is not yet clear whether the bottom of the conduction band is primarily $5d$ or primarily $6s$ in nature, although the fact that the $4f^6 6s$ level is below the $4f^7 5d$ level in the free Eu^+ ion is evidence in favor of the bottom of the conduction band being predominantly $6s$.[203] On the other hand, an explicit band calculation yielded a $5d$ band with a minimum over 1 eV below the bottom of the $6s$ band.[204]

As is true of other strongly correlated materials, the transport properties are dominated not by the intrinsic band structure, but by the electronic structure of the defects. No matter what the nature of the conduction band, there is no doubt that the Fermi energy lies above its minimum at temperatures below the metal–insulator transition discussed previously. The large increase in conductivity as the temperature is lowered through the transition region has been shown to be primarily a change in carrier concentration rather than in mobility.[205] Penney *et al.*[200] have shown that the transition can be accounted for by an occupied defect level rising relative to the conduction band edge, eventually crossing the minimum and yielding a metallic state. The fact that oxygen-rich EuO is insulating at all temperatures is simple to understand: Eu vacancies act as acceptors, since they lead to the formation of two Eu^{3+} ions, which can accept electrons. The holes can be created in either the oxygen $2p$ band or the Eu $4f^7$ band; however, transport in the latter should be quite difficult, so that the observed 0.6-eV activation energy is very likely the energy to create a hole in the $2p$ band. The localized $4f^7$ band is probably located above the top of the $2p$ band, even in oxygen-rich material, in which case it serves to pin the Fermi energy. The onset of ferromagnetism should have no effect on this energy, and no transition is observed. On the other hand, in Eu-rich EuO the defects are oxygen vacancies. These either lead to the formation of bound-Eu^+–vacancy-Eu^+ complexes, which contain two extra electrons and thus act as double donors, or cause an additional two electrons to be trapped at the site of each oxygen vacancy, also creating double donors. Of course, just as in NiO, partial compensation very likely occurs in either situation.

Two major models for the insulator–metal transition have been proposed. One of them[205,206] assumes that the ground state of Eu-rich crystals is one in which both additional electrons are trapped at each oxygen vacancy position, a situation identical to the formation of F' color centers in alkaline earth oxides. The ground state of the heliumlike configuration is a spin singlet, which is independent of magnetic fields. When the temperature is lowered through the Curie point the internal magnetic field set up by the ferromagnetism splits the conduction band into spin-up and spin-down components. At sufficiently intense fields the bottom of the conduction band passes through the defect level, thus inducing an insulator–metal transition. This model is schematically outlined in Figure 21. Recently a model which postulates formation of bound magnetic polarons has been proposed.[201] The basic assumption in this model is that the ground state of an oxygen vacancy has only one electron trapped at the vacancy; the second excess electron then is bound to the six surrounding Eu^{2+} ions. If the hopping time of the electron between Eu^{2+} sites is sufficiently short, the electron can reduce its energy by polarizing the spins of the six ions, forming a local ferromagnetic complex known as a bound magnetic polaron. If the exchange energy decrease from magnetic polaron formation is greater than the binding energy of the second electron to the F' center, then the polaron state is the ground state; otherwise, the heliumlike state is lowest. If the bound magnetic polaron is the ground state, then as the crystal becomes ferromagnetic as a whole the polaronic binding energy decreases and the polaron spreads out into a hydrogenic-type orbit around the singly ionized vacancy. If a sufficient density of vacancies is present, a Mott transi-

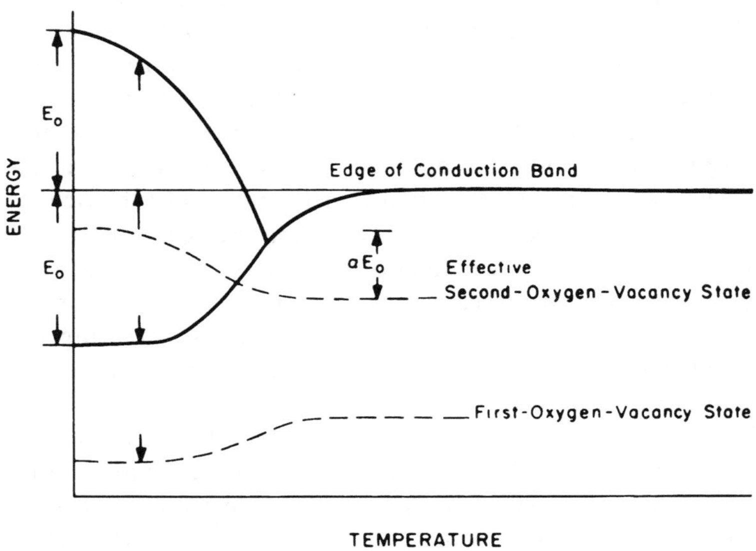

Fig. 21. Sketch of energy levels as a function of temperature near the conduction-band edge of EuO. From Oliver *et al.*[205]

tion in the defect band leads to metallic conduction, but only at low temperatures when the polaron delocalizes. In more stoichiometric samples carrier freezeout at sufficiently low temperatures should exist, in agreement with observations. However, further experiments are necessary before the actual cause of the insulator–metal transition is determined with any degree of certainty.

5.4. Amorphous Semiconductors

The important issues to decide with regard to transport in amorphous semiconductors are: (1) Is there a band tail? (2) If so, does a sharp mobility edge exist? (3) Is conduction predominantly above the mobility edge, in the band tail, in an impurity band, or in the vicinity of the Fermi energy? (4) Is conduction primarily through extended states or by hopping through localized states? (5) Do small polarons form? (6) Where is the Fermi energy relative to the band or mobility edges? (7) Is the Fermi energy pinned, and, if so, what is the magnitude of $g(\varepsilon_F)$? (8) Is conduction predominantly p type or n type? (9) What is the mobility and mean free path of the current carriers? Although it would appear that such questions should be relatively easily answered, in the case of amorphous solids even the simplest experiments often yield ambiguous conclusions or contradictory results. The problems encountered can be exemplified by a consideration of amorphous silicon and germanium. Since the two materials are so similar, we shall concentrate on amorphous silicon, and refer to results on amorphous germanium only when the data on silicon have not yet been reported.

Amorphous Si cannot be prepared in bulk form by quenching from the liquid state, since the liquid has complete different coordination from the solid. Thus only thin films, less than 20 μm in thickness, can be prepared. Even the structure of amorphous Si and Ge is still a matter of lively debate,[207] although the bulk of the evidence at present is in favor of a random network of tetrahedrally coordinated atoms. Small distortions (average of 10°) of the bond angles result in a random network which can be shown to fill space with up to 97% of the density of the crystal[208] and which is in agreement with electron diffraction experiments.[209] Nevertheless, a model of amorphous Si or Ge as a collection of microcrystallites 14 Å in diameter held together by an amorphous matrix[210] cannot yet be completely ruled out.

However, even if we are given the ideal amorphous structure, there is still the problem that most vapor-deposited films (the usual method of preparation for either amorphous Si or Ge) are far from ideal, but, in fact, contain extensive regions of voids, as in a slice of Swiss cheese.[9] These voids lead to interior dangling bonds, which can dominate the transport behavior. There is evidence that the voids tend to coalesce as the films are annealed below the crystallization temperature.[211] There is some evidence that the voids are rod-shaped[212] and other evidence that they are cracklike.[213] Other structured problems that must be faced before attempting to analyze the transport properties are thickness variations and the presence of impurities. Hauser[214] observed that 240-Å films of amorphous Ge are shorted out by planar electrodes, indicative of thickness variations comparable with the film thickness itself. In addition, the presence of oxygen in films of amorphous Si can have an enormous effect on the transport.[215,216] The possibility also exists that annealing of the films, in order to mitigate the effects of the void network, might, in fact, add oxygen impurities to the films in a systematic manner.[215]

A brief review of the early literature on transport properties of amorphous Si or Ge immediately focuses on these problems. The optical band gap of amorphous Si is 0.6,[217] 1.6,[218] or 1.8 eV.[219] The absorption edge is gradual,[220] indicative of disorder-induced band tailing, is sharp even in as-deposited films,[221] or is gradual in as-deposited films but sharp after annealing.[222] The logarithm of the electrical conductivity is a linear function of T^{-1},[223] a linear function of $T^{-1/4}$,[224] the latter at low temperatures, the former at high temperatures,[225] or the latter for fully annealed films, the former for as-deposited films.[226] The Hall effect indicates that conduction is predominantly *n* type,[227] or *p* type,[228] or cannot be detected at all.[229] Thermoelectric power measurements clearly show that either electrons[230] or holes[231] dominate the transport near 300°K. It is rather clear that the films must be carefully characterized and the experimental techniques critically analyzed if any order is to be made out of this apparent chaos.

Some of the contradictory results can be interpreted by noting that two classes of vapor-deposited amorphous Si and Ge exist.[215] One set of films are prepared in ultrahigh vacuums, of the order of 10^{-11} Torr, by slow deposition on heated substrates. We shall call these type I films. Type I films yield sharp optical gaps, of the order of 0.6 eV for amorphous Si, and show little or no effects of annealing. No evidence for any band tails exist[221]; thus, the question of a sharp mobility edge in these films is moot. The type I films exhibit a tempera-

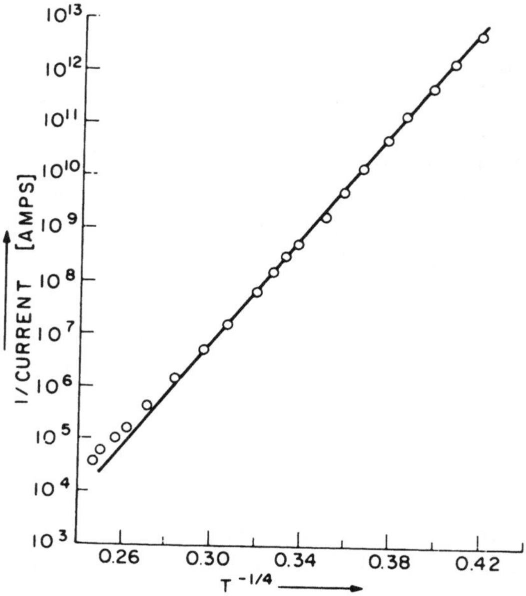

Fig. 22. Resistance as a function of temperature for a type I film of amorphous Ge. From Pollack *et al.*[232]

ture dependence of the electrical conductivity in accord with Eq. (41), thus indicating the predominance of phonon-assisted hopping in states in a band of localized states about 0.3 eV wide surrounding the Fermi energy.[232] Typical results are shown in Figure 22. Similar films of decreasing thickness start to exhibit a linear dependence of the logarithm of the conductivity on $T^{-1/3}$, characteristic of phonon-assisted hopping in two dimensions.[233] From these results a density of states at the Fermi energy $g(\varepsilon_F)$ of approximately $10^{18}\ \text{cm}^{-3}\ \text{eV}^{-1}$ can be estimated. Thus, the picture that emerges for the electronic structure of type I films is one of sharp band edges at a somewhat smaller gap than that of the corresponding crystal, with a relatively wide defect band containing approximately 10^{17} states/cm^3 in the gap. This defect band pins the Fermi energy and controls the transport properties. The defect band most likely arises from well-defined complexes, such as vacancies or divacancies.

A rather different view is obtained from a consideration of films deposited in vacuums of about 10^{-7} Torr at higher deposition rates on room-temperature substrates. These films, which we shall call type II, initially show a gradual optical absorption edge, which could indicate the existence of disorder-induced band tails. Upon annealing, these tails disappear, and the edge shifts to higher energies, about twice the energies of the edges found in type I films. There is evidence that the void structure coalesces and the dangling-bond density decreases upon annealing.[209,211] The electrical conductivity of as-deposited type II films generally obeys a $T^{-1/4}$ law, Eq. (41). Upon annealing, the conductivity decreases, and at higher temperatures a T^{-1} behavior, Eq. (32) is obtained.[234] Typical data are shown in Figure 23. At the highest annealing temperature the activation energy at high temperatures becomes just half the optical energy gap of the films, indicating intrinsic bandlike conduction, Eq. (7). However, since the optical gap is sharp after this anneal, this result does not provide any evidence for a sharp mobility edge, just a sharp band edge. These results plotted on a $T^{-1/4}$ curve are shown in Figure 24. The deviations from Eq. (41) are evident in the thermoelectric-power as well as in the resistivity behavior. Similar linear $T^{-1/4}$ plots for as-deposited material are also obtained from studies of low-bias tunneling into amorphous Si and Ge films.[218] However, careful interpretation of the temperature dependence of the conductivity often yields physically

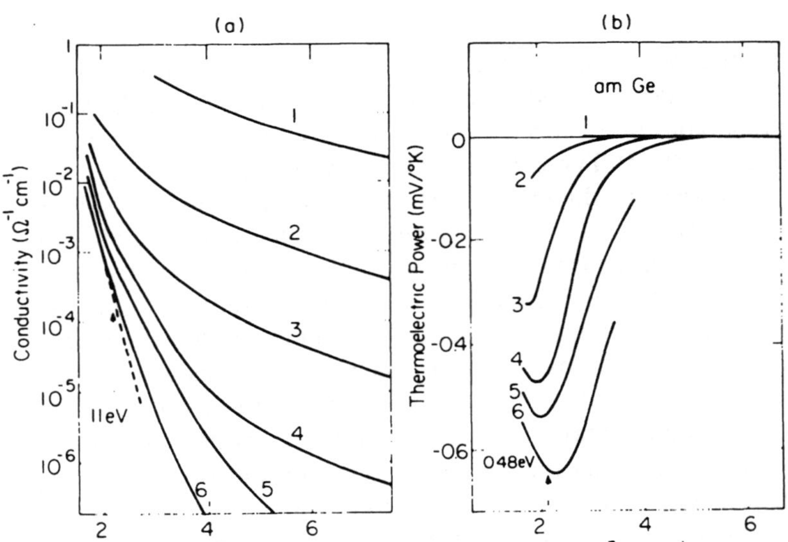

Fig. 23. Temperature dependence of (a) electrical conductivity and (b) thermoelectric power of type II amorphous germanium films for different annealing steps. The higher the number on the curve, the higher the annealing temperature. From Beyer and Stuke.[234]

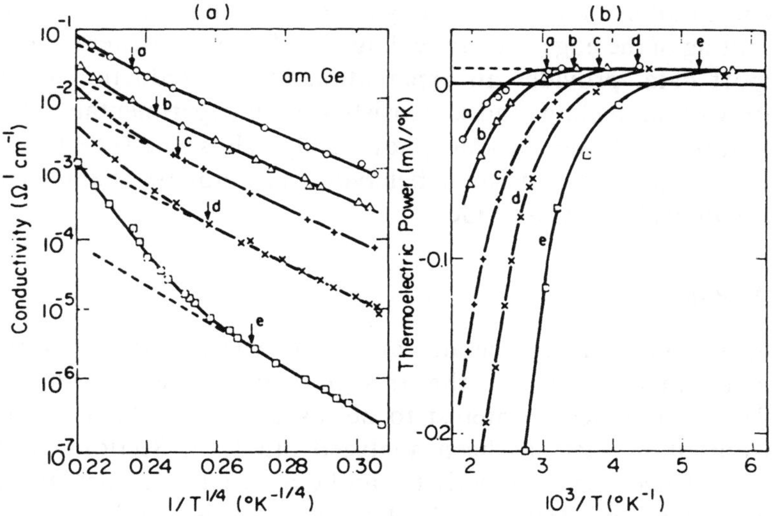

Fig. 24. (a) Electrical conductivity plotted as a function of $T^{-1/4}$ for type II films of amorphous Ge for different annealing steps. (b) Corresponding thermoelectric power curves. From Beyer and Stuke.[234]

unreasonable values for the density of states at the Fermi energy.[226,235] In such films an alternative model to phonon-assisted hopping should be considered,[55] which is bandlike conduction in the absence of a sharp mobility edge. As the material is annealed the optical results suggest a sharpening of the band edge. Percolation calculations indicate that as the band tail abates, conduction passes smoothly from a pseudo-$T^{-1/4}$ region to ordinary T^{-1} behavior. This is not an unreasonable explanation of the data on type II amorphous Si and Ge films. We should note that in neither type of film is there any evidence for the applicability of the CFO model to amorphous Si or Ge.

Finally, the major differences between type I and type II films, particularly with regard to the discrepancies in optical edges, must be explained. Perhaps the simplest resolution of the problem would be the existence of amorphous polytypes, just as there are several allotropic crystalline forms of Si and Ge. But a second possibility that should not be overlooked is that the type II films, prepared in much less of a vacuum than the type I films, contain a significant fraction of oxygen impurities.[215] It has been shown[236] that the presence of oxygen in amorphous Ge films sharpens the optical edge and shifts it to higher photon energies. This suggests that adsorption and subsequent diffusion of oxygen may be responsible for the sharpening and shifting of the edge with annealing.(222) This is consistent with the transport data, provides an explanation for the reduction in spin density with annealing, and is consistent with experimental(237) and theoretical(238) studies of amorphous SiO_x. Thus, the type II films, although exhibiting consistent behavior, may not be characteristic of pure amorphous Si and Ge.

5.5. Metals

Although ordinary metals cannot be regarded in any way as unconventional materials, the transport properties of imperfect metals are of sufficient interest to be discussed as a final example. Transport in metals has been analyzed briefly in Section 3.1; the problem of electrical conductivity can be reduced to a calculation of the relaxation time τ introduced in Eq. (3). Any deviation from periodicity contributes to τ by leading to the scattering of conduction electrons. If the scattering processes are independent, the probability of a collision is the sum of the probabilities due to any of the imperfections. Thus, the effective relaxation time can be written

$$\frac{1}{\tau} = \frac{1}{\tau_p} + \frac{1}{\tau_i} + \frac{1}{\tau_d} + \frac{1}{\tau_m} + \frac{1}{\tau_s} + \cdots \tag{62}$$

where τ_x represents the relaxation time due to any individual deviation from periodicity. In Eq. (62) τ_p represents the relaxation time due to the presence of phonons, τ_i to impurities, τ_d to defects, τ_m to magnetic (spin) disorder, and τ_s to the surface; in addition, contributions could be present from electron–electron interactions, electron–spin interactions, etc. Since the resistivity ρ is proportional to the reciprocal of the relaxation time [see Eqs. (2) and (3)], Eq. (62) implies

$$\rho = \rho_p + \rho_i + \rho_d + \rho_m + \rho_s + \cdots \tag{63}$$

a result known as Matthiessen's rule.(239) Since the number of phonons present is a strong function of temperature, ρ_p can be expected to be very temperature dependent and become small at low temperatures. Alternatively, most of the remaining contributions to the resistivity should be essentially independent of temperature. Thus, at sufficiently high temperatures the resistivity of any metal is dominated by ρ_p and is independent of impurities and defects, but at low temperatures the effects of impurities and defects dominate.

The phonon contribution to resistivity ρ_p was first calculated by Bloch[240]; refinements have been introduced by Sommerfeld and Bethe,[241] Bardeen,[242] and Bailyn,[243] and the final result is indeed complex. However, the basic temperature dependence of ρ_p can be estimated without resort to detailed mathematics. At sufficiently high temperatures the classical limit is quite appropriate, in the sense that the equipartition theorem is applicable. The highest-energy phonons are of the order of $k\Theta$, where Θ is the Debye temperature (e.g., 344°K for Cu). At temperatures large compared to Θ the change in electron energy during a collision is thus small compared to kT, and the collisions can be treated as elastic. The collision probability is then simply proportional to the mean square amplitude of the lattice vibrations, which by the equipartition theorem is itself proportional to T. Thus the resistivity at high temperatures increases linearly with T.

At low temperatures, however, there are only low-frequency phonons present in significant concentration. The number of such phonons varies as T^3. But, in addition, many scattering processes are necessary to restore equilibrium. The low-energy phonons can scatter electrons only through a small angle, of the order of T/Θ. Since the contribution to the resistivity of each scattering event depends on the square of the sine of the scattering angle, an additional factor of T^2 is introduced into ρ_p. Thus, the contribution of phonon scattering to the resistivity of metals at low temperatures is proportional to T^5.

The rapid decrease in ρ_p with decreasing temperature below Θ implies that one of the temperature-independent contributions to the resistivity must dominate at sufficiently low T. The sizes of these contributions depend very much on the physical and chemical purity of the sample. If there is a large concentration of impurities, ρ_i is large and eventually predominates at very low temperatures. In very pure samples the predominant scattering is usually by vacancies, interstitials, or dislocations, and $\rho \simeq \rho_d$ as T approaches zero. In relatively pure, carefully annealed samples ρ_s can even become the major contribution at low temperatures.

Because of the fact that all samples of a particular metal ordinarily have the same resistivity at room temperature ($\rho \simeq \rho_p$) but the higher-quality samples have much lower resistivity as T approaches zero, the ratio of the resistance of a particular sample at room temperature to that at liquid helium temperature,

$$R \equiv R(300°\mathrm{K})/R(4°\mathrm{K}) \tag{64}$$

is an excellent guide to the quality of the sample. R is usually called the resistance ratio and should generally be very large in pure samples. For example, the resistance ratio of ordinary commercial copper wire is of the order of 100, while ultra pure samples of copper have been prepared with values of R near 10,000.[244] Figure 25 shows the low-temperature resistivities of several samples of gold of varying purity.[245] The purest samples exhibit the T^5 behavior expected for ρ_p in this temperature range. All the curves can be represented by such a contribution ρ_p plus a temperature-independent term obtained by the extrapolation of $\rho(T)$ to zero temperature. At higher temperatures all the curves merge into ρ_p.

For several metals the resistance ratios of even pure, annealed samples are quite low; e.g., for Mn typical values of R are of the order

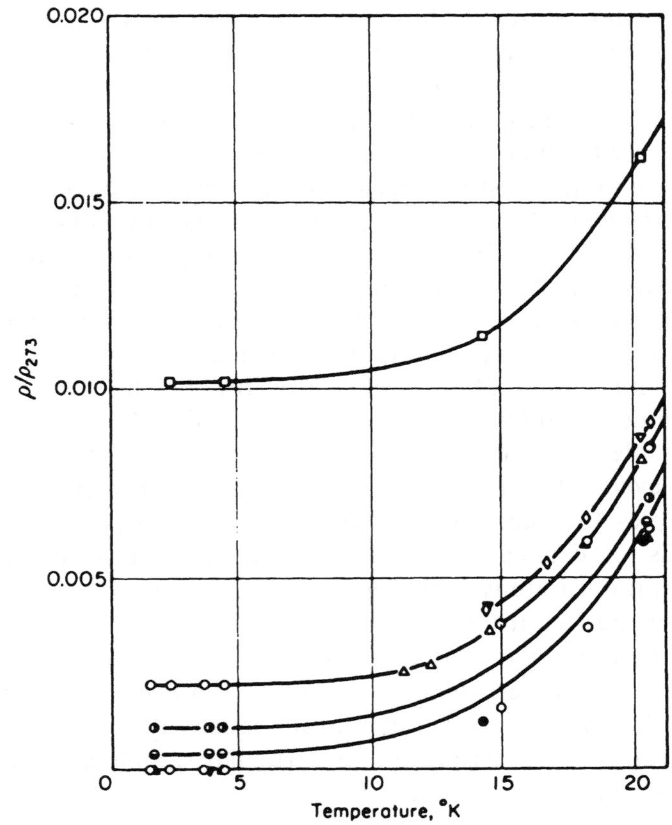

Fig. 25. Electrical resistivity of several samples of Au of various purities at low temperatures. From van den Berg.[245]

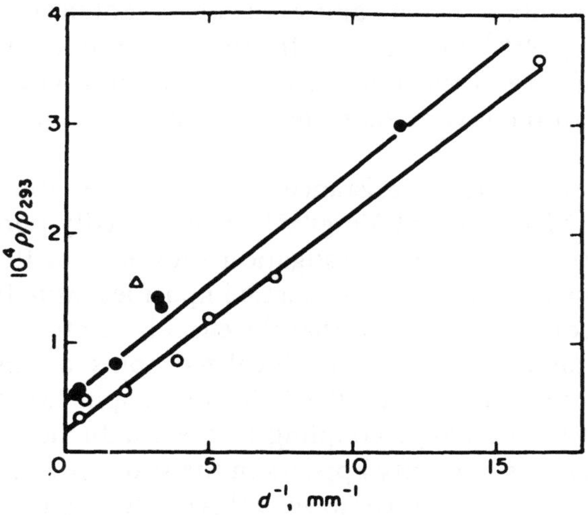

Fig. 26. Electrical resistivity of three different types of pure indium wire as a function of the inverse diameter. From Olsen.[250]

of 15.[246] The resistance, even at the lowest temperatures investigated, does not become temperature independent but varies as T^2. The explanation stems from the fact that these metals exhibit long-range magnetic order at low temperatures; e.g., Mn is antiferromagnetic below 100°K. At zero temperature the magnetic order is perfect, and the spins make no contribution to the resistivity. However, as indicated in Section 1.1, as the temperature increases the spins become disordered, and the resulting deviations from perfect periodicity can scatter conduction electrons. This is called spin-disorder scattering[247] and contributes the term ρ_m in Eq. (63) to the resistivity. At high temperatures when the spins are randomly aligned perfect periodicity is reobtained, and ρ_m becomes very small.

If no localized spins are present and the metal is ultrapure and physically free from defects, the mean free path Λ at low temperatures can become very large; e.g., in metallic (white) tin mean free paths of the order of several mm have been attained.[248] If the sample is a thin film of thickness d or a wire of diameter d, where $d \ll \Lambda$, then surface scattering will limit the actual mean free path to values of the order of d. Since resistivity is inversely proportional to mean free path, for thin films or wires, we expect

$$\rho = \rho_b \alpha \Lambda / d \tag{65}$$

where ρ_b is the bulk resistivity and α is a constant of the order of unity. For wires Dingle[249] finds $\alpha = 1$. In Figure 26 the low-temperature resistivities of several different varieties of indium wire are plotted as a function of the inverse diameter, showing that Eq. (65) is indeed obeyed.[250]

Finally, anomalous resistance minima were observed many years ago by Meissner and Voigt[251] in supposedly pure metals at low temperatures. Further investigation showed that these minima were due to small amounts of magnetic impurities, usually transition metals.[245] Kondo[252] showed that the effect is due to a scattering of the conduction electrons by the localized spins of the magnetic impurities. Because of the fact that a localized spin can be either up or down and the exchange coupling to the conduction electrons is spin dependent, a singularity appears in the scattering probability as T approaches zero; the contribution of this scattering to the resistivity is

$$\rho_{\text{mi}} = K[1 + |J|/\varepsilon_{\text{F}} \ln(\varepsilon_{\text{F}}/kT)] \tag{66}$$

where K is a constant proportional to the magnetic-impurity concentration and J is the exchange energy. It is clear that if the remaining

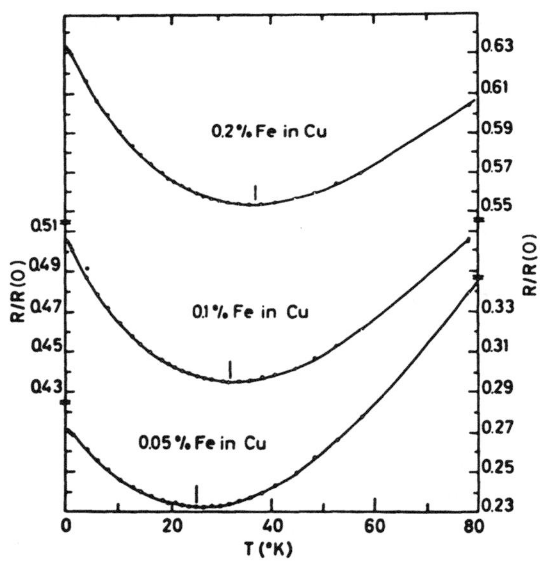

Fig. 27. Relative resistance as a function of temperature for Cu with several different concentrations of Fe impurities, clearly exhibiting the Kondo effect. From Franck *et al.*[254]

contributions to the resistivity at low temperatures are either independent of temperature (e.g., nonmagnetic impurity scattering or defect scattering) or vary as T^5 (phonon scattering at $T \ll \Theta$), the resistivity behaves as

$$\rho = C + C'T^5 - (K|J|/\varepsilon_F) \ln T \tag{67}$$

where C and C' are independent of temperature. Equation (67) exhibits a minimum at a temperature given by

$$T_{min} = (K|J|/C'\varepsilon_F)^{1/5} \tag{68}$$

Since K is proportional to the impurity concentration, the temperature of the resistance minimum should increase as the one-fifth power of the concentration, in agreement with experiment.[253] Typical results are shown in Figure 27.[254] The resistance-minimum phenomenon is now known as the Kondo effect.

Acknowledgments

The original work referred to in this article was supported, in part, by the Advanced Research Projects Agency (monitored by the U.S. Army Research Office—Durham) and in part by the National Science Foundation. I should also like to thank Dee Bowman for typing the manuscript under most adverse conditions.

References

1. J. M. Ziman, *Electrons and Phonons*, Oxford Univ. Press, London (1960).
2. C. E. Moore, Atomic Energy Levels, Nat. Bur. Std. (U.S.) Circ. No. 467, Vol. I–III (1949–58).
3. E. U. Condon and G. H. Shortley, *Theory of Atomic Spectra*, Cambridge Univ. Press (1935).
4. G. F. Koster and J. C. Slater, Simplified impurity calculation, *Phys. Rev.* **96**, 1208–1233 (1954).
5. W. Kohn, in *Solid State Physics*, Vol. 5, pp. 257–320, Academic Press, New York (1957).
6. F. C. Brown, *Physics of Solids*, W. A. Benjamin, New York (1967).
7. D. Adler, in *Physics of Electronic Ceramics*, (L. L. Hensch and D. B. Dove, eds.), Part A, pp. 29–66, Marcel Dekker, New York (1971).
8. D. Weaire, Vol. 1, Chapter 2 of this Treatise.
9. S. C. Moss and J. F. Graczyk, Evidence of voids within the as-deposited structure of glassy silicon, *Phys. Rev. Lett.* **23**, 1167–1171 (1969).
10. D. Adler, *Amorphous Semiconductors*, CRC Press, Cleveland, Ohio (1971).

11. L. Van Hove, The occurrence of singularities in the elastic frequency distribution of a crystal, *Phys. Rev.* **89**, 1189–1193 (1953).
12. F. J. Blatt, *Physics of Electronic Conduction in Solids*, McGraw-Hill, New York (1968).
13. D. K. C. MacDonald, *Thermoelectricity*, Wiley, New York (1962).
14. A. H. Wilson, *Theory of Metals*, Cambridge Univ. Press, London (1953).
15. N. F. Mott, The transition to the metallic state, *Phil. Mag.* **6**, 287–309 (1961).
16. D. Adler and J. Feinleib, Electrical and optical properties of narrow-band materials, *Phys. Rev. B* **2**, 3112–3134 (1970).
17. J. Hubbard, Electron correlations in narrow energy bands, *Proc. Roy. Soc. A* **276**, 238–257 (1963).
18. J. Hubbard, Electron correlations in narrow energy bands. III. An improved solution, *Proc. Roy. Soc. A* **281**, 401–419 (1964).
19. G. Kemeny and L. G. Caron, Self-consistent pair correlations in Mott-type insulators, *Phys. Rev.* **159**, 768–774 (1967).
20. G. W. Pratt and L. B. Caron, Correlation and magnetic effects in narrow-energy bands, *J. Appl. Phys.* **39**, 485–486 (1968).
21. R. A. Bari, D. Adler, and R. V. Lange, Electrical conductivity in narrow energy bands, *Phys. Rev. B* **2**, 2898–2905 (1970).
22. N. Ohata and R. Kubo, Electrical conduction in a narrow band: I. Moment method, *J. Phys. Soc. Japan* **28**, 1402–1412 (1970).
23. W. F. Brinkman and T. M. Rice, Single particle excitations in magnetic insulators, *Phys. Rev. B* **2**, 1324–1338 (1970).
24. N. Ohata, Electrical conduction in a narrow band: II. Effect of randomness in atomic distributions, *J. Phys. Soc. Japan* **29**, 1138–1144 (1970).
25. D. C. Langreth, Hall coefficient of Hubbard's model, *Phys. Rev.* **148**, 707–711 (1966).
26. D. Adler, in *Solid State Physics*, Vol. 21, pp. 1–113, Academic Press, New York (1968).
27. W. F. Brinkman and T. M. Rice, Hall effect in the presence of strong spin-disorder scattering, *Phys. Rev. B* **4**, 1566–1571 (1971).
28. D. M. Edwards and A. C. Hewson, Comment on Hubbard's theory of the Mott transition, *Rev. Mod. Phys.* **40**, 810–811 (1968).
29. W. F. Brinkman and T. M. Rice, Application of Gutzwiller's variational method to the metal–insulator transition, *Phys. Rev. B* **2**, 4302–4304 (1970).
30. M. C. Gutzwiller, Correlation of electrons in a narrow *s* band, *Phys. Rev.* **137**, A1726–A1735 (1965).
31. H. Frohlich, Electrons in lattice fields, *Adv. Phys.* **3**, 325–361 (1954).
32. G. R. Allcock, On the polaron rest energy and effective mass, *Adv. Phys.* **5**, 412–451 (1956).
33. R. P. Feynman, Slow electrons in a polar crystal, *Phys. Rev.* **97**, 660–665 (1955).
34. T. Holstein, Studies of polaron motion: I. The Molecular-crystal model, *Ann. Phys. (N.Y.)* **8**, 325–342 (1959).
35. T. Holstein, Studies of polaron motion: II. The "small" polaron, *Ann. Phys. (N.Y.)* **8**, 353–389 (1959).
36. L. D. Landau, On the motion of electrons in crystal lattices, *Phys. Z. Sowjetunion* **3**, 664–665 (1933).
37. I. G. Austin and N. F. Mott, Polarons in crystalline and non-crystalline materials, *Adv. Phys.* **18**, 41–102 (1969).

38. L. Friedman and T. Holstein, Studies of polaron motion: III. Hall mobility of the small polaron, *Ann. Phys.* (*N.Y.*) **21**, 494–549 (1963).
39. Yu. A. Firsov, Theory of the Hall effect in low-mobility semiconductors, *Soviet Phys.—Solid State* **5**, 1566–1580 (1964).
40. J. Schnakenberg, The Hall coefficient of the small polaron, *Z. Physik* **185**, 123–138 (1965).
41. T. Holstein and L. Friedman, Hall mobility of the small polaron, II. *Phys. Rev.* **165**, 1019–1031 (1968).
42. Yu. A. Firsov, Hall effect in polaron semiconductors, *Soviet Phys.—Solid State* **10**, 2387–2394 (1969).
43. D. Emin, The Hall mobility of a small polaron in a square lattice, *Ann. Phys.* (*N.Y.*) **64**, 336–395 (1971).
44. A. Miller and E. Abrahams, Impurity conduction at low concentrations, *Phys. Rev.* **120**, 745–755 (1960).
45. N. F. Mott and E. A. Davis, *Electronic Processes in Non-Crystalline Materials*, Oxford Univ. Press, London (1971).
46. N. F. Mott, Electrons in disordered structures, *Adv. Phys.* **16**, 49–144 (1967).
47. N. F. Mott, Conduction in non-crystalline systems: I. Localized electronic states in disordered systems, *Phil. Mag.* **17**, 1259–1268 (1968).
48. E. N. Economou and M. H. Cohen, Localization in disordered materials: existence of mobility edges, *Phys. Rev. Lett.* **25**, 1445–1448 (1970).
49. T. P. Eggarter and M. H. Cohen, Simple model for density of states and mobility of an electron in a gas of hard-core scatterers, *Phys. Rev. Lett.* **25**, 807–810 (1970).
50. S. F. Edwards, The localization of electrons in disordered systems, *J. Non-Crystall. Solids* **4**, 417–425 (1970).
51. N. F. Mott, Conduction in non-crystalline systems: IV. The minimum metallic conductivity, *Phil. Mag.* **26**, 1015–1026 (1972).
52. J. M. Ziman, The localization of electrons in ordered and disordered systems: I. Percolation of classical particles, *J. Phys. C* **1**, 1532–1538 (1968).
53. H. L. Frisch, J. M. Hammersley, and D. J. A. Welsh, Monte-Carlo estimates of percolation probabilities for various lattices, *Phys. Rev.* **126**, 949–951 (1962).
54. B. J. Last and D. J. Thouless, Percolation theory and electrical conductivity, *Phys. Rev. Lett.* **27**, 1719–1721 (1971).
55. D. Adler, L. P. Flora, and S. D. Senturia, Electrical conductivity in disordered systems, *Solid State Commun.* **12**, 9–12 (1973).
56. M. H. Cohen, H. Fritzsche, and S. R. Ovshinsky, Simple band model for amorphous semiconducting alloys, *Phys. Rev. Lett.* **22**, 1065–1068 (1969).
57. D. Adler and J. Feinleib, in *Electronic Density of States* (L. H. Bennett, ed.), Nat. Bur. Std. Special Publication No. 323, pp. 493–504, U.S. Government Printing Office, Washington, D.C. (1971).
58. N. F. Mott, Conduction in non-crystalline materials: III. Localized states in a pseudogap and near extremities of conduction and valence bands, *Phil. Mag.* **19**, 835–852 (1969).
59. N. F. Mott, Conduction in glasses containing transition-metal ions, *J. Non-Crystall. Solids* **1**, 1–17 (1968).
60. V. Ambegaokar, B. I. Halperin, and J. S. Langer, Hopping conductivity in disordered systems, *Phys. Rev. B* **4**, 2612–2620 (1971).

61. M. Pollack, A percolation treatment of dc hopping conduction, *J. Non-Crystall. Solids* **8–10**, 486–491 (1972).
62. D. Jerome, T. M. Rice, and W. Kohn, Excitonic insulator, *Phys. Rev.* **158**, 462–475 (1967).
63. B. I. Halperin and T. M. Rice, in *Solid State Physics*, Vol. 21, pp. 115–192, Academic Press, New York (1968).
64. B. I. Halperin and T. M. Rice, Possible anomalies at a semimetal-semiconductor transition, *Rev. Mod. Phys.* **40**, 755–766 (1968).
65. J. Zittarz, Transport properties of the "excitonic insulator": Electrical conductivity, *Phys. Rev.* **165**, 605–611 (1968).
66. D. Adler and H. Brooks, Theory of semiconductor-to-metal transitions, *Phys. Rev.* **155**, 826–840 (1967).
67. D. Adler and J. Feinleib, Semiconductor-to-metal transition in V_2O_3, *Phys. Rev. Lett.* **12**, 700–703 (1964).
68. J. J. Hallers and G. Vertogen, Electronically induced crystallographic transition, *Phys. Rev. B* **4**, 2351–2357 (1971).
69. D. Adler, Mechanisms for metal–nonmetal transitions in transition-metal oxides and sulfides, *Rev. Mod. Phys.* **40**, 714–736 (1968).
70. I. Nebenzahl and M. Weger, Band structure and lattice distortion in V_2O_3, *Phys. Rev.* **184**, 936–941 (1969).
71. D. Adler, in *Critical Phenomena in Alloys, Magnets and Superconductors* (R. E. Mills, E. Ascher, and R. I. Jaffee, eds.), pp. 567–591, McGraw-Hill, New York (1971).
72. W. Paul, The present position of theory and experiment for VO_2, *Mat. Res. Bull.* **5**, 691–702 (1970).
73. C. J. Hearn, The metal–insulator transition in VO_2, *Phys. Lett.* **38A**, 447–448 (1972).
74. D. Adler, Antiferromagnetism in Ti_2O_3, *Phys. Rev. Lett.* **17**, 139–141 (1966).
75. J. C. Slater, Magnetic effects and the Hartree–Fock equation, *Phys. Rev.* **82**, 538–541 (1951).
76. L. N. Bulaevskii and D. I. Khomskii, Insulator–metal phase transitions in antiferromagnets, *Soviet Phys.—Solid State* **9**, 2422–2426 (1968).
77. E. Hanamura, Lattice instability associated with metal–semiconductor transitions, *Rev. Mod. Phys.* **40**, 744–747 (1968).
78. N. Kristoffel and P. Konsin, Displacive vibronic phase transitions in narrow-gap semiconductors, *Phys. Stat. Sol.* **28**, 741–739 (1968).
79. A. G. Aronov and E. K. Kudinov, Phase transition in strong electron–phonon interaction, *Soviet Phys.—JETP* **28**, 704–709 (1069).
80. D. C. Mattis and W. D. Langer, Role of phonons and band structure in metal-insulator phase transition, *Phys. Rev. Lett.* **25**, 376–380 (1970).
81. D. C. Mattis, Role of phonons in metal–insulator phase transition, *J. de Physique* **32**(Suppl. C1), 1086–1089 (1971).
82. A. M. de Graaf and R. Luzzi, Crystallographic distortion, electron–electron interaction and the metal–nonmetal transition, *Helv. Phys. Acta* **41**, 764–766 (1968).
83. J. J. Hallers, Theories on the metal–nonmetal transition, Ph.D. Thesis, Univ. of Groningen (1972).
84. K. Elk, On a criterion for antiferromagnetism connected with the metal–insulator phase transition, *Phys. Stat. Sol.* **45**, 305–309 (1971).

85. D. Pines, in *Solid State Physics*, Vol. 1, pp. 368–450 (1955).
86. H. Frohlich, in *Quantum Theory of Atoms, Molecules, and the Solid State* (P. O. Lowdin, ed.), pp. 465–468, Academic Press, New York (1966).
87. D. Adler, unpublished work, 1964; see discussion to paper by G. J. Hyland, *Rev. Mod. Phys.* **40**, 739–742 (1968).
88. S. Doniach, The insulator–metal transition, *Adv. Phys.* **18**, 819–848 (1969).
89. L. M. Falicov and J. C. Kimball, Simple model for semiconductor–metal transitions: SmB_6 and transition-metal oxides, *Phys. Rev. Lett.* **22**, 997–999 (1969).
90. R. Ramirez, L. M. Falicov, and J. C. Kimball, Metal–insulator transitions: A simple theoretical model, *Phys. Rev. B* **2**, 3383–3393 (1970).
91. B. Alascio, A. Lopez, and V. Grunfeld, A model for the phase diagram of $(V_{(1-x)}Cr_x)_2O_3$, *Solid State Commun.* **9**, 1711–1713 (1971).
92. L. M. Falicov and C. E. T. Goncalves da Silva, Metal–insulator and magnetic phase transitions: A thermodynamic model, *Solid State Commun.* **10**, 455–458 (1972).
93. J. C. Kimball, Magnetic metal–nonmetal transitions: A simple model, *Phys. Rev. Lett.* **29**, 127–130 (1972).
94. P. Soven, Coherent-potential model of substitutional disordered alloys, *Phys. Rev.* **156**, 809–813 (1967).
95. M. Plischke, Coherent potential approximation calculation on the Falicov–Kimball model of the metal–insulator transition, *Phys. Rev. Lett.* **28**, 361–363 (1972).
96. C. E. T. Goncalves da Silva and L. M. Falicov, Metal–insulator transitions: A coherent potential approximation, *J. Phys. C* **5**, 906–913 (1972).
97. J. R. Townsend, Solid-state absorption spectra of Mg and MgO, *Phys. Rev.* **92**, 556–560 (1953).
98. J. Yamashita, Oxygen band in magnesium oxide, *Phys. Rev.* **111**, 733–735 (1958).
99. F. A. Kroger, Point defects and phase stability of transition-metal compounds, *J. Phys. Chem. Solids* **29**, 1889–1899 (1968).
100. R. E. Watson, Iron series Hartree–Fock calculations, *Phys. Rev.* **118**, 1036–1045 (1960).
101. W. Low, Paramagnetic and optical spectra of divalent nickel in cubic crystalline fields, *Phys. Rev.* **109**, 247–255 (1958).
102. G. K. Wertheim and S. Hufner, X-ray photoemission band structure of some transition-metal oxides, *Phys. Rev. Lett.* **28**, 1028–1031 (1972).
103. D. M. Roessler and W. C. Walker, Electronic spectrum and ultraviolet optical properties of crystalline MgO, *Phys. Rev.* **159**, 733–738 (1967).
104. V. A. Fomichev, T. M. Zimkina, and I. I. Zhukova, Investigation of the energy structure of MgO by ultrasoft X-ray spectroscopy, *Soviet Phys.—Solid State* **10**, 2421–2427 (1969).
105. B. Henderson and J. E. Wertz, Defects in the alkaline earth oxides, *Adv. Phys.* **17**, 749–855 (1968).
106. E. Yamaka and K. Sawamoto, Electrical conductivity and thermoelectric motive force in MgO single crystals, *J. Phys. Soc. Japan* **10**, 176–179 (1955).
107. R. Mansfield, The electrical conductivity and thermoelectric power of magnesium oxide, *Proc. Phys. Soc.* (*London*) **B66**, 612–614 (1953).
108. A. Lempicki, The electrical conductivity of MgO single crystals at high temperatures, *Proc. Phys. Soc.* (*London*) **B66**, 281–283 (1953).

109. S. P. Mitoff, Electrical conductivity of single crystals of MgO, *J. Chem. Phys.* **31**, 1261–1269 (1959).
110. S. P. Mitoff, Electronic and ionic conductivity in single crystals of MgO, *J. Chem. Phys.* **36**, 1383–1389 (1962).
111. M. O. Davies, Transport phenomena in pure and doped magnesium oxide, *J. Chem. Phys.* **38**, 2047–2055 (1963).
112. T. J. Lewis and A. J. Wright, The electrical conductivity of magnesium oxide at low temperatures, *Brit. J. Appl. Phys.* (*J. Phys. D.*) **1**, 441–447 (1968).
113. R. Lindner and G. D. Parfitt, Diffusion of radioactive magnesium in magnesium oxide crystals, *J. Chem. Phys.* **26**, 182–185 (1957).
114. S. P. Mitoff, Bulk vs. surface conductivity of MgO crystals, *J. Chem. Phys.* **41**, 2561–2562 (1964).
115. N. A. Surplice, The electrical conductivity of calcium and strontium-oxides, *Brit. J. Appl. Phys.* **17**, 175–180 (1966).
116. W. D. Copeland and R. A. Swalin, Studies on the defect structure of strontium oxide, *J. Phys. Chem. Solids* **29**, 313–325 (1968).
117. C. G. Fonstad and R. H. Rediker, Electrical properties of high-quality stannic-oxide crystals, *J. Appl. Phys.* **42**, 2911–2918 (1971).
118. D. Adler, Electrical and optical properties of transition-metal oxides, *Radiation Effects* **4**, 123–131 (1970).
119. J. B. Goodenough, Metallic oxides, *Progr. Solid State Chem.* **5**, 145–399 (1971).
120. J. A. Wilson, Systematics of the breakdown of Mott insulation in binary transition-metal compounds, *Adv. Phys.* **21**, 143–198 (1972).
121. A. Ferretti, D. B. Rogers, and J. B. Goodenough, The relation of the electrical conductivity in single crystals of rhenium trioxide to the conductivities of Sr_2MgReO_6 and Na_xWO_3, *J. Phys. Chem. Solids* **26**, 2007–2011 (1965).
122. L. F. Mattheiss, Band structure and Fermi surface of ReO_3, *Phys. Rev.* **181**, 987–1000 (1969).
123. L. F. Mattheiss, Crystal-field effects in the tight-binding approximation: ReO_3 and perovskite structures, *Phys. Rev. B* **2**, 3918–3935 (1970).
124. S. M. Marcus, Measurement of the de Haas–van Alphen effect in the transition-metal oxide ReO_3, *Phys. Lett.* **27A**, 584–585 (1968).
125. J. E. Graebner and E. S. Greiner, Magnetothermal oscillations and the Fermi surface of ReO_3, *Phys. Rev.* **185**, 992–994 (1969).
126. J. G. Aiken and A. G. Jordan, Electrical transport properties of single-crystal nickel oxide, *J. Phys. Chem. Solids* **29**, 2153–2167 (1968).
127. C. M. Osburn and W. R. Vest, Defect structure and electrical properties of NiO: II. Temperatures below equilibrium, *J. Phys. Chem. Solids* **32**, 1343–1354 (1971).
128. V. P. Zhuze and A. I. Shelykh, Hall effect in nickel oxide, *Soviet Phys.—Solid State* **5**, 1278–1280 (1963).
129. S. Koide, Electrical properties of $Li_xNi_{(1-x)}O$ single crystals, *J. Phys. Soc. Japan* **20**, 123–132 (1965).
130. S. P. Mitoff, Electrical conductivity and thermodynamic equilibrium in nickel oxide, *J. Chem. Phys.* **35**, 882–889 (1961).
131. I. Bransky and N. M. Tallan, High-temperature defect structure and electrical properties of NiO, *J. Chem. Phys.* **49**, 1243–1249 (1968).
132. Ya. M. Ksendzov and I. A. Drabkin, Forbidden-band width of nickel monoxide, *Soviet Phys.—Solid State* **7**, 1519–1520 (1965).

133. S. Pizzini and R. Morlotti, Thermodynamic and transport properties of stoichiometric and nonstoichiometric nickel oxide, *J. Electrochem. Soc.* **114**, 1179–1189 (1967).
134. A. J. Springthorpe, I. G. Austin, and B. A. Austin, Hopping conduction in $Li_xNi_{(1-x)}O$ crystals at low temperatures, *Solid State Commun.* **3**, 143–146 (1965).
135. Ya. M. Ksendzov, B. K. Avdeenko, and V. V. Marakov, Semiconductor properties of single crystals of nickel oxide, *Soviet Phys.—Solid State* **9**, 828–834 (1967).
136. S. van Houten, Semiconduction in $Li_xNi_{(1-x)}O$, *J. Phys. Chem. Solids* **17**, 7–17 (1960).
137. I. G. Austin, A. J. Springthorpe, B. A. Smith, and C. E. Turner, Electronic transport phenomena in single-crystal NiO and CoO, *Proc. Phys. Soc.* (*London*) **90**, 156–174 (1967).
138. A. J. Bosman, H. J. van Daal, and G. F. Knuvers, Hall effect between 300K and 1000K in NiO, *Phys. Lett.* **19**, 372–373 (1965).
139. H. J. van Daal and A. J. Bosman, Hall effect in CoO, NiO, and α-Fe_2O_3, *Phys. Rev.* **158**, 736–747 (1967).
140. A. J. Bosman and C. Crevecoeur, Mechanism of the electrical conduction in Li-doped NiO, *Phys. Rev.* **144**, 763–770 (1966).
141. W. E. Spear and D. S. Tannhauser, Hole transport in pure NiO crystals, *Phys. Rev. B* **7**, 831–833 (1973).
142. I. G. Austin and N. F. Mott, Polarons in crystalline and non-crystalline materials, *Adv. Phys.* **18**, 41–102 (1969).
143. A. J. Bosman and H. J. van Daal, Small-polaron versus band conduction in some transition-metal oxides, *Adv. Phys.* **19**, 1–117 (1970).
144. S. Kabashima and T. Kawakubo, High-frequency conductivity of NiO, *J. Phys. Soc. Japan* **24**, 493–497 (1968).
145. T. M. Wilson, A study of the electronic structure of the first-row transition-metal compounds, *Intern. J. Quant. Chem.* **IIIS**, 757–774 (1970).
146. L. F. Mattheiss, Electronic structure of the 3*d* transition-metal monoxides: I. Energy-band results, *Phys. Rev. B* **5**, 290–306 (1972); II. Interpretation, *Phys. Rev. B* **5**, 306–315 (1972).
147. D. Reinen, Color and composition in inorganic solids: VII B. Light absorption of divalent nickel in mixed crystals on $Mg_{(1-x)}Ni_xO$ and in tetrahedral coordination, *Ber. Bunsenges. Physik. Chem.* **69**, 82–87 (1965).
148. N. Kawai and S. Mochizuki, Insulator–metal transition in NiO, *Solid State Commun.* **9**, 1393–1395 (1971).
149. G. K. Wertheim and S. Hufner, X-ray photoemission band structure of some transition-metal oxides, *Phys. Rev. Lett.* **28**, 1028–1031 (1972).
150. D. E. Eastman, unpublished data (1973).
151. R. J. Powell and W. E. Spicer, Optical properties of NiO and CoO, *Phys. Rev. B* **2**, 2182–2193 (1970).
152. D. Adler, Band structure of magnetic semiconductors, *IBM J. Res. Develop.* **14**, 261–268 (1970).
153. R. Glosser and W. C. Walker, Electroreflectance observation of localized and itinerant electron states in NiO, *Solid State Commun.* **9**, 1599–1602 (1971).
154. M. A. Kolber and R. K. MacCrone, Bound-polaron hopping in NiO, *Phys. Rev. Lett.* **29**, 1457 (1972).

155. K. H. Johnson, R. P. Messmer, and J. W. D. Connolly, Localized electronic excitations in nickel oxide, *Solid State Commun.* **12**, 313–316 (1973).
156. J. C. Slater and K. H. Johnson, Self-consistent-field $X\alpha$ cluster method for polyatomic molecules and solids, *Phys. Rev. B* **5**, 844–853 (1972).
157. P. D. Dernier and M. Marezio, Crystal structure of the low-temperature antiferromagnetic phase of V_2O_3, *Phys. Rev. B* **2**, 3771–3776 (1970).
158. J. Feinleib and W. Paul, Semiconductor-to-metal transition in V_2O_3, *Phys. Rev.* **155**, 841–850 (1967).
159. R. M. Moon, Antiferromagnetism in V_2O_3, *Phys. Rev. Lett.* **25**, 527–529 (1970).
160. A. S. Barker and J. P. Remeika, Optical properties of V_2O_3 doped with chromium, *Solid State Commun.* **8**, 1521–1524 (1970).
161. M. S. Kozyreva, V. N. Novikov, and B. A. Tallerchik, Optical properties of V_2O_3 in the region of the phase transition, *Soviet Phys.—Solid State* **14**, 639–643 (1972).
162. I. G. Austin and C. E. Turner, The nature of the metallic state in V_2O_3 and related oxides, *Phil. Mag.* **19**, 939–149 (1969).
163. A. Menth and J. P. Remeika, Magnetic properties of $(V_{1-x}Cr_x)_2O_3$, *Phys. Rev. B* **2**, 3756–3762 (1970).
164. D. Adler, J. Feinleib, H. Brooks, and W. Paul, Semiconductor-to-metal transitions in transitional-metal compounds, *Phys. Rev.* **155**, 851–860 (1967).
165. D. B. McWhan and T. M. Rice, Critical pressure for the metal–semiconductor transition in V_2O_3, *Phys. Rev. Lett.* **22**, 887–890 (1969).
166. D. B. McWhan, T. M. Rice, and J. P. Remeika, Mott transition in Cr-doped V_2O_3, *Phys. Rev. Lett.* **23**, 1384–1387 (1969).
167. G. K. Wertheim, J. P. Remeika, H. J. Guggenheim, and D. N. E. Buchanan, Mössbauer effect study of metal–insulator transition in V_2O_3, *Phys. Rev. Lett.* **25**, 94–96 (1970).
168. D. B. McWhan and J. P. Remeika, Metal–insulator transition in $(V_{1-x}Cr_x)_2O_3$, *Phys. Rev. B* **2**, 3734–3750 (1970).
169. T. M. Rice and D. B. McWhan, Metal–insulator transition in transition-metal oxides, *IBM J. Res. Develop.* **14**, 251–257 (1970).
170. T. M. Rice, D. B. McWhan, and W. F. Brinkman, in *Proc. Tenth Int. Conf. on the Physics of Semiconductors* (S. P. Keller, J. C. Hensel, and F. Stern, eds.), pp. 293–300, USAEC Div. Tech. Inform., Oak Ridge, Tennessee (1970).
171. D. B. McWhan, T. M. Rice, and J. P. Remeika, in *Proc. Int. Conf. on the Physics of Solids under Pressure, Grenoble, France, 1979*, pp. 149–156, Editions du CNRS, Paris (1970).
172. T. M. Rice and W. F. Brinkman, in *Critical Phenomena in Alloys, Magnets, and Superconductors* (R. E. Mills, E. Ascher, and R. I. Jaffee, eds.), pp. 593–612, McGraw-Hill, New York (1971).
173. D. B. McWhan, A. Menth, and J. P. Remeika, Metal–insulator transitions in transition-metal oxides, *J. de Physique* **32** (C1), 1079–1085 (1971).
174. D. B. McWhan, A. Menth, J. P. Remeika, W. F. Brinkman, and T. M. Rice, Metal–insulator transitions in pure and doped V_2O_3, *Phys. Rev. B* **7**, 1920–1931 (1973).
175. A. F. Reid, T. M. Sabine, and D. A. Wheeler, Neutron diffraction and other studies of magnetic ordering in phases based on Cr_2O_3, V_2O_3, and Ti_2O_3, *J. Solid State Chem.* **4**, 400–409 (1972).
176. C. Bonnelle, Contribution to the study of transition metals, *Ann. Physique (Paris)*, **1**, 439–481 (1966).

177. D. W. Fischer, Use of soft X-ray band spectra for determining molecular orbital structure: I. Vanadium octahedral and tetrahedral sites, *Appl. Spectry.* **25**, 263–270 (1971).
178. A. Gainotti, C. Ghezzi, and M. Manfredi, Semiconductor-to-metal transition and positron annihilation in V_2O_3, *Nuovo Cimento*, **62B**, 121–129 (1969).
179. A. Greenberger and S. Berko, Angular distribution of 2γ from positron annihilation in V_2O_3, *Bull. Am. Phys. Soc.* **17**, 358 (1972).
180. T. N. Kennedy and J. D. Mackenzie, Suppression of the semiconductor–metal transition in vanadium oxides, *J. Non-Crystall. Solids* **1**, 326–330 (1969).
181. A. Jayaraman, D. B. McWhan, J. P. Remeika, and P. D. Dernier, Critical behavior of the Mott transition in Cr-doped V_2O_3, *Phys. Rev. B* **2**, 3751–3756 (1970).
182. V. P. Zhuze, A. A. Andreev, and A. I. Shelykh, The Hall effect in V_2O_3 single crystals in the metallic conductivity region, *Soviet Phys.—Solid State* **10**, 2914–2916 (1969).
183. E. D. Jones, Temperature dependence of the vanadium NMR frequency shift in V_2O_3, *J. Phys. Soc. Japan* **27**, 1692–1693 (1969).
184. H. Zeiger, Free-energy model of the high-temperature metal–insulator transition in Ti_2O_3 and V_2O_3, *Bull. Am. Phys. Soc.* **18**, 399 (1973).
185. P. G. Dickens and M. S. Whittingham, The tungsten bronzes and related compounds, *Quart. Rev. Chem. Soc.* **22**, 30–44 (1968).
186. A. Narath and D. C. Wallace, Nuclear magnetic resonance in cubic sodium tungsten bronzes, *Phys. Rev.* **127**, 725–729 (1962).
187. W. R. Gardner and G. C. Danielson, Electrical resistivity and Hall coefficient of sodium tungsten bronze, *Phys. Rev.* **93**, 36–51 (1954).
188. L. D. Muhlestein and G. C. Danielson, Seebeck effect in sodium tungsten bronze, *Phys. Rev.* **160**, 562–567 (1967).
189. M. J. Sienko and J. M. Berak, in *The Chemistry of Extended Defects in Non-Metallic Solids* (L. Eyring and M. O'Keeffe, eds.), pp. 541–554, American Elsevier, New York (1970).
190. J. M. Berak and M. J. Sienko, Effect of oxygen deficiency on electrical transport properties of tungsten trioxide crystals, *J. Solid State Chem.* **2**, 109–133 (1970).
191. S. Tanasaki, On the phase transition of tungsten trioxide below room temperature, *J. Phys. Soc. Japan* **15**, 566–573 (1960).
192. A. D. Wadsley, The crystal structure of $Na_{(2-x)}V_6O_{15}$, *Acta Cryst.* **8**, 695–701 (1955).
193. J. H. Perlstein and M. J. Sienko, Single-crystal studies of electrical conductivity, Seebeck effect, and Hall voltage in sodium vanadium bronze and a crystal-field model of electron transport, *J. Chem. Phys.* **48**, 174–181 (1968).
194. J. Graham and A. D. Wadsley, The crystal structure of the blue potassium molybdenum bronze, $K_{0.28}MoO_3$, *Acta Cryst.* **20**, 93–100 (1966).
195. G. H. Bouchard, J. Perlstein, and M. J. Sienko, Solid-state studies of potassium molybdenum bronzes, *Inorganic Chem.* **6**, 1682–1685 (1967).
196. D. S. Perloff, M. Vlasse, and A. Wold, Anisotropic electrical behavior of the blue potassium molybdenum bronze, $K_{0.30}MoO_3$, *J. Phys. Chem. Solids* **30**, 1071–1076 (1969).
197. W. Fogle and J. H. Perlstein, Semiconductor-to-metal transition in the blue potassium molybdenum bronze, $K_{0.30}MoO_3$; example of a possible excitonic insulator, *Phys. Rev. B* **6**, 1402–1412 (1972).

198. S. Methfessel and D. C. Mattis, in *Handbuch der Physik* (S. Flugge, ed.), Vol. 18/1, pp. 387–562, Springer-Verlag, Berlin (1968).
199. M. W. Shafer, J. B. Torrance, and T. Penney, in *Magnetism and Magnetic Materials —1971* (AIP Conf. Proc. No. 5; C. D. Graham and J. J. Rhyne, eds.), pp. 840–844, American Institute of Physics, New York (1972).
200. T. Penney, M. W. Shafer, and J. B. Torrance, Insulator–metal transition and long-range magnetic order in EuO, *Phys. Rev. B* **5**, 3669–3674 (1972).
201. J. B. Torrance, M. W. Shafer, and T. R. McGuire, Bound magnetic polarons and the insulator–metal transition in EuO, *Phys. Rev. Lett.* **29**, 1168–1171 (1972).
202. D. E. Eastman, F. Holtzberg, and S. Methfessel, Photoemission studies of the electronic structure of EuO, EuS, EuSe, and GdS, *Phys. Rev. Lett.* **23**, 226–229 (1969).
203. T. Kasuya, *s–f* Exchange interactions and magnetic semiconductors, *Crit. Rev. Solid State Sci.* **3**, 131–164 (1972).
204. S. J. Cho, Spin-polarized energy bands in Eu chalcogenides by the augmented-plane-wave method, *Phys. Rev. B* **1**, 4589–4603 (1970).
205. M. R. Oliver, J. O. Dimmock, A. L. McWhorter, and T. B. Reed, Conductivity studies in europium oxide, *Phys. Rev. B* **5**, 1078–1098 (1972).
206. S. von Molnar and T. Kasuya, in *Proc. Tenth Int. Conf. on the Physics of Semiconductors, Cambridge, Mass., 1970* (S. P. Keller, J. C. Hensel, and F. Stern, eds.), pp. 233–242, USAEC Div. Tech. Inf., Springfield, Virginia (1970).
207. S. C. Moss and D. Adler, Amorphous silicon and germanium revisited: I. Structural aspects, *Comments on Solid State Physics* **5**, 47–55 (1973).
208. D. E. Polk, Structural model for amorphous silicon and germanium, *J. Non-Crystall. Solids* **5**, 365–376 (1971).
209. S. C. Moss and J. F. Graczyk, in *Proc. Tenth Int. Conf. on the Physics of Semiconductors, Cambridge, Mass., 1970* (S. P. Keller, J. C. Hensel, and F. Stern, eds.), pp. 658–662, USAEC Div. Tech. Inform., Oak Ridge, Tennessee (1970).
210. M. L. Rudee and A. Howie, The structure of amorphous Si and Ge, *Phil. Mag.* **25**, 1001–1007 (1972).
211. M. H. Brodsky, R. S. Title, K. Weiser, and G. D. Pettit, Structural, optical, and electrical properties of amorphous silicon films, *Phys. Rev. B* **1**, 2632–2641 (1970).
212. G. S. Cargill, Anisotropic microstructure in evaporated amorphous germanium films, *Phys. Rev. Lett.* **28**, 1372–1375 (1972).
213. F. L. Galeener, Optical evidence for a network of cracklike voids in amorphous germanium, *Phys. Rev. Lett.* **27**, 1716–1719 (1971).
214. J. J. Hauser, Anisotropic electrical properties of amorphous germanium, *Phys. Rev. Lett.* **29**, 476–479 (1972).
215. D. Adler and S. C. Moss, Amorphous silicon and germanium revisited: II. Electronic structure and transport, *Comments on Solid State Physics*, **5**, 63–72 (1973).
216. S. C. Moss, P. Flynn, and L. O. Bauer, Impurity effects on the structure of amorphous silicon and germanium prepared in various ways, *Phil. Mag.* **27**, 441–456 (1973).
217. D. T. Pierce and W. E. Spicer, Electronic structure of amorphous Si from photoemission and optical studies, *Phys. Rev. B* **5**, 3017–3029 (1972).

218. J. Sauvage, C. J. Mogab, and D. Adler, Temperature-dependent tunnelling into amorphous silicon, *Phil. Mag.* **25**, 1305–1312 (1972).
219. M. H. Brodsky, D. Kaplan, and J. F. Ziegler, in *Proc. Eleventh Int. Conf. on the Physics of Semiconductors, Warsaw, 1972*, pp. 529–535, Polish Scientific Publishers, Warsaw (1972).
220. R. Grigorovici and A. Vancu, Optical constants of amorphous silicon films near the main absorption edge, *Thin Solid Films* **2**, 105–110 (1968).
221. J. E. Fischer and T. M. Donovan, Optical and photoelectric properties of amorphous silicon, *J. Non-Crystall. Solids* **8–10**, 202–208 (1972).
222. M. L. Theye, influence of annealing on the optical properties of amorphous germanium films, *Mat. Res. Bull.* **6**, 103–118 (1971).
223. R. C. Chittick, J. H. Alexander, and H. F. Sterling, The preparation and properties of amorphous silicon, *J. Electrochem. Soc.* **116**, 77–81 (1969).
224. M. Morgan and P. A. Walley, Localized conduction processes in amorphous germanium, *Phil. Mag.* **23**, 661–671 (1971).
225. J. Stuke, in *Conduction in Low-Mobility Materials* (N. Klein, D. S. Tannhauser, and M. Pollack, eds.), pp. 193–206, Taylor and Francis, London (1971).
226. A. Lewis, Evidence for the Mott model of hopping conduction in the anneal stable state of amorphous silicon, *Phys. Rev. Lett.* **29**, 1555–1558 (1972); Erratum, *Phys. Rev. Lett.* **30**, 1238 (1973).
227. A. H. Clark, Electrical and optical properties of amorphous germanium, *Phys. Rev.* **154**, 750–757 (1967).
228. A. Nwachuku and M. Kuhn, Tunneling into amorphous germanium films, *Appl. Phys. Lett.* **12**, 163–165 (1968).
229. H. Piller and S. A. Khan, in *Proc. Tenth Int. Conf. on the Physics of Semiconductors, Cambridge, Mass., 1970* (S. P. Keller, J. C. Hensel, and F. Stern, eds.), pp. 662–666, USAEC Div. Tech. Inform., Oak Ridge, Tennessee (1970).
230. J. Stuke, Review of optical and electrical properties of amorphous semiconductors, *J. Non-Crystall. Solids* **4**, 1–26 (1970).
231. R. Grigorovici, N. Croitoru, and A. Denvenyo, Thermoelectric power in amorphous germanium, *Phys. Stat. Sol.* **16**, K143–K145 (1966).
232. M. Pollack, M. L. Knotek, H. Kurtzman, and H. Glick, DC conductivity of amorphous germanium and the structure of the pseudogap, *Phys. Rev. Lett.* **30**, 856–859 (1973).
233. M. L. Knotek, M. Pollack, T. M. Donovan, and H. Kurtzman, Thickness dependence of hopping transport in amorphous-germanium films, *Phys. Rev. Lett.* **30**, 853–856 (1973).
234. W. Beyer and J. Stuke, Thermoelectric power of amorphous semiconductors, *J. Non-Crystall. Solids* **8–10**, 321–325 (1972).
235. M. H. Broksky and R. J. Gambino, Electrical conduction in evaporated amorphous silicon films, *J. Non-Crystall. Solids* **8–10**, 739–744 (1972).
236. M. Zavetova, S. Koc, and J. Zemek, Steep vs. exponential absorption edge in amorphous germanium: evidence for the effect of oxygen, *Czech. J. Phys.* **22**, 429–431 (1972).
237. H. R. Philipp, Optical and bonding model for non-crystalline SiO_x and SiO_xN_y materials, *J. Non-Crystall. Solids* **8–10**, 627–632 (1972).

238. A. J. Bennett and L. M. Roth, Calculation of the optical properties of amorphous SiO_x materials, *Phys. Rev. B* **4**, 2686–2696 (1971).
239. A. Mattheissen and C. Voigt, On the influence of temperature on the electrical conductivity of alloys, *Ann. Physik Chemie* (*Leipzig*) **122**, 19–78 (1864).
240. F. Bloch, On the quantum mechanics of electrons in crystal lattices, *Z. Physik* **52**, 555–600 (1928).
241. A. Sommerfeld and H. Bethe, Electron theory of metals, *Handbuch der Physik* **2412**, 333–622 (1933).
242. J. Bardeen, Conductivity of monovalent metals, *Phys. Rev.* **52**, 688–697 (1937).
243. M. Bailyn, Transport in metals: II. Effect of the phonon spectrum and Umklapp processes at high and low temperatures, *Phys. Rev.* **120**, 381–404 (1960).
244. J. E. Kunzler, in *Ultra-High-Purity Metals*, pp. 171–200, American Society for Metals, Metals Park, Ohio (1962).
245. G. J. van den Berg, Ph.D. Thesis, Univ. of Leiden, unpublished (1938).
246. G. K. White and S. B. Woods, Conductivity of α-manganese, *Can. J. Phys.* **35**, 346–348 (1957).
247. P. G. deGennes and J. Friedel, Anomalies in the resistivity in certain magnetic metals, *J. Phys. Chem. Solids* **4**, 71–77 (1958).
248. V. B. Zernov and Yu. V. Sharvin, Measurement of the resistance of high-purity tin at helium temperatures, *Soviet Phys.—JETP* **9**, 737–741 (1959).
249. R. B. Dingle, The electrical conductivity of thin wires, *Proc. Roy. Soc.* (*London*) **A201**, 545–560 (1950).
250. J. L. Olsen, *Electrical Transport in Metals*, Interscience, New York (1962).
251. W. Meissner and B. Voigt, Measurements with the aid of liquid helium: XI. Resistance of pure metals at low temperatures, *Ann. Physik* (*Leipzig*) **7**, 761–797 (1930).
252. J. Kondo, Resistance minimum in dilute magnetic alloys, *Progr. Theor. Phys.* (*Kyoto*) **32**, 37–49 (1964).
253. W. B. Pearson, Electron transport in copper and dilute alloys at low temperature: IV. Resistance minimum: temperature of occurrence as a function of solute concentration, *Phil. Mag.* **46**, 920–923 (1955).
254. J. P. Franck, F. D. Manchester, and D. L. Martin, The specific heat of pure copper and of some dilute copper and iron alloys showing a minimum in the electrical resistance at low temperatures, *Proc. Roy. Soc.* (*London*) **A263**, 494–507 (1961).

5

The Imperfect Solid—Semiconductors

Robert F. Brebrick
Metallurgy and Materials Science
Marquette University, Milwaukee, Wisconsin

1. Imperfections in Crystalline Semiconductors

1.1. Electronic Energy States in the Perfect Crystal

In this discussion of semiconductors the band picture will be adopted for the electronic energy states. In a perfect crystalline semiconductor at 0°K the valence electrons completely fill a quasicontinuous band of levels, the valence band, which in turn is separated by an energy gap from the next higher, unoccupied band of levels called the conduction band. The valence electrons are delocalized and belong to the crystal as a whole in the sense that their wave functions possess identical values at equivalent points in every unit cell of the crystal. Since there are no empty energy states accessible at 0°K, the electrons in the filled band cannot receive net energy from an imposed electric field, i.e., a filled band does not contribute to the electrical conductivity. In order for this to occur in either a perfect or imperfect (impure) semiconductor, electrons must be excited into the conduction band, where there are many empty levels, or out of the valence band, leaving empty levels there. In the latter case the net movement of the electrons in the nearly filled valence band is equivalent to that of positive charge carriers, called holes, whose number is equal to that of the empty energy levels. The conductivity is then given in terms of the concentration of electrons n, that of holes p, and the electron and hole mobilities μ_n and μ_p, as

$$\sigma = ne\mu_n + pe\mu_p \tag{1}$$

The mobilities are the average drift velocities attained in an electrostatic field gradient of 1 V/cm. The simplest situation is intrinsic excitation and occurs when a valence-band electron is excited into the conduction band, giving one conduction-band electron and one valence-band hole. For thermodynamic reasons this process will occur to an increasing degree as the temperature is raised. For the perfect, and therefore pure and stoichiometric, semiconductor the numbers of (conduction band) electrons and (valence band) holes at a given temperature are equal (this concentration is called the intrinsic carrier concentration n_i) and are determined primarily by the magnitude of the energy gap. At 300°K n_i is 1.4×10^{10} cm^{-3} for Si, 2.5×10^{13} cm^{-3} for Ge, and 10^{16}–10^{17} cm^{-3} for PbS, PbSe, and PbTe. Typically n_i is below 10^{17} cm^{-3} at 300°K for many semiconductors and is therefore small compared to the 10^{22}–10^{23} cm^{-3} concentration of atoms in a solid. This fact is of central importance in understanding the sensitivity of semiconductor properties to the presence of certain atomic point defects, as discussed in the next section.

In the simple band picture the net resultant of the interactions among the many electrons and nuclei present is the same as though the valence- and conduction-band electrons were noninteracting particles in a constant potential field, with two exceptions: There are energy gaps in the distribution of energy levels and the effective mass of the particles is generally different, and often much smaller, than that of the free-electron mass. The mobilities, for instance, are given in terms of the effective mass and the mean time between collisions as

$$\mu_i = \tau_i/m_i \tag{2}$$

where i is n for electrons and p for holes. At room temperature μ is typically in the range 10^2–10^4 cm/V-sec. Then if n_i is 10^{14}–10^{17} cm^{-3}, the conductivity of the intrinsic semiconductor by Eq. (1) is between 0.03 and 6400 Ω^{-1} cm^{-1}, compared to values of 10^4–10^6 for good metals.

As discussed in more detail in Chapter 2 of Volume 1 and Chapter 4 of Volume 2, the simple *broadband* picture requires modification when transition metal compounds or very ionic compounds are considered. In the former case the d electrons occupy a narrow band of energy levels, and moreover can be localized about the transition element atoms while still being the highest energy electrons present. The transport properties are then very different from those of a simple broadband semiconductor. In the latter case the energy bands tend to be very narrow and conduction-band electrons or valence-band holes tend to polarize the ions about them, the electronic carrier and its neighboring, displaced, ions being called a polaron. This interaction is usually described as one between the electron or hole and the optical vibrational modes of the crystal, but is analogous in its broad features to that giving rise to the Debye–Hückel ionic atmosphere in aqueous salt solutions. The simple band picture is on the whole adequate for the elemental semiconductors Ge, Si, Se, and Te and a number of binary semiconductors such as the zincblende or wurzite structure II–VI's (ZnS, ZnSe, ZnTe, CdS, CdSe, CdTe, HgS), the orthorhombic or rocksalt structure IV–VI's (GeS, GeSe, GeTe, SnS, SnSe, SnTe, PbS, PbSe, PbTe), and the zincblende structure III–V's (AlP, AlAs, AlSb, GaP, GaAs, GaSb, InP, InAs, InSb). Some relatively weak interaction between electrons and holes must be considered to account for all the experimental details, but even this becomes less important the larger the concentrations of mobile, delocalized electrons present. At high carrier concentrations the interaction between charged particles of all types is given by a screened Coulombic potential which is the product of the ordinary expression and a factor $e^{-r/\lambda}$, where r is the radial distance and the Debye length λ depends upon the dielectric constant ε and free-electron concentration n as $\lambda = (\varepsilon kT/4\pi ne^2)^{1/2}$. Thus many interaction phenomena are seen most clearly at low temperatures in semiconductors such as Ge, Si, GaAs, or GaP, which have been prepared in a very pure state and have not been seen at all in the IV–VI, Pb-salt semiconductors, which have abnormally high static dielectric constants and which have only rarely been prepared with carrier concentrations below 10^{16} cm^{-3}.

1.2. Atomic Point Defects

1.2.1. Impurity Levels

Foreign atoms or native atomic point defects present in a crystalline semiconductor can affect the band structure in a significant way. At not too high concentrations the main effect is the introduction of localized "impurity" levels in the energy gap of the electronic energy band structure which are of three types: donor levels, acceptor levels, and isoelectronic levels. For a *donor* defect in a semiconductor at 0°K the valence band is completely filled, the conduction band is completely empty, and a localized level lies in the forbidden energy gap and contains one or more electrons. Shallow donor levels, discussed in more detail later, are separated from the conduction-band edge by an energy which is a small fraction of the energy gap. (See

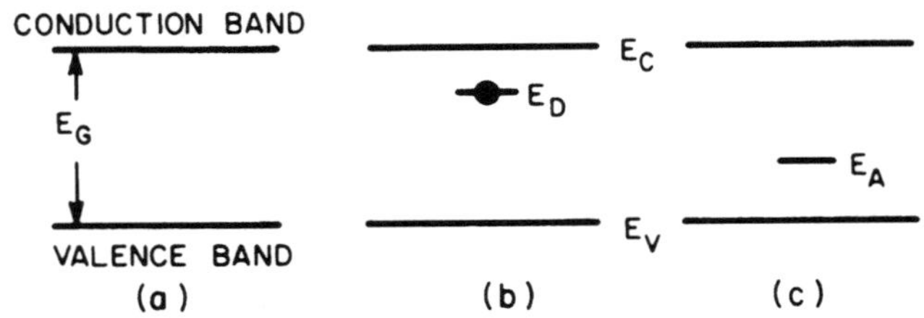

Fig. 1. Occupancy of the electronic energy levels in a semiconductor at 0°K. Energy is plotted vertically, spatial extent of the wave function horizontally. (a) A perfect semiconductor. (b) A single donor is present. (c) A single acceptor is present. A circle on the level indicates it is occupied by an electron at 0°K.

Figure 1b.) For an *acceptor* defect in a semiconductor at 0°K the valence band is completely filled, the conduction band is completely empty, and a localized, unoccupied level lies in the energy gap. Shallow acceptors are displaced from the valence-band edge by an ionization energy which is a small fraction of the energy gap (Figure 1c).

1.2.2. Foreign Atoms as Donors, Acceptors, or Alloy Impurities

Interstitial foreign atoms in either elemental or compound semiconductors are usually observed to be donors or, like H in Ge, to be electrically inactive, i.e., neither donors nor acceptors. The possibility of acceptor behavior exists for an interstitial atom if it can form a closed shell upon ionization of the acceptor, i.e., upon promotion of an electron from the valence band to the acceptor level. Empirically it is found that a substitutional foreign atom in an elemental semiconductor is a donor if it possesses more valence electrons than the host atom replaced, an acceptor if it possesses fewer, and neither if it possesses the same number. In the last case the foreign atom will be called an alloy impurity. Complications exist for this case which are discussed below under isoelectronic traps. Thus P, As, and Sb with five, and Al, Ga, and In with three valence electrons all form substitutional solid solutions in Ge and Si. The first group acts as donors, the second as acceptors. Silicon in Ge, or the converse, behaves as an alloy impurity. Since not all of the outer shell electrons of an atom are always involved in bonding, the empirical rules for donor or acceptor behavior of substitutional foreign atoms in compound semiconductors can be more accurately stated in terms of oxidation numbers. If the predominant oxidation number of an atom is positive, then it will tend to substitute for the host atom in a compound semiconductor with the positive oxidation number, assuming it does not go into solid solution in an interstitial position. The foreign atom will behave as a donor if its oxidation number is larger than that of the host atom replaced, as an acceptor if its oxidation number is smaller, and as an alloy impurity if its oxidation number is the same. Analogously, if a foreign atom whose predominant oxidation number is negative substitutes for a host atom with a negative oxidation number, it will act as a donor if its oxidation number is *less* negative than that of the atom replaced, as an acceptor if it is more negative, and as an alloy impurity if it is the same. Thus in PbTe the oxidation numbers of Pb and Te are respectively $+2$ and -2 although the Pb atom has an outer shell configuration s^2p^2. Aluminum, Ga and In, with a predominant oxidation number of $+3$, and Na, K, Cs, and Ag, with $+1$, all enter PbTe substitutionally for Pb. The former group acts as donors with a single electron in the donor level, (singly ionizable donors), the latter group acts as singly ionizable acceptors. Chlorine, Br and I have a predominant oxidation number of -1 in combination with metallic elements. They enter the PbTe lattice substitutionally for Te and act as singly ionizable acceptors. Bismuth and Sb have five valence electrons but exhibit a predominant oxidation number of $+3$ in combination with less metallic elements. Therefore both of these elements are incorporated on sites in the Pb sublattice of PbTe, i.e., they enter substitutionally for Pb, and they behave as donors. Generally the predominant oxidation number of a nontransition element in Groups I to V of the periodic table tends to drop by two as one proceeds down the table to the heavier elements (the inert *s*-pair effect of inorganic chemistry). Thus while the predominant oxidation number of Ge and Si is $+4$, that of Pb in the same group is $+2$, although Pb does form some compounds, such as lead tetraethyl, and PbO_2, in which its oxidation number is $+4$. Similarly the predominant oxidation number of Tl is $+1$ and not $+3$ like Al, Ga, and In in the same group, and Tl substitutes for Pb in PbTe and acts as an acceptor. Also Tl ought to enter InSb substitutionally for In and behave as an acceptor and, as observed, ought not to form III–V compounds as do Al, Ga, and In. Aluminum and Ga ought to substitute for In and be alloy impurities. As a matter of fact, a complete range of solid solutions is formed between AlSb and InSb and between GaSb and InSb. As the concentrations change, the intrinsic semiconductor properties such as the energy gap and effective masses of electrons and holes change smoothly and as yet there is no evidence for localized levels in the energy gap. It no doubt is apparent that the use of oxidation numbers offers a more generally valid method of predicting donor or acceptor behavior than the use of the number of valence electrons. However, the oxidation number rule itself becomes ambiguous for elements possessing two or more equally prominent oxidation numbers, as is typically the case for the transition elements.

The basic concepts of a crystalline semiconductor described thus far already include electrons and holes as charged particles. Subsequent quantitative analysis will depend in part upon an equation expressing the electrical neutrality of the crystal as a whole. Therefore it is necessary to specify the charges residing on un-ionized donors and acceptors, that for the singly ionized defects being more positive and more negative, respectively, by one unit of the electronic charge. The usual assumption is that un-ionized interstitials have zero charge while substitutional alloy impurities and un-ionized substitutional donors and acceptors have the same charge as the host atom they replace. Thus an un-ionized, substitutional As atom in Ge or Si is a donor whose charge is the same as that on the Ge atom, zero. In a compound semiconductor, whose formula is M_aN_b if the crystal is perfect, there will in general be some element of ionicity in the crystal binding. If all the M atoms occupy sites which are crystallographically equivalent and distinct from the equivalent sites occupied by the N atoms, then all the M atoms must bear the same fractional or integral charge, which is equal in magnitude but opposite in sign from that common to all the N atoms. We suppose the M atoms have a charge $+\delta$ and the N atoms $-\delta$ in units of the magnitude of the charge on a free electron. Then in these units an un-ionized donor atom $D^{\times}$ substituting for an M atom must also have a charge of $+\delta$, while the singly ionized donor has a charge of $+(\delta + 1)$. It should be noted that the charge on a host atom or substitutional impurity is only the same as the oxidation number if the binding is purely ionic. With the above assumption concerning the charge on substitutional impurities, the charge on these and on the host atoms drops out of the electroneutrality equation and neither its magnitude or sign need be known. As discussed immediately in the next section, the sign, but not the magnitude, of the charge on the host atoms must be known in order to assign donor or acceptor character to the native atomic point defects in compound semiconductors.

1.2.3. Native Atomic Point Defects

We assume the M atoms bear a charge $+\delta$ in the ordered semiconductor compound M_aN_b. Generally the identity of the atomic species playing the role of M for a specific compound is intuitively clear from a chemical standpoint, M being the more metallic, less electronegative of the two species. In the III–V compounds the situation is not so clear. The tetrahedrally coordinated structure can be pictured as purely ionic or purely covalent. In either case both types of atom attain a stable, closed *sp* subshell. However, for pure ionic binding the group III atom becomes a cation with charge +3, while for covalent binding it has a formal charge of −1 because of the one coordinate bond formed by each group V element. One would expect the actual situation to lie somewhere in between. Electron density maps[1] constructed from X-ray diffraction of InSb indicate the magnitude of the charge transfer as 0.45*e*, in close agreement with that required to account for ionized impurity scattering and for infrared absorption. Moreover, the excess positive charge resides upon the Sb, so that it plays the role of M. Because of the greater electronegativity as As and P the charge transfer is expected to be different and possibly reversed in InAs and InP. A vacancy in the sublattice of the M atoms, which bear the fractional or integral positive charge, is an acceptor. An N vacancy is a donor. Interstitial M atoms are donors, while interstitial N atoms are acceptors. This empirical assignment is apparently consistent with observed behavior, including that of the ionic alkali halides. The place-exchange or antistructure or native substitutional defect is predominant in Bi_2Te_3[2,3] and Sb_2Te_3 and is less likely to be predominant for more ionic compounds. An N atom on an M site is a donor, while an M

TABLE 1
Electrical Behavior of Native Atomic Point Defects in the Ordered Semiconductor Compound M_aN_b

Point defect	Electrical behavior	When defect can be predominant
N vacancy	Donor	Excess M
M interstitial	Donor	Excess M
M vacancy	Acceptor	Excess N
N interstitial	Acceptor	Excess N
M on N site	Acceptor	Excess M
N on M site	Donor	Excess N

atom on an N site is an acceptor. Table 1 summarizes these assignments and indicates under what conditions of stoichiometry the defect is more likely to be predominant. This table will be referred to again. Similar to the assumption made for impurity atoms, and with the same resultant simplification of the electroneutrality condition, un-ionized vacancies are assumed to bear the same charge as the atom replaced. Un-ionized interstitials are assumed to have zero charge.

Extensive investigations have been carried out on the electrical activity of atomic point defects in Ge and Si and it is established that a vacancy acts as an acceptor.

1.2.4. Ionization Energy for Shallow Donors and Acceptors

The ionization energies of shallow donors and acceptors are calculated using a hydrogenic model. For a singly ionizable donor, the model consists of a single positive charge arising from the atomic core of the donor and a negative charge, whose mass m_e is the semiconductor effective mass, both immersed in a continuum whose dielectric constant ε is that of the semiconductor. A series of localized donor levels arises whose energies relative to the conduction-band edge are

$$E_n\,(\text{eV}) = -13.6(m_n/m_0)/\varepsilon^2 n^2, \qquad n = 1, 2, \ldots \tag{3}$$

where m_0 is the free-electron mass. The value of E_n for $n = 1$ represents the ionization energy required to excite an electron from its ground state when trapped at the donor to the bottom of the conduction band. Values of n greater than one represent excited donor states. An analogous equation holds for singly ionizable acceptors with the energy being measured relative to the top of the valence band. The Bohr radius of the ground state is given by

$$r_1 = \varepsilon h^2/4\pi m_n \tag{4}$$

Because the dielectric constant for broadband semiconductors is often in the range of 5–15 while the effective masses are between 0.05 and 0.5 that for the free electron, the ionization energies are small compared to that for the H atom and the first Bohr radius is large. Typical values are 0.003–0.06 eV for the former and 20–80 Å for the latter. A tabulation of measured impurity ionization energies is reproduced in Table 2 for GaAs, for which $\varepsilon = 12.5$, $m_n = 0.067m_0$, and $m_p = 0.47m_0$. The hydrogenic model predicts a donor level at 0.0058 eV (below the conduction band) and an acceptor level at 0.041 eV (above the valence band), in reasonable agreement with observed values for the shallow levels.

TABLE 2
Impurity Ionization Energies in GaAs

Donors												
Element	S	Se	Te	Si	Ge	Sn	C	O	O			
Level, meV	6.1	5.9	5.8	5.8	6.1	60	6.0	750	400			
Acceptors												
Element	Zn	Cd	Ge	Cu	Cu	Li	C	Si	Cr	Mn	Fe	Co
Level, meV	29	30	38	150	470	23	19	30	790	100	520	160

As the Bohr radius of the ground state given by Eq. (4) decreases and becomes comparable to the lattice parameter, the hydrogenic model is no longer valid and the ionization energies ought to reflect more of the chemical nature of the specific impurity. The actual ionization energy will then be larger than that calculated by Eq. (3). Some impurities with deep levels in GaAs are shown in Table 2.[4] Transition metals typically form deep levels.

Complications arise even in the case of shallow donors and acceptors as their concentrations increase. The ionization energy given by Eq. (3) with $n = 1$ is shifted by the presence of other donors and acceptors. Luminescence spectra[5,6] in GaP, in which an electron trapped at a donor recombines with a hole at an acceptor center, show sharp lines for donor–acceptor pairs up to 68 coordination shells apart, the energy shift being substantially all due to a Coulombic interaction between the impurity atom cores and depending upon their separation r as $e^2/\varepsilon r$. As the impurity concentration increases even further and the ground-state orbits start to overlap, the initially sharp donor and acceptor levels spread out over an energy interval and form impurity bands. Eventually these overlap with the conduction or valence bands. Theoretical analysis[7] of the initial spread of the impurity levels leads to a Gaussian distribution of the density of impurity states. Other theoretical analyses[8–10] have focused upon the behavior of the conduction- and valence-band edges at high electron or hole concentrations and predict the development of an exponential tail in a plot of the density of electronic states vs. energy that extends into what would be the energy gap in the purer material. These effects are discussed again in Section 4.

In Section 1.1 it was pointed out that in a pure, perfect (intrinsic) semiconductor at 300°K the concentration of electrons and holes is typically 10^{17} cm^{-3} and often much less. Here it has been pointed out that the ionization energies for shallow donor and acceptors are often comparable to the thermal energy at 300°K, 0.025 eV. As a consequence, shallow donors and acceptors are ionized to a significant extent, if not completely at room temperature. The electron or hole concentration then can be fixed by the presence of electrically active impurities or native atomic point defects at the level of 10^{17} cm^{-3} or less. Semiconductors therefore are sensitive to extremely small concentrations of these defects. As will be discussed again, this means that the properties of compound semiconductors can be sensitive not only to small concentrations of foreign atoms, but also to very small changes in the composition of the pure compound, changes often well within the precision of quantitative chemical analysis.

1.2.5. Isoelectronic Traps

We have called a substitutional impurity with the same oxidation number as the host atom replaced an alloy impurity and characterized it by an absence of localized levels in the energy gap. The alloy impurities commonly used are those from the same group of the periodic table as the replaced host atom. These often can be incorporated at any concentration up to complete substitution. Recent luminescence measurements(11) show this simple description is not always adequate, however. If in the compound M_aN_b, with M bearing the fractional positive charge, an impurity with the same oxidation number as N, but sufficiently more electronegative, is substituted for N, then a localized level is introduced in the energy gap which is called an isoelectronic trap. At 0°K the valence band is full, the conduction band empty, and the isoelectronic trap empty. This type of isoelectronic impurity is obviously not a donor defect. It is a very deep acceptor level and as a result its ionization energy is not given accurately by Eq. (3). Since it is very deep, its position is near the conduction-band edge. It differs from an ionized donor in that its attraction for an electron is short range compared to the Coulombic attraction of the latter. As a result, there are no long-distance interactions leading to pair spectra involving isoelectronic traps. When the isoelectronic trap formed by a very electronegative atom captures an electron a long-range Coulombic potential attractive to a hole results. The isoelectronic trap is thus an efficient radiative recombination center, i.e., it provides an efficient mechanism by which an electron and hole can recombine with the emission of radiation rather than by a process in which the energy liberated is taken up by the thermal vibration of the lattice. An example of this type of isoelectronic defect is N substituted for P in GaP. The N level is about

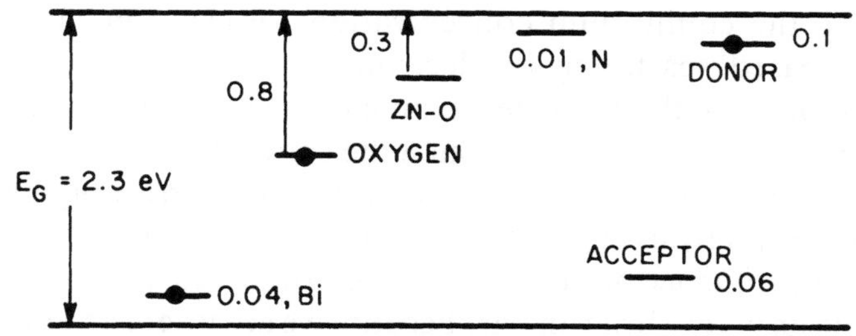

Fig. 2. Positions of some impurity levels in the energy gap of GaP. A circle indicates the occupancy of the level by an electron at 0°K.

2.3 eV above the valence band (0.01 eV below the conduction band), compared to the 0.06 eV level of shallow acceptors. The other simple type of isoelectronic impurity is exemplified[12] by Bi substituted for Ga in GaP. Here an atom with the same oxidation number as M, but more electropositive, substitutes for it. At 0°K the valence band is filled, the conduction band empty, and a deep, filled, localized level is present in the energy gap. The attraction between the electropositive isoelectronic trap and a hole is short range, so that no long-range Coulombic interaction is seen between defect pairs in which one member is the isoelectronic defect. Moreover, the electropositive, isoelectronic trap with a trapped hole exerts a Coulombic attraction for an electron and therefore acts as a radiative recombination center. The filled Bi level is 2.2 eV below the conduction band (0.04 eV above the valence band) compared to the 0.10 level of shallow donors. These levels are shown schematically in Figure 2. For reasons depending upon the band structure of GaP, the Bi isoelectronic trap is not as efficient a center for radiative recombination of a bound electron–hole pair as is N.[6]

1.3. Dislocations, Grain Boundaries, and Surfaces

Atomic point defects are of atomic size in three dimensions and are often characterized as "zero-dimensional" defects. Dislocations, grain boundaries, and surfaces can be idealized as defects disrupting the periodicity of an otherwise perfect crystal that are respectively one-dimensional and two-dimensional. Like atomic point defects, their effects in general are threefold: (a) the introduction of localized levels within the energy gap of the electronic energy band structure, which influences the electrical and optical properties significantly by determining the equilibrium concentrations of electrons and holes, (b) the increased scattering of electrons and holes, resulting in a reduced mobility of these species in an electric field, and (c) the effect on the rates and mechanisms by which the concentrations of electrons and holes return to equilibrium upon perturbation by a high electric field, illumination by light, etc. In addition, the line and surface defects have a strong influence on the mechanical properties. Discussion here of these defects is very brief, in part because we are interested primarily in effects of type (a), and in part because of the tendency in semiconductor technology to use single crystals in which there are at most subgrain (very low angle) boundaries and small dislocation densities. The latter typically are $10^4\ \text{cm}^{-2}$ or below in compound semiconductors and 0–$10^3\ \text{cm}^{-2}$ in Ge and Si.

Every actual semiconductor must of course be finite in size and bounded by external surfaces. The necessary disruption of the periodic structure introduces localized surface states whose nature can be determined by absorbed layers present accidentally or by design. As a result, the position of the conduction- and valence-band edges relative to the Fermi level (electrochemical potential) of the electrons and holes can be very different near the surface than in the bulk. This effect is important when the region influenced by the surface layer is a significant fraction of the total semiconductor thickness or when phenomena such as light absorption occur predominantly in the surface layers.

The effects of grain boundaries are similar to those of external surfaces but can be magnified by decreasing the grain size. The very presence of grain boundaries reflects the nonequilibrium conditions during crystal growth and substantial differences can exist in the chemical composition near and away from grain boundaries in specific cases. These are particularly important when there is an excess of one type of impurity in the bulk (say donors) and an excess of the other type (acceptors) in the grain boundary region.

The two extreme forms of simple dislocations are the edge and screw dislocations. The former can be visualized in a close-packed metal as resulting from the insertion of an extra half plane of atoms into an otherwise perfect crystal. The row of atoms along the edge of the half plane which is within the crystal is the dislocation line. In more complicated structures with more than one atom per primitive unit cell, such as exhibited by ordered binary compounds, two or more adjacent half planes must be inserted so that far enough away from the dislocation line the structure looks perfect on all sides of the dislocation line. Edge dislocation lines lying along the same crystallographic direction can be of two signs depending upon whether the extra half plane is inserted from the top or bottom of the crystal. The crystal is in compression on the side of the dislocation line where the extra half plane or planes lie and in tension on the opposite side. (The strain field has cylindrical symmetry about a pure screw dislocation line.) As a result, edge dislocations interact with themselves and with atomic point defects. An impurity atom whose radius is different than that of the host atoms will tend to segregate on the appropriate side of the edge dislocation line, the resulting partial strain relief being equivalent to a binding energy. Upon heating to temperatures high enough for self-diffusion to be significant, edge dislocations of opposite sign tend to annihilate each other in pairs and the excess of one sign form vertical arrays so that the compressive strain of one dislocation can be partially cancelled by the tensile strain of one adjacent. Such a vertical array of edge dislocations forms a particularly simple kind of grain boundary, called a symmetric tilt boundary, in which one grain is displaced relative to the other by rotation about a single axis parallel to the dislocation lines. By a simple argument the angle of rotation in radians θ and the average spacing of the dislocation lines D are related by

$$D\,(\text{cm}) = b/\theta \qquad (5)$$

where b is the crystallographic repeat distance in a direction perpendicular to the dislocation lines (the magnitude of the Burger's vector). There are other characteristics of dislocations common to all classes of materials that are directly attributable to the fact that dislocations represent a region of misfit in which the energy of the material is increased. Thus dislocation lines, along with grain boundaries, serve as nucleation sites for the precipitation of new phases and in fact have been rendered visible[13] in transparent crystals by exploiting this tendency. Chemical attack of surfaces tends to be most rapid where dislocation lines emerge, and appropriate etches develop pits marking such points.[14] Edge dislocation lines serve as a source and sink of vacancies or interstitials, so that the equilibrium concentrations of these defects are more rapidly attained in their presence. Without dislocations vacancies can be created or destroyed only at an external surface or grain boundary and the diffusion lengths required are roughly half the mean grain diameter. With edge dislocations distributed throughout the bulk the diffusion length is half the average dislocation separation and consequently the equilibration time, dependent on the square of the diffusion length, is significantly reduced.

The specifically electrical effects of dislocations in semiconductors have been most extensively studied in Ge and Si.[15] The row of atoms along an edge dislocation line are pictured as each possessing two broken bonds containing one electron each (dangling bonds) which, similar to a vacancy in these materials, act as acceptor levels. Thus plastic deformation of Ge containing an excess of donor impurities, *n*-type Ge, results in a multiplication of the number of dislocations present, as in metals. In fact, enough dislocations can be generated that the associated acceptor levels outnumber the donor impurities present, thereby converting the initial excess of electrons to an excess of holes (converting the initially *n*-type Ge to *p*-type). At smaller dislocation concentrations the dislocation acceptor levels can result in a cylindrical region of *p*-type material in an otherwise *n*-type matrix. The donor or acceptor behavior of dislocations in compound semiconductors has not been investigated so thoroughly, although some work[16] has been done on InSb. If a dislocation density of 10^5 cm^{-2} is assumed and if one atom in every ten along a dislocation line is active, or about 10^7 per cm of dislocation line, then there are 10^{12} impurity levels per cm^3 of semiconductor. Thus in carefully grown crystals with minimal plastic deformation the impurity levels from atomic point defects will generally far outnumber those from dislocations. Dislocations, like other defects, however, can have significant effects of types (b) and (c) mentioned above, and for this reason it is desirable to control, if not minimize, their concentration. One particular area where this problem arises is in the growth of thin, epitaxial films on substrates of the same structure but different lattice parameter. In the simplest view the lattice mismatch tends to be taken up by the inclusion in the film of a number of edge dislocations proportional to the lattice parameter mismatch.[17]

1.4. Excitation Defects

The defects considered thus far involve a disruption of periodicity by static atomic displacements. A more general view of defects would include dynamic displacements present even in a crystal with no atomic point defects, dislocations, or grain boundaries whenever the temperature is above 0°K. The vibrational modes of a crystal and electrons and holes are examples. At room temperature and above the mobilities of electronic carriers are often determined primarily by the interaction between the vibrational modes and the electrons and holes. This so-called lattice scattering is subdivided into acoustical mode and optical mode scattering and along with scattering from un-ionized or ionized donors and acceptors is discussed in Chapter 4 of this volume. Here we just mention two other dynamic excited states that can be viewed as refinements of simple band theory. Both involve the idea that the interactions between conduction-band electrons and valence-band holes with one another or with the lattice are not completely taken account of by a model in which the electrons and holes are noninteracting particles with an effective mass different than that of the free electron. The polaron mentioned in Section 1.1 and discussed in detail in Chapter 4 is a quasiparticle consisting of the electronic carrier and the local distortion of the lattice it induces. The mass of the polaron governs its response to an applied force and is only slightly larger than the effective mass calculated ignoring this interaction for many broadband, narrow-energy-gap semiconductors with $m_p, m_n \leq 0.1m_0$. On general grounds the interaction is also expected to be less effective as the concentrations of electronic carriers increases. The exciton represents a bound state between an electron and hole in which both move through the lattice as a pair, carrying no net electrical current. In the tight binding limit, representing the opposite extreme of the simple band picture, the exciton can be pictured as the promotion of an electron from an atomic ground state to an atomic excited state. Such a description is accurate for an ionic crystal such as KBr, with the exciton forming on the Br ion, or for molecular crystals such as benzene or anthracene, in which the strong binding is between atoms in distinct molecular units. For broadband semiconductors with effective masses roughly one-tenth or less that of the free electron the exciton is loosely bound and can exist in a set of energy levels given by

$$E_n = E_G - \mu e^4/2h^2\varepsilon^2 n^2, \qquad n = 1, 2, \ldots \tag{6}$$

where E_G is the energy gap and μ is the reduced electron–hole mass. It can be seen that Eq. (6) corresponds to a hydrogenic model similar to that used to calculate the ionization energies of shallow donors and acceptors. The difference $E_1 - E_G$ represents the energy required to dissociate an exciton in its ground state into a free electron and hole and is about 0.01 eV for $\varepsilon = 10$ and $m_p = m_n = 0.1m_0$. More complicated excitations have been used in the interpretation of low-temperature luminescence spectra in the III–V and II–VI compounds. Examples are excitons bound to neutral donors, ionized donors, and isoelectronic traps.[18]

2. Electron-Hole Equilibria

2.1. Conduction-Band Electrons and Valence-Band Holes

The electrons both in the bands and trapped at impurity centers are fermions and are distributed among the various energy levels according to Fermi–Dirac statistics. The concentration of electrons n_E in one of the N_L levels per unit volume at energy E_L is given by

$$n_L/(N_L - n_L) = e^{(E_F - E_L)/kT} \tag{7}$$

where E_F is a parameter called the Fermi level, k is the Boltzmann constant, and T is the absolute temperature. The electronic energy levels are all measured relative to some common, unspecified zero that can be taken as the vacuum level. Equation (7) states that the ratio of occupied to unoccupied states is given by a Boltzmann factor and agrees with classical statistics only when $n_L \ll N_L$. Equation (7) can be rearranged to give the Fermi–Dirac distribution in conventional form:

$$n_L/N_L = [1 + e^{(E_L - E_F)/kT}]^{-1} \tag{8}$$

The right side of Eq. (8) is the probability that one of the N_L levels at energy E_L is occupied by an electron. Consequently, the probability that one of these levels is *not* occupied by an electron is obtained by subtracting n_L/N_L from unity. Therefore the concentration of holes at E_L is given by

$$p_L/N_L = [1 + e^{(E_F - E_L)/kT}]^{-1} \tag{9}$$

in which the right hand side of the equation differs from that of Eq. (8) only in the sign of the argument for the exponential. It can be seen that when E_F is equal to the energy of a particular level then the probability of occupation for that level is one-half. At $T = 0°K$ all the states whose energies are equal to or less than E_F are occupied by electrons, while those with larger energies are unoccupied.

If the levels of the quasicontinuous conduction band are arranged in subgroups so that N_S of them per unit volume occur within a small energy interval about E_S, then the concentration n_S of electrons in each of these subgroups of levels is given by Eq. (8). The total concentration of conduction-band electrons is obtained by summing n_S over all of the subgroups in the band and is given by

$$n = \sum n_S = \sum N_S[1 + e^{(E_S - E_F)/kT}]^{-1}$$
$$= \int_{E_C} N_S[1 + e^{(E_S - E_F)/kT}]^{-1}\, dE_S \qquad (10)$$

where the integral is a valid approximation for the summation because of the quasicontinuous distribution of levels in the conduction band, E_C is the energy of the bottom edge of the conduction band, and the density of states N_S is a function of energy. Similarly, using Eq. (9), the total concentration of valence band holes is

$$p = \sum p_R = \int_{E_V} N_R[1 + e^{(E_F - E_R)/kT}]^{-1}\, dE_R \qquad (11)$$

where E_V is the top edge of the valence band. Thus the concentrations of electrons and holes is determined by T, E_F, and the densities of states in the bands.

In a *nondegenerate* semiconductor the Fermi level E_F is within the energy gap and at least a few kT away from either band edge. This is the usual state of a pure semiconductor at low temperatures and is departed from at high enough T or, more commonly, at high enough impurity concentrations. If the Fermi level lies in the energy gap but within a few kT of either band edge, or if it lies in one of the bands, the semiconductor is said to be *degenerate*. For a nondegenerate semiconductor Eqs. (10) and (11) for n and p simplify because the exponential on the right side of both equations is then large compared to unity, which can be neglected. The concentration of electrons can then be written as

$$n = \int_{E_C} N_S e^{(E_F - E_S)/kT}\, dE_S$$
$$= e^{(E_F - E_C)/kT} \int_0 N_S e^{(E_C - E_S)/kT}\, d(E_S - E_C) = N_C e^{(E_F - E_C)/kT} \qquad (12)$$

where N_C is an effective density of states in the conduction band defined by

$$N_C = \int_0^{\infty} N_S e^{-(E_S - E_C)/kT}\, d(E_S - E_C) \qquad (13)$$

Similarly for a nondegenerate semiconductor the concentration of holes is obtained from Eq. (11) in terms of the effective density of states for the valence band N_V as

$$p = N_V e^{(E_V - E_F)/kT} \tag{14}$$

where

$$N_V = \int_0^\infty N_R e^{(E_R - E_V)/kT}\, d(E_V - E_R) \tag{15}$$

Because the arguments of the exponentials in Eqs. (13) and (15) are negative for energies within the bands, the effective densities of states for the conduction and valence bands depend most heavily upon the values for the densities of states near the respective band edges. For the same reason the upper limits of integration can be taken as infinity.

The allowed energy states in each band are uniformly distributed in electron wave vector or k space. The number of states per unit volume of crystal between wave vector magnitudes k and $k + dk$ is twice the volume of a spherical shell in k space divided by $8\pi^3$, i.e.,

$$N\, dk = k^2\, dk/\pi^2 \tag{16}$$

Evaluation of n or p depends upon the relationship between E and k. A simple case is the parabolic band with an extremum at $k = 0$, for which this relationship for the conduction band is given by

$$E_S = E_C + (h^2 k^2/8\pi^2 m_n) \tag{17}$$

where m_n is the electron effective mass. For the valence band the plus sign is changed to minus and E_C and m_n are replaced by E_V and m_p, respectively. For simple parabolic bands described by Eq. (17) this equation and Eq. (16) can be inserted into Eqs. (13) and (15) for the effective densities of states, giving

$$N_C = 2\left(\frac{2\pi m_n kT}{h^2}\right)^{3/2} \tag{18a}$$

$$N_V = 2\left(\frac{2\pi m_p kT}{h^2}\right)^{3/2} \tag{18b}$$

The concentrations of electrons and holes for a nondegenerate semiconductor are related, as can be seen by multiplying Eqs. (12) and (14):

$$np = N_C N_V e^{(E_V - E_C)/kT} = N_C N_V e^{-E_G/kT} = n_i^2 \tag{19}$$

where the magnitude of the energy gap E_G is recognized as the difference between the edges of the conduction and valence bands, $E_C - E_V$, and n_i is the intrinsic carrier concentration. The right side of Eq. (19) is a function of T and intrinsic semiconductor properties independent of n and p. Thus as n increases, p must decrease proportionately. As long as the condition for nondegeneracy holds, Eq. (19) is valid even when impurity levels are present, although these have not been discussed as yet.

An *intrinsic* semiconductor is one for which $n = p = n_i$. In view of Eq. (19) the total carrier concentration $n + p$ is a minimum when the semiconductor is intrinsic. The position of the Fermi level in an intrinsic semiconductor $E_F{}^i$ can be obtained by setting the right sides of Eqs. (12) and (14) equal to one another. For the general (nondegenerate) case one obtains

$$E_F{}^i = \tfrac{1}{2}(E_C + E_V) - \tfrac{1}{2}kT\ln(N_C/N_V) \tag{20}$$

whereas for simple parabolic bands Eqs. (18a) and (18b) for N_C and N_V give

$$E_F{}^i = \tfrac{1}{2}(E_C + E_V) - \tfrac{3}{4}kT\ln(m_n/m_p) \tag{21}$$

The intrinsic Fermi level is at the midpoint of the energy gap if the effective density of states is the same for the conduction and valence bands. Otherwise it is displaced toward the band edge with the smaller density of states. This displacement is one of the factors governing the position of the homogeneity range of a compound semiconductor relative to the intrinsic composition.[38] The Fermi level can be calculated from Eq. (12) if n is known or from Eq. (14) if p is known. Similarly, the Fermi level can be expressed in terms of the difference $n - p$ by starting with Eq. (19). The result, valid for the nondegenerate case, is

$$E_F = E_F{}^i + kT\sinh^{-1}[(n - p)/2n_i] \tag{22}$$

In actual band structures the variation of E with electron wave vector k is generally different for crystallographically nonequivalent directions and is typically obtained by a combination of theory and experiment. Figure 3 is a schematic showing this variation in two high-symmetry directions for Ge. The maximum energy for the valence band occurs at $k = 0$, where two bands meet and the energy level is degenerate. [The reader will note that the word degenerate is used in two senses in discussing solids, that just employed being the

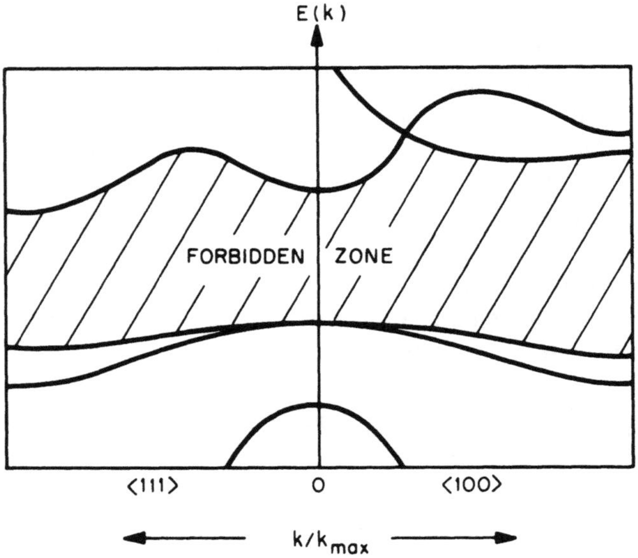

Fig. 3. Schematic energy band structure for Ge. The variation of energy with electron wave vector is shown for two high-symmetry directions.

quantum mechanical usage and referring to an energy level corresponding to two or more independent wave functions. The second usage refers to the Fermi–Dirac distribution function, Eq. (8), and has been defined in Section 2.1.] There are three types of minima in the conduction band. The ⟨111⟩ minimum at the zone edge is the lowest. That at the zone center, ⟨000⟩, and that lying along the ⟨100⟩ direction are respectively 0.15 and 0.18 eV higher. The smallest energy gap, 0.65 eV is therefore between the ⟨000⟩ valence-band maximum and the ⟨111⟩ conduction-band minimum and because the k values are different Ge is called an *indirect-gap semiconductor.* Some semiconductors have a *direct gap*, with the conduction-band minimum and valence-band maximum occurring at the same value of k, zero for GaAs. Only levels within a few kT of the lowest conduction-band minimum and highest valence-band maximum are important in determining n and p at thermal equilibrium. For a nondegenerate extremum E can be expanded in a Taylor series in k about the extremum value E_{ex} at $\langle k_x{}^0 k_y{}^0 k_z{}^0 \rangle$. The linear terms in k will be missing because the expansion is about an extremum point and for energies not too different from that at the extremum the series can be truncated after the quadratic terms. For a cubic crystal we have

$$E = E_{ex} + \frac{h^2}{8\pi^2} \sum_{i,j=x,y,z} \frac{(k_i - k_i^0)(k_j - k_j^0)}{m_{ij}} \tag{23}$$

where the plus sign is changed to minus for a nondegenerate valence-band maximum. In contrast to the case of a simple parabolic band the effective mass is now a tensor with components

$$\frac{1}{m_{ij}} = \frac{4\pi^2}{h^2} \frac{\partial^2 E}{\partial k_i \partial k_j} \tag{24}$$

By a translation of the origin of k space to the extremum point and by a suitable rotation the quadratic form of Eq. (23) can be diagonalized to give

$$E = E_{ex} + \frac{h^2}{8\pi^2}\left(\frac{k_l^2}{m_l} + \frac{k_{t1}^2}{m_{t1}} + \frac{k_{t2}^2}{m_{t2}}\right) \tag{25}$$

where k_l is along the direction in which the extremum occurs and k_{t1} and k_{t2} are in directions at right angles to this and to each other. The constant-energy surfaces are now ellipsoids with principal axes proportional to $m_l^{1/2}$, $m_{t1}^{1/2}$, and $m_{t2}^{1/2}$. m_l is called the longitudinal effective mass and m_{t1} and m_{t2} are called transverse effective masses. If a band extremum for a cubic crystal lies along the $\langle 100 \rangle$ or $\langle 111 \rangle$ directions, the two transverse masses must be equal by symmetry and the constant-energy surfaces are ellipsoids of revolution. If the extremum of one band for a cubic crystal is in the $\langle 111 \rangle$ direction, but within the zone, then there are eight equivalent minima corresponding to the opposite ends of four equivalent $\langle 111 \rangle$ directions. If, however, the extremum occurs at the zone edge as in Figure 3, then there are only four minima. At each of the eight equivalent $\langle 111 \rangle$ points only positive or negative values for k_l in Eq. (25), and therefore only one-half of a constant-energy surface, fall within the zone.

The translation and rotation of the k axes required to express the E–k relationship near a band extremum in the form of Eq. (25), involving no distortion of scale, does not upset the uniform density of allowed electron energy states in k space. Therefore near a band extremum the product of the number of states per unit volume of crystal and unit energy times dE is given by the factor $1/8\pi^3$ times the volume contained between a surface with constant energy E and one with constant energy $E + dE$. If E is measured from the energy at the band extremum, this is

$$N(E)\,dE = 4\pi N_{ex}(2m^*)E^{1/2}\,dE \tag{26}$$

where the density-of-states effective mass m^* is given by

$$m^* = (m_l m_{t1} m_{t2})^{1/3} \tag{27}$$

and N_{ex} is the number of equivalent extrema. This density of states for a so-called nondegenerate, multivalley semiconductor is similar to that for a simple parabolic band, a geometric average of the principal components of the effective mass tensor replacing the scalar effective mass. The effective density of states at the conduction-band edge is then obtained from Eq. (18a) by replacing m_n by m^* above and multiplying by N_{ex}, i.e.,

$$N_c = 2N_{ex}\left(\frac{2\pi m^* kT}{h^2}\right)^{3/2} \tag{28}$$

For Si the lowest conduction-band minima lie inside the zone along the ⟨100⟩ direction and $N_{ex} = 6$.

The E–k relationship is more complicated[19] when two bands meet at an extremum, as is the case for the highest valence-band extremum at $k = 0$ in Figure 3. An approximation adequate for calculating the equilibrium values of n and p consists in describing the maximum as two independent spherically symmetric parabolic bands with effective masses m_{p1} and m_{p2}. (The valence band with the sharper maximum has the smaller effective mass and is called the light hole band. The broader band is called the heavy hole band.) The effective density of states is then given by Eq. (18b) with $m_p = m_{p1} + m_{p2}$.

2.2. Donors and Acceptors

We represent the total concentration of donors as D, that of un-ionized donors as $D^\times$, and that of singly ionized donors as D^+. For simplicity the excited states of the donor are ignored, as are those of the acceptor to be considered. A more complete analysis is given elsewhere (Ref. 20, p. 140). The analogous concentrations for acceptors are A, $A^\times$, and A^-. Thus we have

$$D = D^\times + D^+ \tag{29a}$$

$$A = A^\times + A^- \tag{29b}$$

and an equation for electroneutrality,

$$n + A^- = p + D^+ \tag{30}$$

Therefore in an intrinsic semiconductor for which $n = p$ we must also have $D^+ = A^-$, which is certainly the case in the pure, perfect semiconductor, for which the total concentrations of donors and acceptors are individually zero, $D = A = 0$. Each un-ionized donor can be viewed as an electron trapped about a donor site and each un-ionized acceptor as a hole trapped about an acceptor site. The concentration of former species is given by an equation of the form of Eq. (7), the latter by subtracting Eq. (7) from unity. Suppose that each donor level is doubly degenerate, in that an electron trapped there can have either of two spin states, but the level can only hold one electron, and analogously for the acceptor levels. With a total concentration of D donors and a concentration of $D^\times$ electrons trapped at donor levels, there are $2D - 2D^\times$ unoccupied levels, each electron occupying one level and removing a second. Then using Eq. (7) we have

$$\frac{D^\times}{2(D - D^\times)} = \frac{D^\times}{2D^+} = e^{(E_F - E_D)/kT} \tag{31}$$

where E_D is the position of the donor level and $E_C - E_D$ is therefore the energy required to remove an electron from the donor to the bottom of the conduction band. Similarly

$$A^\times/2A^- = e^{(E_A - E_F)/kT} \tag{32}$$

where $E_A - E_V$ is the energy required to remove a hole from the acceptor level to the top of the valence band (or equivalently, the energy to lift an electron from the top of the valence band to the ground acceptor level). As E_F moves from a few kT below the donor level energy E_D to a few kT above it, the ratio of un-ionized to ionized donors changes from close to zero to a large number. Eliminating $D^\times$ and A^- in favor of the total donor and acceptor concentrations and rearranging gives the concentrations of ionized impurities as

$$D^+ = D[1 + 2e^{(E_F - E_D)/kT}]^{-1} \tag{33}$$

$$A^- = A[1 + 2e^{(E_A - E_F)/kT}]^{-1} \tag{34}$$

For the nondegenerate case Eq. (22) for E_F is still valid and together with the Eqs. (12) and (14) for n and p and Eqs. (33) and (34) fixes the electron–hole distribution over all types of levels if the difference $n - p$ is known. If $n - p$ is not known, the latter four equations for n, p, D^+, and A^- must be substituted into the electroneutrality equation given by Eq. (30). The Fermi level is then in principle fixed and depends upon T, D, and A as well as the semiconductor parameters N_C, N_V,

E_G, $E_C - E_D$, and $E_A - E_V$. The equation, however, is quartic in $\exp[(E_F - E_F{}^i)/kT]$ and must be solved numerically[20] or piecewise by Brouwer's method discussed in Chapter 6 of Volume 1. However, the situation simplifies when there is a range of energies for E_F in which the semiconductor is nondegenerate and simultaneously E_F is at least a few kT below E_D and at least a few kT above E_A. Such a range is possible with shallow donors and acceptors. Then all the donors and acceptors are essentially ionized, so that $n - p = D - A$. The Fermi level can be obtained from Eq. (12) in terms of $D - A$ and n and p subsequently calculated using Eqs. (12) and (14), respectively.

If more than one type of singly ionizable donor is present, then equations analogous to Eqs. (29a), (31), and (33) hold for each. Similarly, equations analogous to Eqs. (29b), (32), and (34) hold for each species of singly ionizable acceptor and in the electroneutrality equation given by Eq. (30) A^- must be replaced by a sum over singly ionized acceptors of all types and D^+ by a sum over singly ionized donors of all types.

The case of multiple ionization can be handled by an extension of the above considerations which will be illustrated for a single species of doubly ionizable donors. The concentrations of un-ionized, singly ionized, and doubly ionized donors are defined as, respectively, $D^\times$, D^+, and D^{++}, so that the total concentration of donors D is given by

$$D = D^\times + D^+ + D^{++} \tag{35}$$

and the electroneutrality condition is

$$n = p + D^+ + 2D^{++} \tag{36}$$

The energy level of the un-ionized donor is taken at E_{D1}, so that the energy required to remove the first electron from the ground donor state to the bottom of the conduction band is $E_C - E_{D1}$. The concentration of donor levels either un-ionized or singly ionized is $D^\times + D^+$, so that the concentration of unoccupied levels for this first electron is $D^\times + D^+ - D^\times$, or D^+. Equation (7) can then be applied to give

$$D^\times/D^+ = e^{(E_F - E_{D1})/kT} \tag{37}$$

Because of the increased positive charge of a singly ionized donor, the removal of a second electron requires more energy and the level of the singly ionized donor E_{D2} lies deeper in the energy gap than E_{D1}. (If it should lie within the valence band, it would effectively be unobservable.) For concreteness we assume this level is twofold spin degenerate. The concentration of electrons trapped in this lower level is D^+ and the total number of un-ionized and ionized states is $D^+ + D^{++}$, so that by Eq. (7) we have

$$D^+/2D^{++} = e^{(E_F - E_{D2})/kT} \tag{38}$$

3. Electrical Properties

Here we discuss only a few of the main electrical properties of broadband semiconductors with the intent of indicating how n and p can be determined experimentally and information thereby obtained concerning the concentrations of donors and/or acceptors. The electrical transport properties are discussed from a more general viewpoint in Chapter 4.

The conductivity has already been given in Eq. (1) when electrons and holes are simultaneously present. When there are appreciable concentrations of carriers in more than one type of conduction or valence band, Eq. (1) must be generalized by writing a term similar to those shown for the carriers in each type of band. If either n or p is much larger than n_i, then by Eq. (19) the other concentration is proportionally reduced so that only one of the two terms of Eq. (1) is necessary. Even then, the conductivity is still the product of a carrier concentration and a mobility, both of which depend upon temperature. A measurement of the Hall coefficient, discussed below, along with the conductivity serves to obtain these factors individually in favorable cases. However, measures of the energy gap E_G have been obtained from the slopes at high temperatures of $\ln(\sigma T^{-3/2})$ vs. $1/T$ plots. The basis for this procedure lies first of all in the fact that at sufficiently high temperatures the carriers resulting from intrinsic excitation across the energy gap outweigh those arising from impurity levels. Consequently, n and p both approach n_i and the high-temperature value of σ is the product of n_i and the sum of the electron and hole mobilities. Second, at room temperature and above the carrier mobilities are determined predominantly by interaction of the carriers with the acoustical and optical vibrational modes of the crystal and depend upon the sum of terms varying as $T^{-3/2}$ and $T^{-1/2}$. Thus the predominant temperature dependence of $\sigma T^{-3/2}$ arises through the exponential dependence of n_i upon $-E_G/2kT$.

When an electrical current flows through a semiconductor with a magnetic field B imposed at right angles a voltage drop transverse to the current flow is developed due to the tendency of the carriers to deflect in the magnetic field.[21] Phenomenologically the Hall coefficient R_H relates this transverse voltage to the current, magnetic field, and geometry of the sample. The general relation between R_H and n and p is complicated, but simplifies for special cases. Only the low-magnetic-field cases when $eB\mu_p$ and/or $eB\mu_n$ are much less than unity are considered here. If a single type of carrier occupies energy levels near a band extremum for which the constant-energy surfaces in k space are ellipsoids of revolution, then the Hall coefficient for electrons is[22]

$$R_H = -(r/ne)\left[\frac{3K(K+2)}{(2K+1)^2}\right] = -(r/ne)S \tag{39}$$

For holes p replaces n, the minus sign is changed to a plus, and r and K are evaluated for holes.

The parameter K is the ratio of the longitudinal effective mass to the transverse effective mass m_l/m_t. For spherical constant-energy surfaces, such as given by Eq. (17), $K = 1$ and the entire factor involving K, the symmetry factor S, reduces to unity. For prolate constant-energy surfaces ($m_l > m_t$) the symmetry factor varies from one to about 0.75 as K increases from one to 50. For oblate constant-energy surfaces ($m_l < m_t$) the symmetry factor varies from unity to about 0.4 as K decreases from unity to 0.1. The former is the case usually encountered, although the valence-band maxima for SnTe appear to have oblate constant-energy surfaces. The scattering factor r depends upon the predominant scattering mechanism and state of degeneracy of the electron distribution. It is equal to 1.18 for lattice scattering, to 1.93 for ionized impurity scattering, and is generally between 1 and 2. When both electrons and holes are present then the Hall constant is given by[21]

$$R_H = (R_n\sigma_n^2 + R_p\sigma_p^2)/(\sigma_n + \sigma_p)^2 \tag{40}$$

where R_n and R_p are calculated from Eq. (39) and its analog for holes, and

$$\sigma_n = ne\mu_n \tag{41a}$$

$$\sigma_p = pe\mu_p \tag{41b}$$

If the symmetry factors are the same for electrons and holes,

$$R_H = -\frac{S}{e}\,\frac{r_n b^2 n - r_p p}{(nb+p)^2} \tag{42}$$

where b is the mobility ratio, μ_n/μ_p. In this case R_H goes to zero when $b^2 n = p$. A schematic of the temperature dependence of the low-field

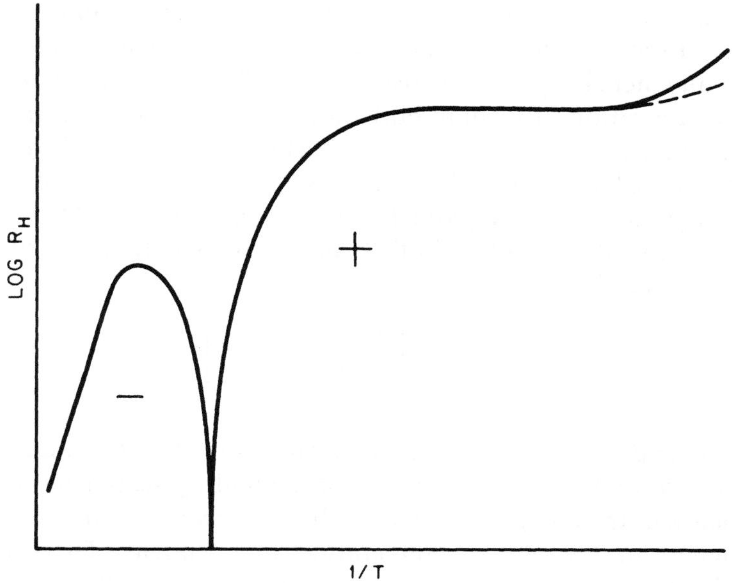

Fig. 4. The Hall coefficient as a function of reciprocal temperature for a semiconductor with an excess of acceptors and a mobility ratio b greater than one. The Hall coefficient is positive to the right of the inverted cusp, negative to the left.

Hall coefficient is shown in Figure 4 for a case that $n - p < 0$, due to an excess of acceptors, and $b > 1$. At the very lowest temperatures R_H is decreasing with increasing temperature due to the ionization of acceptors. If the sample is compensated, i.e., if donors are also present but at a concentration lower than that of the acceptors, the slope of $\ln R_H$ vs. $1/T$ is the ionization energy of the acceptors divided by the Boltzmann constant, i.e., $(E_A - E_V)/k$. This might be expected to be the usual case. If the sample contains only acceptors and is uncompensated, the slope is one-half that for the former case, as indicated by the dashed line in Figure 4. Over the temperature range corresponding to the flat portion of the curve all the acceptors are ionized but no significant intrinsic excitation has yet occurred. Up to this point only holes are present ($p - n \approx p$), R_H is positive, and the one-carrier formula, Eq. (39), is adequate. The net excess of ionized acceptors, $A^- - D^+$, is equal to $p - n \approx p$ by the electroneutrality equation, Eq. (30), and can be obtained from the value of R_H in the flat region. The Pb-salt members of the IV–VI family typically show[23] extended flat regions and in fact have shown no sign of the low-temperature carrier freezeout on impurity levels indicated in Figure 4. In more complicated cases the flat region is not present, R_H continually increasing with decreasing T. As T is increased further the intrinsic ionization furnishing both electrons and holes becomes important and a two-carrier formula such as Eq. (40) or Eq. (42) must be used. Since we have assumed $b > 1$, as is typical, R_H starts to decrease and changes sign, since R_n and R_p are of opposite sign. At the very highest temperatures shown both n and p are essentially equal to n_i and R_H is given by

$$R_H = -\frac{1}{n_i e}\left[\frac{r_n S_n b^2 + r_p S_p}{(b+1)^2}\right] \tag{43}$$

where S_n and S_p are the symmetry factors for electrons and holes, respectively, and r_n and r_p are the corresponding scattering factors. A measure of the energy gap can be obtained by assuming the factor in brackets in Eq. (43) is temperature independent. Then in the intrinsic region the slope of $\ln(R_H T^{-3/2})$ vs. $1/T$ is $E_G/2k$. Difficulties in this procedure can arise when semiconductor compounds are studied if the temperatures required to make the semiconductor intrinsic are also high enough for diffusion to occur rapidly.[24]

When $b^2 n$ is either much larger or much smaller than p, the one-carrier formulas for R_H and σ apply and the Hall mobility, defined as the product $R_H\sigma$, differs from the drift mobility only by the factor rS appearing in Eq. (39). The variation of the Hall mobility with T and impurity level concentration has been used to determine the predominant scattering mechanisms and at about 77°K and below, when ionized impurity scattering is important, to determine $D^+ + A^-$ for Ge and Si and for the III–V and some II–VI compounds.[25] Since $D^+ - A^-$ is obtained from the Hall coefficient, D^+ and A^- can be obtained individually in this case.

When two dissimilar materials a and b are joined and one junction kept at a temperature ΔT higher than the other, an emf develops that is proportional to ΔT and a material property $\alpha_a - \alpha_b$, where α_a and α_b are the Seebeck coefficients of the materials a and b. At room temperature the Seebeck coefficients of semiconductors are in the mV/°C range, while those for metals are in the μV/°C range. When both electrons and holes are present in a semiconductor the Seebeck coefficient is given by[21]

$$\alpha = (\alpha_n\sigma_n + \alpha_p\sigma_p)/(\sigma_n + \sigma_p) \tag{44}$$

where σ_n and σ_p are the partial conductivities defined by Eqs. (41a) and (41b). For a nondegenerate semiconductor in which the relaxation time for both electrons and holes depends upon the energy of the carrier as E^{-s}

$$\alpha_n = -(k/e)\{(E_C - E_F)/kT + (\tfrac{5}{2} - s)\} \tag{45}$$

$$\alpha_p = (k/e)\{(E_F - E_V)/kT + (\tfrac{5}{2} - s')\} \tag{46}$$

These simple equations are inappropriate at low temperatures where the behavior of the relaxation times are more complicated, but they do serve to indicate the temperature and impurity dependences at about room temperature and above. The relaxation time index s is $\frac{1}{2}$ for acoustic mode scattering, $-\frac{1}{2}$ for optical mode scattering well above the Debye temperature, and $-\frac{3}{2}$ for ionized impurity scattering. With these scattering modes predominant α_n is negative and α_p positive, so that α at a given T for samples of a semiconductor with different donor and acceptor concentrations, i.e., different values for $n - p$, changes sign. This occurs at the intrinsic concentration where $n = p$ only if the mobility ratio b is unity. If $b > 1, \alpha = 0$ at a negative value of $n - p$. Because the two-carrier Hall coefficient given by Eq. (40) is a different weighted average of the partial Hall coefficients, the zero in the Hall coefficient occurs a value of $(n - p)$ further from zero than that in the Seebeck coefficient.

For a nondegenerate semiconductor the Fermi level in Eq. (45) can be replaced by n using Eq. (12) and the Fermi level in Eq. (46) by p using Eq. (14) to give

$$\alpha_n = -(k/e)\{\ln(N_C/n) + (\tfrac{5}{2} - s)\} \tag{47}$$

$$\alpha_p = (k/e)\{\ln(N_V/p) + (\tfrac{5}{2} - s')\} \tag{48}$$

If either carrier is in excess, so that nb is either much smaller or larger than p, then α is equal to the partial Seebeck coefficient of the carrier in excess. The Seebeck coefficient at fixed T is shown schematically as a function of $(n - p)/2n_i$ in Figure 5 for the case that the mobility ratio b is greater than unity. Because $b > 1$ the Seebeck coefficient is zero at a negative value of $n - p$. Both electrons and holes are contributing significantly to the value of α in the region about $\alpha = 0$ and between the maximum and minimum. On either side of the extrema, one type of carrier is predominant and σ varies linearly with $\ln n$ or $\ln p$ with a slope of $\pm k/e = \pm 86\ \mu V/°C$. At the highest values of

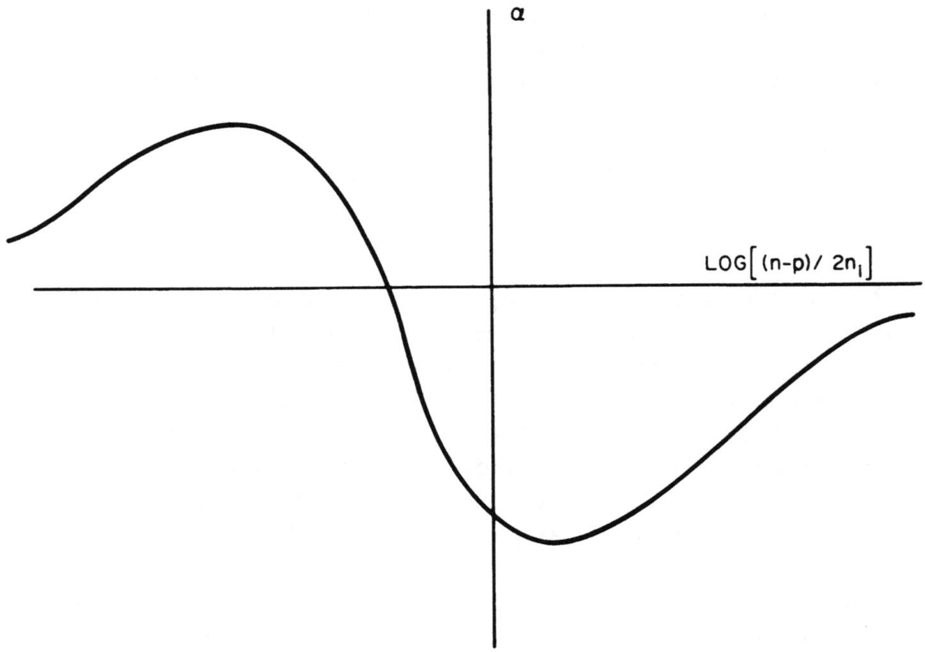

Fig. 5. The thermoelectric power as a function of reduced net electron concentration. The mobility ratio b is greater than one.

$|n - p|$ the semiconductor is degenerate, the above equations are no longer valid, and α varies less rapidly with $n - p$.

4. Optical Properties

4.1. Intrinsic Behavior

A number of review articles discuss absorption and luminescence in the III–V[4,6,18] and II–VI[26,27] compounds and a recent book by Pankove[28] serves as an excellent introduction to optical processes in solids. Two primary optical absorption mechanisms that are independent of atomic point defects are the excitation of the optical modes of the lattice vibrations and the excitation of electrons from the valence to the conduction bands. Lattice absorption is not expected in elemental substances on the simplest theoretical grounds, but has been seen in Si. This absorption occurs within roughly the 20–200 μm range and, in an isomorphous series of compounds, shifts to longer wavelengths within this range with increasing molecular weight. The spectra show complex structure due to the various combinations of two or more phonons that are necessarily excited to conserve momentum. Impurities can introduce localized modes into the vibrational spectrum. An example[29] is O in Si, which produces an absorption band at 9 μm due to the Si–O bond. The donor level that would be expected when O substitutes for Si apparently lies deep in the valence band and produces no significant influence on the concentrations of electrons and holes.

The band-to-band electronic excitation produces an absorption band starting close to the frequency corresponding to the semiconductor energy gap. By the Planck relation the energy of a photon of frequency ν is given by $h\nu$. In the simplest case this energy just matches that required to excite an electron from the top of the valence band to the bottom of the conduction band, i.e., E_G. If $E_G = 1$ eV, the fundamental (absorption band) edge begins near 1.2 μm. At energies 0.1–1 eV greater than E_G the absorption coefficient is in the 10^3–10^4 cm^{-1} range, compared to a value of the order of 10 cm^{-1} for lattice absorption. Two general restrictions govern the absorption process; the conservation of energy and a selection rule relating the wave vectors of the photon, electron, and any phonons involved. For all practical purposes the latter restriction can be interpreted as an equation for the conservation of momentum. Transitions meeting the constraints of constant energy and momentum can still be allowed or forbidden depending upon parity of the wave functions in the initial and final states. As already discussed in connection with Figure 3, the energy gap in Ge is an indirect gap between the valence-band extremum at $k = 0$ and a conduction-band minimum at the zone edge in the $\langle 111 \rangle$ direction. When an electron is excited across this gap its associated crystal momentum changes from zero to $h/2a_0$, where a_0 is the lattice parameter. The momentum of a photon with an energy corresponding to E_G is hc/ν (c is the speed of light) and is much smaller than this. Therefore a phonon must be created or destroyed in order to conserve momentum. The slightly higher-energy, direct transition from the valence-band maximum to the conduction-band minimum at $k = 0$, requiring no change in k, need involve no phonons.

For a direct transition the absorption coefficient at a frequency ν is given by

$$\alpha(h\nu) = A \sum P_{if} n_i n_f \tag{49}$$

where n_i is the concentration of electrons in an initial state, n_f is the concentration of empty energy levels in a final state at energy $h\nu$ higher, P_{if} is a transition probability, and the sum is over all initial and final states. For simple parabolic bands, as given by Eq. (17), and an allowed transition P_{if} is independent of photon frequency, giving

$$\alpha(h\nu) = e^2 \frac{\{2[1/m_n + 1/m_p]\}^{3/2}}{n_R c h^2 m_n} (h\nu - E_G)^{1/2} \tag{50}$$

where n_R is the index of refraction. Thus a plot of α^2 vs. $h\nu$ extrapolates to a zero value of α at a value of $h\nu$ equal to E_G. For a forbidden direct transition P_{if} is proportional to $(h\nu - E_G)$ and by the same arguments as above one obtains

$$\alpha(h\nu) = A(h\nu - E_G)^{3/2}/h\nu \tag{51}$$

where A depends upon the effective masses and the index of refraction.

Indirect transitions may occur at photon energies slightly less than the energy change of the electron if a phonon is also absorbed or, at photon energies slightly higher, if a phonon is emitted. In the former case the absorption coefficient depends on the number of phonons of the requisite energy and so becomes small at low temperatures. If E_p is the energy of the phonon, the absorption coefficient depends upon frequency as

$$\alpha(h\nu) = A(h\nu - E_G - E_p)^2 \tag{52}$$

for $h\nu > E_G - E_p$ and as

$$\alpha(h\nu) = A(h\nu - E_G - E_p)^2 + B(h\nu - E_G + E_p)^2 \tag{53}$$

for $h\nu > E_G + E_p$ with the parameter A approaching zero and the parameter B approaching a constant value as the temperature approaches 0°K. The absorption coefficients for indirect transitions are typically 10^{-3} those for direct transitions. Of the III–V compounds listed in Section 1, the Al compounds and GaP are indirect-gap semiconductors; the others have direct gaps.

When the equilibrium concentrations of electrons and holes are perturbed by some means, they tend to return to their equilibrium values by radiative mechanisms, in which some or all of the excitation energy is carried off in a photon, or by nonradiative mechanisms, which increase the thermal energy of the lattice. At small departures from equilibrium the relative importance of the radiative mechanisms is dependent upon the magnitude of the absorption coefficient and hence tends to be small for indirect-gap semiconductors, unless appropriate atomic point defects are introduced as discussed below. This can be seen as follows. At thermal equilibrium the rate G at which electron–hole pairs are optically generated by absorption of the blackbody radiation field must equal the rate at which electron–hole pairs recombine to emit a photon. The latter rate is assumed to be the product of np and a recombination rate constant R_r. The generation rate is

$$G = \int_0^\infty \frac{8\pi n_R{}^2 h^2 \nu^2}{h^3 c^2 (e^{h\nu/kT} - 1)} \alpha(h\nu)\, d(h\nu) \tag{54a}$$

$$\simeq C \int_0^\infty \alpha(h\nu)\, e^{-h\nu/kT}\, d(h\nu) \tag{54b}$$

where the term in parentheses in Eq. (54a) gives the number of photons of frequency ν and where unity in the denominator has been neglected in obtaining Eq. (54b). Equating the generation rate given by Eq. (54b) to the radiative recombination rate $R_r np$, one can obtain R_r. At small departures from equilibrium the radiative recombination rate is still given by the same expression with the prevailing values for n and p inserted,

$$\mathrm{R}_r np = (Cnp/n_i{}^2) \int_D^\infty \alpha(h\nu) e^{-h\nu/kT}\, d(h\nu) \tag{55}$$

The formation of excitons appears as narrow peaks in α near the absorption edge of direct-gap materials or as steps in the absorption edge of indirect-gap semiconductors. In both cases these features tend to be broadened at higher temperature. At high enough optical frequencies transitions involving deeper valence band and/or higher conduction bands commence and produce more or less sharp changes in the dependence on α upon ν.

4.2. Influence of Atomic Point Defects

Atomic point defects electrically active as donors, acceptors, or isoelectronic traps influence optical absorption in a number of ways. Absorption is introduced at photon energies less than that corresponding to E_G by a number of mechanisms. Some of these are: (1) the excitation or ionization of initially un-ionized donors or acceptors, (2) formation of excitons bound at isoelectronic traps, (3) formation of excitons bound to donor–acceptor pairs, and (4) excitation of conduction-band electrons or valence-band holes within their bands, i.e., free-carrier absorption. Absorption due to 1, 2, and 3 is only seen at temperatures low enough that the concentration of the initial species in the transition is significant. The free-carrier absorption is stronger the larger the concentration of electrons or holes and therefore the larger the excess concentration of donor or acceptors, provided the temperature is high enough to ionize these.

In addition to the above the absorption spectrum near the fundamental edge itself (corresponding to the excitation of an electron across the energy gap) can be affected. If E_G is a few tenths of an eV and the free-carrier concentration is high, absorption by the latter can overlap that at the fundamental edge.

When the electron or hole concentration is high enough that all the energy states near a band edge are occupied, then absorption involving an electron transition from the highest filled valence-band states to the lowest empty conduction-band states must necessarily begin at photon energies greater than E_G. The fundamental edge is therefore shifted to energies larger than that observed for purer materials (Burstein–Moss shift). An idea of the carrier concentration required to produce a given shift can be obtained by integrating the density of states given by Eq. (26) to obtain the number of states between the band edge and some energy E. The result is

$$N = N_{ex}(8\pi/3)(2m^*E/h^2)^{3/2} \tag{56}$$

where N_{ex} is the number of equivalent extrema. For $m^* = 0.1m_0$ and $N_{ex} = 1$ there are $10^{19}\,\text{cm}^{-3}$ states within 0.1 eV of the band edge and a carrier concentration equal to $10^{19}\,\text{cm}^{-3}$ produces roughly a 0.1-eV shift in the absorption edge.

Finally, the Coulombic fields associated with ionized donors or acceptors result in a fluctuating, internal electric field that perturbs the band edges and causes some of the band states to shift. An exponential density of states develops which extends into the band gap from the positions of the band edges when the donors and acceptors are absent or un-ionized.[30] The absorption coefficient then depends exponentially upon photon frequency near the fundamental edge and drops less steeply with decreasing frequency than predicted by Eq. (50), (51), or (53).

Free-carrier absorption corresponds to an electron or hole changing energy states within its band. These transitions are necessarily indirect and require phonon absorption or emission or interaction with impurities in order to conserve momentum. The absorption coefficient is found to increase as λ^p, where λ is the photon wavelength and p is between 1.5 and 3.5, depending upon whether the predominant scattering mechanism is via acoustical phonons, optical phonons ($p = 2.5$), or ionized impurities ($p = 3.5$). The absorption coefficient also depends (for excess electrons) upon the electron concentration n and effective mass m_n and the relaxation time τ. The classical relation is

$$\alpha = ne^2\lambda^2/8\pi^2 m_n n_R c^3 \tau \tag{57}$$

Here n_R is the index of refraction and c is the speed of light.

An electronegative isoelectronic trap, such as N substituting for P in GaP, attracts an electron via a short-range, non-Coulombic potential. The negatively charged complex then can trap a hole via a Coulombic interaction to form an exciton bound to an isoelectronic trap. Thus the absorption of radiation can form a free electron and hole, but at a slightly lower photon energy can also form an exciton (electron–hole pair) bound to the isoelectronic trap. Moreover, pairs of N atoms interact[11] to form new levels for their bound excitons, the excitons being held together most strongly for nearest-neighbor N atoms and less strongly for more distant pairings, as shown schematically in Figure 6. Since only discrete pair separations are allowed in the crystal structure, a number of discrete lines are seen. The notation NN_k indicates the absorption due to excitons bound to N atom pairs that are in the kth nearest-neighbor positions. The lines crowd together and approach the position for an isolated N atom as the pair separation increases. This high-energy limit is near the fundamental edge since the binding energy of an exciton to an isolated N atom relative to that of a free electron and hole is small. Therefore the energy required to create such an exciton by exciting an electron from the valence-band is not much less than that required to create a free electron and hole. The inherently low absorption coefficient of indirect-gap semiconductors is increased by the inclusion

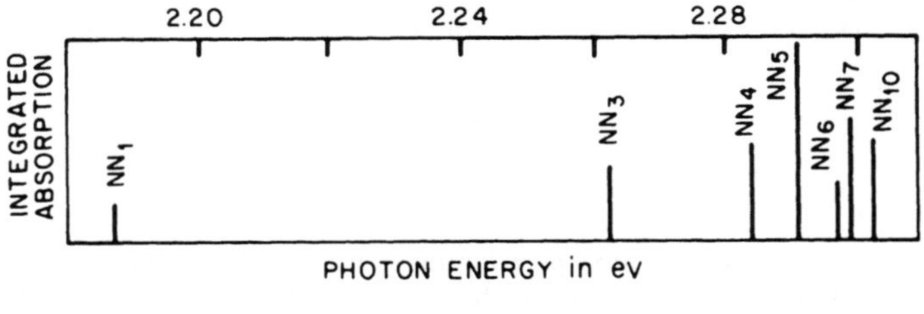

Fig. 6. Schematic absorption spectrum due to isolated and paired N isoelectronic traps in GaP at 1.6°K. Ref. 5.

of a sufficient concentration of isoelectronic traps and consequently the probability of radiative recombination is also increased, as shown by Eq. (55). Momentum is conserved in the radiative transfer by coupling with the impurity. Oxygen substituting for Te in ZnTe is also thought to be an isoelectronic trap. It appears that when a substitutional impurity, with the same oxidation number as the atom replaced, is sufficiently different in electronegativity to form an isoelectronic trap, then the solubility of the impurity is limited and it does not form a complete range of solid solution.

The behavior of donor and acceptor pairs has been investigated in most detail in GaP using luminescence studies.[5,31] For distant, as opposed to nearest-neighbor, pairs an electron trapped at a donor level with an ionization energy $E_C - E_D$, and a hole trapped at an acceptor with an ionization energy $E_A - E_V$, recombine, with the emission of a photon of energy $h\nu$ given by

$$h\nu = E_G - (E_C - E_D) - (E_A - E_V) + (e^2/\varepsilon r) \quad (58)$$

$$= E_D - E_A + (e^2/\varepsilon r) \quad (59)$$

where r is the separation between the donor and acceptor atoms. A schematic of one type of spectrum (type 2) observed in GaP at 1.6°K is shown in Figure 7. Here the donors and acceptors occupy sites in different sublattices and the pair separation r takes discrete values $[(8m - 5)/16]^{1/2}a_0$, where a_0 is the lattice parameter and m is an

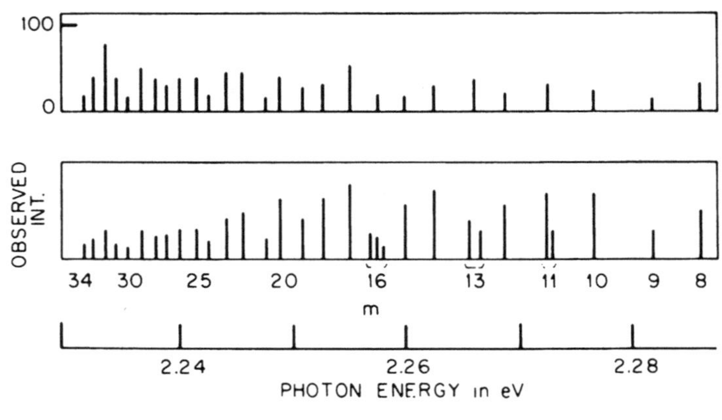

Fig. 7. Donor–acceptor pair luminescence in GaP. Upper figure shows the number of sites in the mth coordination shell when the donor and acceptor occupy sites in different sublattices of the zincblende structure. Lower figure shows the observed intensities.

integer, certain values of which are not allowed. The value of m is shown beneath the corresponding emission lines. Just above the spectrum the number of sites occurring at a given value of m is also plotted. It is seen that the relative intensities observed for close values of m vary in the same way as the number of sites. Thus a random distribution of donor–acceptor pairs and an electron–hole recombination probability that varies slowly with r account for the observed intensities. When the donor and acceptor are on sites in the same sublattice (type 1 spectrum) r takes the values $(\frac{1}{2}m)^{1/2}a_0$, with $m = 14$, 30, 46, 56, 62,... not occurring. The spacing and number of sites for given value of m are distinctly different from the type 2 spectrum. Utilizing these differences and using isotopes, the identities of various donor and acceptors have been established and accurate values obtained for the sum of the donor and acceptor ionization energies in some cases. These include C and O, Zn and O, and Cd and O in GaP.

When a singly ionizable donor and acceptor are on nearest-neighbor sites the localized levels are removed from the energy gap for a shallow donor and acceptor, or an isoelectronic trap may be formed when one type of level is deep. The interstitial Li-substitutional B pair in Ge is an example of the former.[32] The pairing is detected by the decrease in the Li diffusion coefficient, by the enhancement of the solubility of Li over that expected from a shift in the Fermi level alone, and by a decrease in the ionized impurity scattering of holes from that expected for the unpaired ionized centers. An example of the latter[4,6] is the Cd–O nearest-neighbor pair in GaP. Experiments indicate that not only can an electron–hole pair trapped at the Cd–O complex recombine, but also an electron trapped at the Cd–O complex can recombine with a hole trapped at a Cd acceptor. This type of pair spectrum does not show sharp lines, but time-resolved spectroscopy indicates that the Coulombic interaction energy included in Eq. (59) is absent. Thus the Cd–O complex attracts an electron via a local, non-Coulombic potential characteristic of an isoelectronic trap. It also appears that donor–acceptor pairs are the activator–coactivator species necessary for efficient luminescence in ZnS phosphors.[27]

5. The Chemical Potentials in Elemental Semiconductors

5.1. Nonassociated Defects

The equations in Section 2 describing electron–hole equilibria in a semiconductor pertain to a partial equilibrium in which the total concentration for each type of donor and acceptor is fixed. These equations are therefore insufficient to describe situations of practical interest in which the donor and acceptor concentrations can be changed by interaction with another phase. Now in a multicomponent, multiphase system at fixed temperature and pressure the condition for equilibrium is that the chemical potential for each component be the same in every phase. Therefore what are needed are equations defining the functional form of the chemical potentials in the semiconductor phase. These statements apply equally to elemental and compound semiconductors. However, in the remainder of this section attention will be focused on elemental semiconductors.

The concentrations of foreign impurities used in semiconductors and indeed, often their solubilities,[33] are generally well below 1 at. %. To a good approximation the chemical potential of the host atoms is then a function of temperature and pressure alone. If it is assumed that the atomic point defects are randomly distributed over an appropriate set of sites, if the excess Gibbs free energy of the semiconductor is changed by a constant amount per un-ionized defect, and if the positions of the bands and the localized levels are independent of the concentrations, then a statistical thermodynamic analysis yields simple forms for the chemical potentials. Consider an atomic point defect which can exist substitutionally as a doubly ionizable donor or interstitially as a singly ionizable donor, whose donor level is at an energy E_{ID}. The total donor concentration D is then given by

$$D = D^{\times} + D^{+} + D^{++} + I_D{}^{\times} + I_D{}^{+} \qquad (60)$$

where $I_D{}^{\times}$ is the concentration of the un-ionized, interstitial form, $I_D{}^{+}$ is the concentration of the singly ionized interstitial form, and the D's are the concentrations of the various substitutional forms as defined in Section 2.2. If S is the concentration of normal sites and αS the concentration of interstitial sites, then the chemical potential of the species D is given by each of the following equations:

$$\mu_D = kT\ln(D^{\times}/S) + \mu_D{}^{\circ} \qquad (61)$$

$$= kT\ln(D^{+}/S) + (E_{\mathrm{F}} - E_{D1}) + \mu_D{}^{\circ} \qquad (62)$$

$$= kT\ln(2D^{++}/S) + (2E_{\mathrm{F}} - E_{D1} - E_{D2}) + \mu_D{}^{\circ} \qquad (63)$$

$$= kT\ln(I_D{}^{\times}/\alpha S) + \mu_{ID}^{\circ} \qquad (64)$$

$$= kT\ln(I_D{}^{+}/\alpha S) + (E_{\mathrm{F}} - E_{ID}) + \mu_{ID}^{\circ} \qquad (65)$$

In writing these equations, it has also been assumed that the total defect concentration is small compared to S, which is then a function of temperature and pressure only.

We note first that if the various expressions for μ_D are equated, equations for the distribution of electrons over the various donor levels are obtained, in particular Eqs. (37) and (38). Therefore Eqs. (61)–(65) along with the analogous expressions to be presented for acceptors would yield all the electron–hole distribution equations except those for conduction-band electrons and valence-band holes.

The standard excess Gibbs free energies $\mu_D{}^\circ$ and μ_{ID}° are composition-independent terms that represent the increase in the excess Gibbs free energy of the semiconductor per un-ionized defect added. These both can be decomposed in the manner illustrated below for $\mu_D{}^\circ$, i.e.,

$$\mu_D{}^\circ = \varepsilon_D{}^\circ - Ts_D{}^\circ + Pv_D{}^\circ \tag{66}$$

where $\varepsilon_D{}^\circ$, $s_D{}^\circ$, and $v_D{}^\circ$ are the increment per un-ionized defect of, respectively, the internal energy, excess entropy, and volume of the semiconductor crystal and are independent of the defect concentrations. In the simplest case $\varepsilon_D{}^\circ$, $s_D{}^\circ$, $v_D{}^\circ$, and the analogs for the interstitial form, ε_{ID}°, s_{ID}°, and v_{ID}°, are constants and the temperature and pressure P dependences of $\mu_D{}^\circ$ and μ_{ID}° are the explicit dependences shown in Eq. (66) for $\mu_D{}^\circ$.

If $\mu_D{}^\circ$ is much larger than μ_{ID}°, then Eqs. (61)–(66) show that the interstitial forms are predominant and the substitutional form can be neglected. Conversely if $\mu_D{}^\circ$ is much smaller than μ_{ID}°, the substitutional forms are predominant. As discussed in Section 2.2, the level E_{D2} for the singly ionized substitutional donor must lie below that, E_{D1}, for the un-ionized form. If E_{D2} lies within the valence band, the substitutional donor will never be doubly ionized in any practical situation and D^{++} is essentially zero.

Although the chemical potential is a simple function of each species of the donor defect, it is not in general a simple function of the total donor concentration D. Suppose for simplicity that $E_{D2} < E_V$, so that the substitutional form is only singly ionizable and $\mu_{ID}^\circ \gg \mu_D{}^\circ$, so that the concentrations of the interstitial forms are negligible compared to that of the substitutional form of the donor. Moreover, suppose the intrinsic Fermi level, given by Eq. (20) or (22), is near the middle of the energy gap while the donor level E_{D1} is near the conduction band edge E_C. Then for the first small additions of donor to the semiconductor the donor will be essentially all ionized and the chemical potential will be given by Eq. (62) with $D^+ = D$ and $E_F = E_F{}^i$. Upon further addition of donors the Fermi level will rise, but so long as E_F is at least a few kT below E_{D1}, the donors will essentially all be ionized. Then the Fermi level is given by Eq. (22) with the net electron concentration $n - p$ equal to the total donor concentration D, i.e.,

$$\mu_D = kT \ln(D/S) + E_F{}^i - E_{D1} + kT \sinh^{-1}(D/2n_i) + \mu_D{}^\circ \quad (67)$$

The last term is a weak function of composition until D becomes comparable to or larger than the intrinsic carrier concentration n_i. Then it contributes an increment of $2.303kT$ for every decade increase in $D/2n_i$ and this situation persists until E_F is near E_{D1} and E_C. Since n_i is generally small compared to S, the concentration of normal lattice sites, the presence of the Fermi level term can make the chemical potential a sensitive function of the donor concentration.

For brevity an acceptor defect will be considered only for the case of a singly ionizable substitutional form. Under the same assumptions as made for donors the chemical potential of the acceptor is given by

$$\mu_A = kT \ln(A^\times/S) + \mu_A{}^\circ \quad (68)$$

$$= kT \ln(A^-/S) - (E_F - E_{A1}) + \mu_A{}^\circ \quad (69)$$

The significant difference between Eq. (69) for μ_A in terms of the concentration of the singly ionized substitutional acceptors and Eq. (62) for μ_D in terms of the concentration of the singly ionized donors is in the sign before the Fermi level E_F. With this in mind the equations for μ_A in terms of the concentration of a possible doubly ionized substitutional form or in terms of those for un-ionized and ionized interstitial forms of the same foreign atom can be written by analogy with Eqs. (61)–(65).

An increase in the total concentration of a donor tends to raise the Fermi level, while an increase in that of an acceptor does the opposite. Therefore μ_D for a given total donor concentration is decreased when the acceptor concentration is increased and μ_A for a given total acceptor concentration is decreased when the donor concentration is increased. All donors and acceptors interact, therefore, through their influence on the Fermi level.

If the predominant native atomic point defect in the elemental semiconductor is a singly ionizable, interstitial donor, then Eqs. (64) and (65) apply with μ_D being the chemical potential of the host atom, which, as noted earlier, is to a good approximation a function of T

and P alone for small defect concentrations. If the predominant native atomic point defect is a vacancy that behaves as an acceptor, Eqs. (68) and (69) do not apply, but rather

$$\mu_H = kT \ln(S/V^\times) + \mu_H^\circ \tag{70}$$

$$= kT \ln(S/V^-) - (E_F - E_{AV}) + \mu_H^\circ \tag{71}$$

where μ_H is the chemical potential of the host atom, $V^\times$ and V^- are the concentrations of, respectively, un-ionized and ionized vacancies, and E_{AV} is the energy of the acceptor level associated with the vacancy. Note that the argument of the logarithmic term is inverted from that when the acceptor is a substitutional foreign atom. Since μ_H is essentially a function of T and P alone, so is $V^\times$. However, the concentration of singly ionized vacancies, and hence the total vacancy concentration, depends on E_F also. Considerable work[34,50] has been done on the native atomic point defects obtained in Ge upon quenching and upon plastic deformation and double-vacancy acceptors as well as single-vacancy acceptors and interstitial donors have been postulated.

5.2. Associated Defects

In addition to the noninteracting defects discussed above, a number of associated species have been detected in Si and Ge. Most of these are donor–acceptor pairs and many involve Li, which is a fast-diffusing, singly ionizable interstitial donor.[32] When the binding of a nearest-neighbor pair is local rather than long-range Coulombic and the concentration of pairs is small compared to that of lattice sites the usual approach is to consider the pairs and single defects as independent species. Assume the localized levels of the impurities are removed upon forming a nearest-neighbor pair. The additional equation required is then the chemical potential of the donor–acceptor pair, which under the above assumptions is given by

$$\mu_{DA} = kT \ln[(DA)/S] + \mu_{DA}^\circ \tag{72}$$

where (DA) is the concentration of pairs and where

$$\mu_{DA} = \mu_D + \mu_A \tag{73}$$

By substituting Eqs. (61), (68), and (72) into Eq. (73) one obtains the mass action law expression of the form

$$(DA)/D^\times A^\times = K(T) \tag{74}$$

Since pairs are inactive as donors or acceptors, the pair concentration does not enter the electroneutrality condition, but it of course must be taken into account in obtaining the total concentration of the donor or acceptor atoms, e.g., if D is the total concentration of a singly ionizable donor, then

$$D = D^{\times} + D^{+} + (DA) \tag{75}$$

Alternatively, the ionized donors and acceptors can be considered to interact like point charges in a dielectric continuum. One then has not only nearest-neighbor pairs but also pairs that are next-nearest neighbors, etc. The basic idea has already been discussed in Section 4.2 in connection with the spectra of *distant* donor–acceptor pairs. There the assumption of a random distribution of pair separations consistent with the lattice geometry was made. The quantitative agreement with observed line positions confirms the accuracy of the interaction energy given by Eq. (59). The qualitative agreement with line intensities is consistent with a random distribution for the distant pairs. Here one is concerned with the concentrations of near-neighbor pairs. If the donors and acceptors are both shallow and singly ionizable, their localized levels are removed from the gap when they are nearest neighbors. It is assumed the same is true for all *close* pairs, where the term *close* pairs is defined below. The analysis is based on the ion-pair concept developed by Bjerrum and modified by Fuoss for electrolyte solutions and subsequently refined by Reiss.[35] The concentrations of ionized donor and acceptors are assumed to be equal. If two oppositely charged ions (ionized impurities) are close enough together, their Debye–Hückel clouds superimpose and approximately cancel so that the ions are subject only to the Coulombic field of their partner. The probability that the nearest neighbor of one ion is another, oppositely charged, ion at a distance between r and $r + dr$ is $g(r)$, where

$$g(r) = h_1 4\pi r^2 \exp(q^2/\varepsilon kTr - 4\pi r^2 N/3) \tag{76}$$

The concentrations of cations and anions are both N, q is the magnitude of the charge on each ion, and h_1 is a normalization constant obtained from the requirement that the integral of the $g(r)$ over all r between the distance of closest approach a and infinity must equal unity. The normalization constant h_1 therefore depends upon the ion concentration N and a. The latter serves as an adjustable parameter. As a function of r, the pair distribution function $g(r)$ falls exponentially from its value at $r = a$ to a long minimum and then rises to a maximum near a value of r characteristic of a random distribution before dropping toward zero as r approaches infinity. All those ions with oppositely charged partners at distances between a and that corresponding to the minimum in $g(r)$ are considered *close* pairs. More distant ions are considered unpaired. Neglecting the small correction involved in the Debye–Hückel activity coefficient for the unpaired impurities, the chemical potentials of the donor and acceptor are given by Eqs. (62) and (69), respectively, if the concentrations of singly ionized species appearing are corrected by subtracting that portion involved in *close pairs*, i.e., one replaces D^+ in Eq. (62) by

$$D^+(1 - \int_a^b g(r)) \, dr \tag{77}$$

where b is the value of r for which $g(r)$ is a minimum. The experimental data on Li acceptor pairs in Ge can be adequately fit using the above approach and reasonable values are obtained for a, the distance of closest approach.(32)

A modification(36) of Eq. (76) appropriate for substitutional impurities confined to discrete sites in one sublattice can be obtained by realizing that the factor $4\pi r^2 \, dr$ in a continuous distribution is equivalent to the number of sites in the kth coordination shell z_k divided by the total number of substitutional sites per unit volume, S. Then the fraction of ionized donors and acceptors that are kth nearest neighbors is $g(r_k)$, with

$$g(r_k) = A(z_k/S)\exp(q^2/\varepsilon kTr_k)\exp\left[-(N/S)\sum_{j=1}^{k-1} z_j\right] \tag{78}$$

where A is a normalization constant and N is the concentration of both ionized donors and ionized acceptors. As is the case in the electrolyte theory, Eq. (78) is not expected to be valid at values of N/S too close to the maximum value of $\frac{1}{2}$. Numerical calculations performed for cubic ZnS show that for $-N/S = 10^{-4}$ about 12% of the ionized donors and acceptors are nearest neighbors at 1000°K and 90% at 500°K. Moreover, the percentage that are second through fifth nearest neighbors is considerably less, so that an analysis that considers only nearest-neighbor impurities as paired and the remainder unpaired(37) yields essentially the same results in this case.

6. The Chemical Potentials for Binary Semiconductor Compounds

6.1. Thermodynamic Considerations

We are interested in obtaining the chemical potentials of impurities and of native defects in semiconductor compounds in order to describe the complete equilibrium which the compounds attain in contact with other phases. First, however, some thermodynamic aspects of the problem are briefly considered.

A useful function for discussing phase equilibria in a binary system M–N is the Gibbs free energy per gram-atomic weight, defined by

$$G = x_M\mu_M + x_N\mu_N \tag{79}$$

where a G is written for every phase, μ_M and μ_N are the chemical potentials, and x_M and x_N are the atomic fractions for the thermodynamic components M and N. At constant T and P the composition

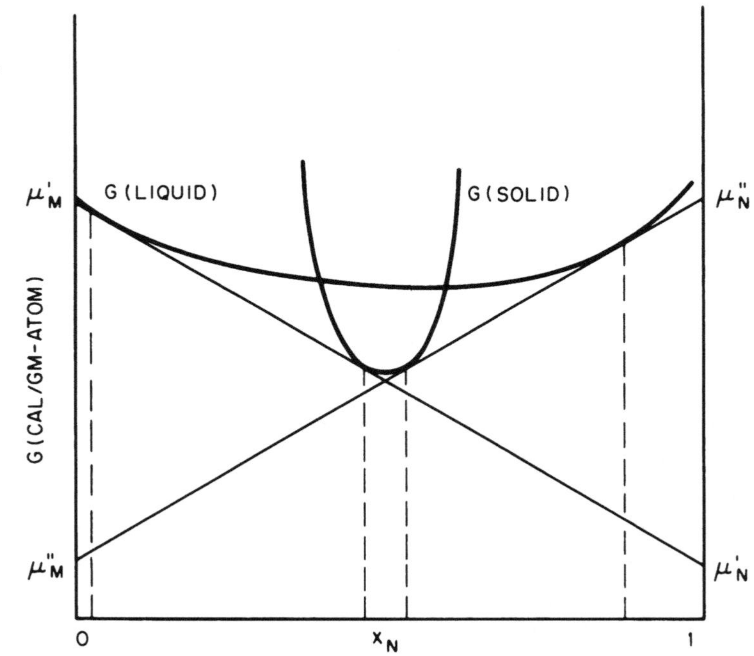

Fig. 8. Isothermal isobaric variation of the Gibbs free energy per gram-atom for a liquid phase and for a solid phase in the binary M–N system. x_N is the atom fraction of component N.

derivative of G, using the Gibbs–Duhem relation, is

$$(\partial G/\partial x_N)_{T,P} = \mu_N - \mu_M \tag{80}$$

The Gibbs free energy per gram-atomic weight at constant T and P is shown schematically in Figure 8 for a solid phase and a liquid phase as a function of x_N. The Gibbs free energy isotherms for any other possible phases are supposed to lie at higher values of G. From Eq. (80) the tangent to G for either phase is $\mu_N - \mu_M$ for the composition at the tangent point and so the intercepts of the tangent with the $x_N = 0$ and $x_N = 1$ give the individual values of μ_M and μ_N, respectively. Thus a straight line simultaneously tangent to two G isotherms represents equality of the chemical potentials for two different phases and equilibrium coexistence of these phases. Two such simultaneous tangents are shown in Figure 8. For either tangent, in the composition interval between the points of tangency G is lowest if the system consists of a mechanical mixture of liquid and solid. As the composition is varied within this interval the relative proportions of the two phases change, but all of the intensive properties (including the compositions of the coexisting phases) remain fixed. At the T and P in question the composition range over which the solid is the stable phase, the so-called homogeneity range, lies between the two tangent points on the G isotherm for the solid. Two important points emerge from Figure 8. The homogeneity range of the solid is expected to be nonzero in general. Even if it were zero at some T, it would not in general remain so over a range of T since the G isotherm for each phase shifts according to

$$-(\partial G/\partial T)_{x_N,P} = x_M\bar{S}_M + x_N\bar{S}_N = S \tag{81}$$

and the partial molar entropies $\bar{S}_M$ and $\bar{S}_N$ and the entropy per gram-atomic weight S are in general different for each phase. Second, the homogeneity limits of the solid phases at a given T and P are each associated with homogeneity limits for another condensed phase, here a liquid phase, and each depend upon the G isotherms for both coexisting phases. It is sometimes convenient to discuss the homogeneity limits of the solid in terms of two general factors, one being the Gibbs free energy isotherm of the solid phase and the other being the values of the chemical potentials at the homogeneity limits (μ_M', μ_M'', μ_N', and μ_N'' in Figure 8) for which experimental values are often available.[38]

It is worth mentioning that in the above discussion the solid phase has in effect been treated thermodynamically as a liquid. In an actual solid dislocations are usually present, except in special circumstances, and grain boundaries can be present. These defects do not reach unique equilibrium states in reasonable laboratory times but do affect the values of the thermodynamic functions. However, the variation in the values of the thermodynamic functions is generally small compared to the changes associated with phase transitions and can be neglected as long as the mean grain diameter is about 1 μm or larger and the dislocation density is about 10^6 cm^{-2} or smaller. A more sophisticated analysis is required in certain instances in which different solid phases with related structures and nearly equal Gibbs free energies occur at compositions close to one another, e.g., the Magneli phases in the Ti–O system.[39]

Figure 8 gives four points on the temperature–composition or T–x projection of the M–N phase diagram. The complete T–x projection is shown schematically in Figure 9. In the arguments presented here the T–x diagram holds for some fixed value of P. However, one can consider the system to be confined by rigid walls and P to be the total vapor pressure at any temperature and composition and obtain

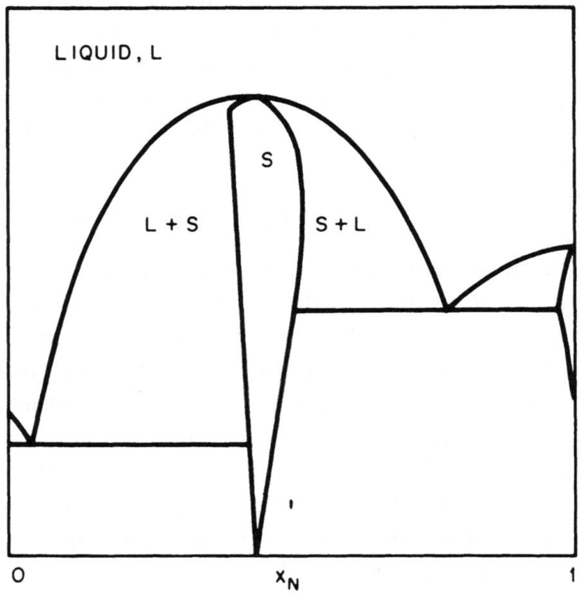

Fig. 9. Temperature–composition projection of the phase diagram for the M–N system, showing an intermediate solid compound.

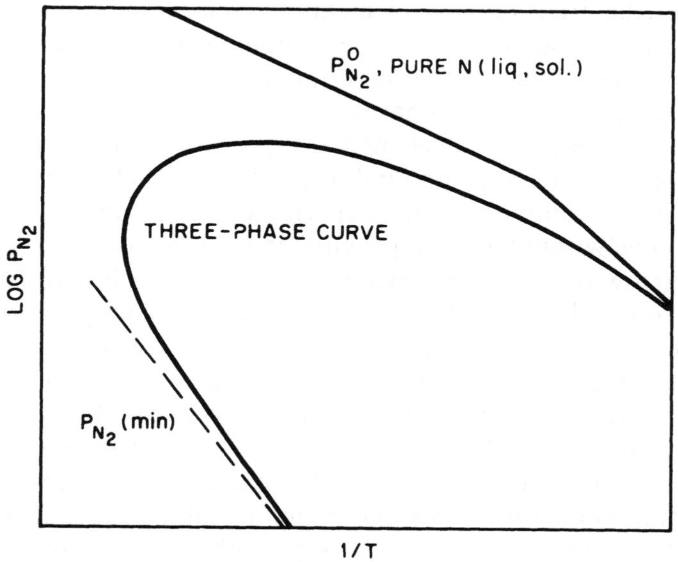

Fig. 10. The partial pressure projection of the phase diagram corresponding to the T–x projection shown in Fig. 9.

an equally valid T–x diagram which corresponds more often to experiment for semiconductor compounds. The two types of T–x projection are different in principle but essentially the same in practice unless the maximum pressure exceeds some tens of atmospheres. Corresponding to the T–x diagram, a P–T projection can also be drawn. However, it is more informative from a thermodynamic point of view to show the various partial pressures instead, with P taken as the total vapor pressure, as shown schematically in Figure 10 for $N_2(g)$. Since the chemical potential of N in a stable phase must decrease monotonically with decreasing x_N, the partial pressure of $N_2(g)$, P_{N_2}, at a given T is less at the M-rich homogeneity limit of the solid than at the N-rich limit, and less at the N-rich limit than for pure N. The P_{N_2}–$10^3/T$ loop in Figure 10 is the three-phase curve along which the solid, liquid, and vapor phase coexist. The area within the three-phase curve specifies those values of P_{N_2} and T for which the solid is a stable phase. For semiconductor compounds the upper and lower branches of the three-phase curve can be separated at a given T by several orders of magnitude or more in P_{N_2}, even when the width of the homogeneity range is very narrow and under 1 at.% wide. Corresponding to the P_{N_2} three-phase curve there will be analogous curves for the other possible elementary vapor species, such as N(g), N_4(g), M(g), etc., as well as for the compound vapor species such as MN(g). General thermodynamic arguments[38] show that the upper and lower legs of the three-phase curves are further apart when given as the partial pressure of an elementary species than when given as the partial pressure of a compound vapor species. The separation is smallest for a compound vapor species whose composition corresponds to one within the homogeneity range of the solid. The position of the narrow three-phase curve in terms of P_{MN} in Figure 10 would depend primarily upon the dissociation constant for MN(g). If the P_{MN} three-phase curve is well above that for any elementary species, the solid compound with the 50 at. % composition within its homogeneity range would appear to sublime congruently. If, on the other hand, the P_{N_2} three-phase curve, or others for elementary species containing only N atoms, occurs at the highest partial pressures, the solid compound would dissociate upon sublimation. The latter behavior is observed with the III–V phosphides and arsenides. The more complicated intermediate behavior in which the partial pressure of no elementary or compound species is predominant over the entire temperature and solid composition range is observed for the IV–VI and II–VI compounds.

Among semiconductor compounds the width of the homogeneity ranges observed varies considerably. For SnTe it is about 0.7 at. % wide at maximum and has been determined directly in terms of atomic percent.[40,41] For a number of IV–VI compounds it is of the order of 0.01 at. % wide and has been indirectly inferred from conductivity and Hall coefficient measurements on samples that were equilibrated at high temperature and then quenched.[42–44] For CdTe and ZnTe[46] it has been inferred from these same electrical measurements made at temperature and under controlled partial pressures. For the III–V compounds, with the exception of GaSb, the width of the homogeneity range appears to be so narrow as to be masked by the donor or acceptor behavior of foreign impurities. However, the P–T projections show the anticipated behavior.[47,48]

6.2. General Characteristics of the Model for a Semiconductor Compound

In the simple view a semiconductor compound M_aN_b is a thermodynamic phase characterized by a particular structure in which the M atoms and N atoms are essentially ordered upon their own sublattices. A perfect structure is possible only for the stoichiometric composition, e.g., $a = b = 1$ for the rocksalt and zincblende structures; $a = 1$, $b = 2$ for the fluorite structure. For other than the stoichiometric composition defects must necessarily be present, which we assume to be atomic point defects. As discussed in Chapter 6 of Volume 1, a general statistical mechanical argument leads to the conclusion that above 0°K even a crystal with the stoichiometric composition will be imperfect and will contain some atomic point defects. The same type of argument leads to the conclusion that as the composition is gradually changed the Gibbs free energy will also change smoothly, although possibly strongly. Therefore the general conclusions drawn from Figure 8 are valid. We assume that at 0°K the pure stoichiometric compound is a semiconductor with no localized levels in the energy gap and that native atomic point defects as well as certain impurities introduce donor or acceptor levels, as discussed in Section 1 and as shown in Table 1. A pure binary semiconductor compound in equilibrium with its own vapor has two degrees of freedom, by the Gibbs phase rule. If the temperature and composition, or more practically, the temperature and one partial pressure, are fixed, then all of the intensive properties of the compound are fixed at equilibrium. These include the concentrations of native atomic point defects of every kind and therefore the concentrations of donor and acceptor levels of each kind (excluding those associated with grain boundaries or dislocations). Control of the electrical properties then reduces to a precise control of composition for which the T–x and p–T projections given in Figures 9 and 10 are relevant. From the general considerations presented this far the homogeneity range of the compound may not even include the stoichiometric composition, though it is expected to be close to it. Furthermore the composition of the compound at the maximum melting point, which is the easiest to obtain by crystal growth since the melting is congruent, need not be the stoichiometric composition.

Samples of the semiconductor compound with compositions ranging from the M-rich homogeneity limit to the N-rich limit can be prepared and ordered by equilibration under increasing values of P_{N_2} or decreasing values of P_M. By reference to Table 1, the concentrations of donor levels will decrease relative to those of acceptor levels unless the place-exchange defect is predominant, in which case the reverse behavior will be seen. Thus the predominance of place-exchange defects can be readily distinguished for the predominance of the various vacancy–interstitial defect combinations. If the samples prepared as described go from n type to p type, the homogeneity range includes the stoichiometric composition; otherwise it lies entirely on one side at the temperature of equilibration. In practice these simple considerations can be complicated by the presence of foreign impurities, or, if the electrical measurements are made on quenched samples, by the inability to freeze in the high-temperature compositions. The latter situation results in an internal micro-precipitate and can usually be detected by a partial pressure or temperature independence of the electron and hole concentrations.

6.3. Chemical Potentials for a Binary Semiconductor Compound

6.3.1. Basic Equations for a Pure Semiconductor

In the simplest model of a semiconductor compound the electronic energy band picture and the association of "impurity" levels with atomic point defects given in Section 1 is adopted and the general assumptions made in Section 6.2 are accepted. In addition, it is assumed that each defect is distributed at random on the sites of an appropriate sublattice, the excess Gibbs free energy of the crystal is increased by a constant amount for each type of un-ionized defect, and the positions of the band levels and "impurity" levels are independent of the defect concentrations. The Gibbs free energy of the crystal is written as an *excess* Gibbs free energy depending linearly upon the number of defects of each kind minus the product of the temperature and the configurational entropy. This is illustrated in Chapter 6 of Volume 1 for a case in which electrons and holes are not considered, and is considered elsewhere for the semiconductor case.[49] The Gibbs free energy is then minimized subject to the constraints of constant T and P, constant numbers of atoms of each kind, a constant ratio of sites of each type, and the electroneutrality condition.

Here we give only the results for a nondegenerate semiconductor compound with the stoichiometric composition MN and with singly ionizable M-vacancy acceptors and N-vacancy donors. The concentrations of M atoms and of N atoms are given by

$$C_{\mathrm{M}} = C_S - V_{\mathrm{M}}^{\times} - V_{\mathrm{M}}^{-} \tag{82}$$

$$C_{\mathrm{N}} = C_S - V_{\mathrm{N}}^{\times} - V_{\mathrm{N}}^{+} \tag{83}$$

where C_S is the concentration of sites in each sublattice, $V_{\mathrm{M}}^{\times}$ and V_{M}^{-} are the concentrations of, respectively, un-ionized and ionized M vacancies, and $V_{\mathrm{N}}^{\times}$ and V_{N}^{+} are the concentrations of, respectively, un-ionized and ionized N vacancies. The charge on the M atoms is supposed to be $+\delta e$, so that, following the discussion in Section 1.2.3, the electroneutrality condition is

$$C_{\mathrm{M}}\delta + V_{\mathrm{M}}{}^{\times}\delta + V_{\mathrm{M}}{}^{-}(\delta - 1) + \sum_{\mathrm{vb}} p_R$$
$$= C_{\mathrm{N}}\delta + V_{\mathrm{N}}{}^{\times}\delta + V_{\mathrm{N}}{}^{+}(\delta - 1) + \sum_{\mathrm{cb}} n_S \tag{84}$$

Similar to the definitions in Section 2.1, p_R is the concentration of holes in the valence band at energy E_R and the sum over p_R is for all energy levels of the valence band, and n_S is the concentration of electrons in the conduction band at energy E_S and the sum over n_S is for all energy levels of the conduction band. Using Eqs. (82) and (83) in (84), the electroneutrality condition simplifies to

$$V_{\mathrm{N}}{}^{+} + \sum_{\mathrm{vb}} p_R = V_{\mathrm{M}}{}^{-} + \sum_{\mathrm{cb}} n_R \tag{85}$$

Thus the sign of the charge on the M atoms and N atoms must be known in order to associate donor or acceptor behavior with the associated vacancies, but the magnitude of the charge does not enter equilibrium equations. The equations resulting from the minimization of G are

$$\mu_{\mathrm{M}} + \mu_{\mathrm{N}} = g_0 - kT\ln(C_S{}^2/C_{\mathrm{M}}C_{\mathrm{N}}) \tag{86}$$

$$\mu_{\mathrm{M}} = -\,\varepsilon_{\mathrm{M}} + kT\ln(2C_{\mathrm{M}}/V_{\mathrm{M}}{}^{\times}) \tag{87}$$

$$\mu_{\mathrm{M}} = -\varepsilon_{\mathrm{M}} + (E_{\mathrm{F}} - E_A) + kT\ln(C_{\mathrm{M}}/V_{\mathrm{M}}{}^{-}) \tag{88}$$

$$\mu_{\mathrm{N}} = -\varepsilon_{\mathrm{N}} - kT\ln(2C_{\mathrm{N}}/V_{\mathrm{N}}{}^{\times}) \tag{89}$$

$$\mu_{\mathrm{N}} = -\varepsilon_{\mathrm{N}} - (E_{\mathrm{F}} - E_D) + kT\ln(C_{\mathrm{N}}/V_{\mathrm{N}}{}^{+}) \tag{90}$$

$$E_{\mathrm{F}} = E_S - kT\ln[(N_S - n_s)/n_S] \tag{91}$$

$$E_{\mathrm{F}} = E_R - kT\ln[(N_R - p_R)/p_R] \tag{92}$$

The assumption of nondegeneracy has resulted in certain terms dropping out of Eq. (86), identical terms dropping out of Eqs. (87) and (88) for μ_{M}, and identical terms dropping out of Eqs. (89) and (90) for μ_{N}.[49] Equations (91) and (92) are unaffected.

The equations necessary to describe electron–hole equilibria and given in Section 2 are contained as a subset of Eqs. (86)–(92). Equations (91) and (92) are identical in form to Eqs. (8) and (9) and, following the arguments in Section 2.1, the total concentration of electrons in the conduction band n or of holes in the valence band p can be obtained.

For the nondegenerate case these are given by Eqs. (12) and (14),

$$n = N_C e^{(E_F - E_C)/kT}, \qquad p = N_V e^{(E_V - E_F)/kT}$$

By equating Eqs. (87) and (88), Eq. (32) is obtained with the acceptor defect identified here as an M vacancy. Similarly by equating Eqs. (89) and (90), Eq. (31) is obtained with the donor defect being an N vacancy. However, what emerges here is that the composition-independent quantities E_A, E_D, ε_M, ε_N, and g_0 are, strictly speaking, excess Gibbs free energies and not energies. Thus the so-called energy gap $E_G = E_C - E_V$ which appears in Eq. (19) for the *np* product of a nondegenerate semiconductor is the excess Gibbs free energy for the creation of an electron–hole pair. The derivative of E_G/T with respect to $1/T$ is then the enthalpy of creation of an electron hole pair.

In Eqs. (86)–(90) g_0 is the Gibbs free energy per MN atom pair in the perfect crystal; ε_M is the increment in the excess Gibbs free energy of the crystal per un-ionized M vacancy; $\varepsilon_M + E_A$, that per ionized M vacancy. Analogously, ε_N is the increment in the Gibbs free energy of the crystal per un-ionized N vacancy; $\varepsilon_N - E_D$, that per ionized N vacancy. These quantities all depend upon temperature and pressure in the same way as shown by Eq. (66) for a foreign donor. The reference levels from which these excess Gibbs free energies are measured is arbitrary and is set by the reference levels chosen for the chemical potentials μ_M and μ_N. Thus one might choose the chemical potential of M(g, 1 atm) and that of N_2(g, 1 atm) or alternatively that of the isolated atoms in their ground states.

6.3.2. The Integral Thermodynamic Properties

The sum of the chemical potentials given by Eq. (86) is obtained by minimizing the Gibbs free energy of the crystal with respect to the number of lattice sites. It can be seen to be a weak function of composition and equal to g_0 and independent of composition for small vacancy concentrations to within terms of the order of $(V_N^{\times} + V_N^{+})/C_S$, etc. If a composition parameter δ is defined as

$$x_N = \tfrac{1}{2} + \delta \tag{93}$$

then the Gibbs free energy per gram-atomic weight of the compound defined by Eq. (79) can be written as

$$G = \tfrac{1}{2}(\mu_M + \mu_N) + \delta(\mu_N - \mu_M) \tag{94}$$

For small departures from stoichiometry δ is small, the last term in Eq. (94) can be shown to be negligible, and therefore G and the other integral thermodynamic functions are approximately independent of composition. Equations (86) and (94) can be combined and the chemical potentials written in terms of the partial pressures of an ideal vapor phase to give

$$2G = g_0 = RT \ln p_M p_{N_2}^{1/2} \tag{95}$$

Here G can be identified as the Gibbs free energy of formation per gramatomic weight of the semiconductor compound from M(*g*, 1 atm) and N_2(g, 1 atm) each at the temperature in question. The graphical implication of the above is that the Gibbs free energy isotherm for the compound, shown schematically in Figure 8, has a very large curvature. Very small changes in composition result in very small changes in G but extremely large changes in μ_M and μ_N. Equation (95) also has a bearing on the p–T projection shown in Figure 10. For the situation considered there the phase coexisting with the compound at its M-rich homogeneity limit approaches pure M at low temperatures and the corresponding partial pressure of M(g) approaches that over pure M, $p_M{}^\circ$. If $p_M{}^\circ$ is inserted for p_M in Eq. (95), the value obtained for p_{N_2} is the minimum value possible over the compound at a particular temperature and is indicated by the dashed line in Figure 10.

The result that the integral molar thermodynamic functions are independent of composition for small departures from stoichiometry is a general one and not dependent upon the details of the specific model considered here.

6.3.3. Chemical Potentials in Terms of Net Carrier Concentration

For narrow-homogeneity-range semiconductor compounds the composition is in general inaccessible to direct measurement, whereas the net electron concentration $n - p$ generally is. Fortunately, the chemical potentials can be expressed entirely in terms of this latter variable and certain material properties for the case of a nondegenerate semiconductor with small departure from stoichiometry. One of the material properties required is the Schottky constant for ionized vacancies, k_S. If Eqs. (88) and (90) are added, equated to Eq. (86), and rearranged, one can obtain

$$V_M{}^- V_N{}^+ = C_M C_N \exp[-(g_0 + \varepsilon_M + \varepsilon_N + E_A - E_D)/kT] \tag{96}$$

For small departures from stoichiometry C_M and C_N can each be replaced by C_S, which in turn is essentially independent of composition. The right hand side of Eq. (96) is then called the Schottky constant for ionized vacancies, i.e.,

$$V_M^- V_N^+ = k_S \tag{97}$$

Substituting this into the electroneutrality condition given by Eq. (85), V_N^+ can be obtained in terms of $n - p$ as

$$\ln(V_N^+/C_S) = \tfrac{1}{2}\ln k_S + \sinh^{-1}[(n-p)/2k_S^{1/2}] \tag{98}$$

Using Eqs. (97) and (98) and the Fermi level given by Eq. (22), the chemical potentials given by Eqs. (88) and (90) can be rewritten as

$$\begin{aligned}\mu_M = {} & -\varepsilon_M + (E_F{}^i - E_A) - \tfrac{1}{2}kT\ln k_S \\ & + kT\sinh^{-1}[(n-p)/2n_i] + kT\sinh^{-1}[(n-p)/2k_S^{1/2}]\end{aligned} \tag{99}$$

$$\begin{aligned}\mu_N = {} & -\varepsilon_N + (E_D - E_F{}^i) - \tfrac{1}{2}\,kT\ln k_S \\ & - kT\sinh^{-1}[(n-p)/2n_i] - kT\sinh^{-1}[(n-p)/2k_S^{1/2}]\end{aligned} \tag{100}$$

The last two terms on the right hand side of Eqs. (99) and (100) are the composition-dependent terms. The first three terms in each equation give the values of the chemical potential for the intrinsic compound, $n - p = 0$. In principle the dependence of the chemical potentials upon the departure from stoichiometry is more complicated. However, when the donor and acceptor levels are shallow essentially all the vacancies are ionized when the semiconductor is nondegenerate and $n - p$ in Eqs. (99) and (100) can be replaced by $C_M - C_N$ or, equivalently, by $2C_S(x_N - \tfrac{1}{2})$.

Since n_i and $k_S^{1/2}$ are generally small compared to the 10^{22}–$10^{23}\ \mathrm{cm}^{-3}$ value of C_S, the chemical potentials are strong functions of composition. Once $|x_N - \tfrac{1}{2}|$ is about three times as large as n_i or $k_S^{1/2}$ every subsequent increase by a factor of ten contributes an increment of 2.3 kT energy units/atom to the chemical potentials.

The difference in the chemical potentials at the intrinsic composition can be obtained from Eqs. (99) and (100). Using the definition of k_S from Eqs. (96) and (97), this difference is

$$\mu_N(\mathrm{int}) - \mu_M(\mathrm{int}) = \varepsilon_M - \varepsilon_N - (2E_F{}^i - E_D - E_A) \tag{101}$$

It can be seen to depend upon two factors: (1) the difference in the excess Gibbs free energy of creation of the un-ionized vacancies and (2) an electronic factor that is zero if the instrinsic Fermi level $E_F{}^i$

coincides with the average position of the donor and acceptor levels associated with the vacancies. This difference is the slope of the G isotherm of the compound at the intrinsic composition by Eq. (80) and is important in determining the position of the homogeneity range relative to the intrinsic composition.[38]

6.3.4. Complete Solution

The pure binary semiconductor compound in equilibrium with its vapor has two degrees of freedom, so that a specification of T and p_{N_2} should be sufficient to fix all of the intensive properties. This means the site fractions $V_M{}^\times/C_S$, $V_M{}^-/C_S$, $V_N{}^\times/C_S$, $V_N{}^+/C_S$, n/N_C, and p/N_V [the effective densities of band states N_C and N_V are defined in Eqs. (13) and (15)], the Fermi level E_F, and the total pressure are all functions of T and p_{N_2}. An exact analytical solution can be obtained if a nondegenerate semiconductor and small departures from stoichiometry are assumed. Although an approximate, piecewise solution has been given in Chapter 6 of Volume 1 and must be resorted to when foreign donors and acceptors are present, we believe the exact solution to be of interest. Under the assumption of small departures from stoichiometry C_S can replace C_M or C_N whenever they occur in Eqs. (86)–(90). For a given T and p_{N_2} Eq. (86) can be used to obtain μ_M and hence p_M. The partial pressures of other vapor species are then all fixed through gas-phase equilibrium constants not considered explicitly here and so, therefore, is the total pressure which appears implicitly in the various excess Gibbs free energies such as g_0, etc.

First a Schottky constant for un-ionized vacancies, k_{S1}, is defined. Equating the sum of Eqs. (87) and (89) to Eq. (86) and rearranging gives

$$V_M{}^\times V_N{}^\times = 4C_S{}^2 \exp[-(g_0 + \varepsilon_M + \varepsilon_N)/kT] \equiv k_{S1} \tag{102}$$

This Schottky constant is related to that for ionized vacancies by

$$k_{S1} = 4k_S \exp[(E_D - E_A)/kT] \tag{103}$$

While the concentrations of ionized vacancies are necessarily equal in the intrinsic compound by Eq. (85) for electroneutrality, from Eqs. (87) and (89) those for the un-ionized vacancies are not equal unless an electronic displacement factor d is unity, where

$$d = \exp[(E_A + E_D - 2E_F{}^i)/2kT] \tag{104}$$

The concentrations of the vacancies and of electrons and holes are then obtainable from Eqs. (86)–(92) in terms of two dimensionless variables

$$z \equiv \exp[(E_F - E_F{}^i)/kT] \quad (105)$$

and

$$Z \equiv \exp\{[\mu_N - \mu_N(\text{int})]/kT\} = [p_{N_2}/p_{N_2}(\text{int})]^{1/2} \quad (106)$$

which are both equal to unity for the intrinsic compound. The equations are

$$n = n_i z \quad (107)$$

$$p = n_i z^{-1} \quad (108)$$

$$V_M{}^\times = k_{S1}^{1/2}(dZ) \quad (109)$$

$$V_N{}^\times = k_{S1}^{1/2}(dZ)^{-1} \quad (110)$$

$$V_M{}^- = k_S{}^{1/2} zZ \quad (111)$$

$$V_N{}^+ = k_S{}^{1/2}(zZ)^{-1} \quad (112)$$

Equations (107), (108), (111), and (112) are substituted into Eq. (85), which, when solved for z, gives

$$z = \left(\frac{1 + \alpha Z^{-1}}{1 + \alpha Z}\right)^{1/2} \quad (113)$$

where α is a temperature-dependent material property defined by

$$\alpha = k_S{}^{1/2}/n_i \quad (114)$$

For compounds with a small energy gap n_i is expected to be large compared to $k_S{}^{1/2}$ and α is less than unity. The intrinsic electronic disorder then is greater than the intrinsic atomic disorder. For insulators the opposite is true and α is much greater than unity.

Equations (107)–(113) represent the desired complete solution. The essential difference in the behavior of materials with small energy gaps and those with large energy gaps resides in the behavior of z near $z = 1$. If m is a positive number and $\alpha = 10^{-m}$, then $z = 1$ to within $\pm 5\%$ over a range of log Z centered at $Z = 1$ and $2(m - 1)$ decades wide. On the other hand, if $\alpha = 10^m$, then z is proportional to Z^{-1} over the same range in log Z. In either case z is proportional to $Z^{-1/2}$ outside this central range.

6.3.5. Chemical Potentials of Impurities

If singly ionizable foreign donors and acceptors are assumed to be present, the electroneutrality condition is changed to

$$n - p = V_N^+ - V_M^- + D^+ - A^- \tag{115}$$

Equations (86)–(92) remain unchanged, and four new equations are added. If the foreign donors occupy interstitial sites, the chemical potential of the donor atom is given by the same expressions as hold for foreign donors in an elemental semiconductor, Eqs. (64) and (65). [In Eqs. (64) and (65) the total concentration of substitutional sites is represented by S, in this section by C_S.] If the foreign acceptors occupy interstitial sites, then Eqs. (68) and (69) apply if S is replaced by αC_S. However, if the foreign donor occupies a site in the M sublattice in the compound MN, then the chemical potential of the donor is

$$\begin{aligned} \mu_{D(M)} &= \mu_M + kT\ln(D^\times/C_S) + \mu_D^\circ \\ &= \mu_M + kT\ln(D^+/C_S) + (E_F - E_{D1}) + \mu_D^\circ \end{aligned} \tag{116}$$

The appearance of μ_M is independent of the assumption that vacancies are the predominant native defects in pure MN. If the donor atom occupies a site in the N sublattice, μ_M is replaced by μ_N. Since μ_M and μ_N are strong functions of composition in general, the appearance of these terms is significant. As an example, the homogeneity range for GaAs is apparently quite narrow, not yet being detected above the accidental background of foreign atoms electrically active as donors or acceptors. Nevertheless, the chemical potential of As and of Ga vary strongly across this homogeneity range, as indicated by the experimental p–T projection.[48] Therefore the chemical potential of a substitutional donor for a fixed value of $D^\times$ can be changed greatly by changing the value of P_{Ga}.

For substitutional foreign acceptors Eqs. (68) and (69) hold if the chemical potential of the host atom replaced is added to the right hand side of each equation.

6.4. Variations of the Basic Model

The detailed discussion in Sections 6.3.1–6.3.5 is for a semiconductor compound whose stoichiometric formula is MN and whose predominant native point defects are singly ionizable M-vacancy acceptors and N vacancy donors. Singly ionizable M-vacancy acceptors and singly ionizable M-interstitial donors, or singly ionizable N-vacancy donors and singly ionizable N-interstitial acceptors, are other simple choices for the predominant defects that could have been made. These different choices yield equations for the chemical potentials that are different and which can be used to distinguish among the defect choices when the departures from stoichiometry are large.[51] However, when this is not the case, i.e., when the defect concentration is 1% or less of the site concentration, the differences in the expressions for the chemical potentials are too small to detect and the equations for the chemical potentials in terms of $n - p$ are of the same functional form as Eqs. (99) and (100).[52] Thus for narrow-homogeneity-range compounds when one relies upon the experimental dependence of $n - p$ upon T and a partial pressure, the predominant defect is not specifically identified. Rather, in the simplest case one obtains a broader characterization typified by the statement: The predominant native defect in M-saturated MN is a singly ionizable acceptor. The acceptor ionization energy may be determined and the enthalpy and excess entropy of creation of the acceptor defect can be established. The equations for the chemical potentials are changed if the place-exchange defects are predominant, as can be seen in Chapter 6 of Volume 1.

If the point defects are considered to be doubly ionizable, an additional equation is required for each defect which expresses the chemical potential of the host atom involved in terms of the concentration of the doubly ionized defect.[38,53] The form of the equation can be inferred from Eqs. (38) and (87)–(90).

Some care must be exercised when compounds with stoichiometric formulas other than one to one are considered.[54] Thus in Bi_2Te_3, with place-exchange defects dominant, only three acceptor Bi atoms on Te sites are obtained for every five excess Bi atoms. Two of the five extra Bi atoms must reside on normal sites unless Bi vacancies are to be also present.

6.5. Refinements

Three very specific assumptions characterize the models for a semiconductor discussed above and in Section 5: (1) Each type of atomic point defect is distributed at random on the sites of an appropriate sublattice, (2) the excess Gibbs free energy of the semiconductor increases linearly with the number of un-ionized defects of each kind, and (3) the positions of the impurity energy levels and band levels are independent of the concentrations of atomic defects, electrons, or holes. These restrictive assumptions have been lifted to various degrees in a number of analyses. One relatively simple approximate approach, involving pairing of ionized donors and acceptors present in equal concentrations, has already been discussed in Section 5.2.

Other analyses have been made which retain assumption 1 but reject 3.

Harvey[55,56] has adapted the Debye–Hückel theory to obtain shifts of the electronic energy levels proportional to the potential ψ', given by

$$\psi' = z_i e/\varepsilon\lambda(1 + a\lambda^{-1}) \tag{117}$$

where a is the distance of closest approach of two oppositely charged defects, ε is the dielectric constant, and λ is the Debye length given by

$$\lambda = (\varepsilon kT/4\pi e^2 \sum n_j z_j^2)^{1/2} \tag{118}$$

The sum in Eq. (118) is over all charged species, n_j and $z_j e$ being the concentration and charge of species j. The positions of the conduction-band edge E_C, the valence-band edge E_V, a donor level E_D, and an acceptor level E_A are then given by

$$E_C = E_C^0 - \tfrac{1}{2}e\psi' \tag{119a}$$

$$E_V = E_V^0 + \tfrac{1}{2}e\psi' \tag{119b}$$

$$E_D = E_D^0 + \tfrac{1}{2}e\psi' \tag{119c}$$

$$E_A = E_A^0 - \tfrac{1}{2}e\psi' \tag{119d}$$

The quantities with zero superscripts are the positions of the energy levels for vanishingly small defect concentrations. With increasing defect concentrations the conduction-band edge and the acceptor level are moved downward relative to some fixed reference level, while the valence-band edge and the donor level are moved upward, leading to a decrease in the energy gap and the donor and acceptor ionization energies.

In a more detailed analysis Morgan[7] has assumed a random distribution of the ionized impurities and calculated the potential energy of an electron due to the additive, screened-Coulomb potentials of all the impurities. The impurity levels, which are located at sharply defined energies at low defect concentrations, at higher defect concentrations are not only shifted, as indicated in Harvey's treatment, but also spread out in energy. The density of states in each impurity band is in general skewed, but when the concentrations of ionized donors and acceptors are approximately equal it is given approximately by a Gaussian distribution

$$N_I(E) = C_I(2\pi)^{-1/2}\sigma^{-1}\exp(-E^2/2\sigma^2) \quad (120)$$

where C_I is the total concentration of ionized impurities of a given species and where

$$\sigma^2 = (2\pi e^4/\varepsilon^2)\lambda \sum n_j z_j^2 e^{-2a/\lambda} \quad (121)$$

In interpreting their results on the solubility of Zn in GaAs at 1000°C Panish and Casey[57] have taken an alternative approach. At low concentrations the chemical potential of the Zn acceptor, which substitutes for Ga, is taken as the analog of Eq. (116) for a donor, i.e.,

$$\mu_{\mathrm{Zn(Ga)}} = \mu_{\mathrm{Ga}} + kT\ln(\mathrm{Zn}^-/C_S) - (E_\mathrm{F} + E_{\mathrm{Zn}}) + \mu^\circ \quad (122)$$

where E_{Zn} is the position of the Zn-acceptor level. For Zn concentrations below about $5 \times 10^{18}\,\mathrm{cm}^{-3}$ the composition dependence of $\mu_{\mathrm{Zn(Ga)}}$ is calculated by using Eq. (122) and by assuming a sharply defined, composition-independent value for E_{Zn} and the valence band edge. For Zn concentrations above $5 \times 10^{18}\,\mathrm{cm}^{-3}$ Eq. (122) is still used but E_{Zn} is set equal to zero and the Fermi level is calculated assuming an exponential density of states at the valence-band edge.

It appears to this author that the shifts of the electronic energy levels at high concentrations of ionized donors and/or acceptors are fairly well understood, so that electron–hole equilibria can be discussed with some confidence in highly doped semiconductors. However, it remains to be demonstrated that the chemical potentials necessary for describing equilibria with external phases does indeed remain unchanged in functional form.

References

1. A. E. Attard and L. V. Azaroff, *J. Appl. Phys.* **34**, 774 (1963).
2. C. B. Satterthwaite and R. W. Ure, Jr., *Phys. Rev.* **108**, 1164 (1957).
3. R. F. Brebrick, *J. Phys. Chem. Solids* **30**, 719 (1969).
4. H. C. Casey and F. A. Trumbore, *Mat. Sci. & Eng.* **6**, 69 (1970).
5. J. J. Hopfield, D. G. Thomas, and M. Gershenzon, *Phys. Rev. Letters* **10**, 162 (1963).
6. D. G. Thomas, *Brit. J. Appl. Phys.* **2**, 637 (1969).
7. T. N. Morgan, *Phys. Rev.* **139**, A343 (1965).
8. D. Redfield, *Phys. Rev.* **130**, 916 (1963).
9. E. O. Kane, *Phys. Rev.* **131**, 79 (1963).

10. B. I. Halperin and M. Lax, *Phys. Rev.* **148**, 77 (1966).
11. D. G. Thomas, J. J. Hopfield, and C. J. Frosch, *Phys. Rev. Letters* **15**, 857 (1965).
12. F. A. Trumbore, M. Gershenzon, and D. G. Thomas, *Appl. Phys. Letters* **9**, 4 (1966).
13. W. C. Dash, *J. Appl. Phys.* **29**, 705 (1958).
14. W. D. Lawson and S. Nielsen, *Preparation of Single Crystals*, Butterworth, London (1958).
15. H. G. Van Bueren, *Imperfections in Crystals*, North-Holland, Amsterdam (1961).
16. J. J. Dugga, *J. Appl. Phys.* **33**, 169 (1962).
17. M. S. Abrahams, L. R. Weisberg, C. J. Buiocchi, and J. Blanc, *J. Mat. Sci.* **4**, 223 (1969).
18. P. J. Dean, *Trans. Met. Soc. AIME* **242**, 384 (1968).
19. G. Dresselhaus, A. F. Kip, and C. Kittel, *Phys. Rev.* **95**, 568 (1954).
20. J. S. Blakemore, *Semiconductor Statistics*, Pergamon Press (1962).
21. R. A. Smith, *Semiconductors*, Cambridge University Press (1959).
22. C. Herring, *Bell System Tech. J.* **34**, 237 (1955).
23. R. S. Allgaier and W. W. Scanlon, *Phys. Rev.* **111**, 1029 (1958).
24. W. W. Scanlon, *Phys. Rev.* **92**, 1573 (1953).
25. H. Ehrenreich, *Phys. Rev.* **120**, 1951 (1963).
26. M. R. Lorenz, *Trans. Met. Soc. AIME* **245**, 539 (1969).
27. F. Williams, *Phys. Stat. Sol.* **25**, 493 (1968).
28. J. I. Pankove, *Optical Processes in Semiconductors*, Prentice-Hall (1971).
29. H. J. Hrostowski, in *Semiconductors* (N. B. Hannay, ed.), Reinhold (1959).
30. D. Redfield and M. A. Afromowitz, *Appl. Phys. Letters* **11**, 138 (1967).
31. D. G. Thomas, M. Gershenzon, and F. A. Trubmore, *Phys. Rev.* **133**, A269 (1964).
32. H. Reiss, C. S. Fuller, and F. J. Morin, *Bell System Tech. J.* **35**, 535 (1956).
33. F. A. Trumbore, *Bell System Tech. J.* **39**, 205 (1960).
34. H. G. Van Bueren, *Imperfections in Crystals*, p. 636, North-Holland, Amsterdam (1961).
35. H. Reiss, *J. Chem. Phys.* **25**, 400 (1956).
36. J. S. Prener, *J. Chem. Phys.* **25**, 1294 (1956).
27. A. B. Lidiard, *Phys. Rev.* **101**, 1427 (1956).
38. R. F. Brebrick, *Progress in Solid State Chemistry* (H. Reiss, ed.), Vol. 3, Pergamon Press (1966).
39. L. A. Bursill and B. G. Hyde, *Progress in Solid State Chemistry* (H. Reiss and J. O. McCaldin, eds.), Vol. 7, Pergamon Press (1972).
40. R. F. Brebrick, *J. Phys. Chem. Solids* **24**, 27 (1963).
41. R. F. Brebrick and A. J. Strauss, *J. Chem. Phys.* **41**, 197 (1964).
42. R. F. Brebrick and E. Gubner, *J. Chem. Phys.* **36**, 1283 (1962).
43. A. R. Calawa, T. C. Harman, M. Finn, and P. Youtz, *Trans. Met. Soc. AIME* **242**, 374 (1968).
44. J. Bloem and F. A. Kroger, *Z. Phys. Chem.* (*Frankfurt*) **7**, 1 (1956).
45. F. T. J. Smith, *Met. Trans.* **1**, 617 (1970).
46. F. T. J. Smith, *Solid State Commun.* **7**, 1757 (1969).
47. J. van den Boomgaard and K. Schol, *Phillips Res. Rep.* **12**, 127 (1957).
48. J. R. Arthur, *J. Phys. Chem. Solids* **28**, 2257 (1967).
49. R. F. Brebrick, *J. Solid State Chem.* **1**, 88 (1969).

50. R. R. Hasiguti (ed.), *Lattice Defects in Semiconductors*, University of Tokyo Press and The Pennsylvania State University Press (1968).
51. G. G. Libowitz and J. B. Lightstone, *J. Phys. Chem. Solids* **28**, 1145 (1967).
52. R. F. Brebrick, *J. Appl. Phys.* (suppl.) **33**, 422 (1962).
53. F. A. Kroger, *Chemistry of Imperfect Crystals*, North-Holland, Amsterdam (1964).
54. F. A. Kroger, *J. Phys. Chem. Solids* **7**, 277 (1958).
55. W. W. Harvey, *J. Phys. Chem.* **71**, 2031 (1967).
56. W. W. Harvey, *Phys. Rev.* **123**, 1666 (1961).
57. M. B. Panish and H. C. Casey, Jr., *J. Phys. Chem. Solids* **28**, 1673 (1967).

6

The Imperfect Solid—Magnetic Properties

E. M. Gyorgy
Bell Laboratories
Murray Hill, New Jersey

1. Introduction

The impact of imperfections on the properties of magnetic materials is of great commercial as well as scientific importance. The vast literature in this field makes a complete survey impractical, and so the foremost problem in attempting a review is the selections of subjects to be included. Because of the natural division of materials into two groups—(1) technical magnetic materials, which are almost invariably polycrystalline, and (2) scientifically interesting materials, which are generally single crystals—adequate review must refer to both groups. Therefore the author has selected for discussion some selected topics which are important with respect to each group of materials.

The theory of magnetism is covered in a variety of textbooks and a very convenient survey of ferromagnetic domains is given in Ref. 1. However, in order to make this chapter somewhat self-contained, a few definitions and calculations pertinent to the subsequent discussion will be very briefly sketched in this section.

Figure 1 shows a schematic magnetization curve which gives the response of a multidomain toroid of magnetic material to a circumferentially applied field. If the material is in the demagnetized state ($M = 0, H = 0$), the initial slope of M versus H is the susceptibility χ.* With a further increase in H, technical saturation is reached where the

* The usual relations between M, H, and B are used; $M = \chi H$, $B = \mu H = (1 + 4\pi\chi)H$.

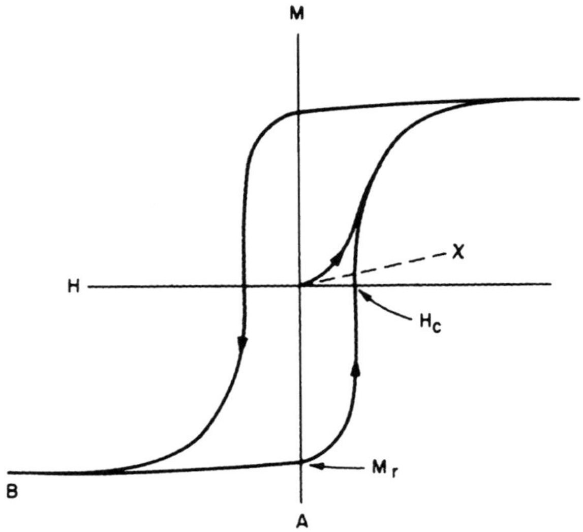

Fig. 1. A schematic M–H (hysteresis) loop.

domains have disappeared and all of the magnetization is parallel to H. A cyclic variation of H takes M around the M–H (hysteresis) loop as shown. The field that produces zero M is the coercive force H_c. The magnetization in zero field following technical saturation is the remanent magnetization M_r.

In Figure 2 we show the magnetization vector $\mathbf{M}$ in a domain wall. The wall in the x–z plane, centered at $y = 0$, separates a domain with $\mathbf{M}$ up on the left from a domain with $\mathbf{M}$ down in the right. The material is taken to have a uniaxial magnetic anisotropy, with the easy direction of magnetization parallel to the z axis. In the domain wall the angle θ gradually changes from zero to π. In order that no free poles are formed on the wall ($\nabla \cdot \mathbf{M} = 0$), the angle ϕ in a static wall is zero. In the wall $\mathbf{M}$ is pointed away from the easy direction, so that there is some anisotropy energy connected with the wall (Section 2.3). This energy term favors a wall of zero thickness. However, in the wall, neighboring spins are not parallel and so exchange energy (Section 2.2) is contained in the wall. The exchange energy favors a wall of infinite thickness. Minimizing both energies leads to a wall thickness of approximately $3.5(A/K_u)^{1/2}$ [1] Here A is the exchange constant [see Eq. (1)] and K_u the uniaxial anisotropy constant. Typically the wall thickness is of the order of 10^{-5} cm. Associated with the wall is a wall energy given by $4(K_u A)^{1/2}$ which generally is between 0.1 and 1.0 ergs/cm^3. The wall energy is denoted by σ_w.

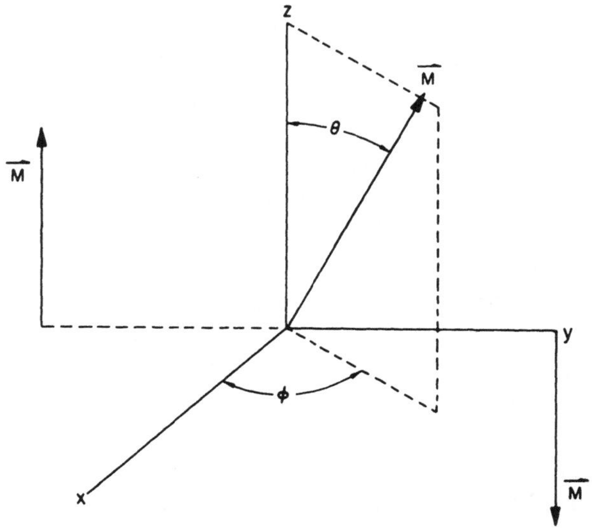

Fig. 2. The orientation of the magnetization vector **M** in a domain wall. The wall in the x–z plane separates a domain with **M** up on the left from a domain with **M** down on the right.

In the following sections on intrinsic properties, impurities, and structure-sensitive effects some of the classifications are of necessity somewhat vague. As will become apparent, some of the intrinsic properties are quite sensitive to impurities. Also the definition of an impurity is somewhat vague since much of the work was performed with ultrapure materials purposely doped. In general, however, the subjects considered under impurities are in large part studied using crystals and the primary emphasis has been on chemical composition. The properties listed under structure-sensitive effects are, almost without exception, investigated using polycrystalline samples and the polycrystalline nature of the materials plays a dominant role in determining the observed properties.

2. Intrinsic Properties

2.1. Saturation Magnetization

The saturation magnetization M_s is the net sum of the individual atomic magnetic moments per unit volume. The atomic moments arise both from the intrinsic moment of any unpaired electrons and from a contribution due to the orbital angular momentum. The unit

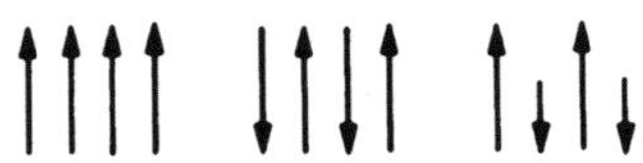

Fig. 3. Possible ordered spin arrangements; (A) ferromagnetic, (B) antiferromagnetic, and (C) ferrimagnetic.

of the spin moment is the Bohr magneton μ_B. The saturation moment in terms of the atomic moments is given by $M_s = n_B \mu_B N D (M_w)^{-1}$, where n_B is the number of Bohr magnetons per molecule, M_w the molecular weight, N Avogadro's number, and D the density. Three distinct spin arrangements are possible (Figure 3). In Figure 3(A) the arrangement is ferromagnetic (all spin parallel), in Figure 3(B) the arrangement is antiferromagnetic (spins of equal size and antiparallel), and in Figure 3(C) the ferrimagnetic configuration is shown (spins antiparallel but of unequal amplitude). Other configurations where the spins are neither parallel nor antiparallel, but rather canted, are also possible. However, since the canted spin arrangement does not appear in the materials which are discussed in this chapter, we will not consider them here. The antiferromagnetic and ferrimagnetic structures shown in Figure 3 may be considered to consist of two antiparallel ferromagnetic sublattices. It is clear that only the ferromagnetic and ferrimagnetic cases will exhibit a net moment.

As shown in Figure 3, the spins appear localized. This description is quite good for ionic compounds but obviously is only an approximation for ferromagnetic metals such as Ni and Fe, which are better described by an electron band picture. A consequence of the nonlocal nature of the wave functions in a metal is that the number of spins per atom may be noninteger.[2]

In ionic crystals, if, as is usually the case, the orbital contribution is small, a knowledge of the spin configuration can be used to interpret the change in M_s produced by compositional changes. Of particular importance among the ionic crystals are oxides with the spinel structure.* The spinel structure may be represented by $Me^{2+}_y Fe^{3+}_{1-y} \times [Me^{2+}_{1-y} Fe^{3+}_{1+y}]O_4$, where the ions in octahedral sites are inside the brackets and the ions in tetrahedral sites are in front of the brackets. Here Me denotes any metal ion. For $y = 1$ the spinel is normal and for $y = 0$ the spinel is inverse. The spinels have the ferrimagnetic spin arrangement (Figure 3C) with the ions on the octahedral (B) sites

* The following discussion is based in large part on the discovery and study of ferrimagnetism by Néel. Reviews of the subject appear in a number of places, for example, Refs. 2 and 3.

forming one of the ferromagnetic sublattices and the ions on the tetrahedral (A) sites forming the other. Thus, if Me has m unpaired electrons, the net moment in Bohr magnetons per formula unit is $[(1 - y)m + (1 + y)5] - [ym + (1 - y)5] = 10y - 2ym + m$. The value of M_s is then seen to depend on the site preference parameter y. In some cases y depends on the thermal history of the material. High temperatures tend to produce a random distribution among the two types of sites ($y = 1/3$). If the material is quenched rapidly to a temperature where diffusion is slow, this high-temperature distribution is frozen in.

Similar arguments apply to material with the garnet structure. In fact, with $Y_3(FeAl)_5O_{12}$, used for microwave devices, thermal anneals have been employed as a means for fine adjustment of M_s.[4] The moment of this material depends on the distribution of the Fe and Al ions over the octahedral and tetrahedral sites.

A further interesting consequence of site preference is the case of NiZn ferrite (spinel) where the addition of a nonmagnetic ion actually increases M_s. The Zn occupies the A sites while $Ni^{2+}(d^8)$ is on the B sites, so we have $Zn_y^{2+}Fe_{1-y}^{3+}[Ni_{1-y}^{2+}Fe_{1+y}^{3+}]O_4$. The net moment per formula unit is $[2(1 - y) + 5(1 + y)] - [5(1 - y)] = 8y + 2$, which increases with the Zn content y.

So far we have tacitly assumed that the material is at absolute zero of temperature and that all the spins are aligned. As the temperature is increased, thermal agitation will tend to misalign the spins, decreasing the average moment. The temperature at which M_s falls to zero is the Curie temperature T_C. Since the two sublattices in a ferrimagnetic material may have different temperature dependences, in some cases the two sublattices may cancel at a temperature below T_C as shown in Figure 4. The temperature at which the cancellation occurs (compensation temperature) can be varied by the substitution of nonmagnetic ions on one of the sublattices.

To conclude this brief introduction to the saturation magnetization, we will consider one case that demonstrates a number of the points just considered. In Fig. 5 the domain structure of a plate of the mixed garnet $(YGdTm)_3(FeGa)_5O_{12}$ is shown. The material is uniaxial (Section 5.1) with easy axis normal to the plate. The sample was grown epitaxially at about 900°C and the Fe–Ga distribution over the octahedral and tetrahedral sites is determined by the growth conditions. The semicircular region was then coated with a thin layer of Si and the sample heated to 650°C. At this temperature diffusion in stoichiometric garnet is quite slow and so the region not coated with Si retains the high-temperature ionic configuration. However, the Si

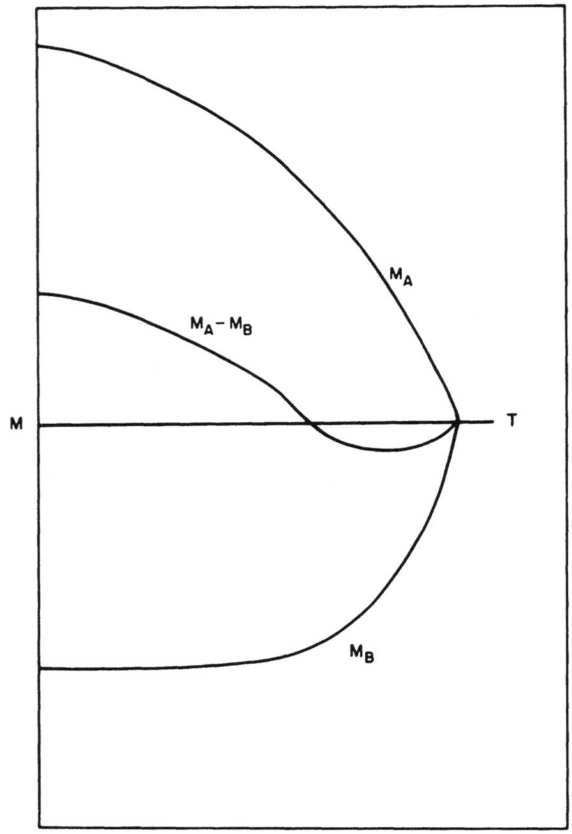

Fig. 4. The moment of the two ferromagnetic sublattices and the net moment of a ferrimagnetic material as a function of temperature. The temperature dependences of M_A and M_B were chosen to show a compensation point: $|M_A| - |M_B| = 0$.

reduces the material under it and the defects produced (probably oxygen vacancies) enhance the diffusion rate. As a result, the Si-covered region reaches an ionic arrangement appropriate to the lower temperature. The change in M_s is quite large; 82 G in the semicircular region and 125 G outside.[5] Both moments are low since the Ga content was adjusted to have the compensation temperature near room temperature.

2.2. Exchange Constant

Of the four "intrinsic" magnetic properties considered in this section, the exchange constant is the least affected by defects and

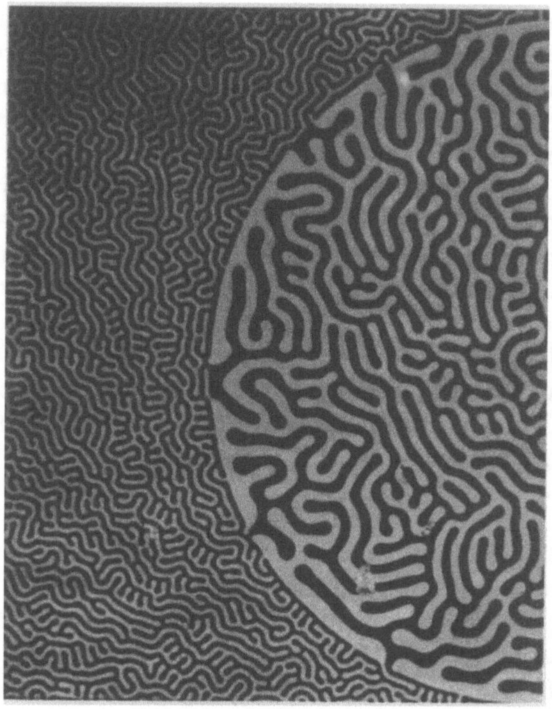

Fig. 5. The domain pattern, made visible by the Faraday effect, of a 6-μm-thick plate of epitaxially grown mixed garnet. The Si coating that covered the semicircular region during heat treatment has been removed. (Unpublished data supplied by R. C. LeCraw and R. Wolfe.)

impurities. Therefore, only a very cursory review of the origins and some convenient expressions of the exchange energy will suffice for our purposes.

The exchange energy is quantum mechanical in origin. The Pauli exclusion principle requires that if the relative directions of two electron spins are changed, the electronic wave functions must also be changed. Thus the interactions are electrostatic rather than magnetic.[2]

The fundamental equation of exchange coupling is $V_{ij} = -2J_{ij}\mathbf{S}_i \cdot \mathbf{S}_j$, which gives the potential energy caused by the exchange effect between atoms i and j. If J_{ij} is positive, the interaction is ferromagnetic; that is, a parallel spin arrangement is favored. If J_{ij} is negative, an antiparallel arrangement is favored and the interaction is antiferromagnetic. In either case, of course, the exchange coupling favors an ordered spin configuration. With increasing temperature, thermal agitation will decrease this order. Very roughly, the spins can reverse when the thermal energy kT is equal to the exchange energy JS^2. If the spins can reverse, the average of the net magnetic moment is zero and so the Curie temperature is given semiquantitatively by $kT_C = JS^2$. No attempt has been made here to correctly account for the crystal structure and the number of interactions involved. A detailed calculation gives $kT_C = 0.28zJ$ for sc structures; $kT_C = 0.325zJ$ for bcc; and $kT_C = 0.346zJ$ for fcc. The value of s has been taken as $\frac{1}{2}$ and z is the number of nearest neighbors.[2]

To obtain the exchange energy density, V_{ij} must be summed over all pairs. To avoid this complication, we will use the form of the exchange energy proposed by Herring and Kittel,[6]

$$E_{ex} = A[(\nabla\alpha_x)^2 + (\nabla\alpha_y)^2 + (\nabla\alpha_z)^2] \tag{1}$$

where α_n are the directions cosines of $\mathbf{M}$ with respect to the crystal axes. Equation (1) does not explicitly depend on a model, but rather is the most general expression possible for the exchange energy density in a cubic crystal. As was indicated in the introduction, the form of Eq. (1) is particularly convenient for the description of gradual variations in the direction of $\mathbf{M}$, such as occur in domain theory. Typical values of A are 2×10^{-6} erg/cm for iron[1] and 4.3×10^{-7} erg/cm for yttrium iron garnet (YIG).[7]

Since the value of A is not often available as a function of temperature, it is convenient to write A in terms of M_s and J_{ij}. In simple ferromagnetic metals, A is proportional to J and M_s^2. In ferrimagnets this relation depends in detail on the sublattice configuration. For a garnet, if the two ferromagnetic sublattices are denoted by o (octahedral) and t (tetrahedral), an unpublished calculation by L. R. Walker gives that A is proportional to $(5/16)\,J_{ot}M_oM_t - 2J_{oo}M_o^2 - (1/3)J_{tt}M_t^2$.

2.3. Anisotropy Constants

The exchange energy (Section 2.2) depends only on the angle between spins and not on the angle with respect to the crystal axes. Thus the whole spin system is free to rotate with respect to the crystal, which is contrary to fact. The required interaction of the spins with the lattice (presumably through spin-orbit coupling) is given by the anisotropy energy. This energy expression must reflect the crystal and time-reversal symmetry. (Time-reversal symmetry here requires that the energy be invariant when **M** is changed to −**M**.) The two lowest-order terms satisfying these conditions for a cubic crystal are[1]

$$E_K = K_1(\alpha_1{}^2\alpha_2{}^2 + \alpha_1{}^2\alpha_3{}^2 + \alpha_2{}^2\alpha_3{}^2) + K_2(\alpha_1{}^2\alpha_2{}^2\alpha_3{}^2) \tag{2}$$

where K_1 and K_2 are the first- and second-order anisotropy constants, respectively. For $\boldsymbol{\alpha} = \langle 100\rangle$, $E_K = 0$; for $\boldsymbol{\alpha} = (1/\sqrt{2})\langle 110\rangle$, $E_K = K_1/4$; and for $\boldsymbol{\alpha} = (1/\sqrt{3})\langle 111\rangle$, $E_K = (K_1/3) + (K_2/27)$. If, as is usually the case, $K_1 \gg K_2$, then for a positive K_1 the energy is minimum for $\boldsymbol{\alpha} = \langle 100\rangle$. This direction is then the easy direction of magnetization. With K_1 negative, $\langle 111\rangle$ is the easy direction.

For a uniaxial (and hexagonal) crystals the two lowest-order terms are

$$E_K = K_u \sin^2\theta + K_b \sin^4\theta \tag{3}$$

where θ is the angle between **M** and the unique (hexagonal) axis.[1]

The theory of the anisotropy in ferromagnetic metals where the electrons are not well localized is not definitely settled[8] and will not be considered here. We only remark, since it will be important later, that Fe has a positive K_1 ($4 \times 10^{+5}$ ergs/cm^3) and Ni a negative K_1 ($-6 \times 10^{+4}$ ergs/cm^3); and so it is not too surprising that there is a Ni–Fe alloy with $K_1 = 0$.

With ionic crystals the situation is simpler, since the anisotropy energy can be approximated by the sum of the anisotropy energies of the individual ions in the lattice.[8] The single-ion contribution to the anisotropy of the 3*d* transition metal ions is given in Ref. 9 and the contribution due to 4*d* and 5*d* ions is reviewed in Ref. 10. The most commonly encountered 3*d* ion is Fe^{3+} (d^5), which has a spherically symmetric half-filled shell. Consequently the spin–orbit coupling and hence K_1 is expected to be small. This is indeed the case; the anisotropy due to Fe^{3+} in either octahedral or tetrahedral sites is small and negative. The Fe^{2+} and Co^{2+} ions in the 3*d* series are a special case inasmuch as they have a large, positive contribution to K_1 when they occupy octahedral sites. In fact, 0.01 Co^{2+} ions per formula unit substituted in Fe_3O_4 changes the room-temperature K_1 from -1.2×10^5 ergs/cm^3 to approximately zero.[11] The large Fe^{2+} contribution is especially significant because the Fe^{2+} content depends on the oxygen stoichiometry. For example, in Fe_3O_4 the room-temperature K_1 value changes by about 20% as the oxygen content is varied by a heat treatment in different oxygen partial pressures.[12]

2.4. Magnetostrictive Constants

Since the exchange integral J_{ij} (Section 2.2) depends on the interatomic distance, in the presence of an exchange interaction the crystal spacing will change (from a hypothetical unmagnetized state) so as to minimize the sum of the exchange and elastic energies. The change in volume produced (volume striction) does not depend on the direction of the magnetization **M**, nor in a cubic crystal does it result in a lowered symmetry. This volume effect is experimentally significant primarily at the Curie temperature, where it produces an anomaly in the thermal expansion coefficient.[1,13]

The anisotropy energy may also depend on the state of strain of the crystal. If it does, the crystal will distort so as to minimize the anisotropy and elastic energies. The distortion, in this case, does depend on the direction of **M**. The expression giving the elongation associated with this distortion must, as was the case for the anisotropy energy, conform to the crystal and time-reversal symmetry. For a cubic crystal the elongation in the direction specified by the unit vector **β** is given to second order by [1,13]

$$\begin{aligned}\delta l/l &= \tfrac{3}{2}\lambda_{100}(\alpha_1{}^2\beta_1{}^2 + \alpha_2{}^2\beta_2{}^2 + \alpha_3{}^2\beta_3{}^2) \\ &\quad + 3\lambda_{111}(\alpha_1\alpha_2\beta_1\beta_2 + \alpha_1\alpha_3\beta_1\beta_3 + \alpha_2\alpha_3\beta_2\beta_3)\end{aligned} \tag{4}$$

where α_n and β_n are the direction cosines of **M** and **β** with respect to the crystal axis. Here λ_{100} and λ_{111} are the linear magnetostrictive constants. For simplicity, or because the exact values are not known, it is often assumed that $\lambda_{100} = \lambda_{111} = \lambda$. Then Eq. (4) reduces to $\delta l/l = \frac{3}{2}\lambda \cos^2\theta$, where θ is the angle between **M** and **β**. In many cases λ_{100} and λ_{111} are of opposite sign and this approximation may be quite misleading.

The derivation of Eq. (4) assumes that the crystal is technically saturated; that is, free of domains. At fields higher than those required for technical saturation, further field-dependent strains (forced magnetostriction) occur. However, forced magnetostriction does not play a significant role in the commonly observed magnetic properties and hence will be neglected here.

Since the linear magnetostriction does not depend directly on the anisotropy energy but rather on the strain dependence of this energy, calculation of λ_{100} and λ_{111} from detailed atomic models is exceedingly difficult. A summary of these calculations will not be attempted here.

For many applications small magnetostrictive coefficients are necessary (Section 4.1). In practice this is generally accomplished by combining two or more materials having magnetostrictive coefficients of opposite sign. In metals it is an alloying process and in ceramics a solid-solution process. One example is the Ni–Fe alloy system, where the positive λ_{111} of the Ni-rich alloys is balanced against the negative λ_{111} of Fe. With only two constituents, λ_{111} and λ_{100} cannot usually both be made simultaneously zero. In another example, important in domain propagation devices, it has been possible by using a solid solution of three rare earth ion garnets to minimize both λ_{111} and λ_{100}.[14]

3. Effect of Impurities

3.1. Microwave Loss

The discussion of microwave loss is most conveniently started by considering the response of a single-crystal ellipsoidal sample to an rf field. It is also assumed that a static field **H** sufficient to saturate the magnetization is applied to the sample. Then the magnetization is everywhere uniform and essentially parallel to **H**. To avoid the complications introduced by eddy currents, only high-resistivity crystals are considered.

The dynamic behavior of the magnetization vector **M** under these conditions was originally given by Landau and Lifshitz.[15] A slight modification of the original equations which clarifies the viscous nature of the damping is given by

$$\dot{\mathbf{M}} = -\gamma[\mathbf{M} \times \mathbf{H}_t] + [\mathbf{M} \times \dot{\mathbf{M}}](\alpha/M_s) \tag{5}$$

where γ, the gyromagnetic ratio, is taken as positive and α is the damping constant. $\mathbf{H}_t$ is the total magnetic field acting on **M** and includes the applied field **H**, the demagnetizing field $\mathbf{H}_d$, the anisotropy field $\mathbf{H}_K$, and the microwave field $\mathbf{H}_{rf}$. The first term in (5) describes the precession of **M** about $\mathbf{H}_t$ at a constant angle and with an angular frequency equal to $\gamma\mathbf{H}_t$. The second term is the damping torque, which tends to spiral **M** in toward $\mathbf{H}_t$. Calculation of $\mathbf{M} \cdot \dot{\mathbf{M}}$ using (5) shows that the magnitude of **M** stays constant. The complete solution to (5) is given in a number of places, for instance, Ref. 16. We will only consider the results obtained for the usual experiment, where the rf frequency is fixed, the applied field is varied, and the microwave power absorbed by the sample is measured as a function of the applied field. Resonance is observed when the applied field is such that the precessional frequency γH_t is equal to the microwave frequency. The width of the resonance

$$\Delta H/H = 2\alpha \tag{6}$$

is an experimental measure of the damping present. In Eq. (6), ΔH is the full linewidth. The microwave power dissipated is absorbed by the uniform precession of **M** (we assume that the sample size is small compared to the rf wavelength). The uniform precession can couple the absorbed power either directly to the lattice or to spacially non-uniform magnetic configurations (spin waves).

The most important sources of direct coupling between the uniform precession and the lattice are the presence of slowly relaxing ions and the possibility of valence exchange. In both cases electrons can move to different energy levels during the precession and therefore the analysis of the two processes is quite similar.

A slowly relaxing impurity is a paramagnetic ion which has low-lying energy levels whose splitting is anisotropic; i.e., the energy splitting depends on the direction of **M**. As **M** precesses, the population of the upper levels is not in instantaneous thermal equilibrium since the relaxation time τ required for an electron to reach equilibrium is taken as finite. The resultant redistribution among the various energy levels leads to a net loss of energy to the lattice.

The linewidth produced by this loss mechanism is given by[17,18]

$$\Delta H = -\frac{1}{M_s}\sum_i \left[\frac{\partial n_i}{\partial \theta}\frac{\partial E_i}{\partial \theta} + \frac{\partial n_i}{\partial \phi}\frac{\partial E_i}{\partial \phi}\right]\frac{\omega\tau_i}{1 + (\omega\tau_i)^2} \tag{7}$$

where n_i is the equilibrium population density of the energy level E_i. The angles θ and ϕ are the angular coordinates of **M** measured from the equilibrium position of **M**, ω is the microwave frequency, and τ_i is the relaxation time of the ith level to thermal equilibrium. The summation is over all the levels in the material. Substantial simplification is obtained if we assume all τ_i are equal and also that the terms in θ and ϕ are the same. With these assumptions (7) reduces to

$$\Delta H = -\frac{2\omega\tau}{M_s[1 + (\omega\tau)^2]}\sum_i \frac{\partial n_i}{\partial \theta}\frac{\partial E_i}{\partial \theta} \tag{8}$$

If the derivatives in (8) are written as $(\partial n_i/\partial E_i)(\partial E_i/\partial \theta)^2$ and if we consider only two levels, the term $\partial n/\partial E$ has a simple form. Under these conditions the linewidth is

$$\Delta H = \frac{n_0}{M_s}\left[\frac{\omega\tau}{1 + (\omega\tau)^2}\right]\left[\frac{1}{kT}\left(\frac{\partial E}{\partial \theta}\right)^2 \operatorname{sech}^2\left(\frac{E}{2kT}\right)\right] \tag{9}$$

where n_0 is the density of the relaxing ions. Equation (9) readily demonstrates the behavior of the slow relaxation mechanism. At low temperatures ($E \gg kT$) the upper level is only slightly populated and ΔH will be small. At high temperatures ($kT \gg E$) the population of the upper state is only slightly altered as **M** precesses and again ΔH is small. Thus ΔH, essentially independent of ω and of the temperature behavior of τ, passes through a peak as the temperature increases from zero. The linewidth can also peak if τ varies such as to make $\omega\tau = 1$. This latter linewidth maximum is the one usually discussed in connection with relaxation losses and is contained in the first set of brackets of (9).

Valence exchange loss is possible if an ion occurs in two valence states in the crystal. For the sake of definiteness we will consider a material where the magnetic lattice contains predominately Fe^{3+} with some Fe^{2+} as an impurity. As discussed in Section 2.3, the energy of an Fe^{2+} ion will in general depend on the direction of **M** with respect to the crystallographic axes. Therefore, for a specified direction **M** a specific Fe^{2+} ion may be in a high-energy state. By the transfer of an electron to an Fe^{3+} ion on another site, the Fe^{2+} ion can move to a lower-energy site. Thus a loss process similar to the one discussed in connection with slowly relaxing ions takes place; the energy of a Fe^{2+} ion is anisotropic and thermal equilibrium can be approached by electron transfer (hopping) from one site to another. (Note that in the valence exchange mechanism the energy of the ion rather than the energy splitting is anisotropic.) As a result, as is shown in Ref. 17, Eq. (7) gives the linewidth for the valence exchange mechanism if τ_i is taken as the electron hopping time. Since the loss does not depend on intraatomic transitions, there are no upper levels to depopulate with decreasing temperature. The temperature dependence of ΔH is then given by $(\omega\tau)\{kT[1 + (\omega\tau)^2]\}^{-1}$ in contrast to the more complicated expression shown in (9) for slowly relaxing ions.

Many experiments have been carried out to study the behavior of slowly relaxing impurities. Much of the effort has been concentrated on investigating the effects of rare earth additions to YIG. With the exception of Gd, all the rare earth ions have anisotropic splitting of low-lying levels and the experimental observations are well described by Eq. (7). In those special cases which are adequately approximated by considering only a low-lying doublet, Eq. (9) gives a good description. The experimental situation has been reviewed by Sparks[(19)] and a vast amount of data has been collected by Von Aulock.[(20)] An observation, important in the discussion of domain wall motion

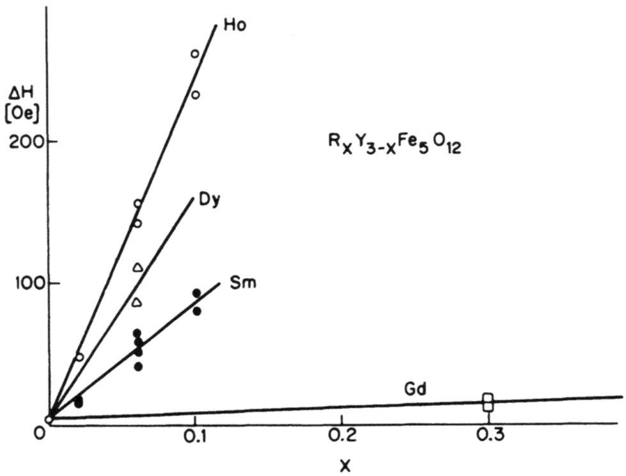

Fig. 6. The room-temperature linewidth of spherical single crystals of YIG containing rare earth substitutions. The data were obtained at 9.25 GHz. (After Ref. 21.)

(Section 3.2), is that, as is implied in (7) and is explicit in (9), the losses are additive. That is, for constant M_s the linewidth increases linearly with number of impurity atoms added to YIG. This linear behavior is illustrated in Figure 6, where ΔH is plotted as a function of rare earth addition.[21]

Representative temperature behavior for a slowly relaxing impurity ion is shown in Figure 7. The linewidth for a Yb-doped YIG sample is plotted as a function of temperature for two rf frequencies (9.3 and 16.8 GHz). One peak, somewhat poorly resolved, is observed at about 20°K independent of frequency. This peak is attributed to a maximum in the population term, the expression given in the second set of brackets in (9). The second peak ($\sim$100°K at 9.3 GHz and $\sim$150°K at 16.8 GHz) arises from the peak in the relaxation loss when $\omega\tau = 1$ and is given by the first set of brackets in (9). It is assumed here, as is generally the case, that τ decreases with increasing temperature. If the behavior of τ as a function of temperature can be calculated and if the angular dependence of the energy splitting E is known, Eq. (9) gives a remarkably good fit to the experimental data. See, for example, Ref. 22.

Depending on the positions of the two types of peaks, a number of different forms of the dependence of ΔH on temperature can be observed. Furthermore, in some cases the term $\partial E/\partial\theta$ in Eq. (9) can

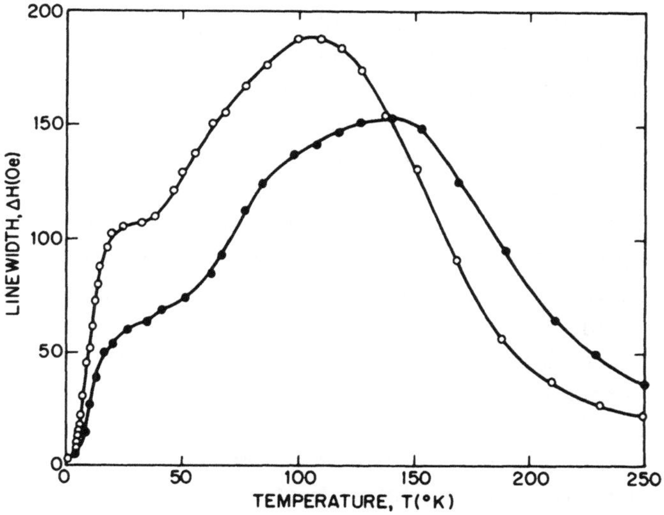

Fig. 7. The linewidth as a function of temperature for Yb-doped YIG at two frequencies with the applied field along the [111] direction. (After Ref. 18).

have a quite complicated angular dependence resulting in sharp linewidth peaks as the static applied field is rotated between different crystallographic directions. A more complete discussion of these observations and further references are given by Sparks.[19,23]

The valence exchange mechanism is more important in the spinel structure than in the garnet structure because the spinel lattice stabilizes with the presence of both divalent and trivalent ions while the garnet lattice is stable with only trivalent ions present. Iron-rich nickel ferrite ($Ni_{1-x}Fe_{2+x}O_4$), for instance, must contain both Fe^{2+} and Fe^{3+}. In the spinel lattice the Fe^{2+} ions occupy octahedral sites. There are four types of nonequivalent octahedral sites each associated with a given $\langle 111 \rangle$ direction (body diagonal).[24] Then we may write that the energy of the ith site containing Fe^{2+} is $E_i = w \cos^2 \theta_i$, where θ_i is the angle between **M** and the $\langle 111 \rangle$ direction of the ith site. The constant w gives the magnitude of the energy and may depend on temperature.

By expressing E_i in terms of θ and ϕ [Eq. (7)], Clogston[17] has found that ΔH is maximum with **H** in the $\langle 100 \rangle$ direction and minimum with **H** in the $\langle 111 \rangle$ direction. Surprisingly, for $Ni_{0.75}Fe_{2.25}O_4$ this is not the case but rather the maximum ΔH occurs in the $\langle 111 \rangle$ direction.[25] This observation implies that there are three types of sites with the $\langle 100 \rangle$ direction unique. The nature of these sites is unknown at present. The peak linewidth at 9.2 GHz occurs at about 100°K[25] and for a frequency of 24 GHz at 150°K.[26] From the relation $\tau = \tau_0 e^{\Delta E/kT}$, expected for valence exchange (ΔE is the barrier for electron hopping), and the fact that $\omega\tau = 1$ at the loss peak, ΔE is found to be 0.03 eV. This value is consistent with the activation energy for conductivity in $NiFe_2O_3$.

An analogous valence exchange loss peak has been observed in $Y_3Fe_{5-\delta}Si_\delta O_{12}$.[27] The Fe ions in YIG are normally trivalent and Fe^{2+} is generally obtained by the substitution of a quadrivalent ion for some of the Fe^{3+}. At 13.4 GHz this loss peak occurs at 380°K. For YIG:Si the loss is a maximum in the $\langle 100 \rangle$ direction, implying that in this case there are four types of sites, each associated with a body diagonal. These sites are presumably the four nonequivalent octahedral sites of garnet lattice. The behavior of octahedral Fe^{2+} will be discussed in more detail in Section 5.2.

So far we have only considered the coupling of energy from the uniform precession to the lattice. Now we will discuss the scattering of energy from the uniform precession into other, spatially nonuniform, modes of magnetization (spin waves). This scattering may be produced either by the presence of impurity ions or by defects such as voids.

Nonmagnetic ions such as Al or Ga in garnet do not appear to scatter appreciable energy from the uniform precession. As shown in Section 3.2, if the energy dissipated per unit volume is independent of M_s, α and therefore ΔH will vary as $1/M_s$. This is just a consequence of the way Eq. (5) is written and is consistent with the form of Eqs. (7)–(9). Then, since M_s is decreased by the addition of Al or Ga and these additions produce no additional losses, ΔH should increase as $1/M_s$. This, for at least $Yb_3Fe_{5-x}Ga_xO_{12}$ and $Y_3Fe_{5-x}Ga_xO_{12}$, is indeed the case.[28] These experiments must be performed on very well-polished samples since the linewidth due to surface pits varies linearly with M_s. This variation may mask the effect of the additional nonmagnetic ions.

The experimental situation is not as clear in the case of spinels. However, since quite narrow linewidths (10 Oe) have been observed in $Ni_{0.36}Zn_{0.64}Fe_2O_4$,[29] it would appear that also in the spinel lattice the scattering from nonmagnetic ions is not an important source of loss.

The magnetic impurity that has received the most attention is Co^{2+}. Haas and Callen[30] predicted a linewidth due to scattering which is proportional to M_s^3. Their calculations give a low-temperature linewidth of 7 Oe for a concentration of 10^{19} Co^{2+} ions/cm^3 in YIG. Experimentally, only 0.1 Oe of the measured linewidth can be attributed to the presence of Co^{2+}.[31] The estimate of the linewidth in Ref. 30 may be too high and scattering may actually take place. Nevertheless the narrow linewidth observed indicates that scattering from Co^{2+} is not an important loss mechanism. Since Co^{2+} has a very large spin–orbit coupling compared to most other magnetic ions, it is probably correct to state that scattering from all magnetic impurities is unimportant.

It has also been suggested by a number of authors that scattering from atomic disorder may play an important part in the loss mechanisms in spinels.[32,33] By atomic disorder is meant the random distribution of the magnetic ions over the various possible crystallographic sites such as are present in inverse or partially inverse spinels. However, experiments show that this type of scattering mechanism is not an appreciable source of loss. Measurements on a well-polished crystal of the nominal composition $MnFe_2O_4$ gave a room-temperature linewidth of less than 4.0 Oe at 13.5 GHz.[34] Manganese ferrite is partially inverse and may be represented by $Mn_{0.8}Fe_2[Mn_{0.2}Fe_{1.8}]O_{12}$, where the bracketed term represents the octahedral sites.[24]

The effect of voids in single crystals has not been extensively studied because of the experimental difficulties involved. In one investigation it was shown that very narrow linewidths ($\Delta H = 0.9$ Oe at room temperature) can be obtained in lithium ferrite having near-zero lattice defect concentration.[35] However, surface roughness, which is, in a sense, equivalent to voids, has been thoroughly investigated.[29,36] The losses due to scattering at the surface add to the losses (absorption, etc.) of the bulk so that $\Delta H = \Delta H_{\text{bulk}} + \Delta H_{\text{surface}}$. Analysis of the scattering losses depends on the exact nature of the model used to describe the surface pits. However, the relation

$$\Delta H_{\text{surface}} = SM_s r/R \tag{10}$$

with S a constant of proportionality, r the radius of the pits, and R the sample radius, does not appear to be very model sensitive. While it is true that finer polishes (decreasing r) decrease ΔH, only the $1/R$ dependence has been experimentally verified unambiguously. As a typical example demonstrating the size of the effect, changing the mean grit size used to polish the YIG spheres from 5 to 15 μm increases the linewidth from 2.5 to 6.5 Oe.[36] Thus, in summary, we see that in single crystals the most prominent scattering mechanism is due to surface pits.

So far we have only considered single crystals. Since most uses of microwave ferrites employ polycrystalline samples, the behavior of polycrystalline samples is of great practical importance. All the loss mechanisms that occur in single crystals, of course, also are relevant to polycrystals. Moreover, in polycrystalline samples the field $\mathbf{H}_t$ [Eq. (5)] acting on $\mathbf{M}$ is not uniform throughout the sample. Thus the resonance response may be broadened even without the introduction of any additional loss.

The variation of $\mathbf{H}_t$ is due to the demagnetizing fields produced by the presence of pores and by the change in the effective anisotropy direction from one grain to another. To continue the discussion, we will first consider only the influence of the anisotropy field. The maximum variation of the anisotropy field H_K is obviously equal to the magnitude of this field, $|K_1/M_s|$. Thus, if the individual grains are independent, $\Delta H_{ans} \approx |K_1/M_s|$. Here ΔH_{ans} is the linewidth contribution arising from the spatial variation of the anisotropy field. However, the grains do not act independently but rather they are coupled by demagnatizing effects. If individual grains precessed at different frequencies, internal discontinuities of the magnetization at the grain boundaries would produce demagnetizing energy. To minimize this demagnetizing energy, $\mathbf{M}$ tends to stay uniform over the sample. The strength of the coupling which tends to keep $\mathbf{M}$ in the various grains parallel depends on the ratio of M_s to H_k. For $M_s \gg H_K$, strong coupling analysis[37] shows that for spherical samples $\Delta H_{ans} \sim 2H_K^2(4\pi M_s)^{-1}$. As stated above, for weak coupling ($H_K \gg M_s$), $\Delta H_{ans} \sim 0.9H_K$.

The contribution to line broadening from the presence of pores arises from the nonuniform demagnetizing fields produced by the discontinuity of $\mathbf{M}$ at the voids. These fields will depend linearly on M_s. Also the linewidth due to pores will be a function of the porosity p. Here p is the fraction of the volume occupied by pores. Consequently ΔH_{pores} may be written as $J'(4\pi M_s)p$, where J' depends on the shape of the sample and the ratio of H_K to M_s.[38–40]* As an example to show the relative magnitude of these two contributions to the line width, Seiden *et al.*[39] considered the case of polycrystalline YIG at room temperature ($|K_1/M_s| = 40$ Oe, $M_s = 140$ G). For $p = 1.7 \times 10^{-2}$ they find $\Delta H_{ans} \sim 5$ Oe and $\Delta H_{pores} \sim 35$ Oe.

In microwave devices the value of M_s is usually specified by some design considerations. As a result, the observed linewidth can only be decreased by decreasing p or $|K_1|$. Several ceramic techniques for

* Implied in the constant J' are both the shape term J of Ref. 39 and a term describing the transition from weak to strong coupling.

obtaining dense samples are given in Ref. 41. The value of $|K_1|$ can be decreased by the addition of certain ions as discussed in Section 2.3. Care must be taken to avoid adding ions which produce microwave losses at the temperature and frequency at which the device is intended to operate. A very successful example of doping to obtain a narrow linewidth is (Ca, In, Ge)-doped YIG ($Y_{2.6}Ca_{0.4}Fe_{3.9}In_{0.7}Ge_{0.4}O_{12}$). Much of the decrease in $|K_1|$ here is attributed to a lowering of T_C. The observed room temperature linewidth of 2.0 Oe is divided as follows: $\Delta H_{ans} = 0.5$ Oe, $\Delta H_{pores} = 0.85$ Oe, ΔH due to rare earth impurities is 0.35 Oe, and unaccounted for are 0.75 Oe.[42]

3.2. Domain Wall Motion

The discussion of domain wall motion will be limited here to high-resistivity single crystals. Also, since the pertinent principles are all demonstrated by cubic crystals, only this type will be considered. Wall motion in polycrystals is discussed in the section covering the permeability. Also, the effect of crystal imperfections will for the most part be neglected. The role of imperfections is neglected solely because almost nothing is known about the interaction of a moving wall and a crystal defect. The interaction of defects on a quasistatic wall in part determines the coercive force and permeability and is discussed in the appropriate sections. Limiting the discussion to almost perfect single crystals still makes it possible to consider much of the available experimental data. This is, of course, a consequence of the fact that in experiments dealing with the fundamentals of wall dynamics the availability of good crystals determined, in part, which materials were studied. However, some recent data concerned with materials of technological importance seem to indicate that in a number of cases unspecified defects play the dominant role in determining the domain wall mobility.

The first domain wall motion experiments in ferromagnetic oxides were performed by Galt,[43] who measured the mobility in Fe_3O_4. These studies were followed by investigations of mobility in $Mn_{1.4}Fe_{1.6}O_4$,[44] YIG,[45,46] YIG with Si additions,[47] and YIG doped with Yb.[48] The samples used for these experiments were "window frame" cores cut from single crystals. For all the materials just listed, $\langle 111 \rangle$ is the easy direction of magnetization (K_1 negative) and therefore the sides of the "window frame" are cut parallel to $\langle 111 \rangle$ as shown in Figure 8. A core of this configuration can support a single domain wall as indicated in the figure. The wall moves under the influence of a circumferentially applied field, and the wall velocity is

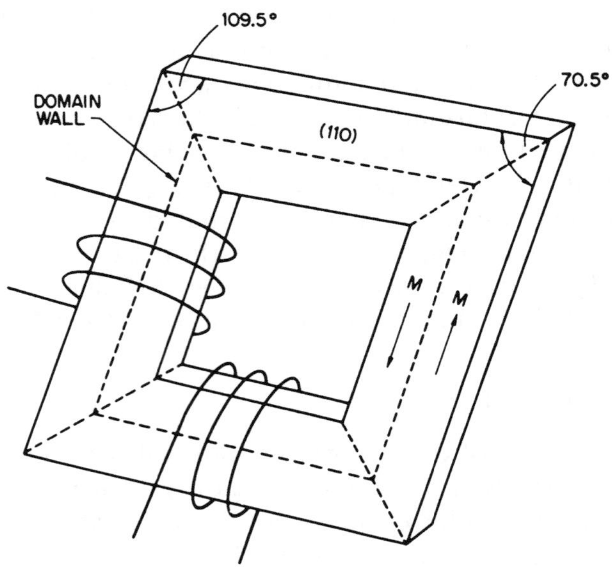

Fig. 8. A "window frame" single-crystal sample used to measure the domain wall mobility. The windings that supply the drive field and the sense of the flux change are shown schematically.

deduced from the voltage induced in a pickup coil wound around one side of the core. The outside edges of the core are carefully polished to minimize any surface pinning or retardation of the moving wall.

With the discovery of mixed garnets with a uniaxial growth-induced anisotropy[49] a number of different techniques for measuring wall motion came into use. These techniques, in general, use samples in the form of thin plates with the easy direction of magnetization normal to the plate. The methods that make use of the optical transparency of thin plates are reviewed by Shumate.[50] The domain wall mobility can also be determined from the magnetic response of a multidomain plate to a high-frequency applied field. In this case the plate is placed in a solenoid (with the normal of the plate parallel to the solenoid axis) and the complex inductance is measured as a function of frequency.[51] We will refer to this technique as the complex susceptibility method.

The phenomenological discussion of wall motion starts with modified Landau–Lifshitz equation (5). In many instances the damping term $(\alpha/M_s)[\mathbf{M} \times \dot{\mathbf{M}}]$ is replaced by $(\lambda/M_s^2)[\mathbf{M} \times (\mathbf{M} \times \mathbf{H})]$, where λ is the Landau–Lifshitz damping term. Simple vector manipulation shows the two damping terms are equivalent if $\alpha \ll 1$ and if we set $\lambda = \alpha\gamma M_s$. The energy dissipation per unit volume (for $\alpha \ll 1$) using the modified term is $[\alpha/(\gamma M_s)][\dot{\mathbf{M}}]^2$, while for the original Landau–Lifshitz term the dissipation is $(\lambda/\gamma^2)[\dot{\mathbf{M}}/M_s]^2$. The first term depends linearly on the magnitude of **M** and the second term is independent of $|\mathbf{M}|$. Due to the ad hoc nature of the damping constant, which form is used is only a matter of convenience and individual preference.

The analysis of a planar wall such as is schematically shown in Figure 2 has been given by Kittel and Galt.[52] By making suitable approximations, they find that with an applied field H along the positive z direction the wall moves in the positive y direction with the velocity given by

$$v = (C\gamma^2 M_s/\lambda)(A/K)^{1/2}H \tag{11}$$

Equation (11) has been derived using the original Landau–Lifshitz term; using the modified damping term, we obtain

$$v = (C\gamma/\alpha)(A/K)^{1/2}H \tag{12}$$

The constant C in Eqs. (11) and (12) is of the order of one; the exact value depends on the details specifying the anisotropy energy (i.e., uniaxial, cubic, etc.). For uniaxial anisotropy $C = 1$. Again we see that the two forms of damping constants imply a different dependence of the velocity on the magnitude of M_s. In a recent review article Hagedorn[53] quotes experimental results for $(Y_xGd_{1-x})_3Fe_5O_{12}$ where the velocity varies linearly with M_s and Eq. (11) appears to be the better description of the domain wall velocity. That is, λ is independent of M_s; if Eq. (12) were used, the constant α would be proportional to M_s^{-1}. In the same article he also gives examples ($Eu_1Er_2Fe_{3.3}Ga_{1.7}O_{12}$) where Eq. (12) appears to be more applicable. The magnetic moment for this composition is, as was discussed in Section 2.1, changed by altering the relative distribution of Ga and Fe on the octahedral and tetrahedral sites. However, since there is no *a priori* reason to expect either α or λ to have any specific dependence on M_s, we will use Eq. (12) in future discussions.

The assumptions used to derive Eqs. (11) and (12) are only valid for low velocities, that is, small applied fields. A solution valid at higher applied fields is given in an unpublished calculation by L. R. Walker. The results are summarized in a review article by Dillon.[54] Walker found an exact particular steady-state solution to Eq. (5) for a planar wall in an infinite medium limited to the case of uniaxial anisotropy. The velocity resulting from his analysis is

$$v = \frac{\gamma H}{\alpha}\left(\frac{A}{K}\right)^{1/2}\left(1 + \frac{\pi M_s^2}{K}\left\{1 - \left[1 - \left(\frac{H}{2\pi M_s\alpha}\right)\right]^{1/2}\right\}^2\right)^{-1/2} \tag{13}$$

It is easily seen that for small applied field, or more particularly $H/2\pi M_s\alpha \ll 1$, the solution approaches that given by Eq. (12). For $H/2\pi M_s\alpha > 1$ the solution becomes complex and is no longer valid. The maximum velocity, which occurs for $H/2\pi M_s\alpha$ slightly less than one, is

$$v_{\max} = 2\pi\gamma M_s(A/K)^{1/2}(2^{1/2}/B)[1 + B - (1 + 2B)^{1/2}]^{1/2} \tag{14}$$

where $B = \pi M_s^2/K$. For YIG at 300°K, $v_{\max} = 4.7 \times 10^4$ cm/sec.

So far we have only discussed the theory of steady-state wall motion under the influence of a constant applied field. We will now consider oscillatory behavior. A wall driven by an oscillating field is important in the discussion of permeability (Section 4.2). To describe oscillatory behavior, the wall equation of motion must include a mass term and possible restoring or spring terms. The existence of a wall mass was first shown by Doring and a complete derivation is given in Ref. 52. Essentially the mass term arises from the fact that in a moving wall the angle ϕ in Figure 2 is no longer equal to zero in the wall. The component of M along y leads to a demagnetizing field H_y. To a good approximation v is proportional to H_y. Consequently the demagnetizing energy associated with H_y^2 may be written as $\frac{1}{2}mv^2$, where m is a constant of proportionality and is associated with an effective mass. The value of m is given by

$$m = (1/4\pi C\gamma^2)(K/A)^{1/2} \tag{15}$$

where C is defined in connection with Eqs. (11) and (12). In general usage the restoring or spring constant is denoted by α. However, since this may cause some confusion with the term α in Eq. (5), we will use κ for the spring constant. The value of κ depends on the exact details of the domain configuration and must be determined for each case. The motion in the y direction of a wall such as illustrated in Figure 2 is then described by

$$m\ddot{y} + \beta\dot{y} + \kappa y = 2M_sH \tag{16}$$

If κ is taken as zero and H is a static field, we see that in the steady state $\beta v = 2M_sH$. Thus $\beta = 2M_s(C\gamma)^{-1}(K/A)^{1/2}\alpha$, where we have used Eq. (12) for v. Depending on the relative values of the constants, Eq. (16) describes either a resonance or a relaxation behavior.

We are now ready to describe the experimental situation, limiting ourselves initially to low applied fields. Interpreting these experiments reduces, in a sense, to finding a theoretical basis for the value of the damping constant α. A large class of experiments has been concerned with the additions of a slowly relaxing impurity into a host lattice, usually a rare earth garnet. Another important class of experiments deals with the losses associated with valence exchange. In both cases the discussion of these loss mechanisms in Section 3.1 (microwave loss) is applicable here.

The domain wall velocity can be derived using arguments similar to those used to derive Eq. (8). However, for our purposes it is sufficient to make a few simplifying assumptions[55,56] and to use Eq. (6) to connect Eq. (8) and Eq. (12). From Eqs. (6) and (8) we find that the damping constant appropriate to the microwave data is

$$\alpha_{rf} = -\frac{\gamma}{M_s}\frac{\tau}{1+(\omega\tau)^2}\sum_i \frac{\partial n_i}{\partial \theta}\frac{\partial E_i}{\partial \theta} \tag{17}$$

For the case of domain wall motion (17) applies if we associate ω with the rate of rotation of **M** in the center of the wall ($d\theta/dt \equiv \omega_{dm}$). For low domain wall velocities $\omega_{dm}\tau \ll 1$. Thus the ratio of α_{dm}, the damping applicable to wall motion, to α_{rf} is

$$\alpha_{dm}/\alpha_{rf} = 1 + (\omega\tau)^2 \tag{18}$$

where ω is, of course, the microwave frequency. Then, using (12), we have

$$\mu \equiv v/H = -\frac{CM_s}{\tau}\left(\frac{A}{K}\right)^{1/2}\left(\sum_i \frac{\partial n_i}{\partial \theta}\frac{\partial E_i}{\partial \theta}\right)^{-1} \tag{19}$$

where μ is the mobility.

The most complete investigation of the effect of a slow relaxing impurity in garnet was carried out by Harper.[48] Using a "window frame" sample of $Yb_{0.15}Y_{2.85}Fe_5O_{12}$ and an undoped sample of $Y_3Fe_5O_{12}$, he deduced the contribution of the Yb ions to the domain wall damping as a function of temperature. Since the wall velocity depends on the reciprocal of the damping and since the losses are, to a good approximation, additive, it is convenient to continue this discussion using the reciprocal of the domain wall mobility (μ^{-1}). Thus

$$\mu^{-1} = \mu_0^{-1} + \mu_1^{-1} \tag{20}$$

where μ_0 is the mobility of the pure sample and μ_1 refers to the

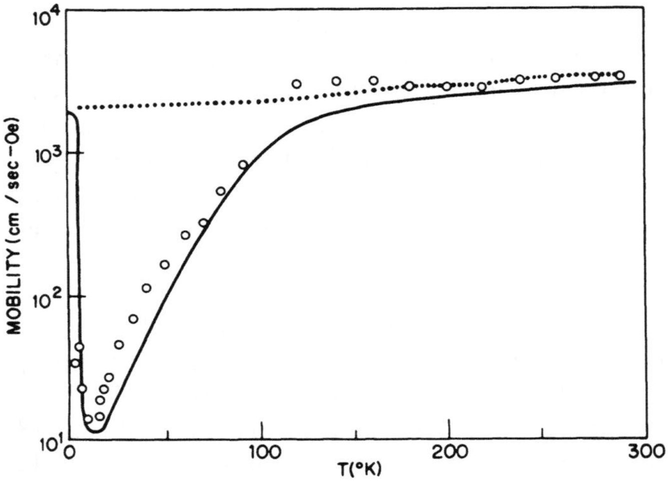

Fig. 9. The measured domain wall mobility for YIG (dotted curve) and for $Yb_{0.15}Y_{2.85}Fe_5O_{12}$ (open circles) as a function of temperature. The solid curve is calculated on the basis of a slow relaxation model. The data in this figure are from Ref. 48.

mobility determined by the losses introduced by the Yb. The experimental results are given in Figure 9, where the mobility of both samples is plotted as a function of temperature. The Y^{3+} ion has the electronic configuration of krypton (completely filled shell) and therefore is, of course, nonmagnetic. Thus the Y^{3+} ions do not contribute to the damping and the mobility shown in Figure 9 for YIG is due to the Fe^{3+} lattice only. The contribution of the Yb^{3+} ions peaks at about 10°K and is quite small at room temperature. The solid curve is a plot of Eq. (19) with the temperature dependence of τ, n_i, and E_i obtained from a number of other independent experiments. The details are given in Ref. 48. The fit of the theoretical expression is very good. The physical reason for the peak in the loss attributed to Yb is the same as that discussed in Section 2.1. At high temperatures the relaxation time τ is very short and in addition the population of the low-lying levels stays essentially constant as **M** rotates since $E_i \ll kT$. Therefore the loss will be small. At very low temperatures only the ground state is populated and again the loss is low even though τ is quite long.

Similar analyses have been carried out for erbium-doped YIG[57] and for manganese ferrite.[55] In the latter experiment the low-temperature loss studied is attributed to the low-lying levels of the Mn^{3+} ions in octahedral sites.

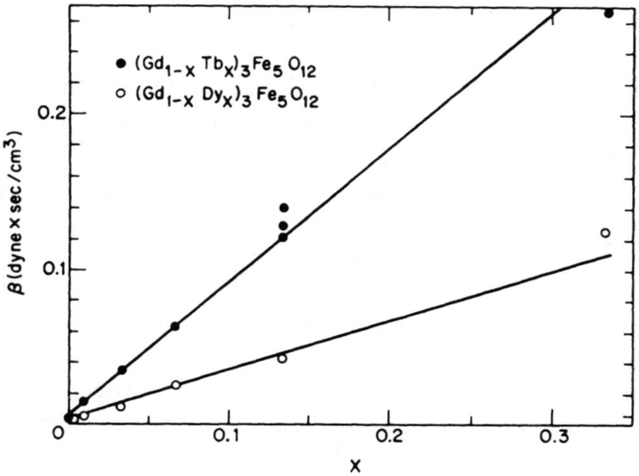

Fig. 10. The room-temperature damping parameter $\beta = 2M_s\mu^{-1}$ as a function of Tb or Dy content in $Gd_3Fe_5O_{12}$. (After Ref. 58.)

The mobility μ at room temperature is an important parameter in the design of domain propagation (bubble) devices.[(49)] In bulk, uniaxial, mixed rare earth iron garnets used for these devices, the mobility at room temperature is, in large part, determined by the high-temperature tails of the relaxation losses just discussed. As stated earlier (Section 3.1), these losses are additive. This additive feature is demonstrated in Figure 10, where $\beta\,(= 2M_s\mu^{-1})$ for the mixed GdTb and GdDy garnets is plotted as a function of Tb and Dy concentration x. These data were obtained with the complex susceptibility method.[(58)] The value of β is plotted rather than the value of μ^{-1} since (19) shows that for the loss processes active here, β is independent of M_s and depends only on the details of the damping and the constants A and K. The value of $(A/K)^{1/2}$ is relatively insensitive to x, while M_s changes by more than a factor of two as x varies between 0.0 and 0.3[(58)] At room temperature, where $\omega\tau \ll 1$, from (18) we would expect the mobilities at the end points to be those deduced from the microwave damping constant α_{rf} of the garnets containing only one rare earth. For all the cases investigated to date this prediction is correct to within a factor of two, with the exception of $Y_3Fe_5O_{12}$. The mobility of YIG will be discussed separately. Further experimental results and the values of α_{rf} for most of the rare earths are given in Ref. 59. The ability to predict the mobility and control the magnetostriction (Section 2.4) and the induced anisotropy (Section

5.1) makes it possible to select compositions suitable for domain propagation devices.

Studies of the effect of the valence exchange mechanism on domain wall motion have been less extensive. The most complete study, carried out on $Ni^{2+}_{0.75}Fe^{2+}_{0.25}Fe^{3+}_{2}O_4$, is reviewed in Ref. 52. The mobility is shown to decrease exponentially with temperature. This behavior is consistent with Eq. (19) since for electron hopping $\tau = \tau_0 \, e^{\Delta E/kT}$, where ΔE is the energy barrier for the transfer of an electron from one octahedral site to another. It should be noted, however, that as τ becomes very large, Eq. (19) is no longer valid since the assumption that $\omega_{dm}\tau \ll 1$ is no longer satisfied. Also, as discussed in Section 3.1, the term $\sum_i(\partial n_i/\partial\theta)(\partial E_i/\partial\theta)$ is expected to have a relatively small temperature dependence since the average of E_i with respect to θ is zero. That is, for the hopping mechanism, in contrast to the situation present with an impurity ion, there are no low-lying levels to depopulate with decreasing temperature. The value of ΔE determined from the domain mobility is in fair agreement with the activation energy found for conductivity in these materials. This is not unexpected because conductivity is, of course, intimately connected with electron hopping.

So far we have not discussed the value of μ_0, the mobility of undoped YIG, shown in Figure 9. The preceding discussion relating α_{rf} to α_{dm} is not valid here since nominally pure YIG presumably does not contain any slowly relaxing impurities. (A large number of experiments show that this is essentially true.) Therefore, without a specific model for guidance, we will take $\alpha_{dm} = \alpha_{rf}$. If the damping constant obtained from microwave measurements ($\Delta H \sim 0.5$ Oe at 9.3 GHz)[60] is used in (12) to evaluate the mobility, we obtain $\mu_0 = 2 \times 10^6$ cm/sec-Oe, three orders of magnitude larger than the observed value. Since this mobility is observed for values of H such that $H/2\pi M_s\alpha < 1$, the limitations discussed in connection with (13) do not apply here. Two earlier investigations[45,46] resulted in mobilities that all agree to within a factor of five. Thus it seems unlikely that gross defects, which would be expected to vary from sample to sample, can account for the very large discrepancy between theory and experiment. The possibility remains, since all the crystals used were grown by essentially the same flux method, that small imperfections inherent in the growth process are present in all the samples and that these tend to limit the wall mobility. To date, little if anything is known about the interaction of dynamic walls and crystal defects and in addition the experimental samples have not been characterized as to defect concentration. As a

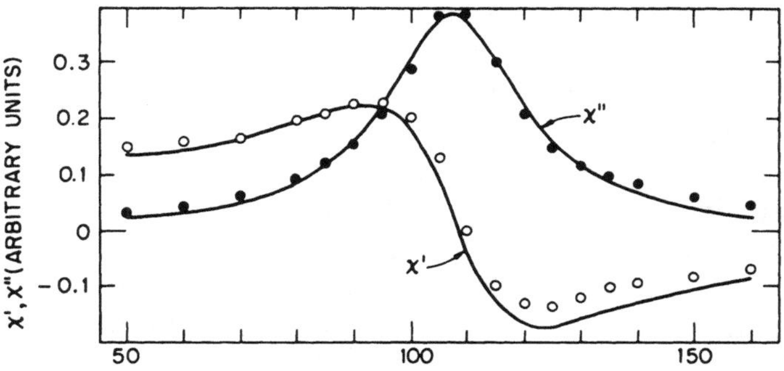

Fig. 11. The complex susceptibility for a 0.1-cm-thick sample of $Y_{1.2}Gd_{1.8}Fe_5O_{12}$. The measurements were taken at room temperature. (From Ref. 51.)

result, the possibility that the mobility is limited by defects cannot, with any certainty, be eliminated. Some theoretical studies, which tacitly assume the measured μ_0 is intrinsic to the perfect material, have been undertaken. An examination of these studies is beyond the scope of this chapter and furthermore none has yet led to a satisfactory explanation.

The experimental data considered so far applied to uniform or to highly damped oscillatory wall motion. Now we will briefly consider experimental data at high frequencies for materials with low loss. Under these conditions the mass term in (16) plays an important part in the domain wall response, and wall resonance, which is important in the discussion of permeability, is observed. Resonance has been observed by a number of investigators, but the clearest results to date have been obtained with uniaxial mixed garnets. The measurements with the uniaxial garnets are made using the complex susceptibility method. The susceptibility is related to the wall displacement y by $\chi = 2My/Hd$, where d is the width of the stripe-domain spacing in zero applied field.[51] The restoring constant κ arises from the magnetostatic demagnetizing energy that determines the stripe spacing. By measuring optically the quasistatic change of d with field, the value of κ can be obtained. Typical experimental results which clearly demonstrate the resonant behavior are shown in Figure 11. The composition of the garnet used was $Y_{1.2}Gd_{1.8}Fe_5O_{12}$. The values of m and β are determined by the best fit to the χ'' curve. Using (15), the calculated value of m is 7.0×10^{-11} g/cm^2, which may be compared to 6.9×10^{-11} g/cm^2 determined from the data.[51] In passing we note that Gd does not substantially add to the loss of YIG since

Gd^{3+}, with a spherically symmetric, half-filled $4f$ shell, has its first excited state more than $10^4\ cm^{-1}$ above the ground state. Thus at room temperature the excited states are not heavily occupied and the relaxation losses will be small.

As the final topic concerning wall motion, the effects observed at high applied fields will be briefly discussed. Since at high fields complicated domain configurations, rather than a single wall, occur using a "window frame," this method is not applicable. Experimental difficulties also limit the complex susceptibility technique to relatively low fields. Thus much of the high-field data, limited here to uniaxial garnets, is obtained by measuring the response of a circular magnetic domain either to an additional applied field normal to the platelet or film (bubble collapse method), or to a field gradient in the plane of the sample (bubble velocity method). These measuring techniques are rather specialized and for the details Refs. 61 and 62 may be consulted.

The most important experimental feature observed at high applied fields is the marked deviation from linearity of the velocity as a function of field. Calhoun and co-workers[63] found that for $Y_3Ga_{1.5}Fe_{3.5}O_{12}$ the velocity measured by the bubble collapse technique varies as $\sqrt{H}$. With some epitaxial garnet films a limiting velocity is reached.[64] The maximum velocity observed is always well below the one predicted by Eq. (14).

At present it is not possible to interpret these high-field results. In fact the velocity measured on different regions of the same epitaxial

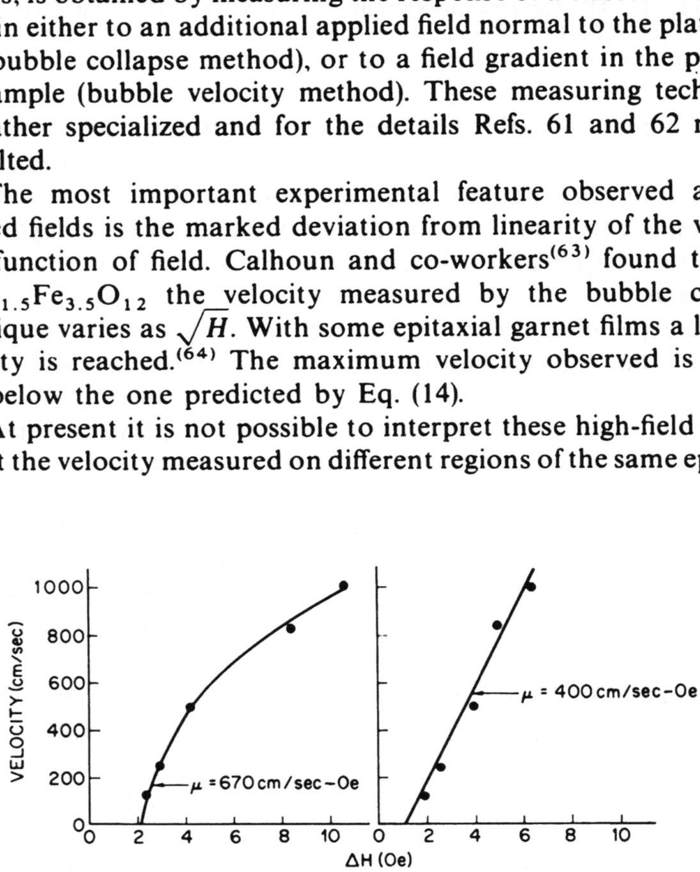

Fig. 12. The bubble velocity as a function of field gradient for two different regions of the same $Y_1Eu_{1.85}Yb_{0.15}Al_{1.1}Fe_{3.9}O_{12}$ epitaxial film. The mobility shown is deduced from the bubble diameter and the measured velocity. (Unpublished data of G. P. Vella-Coleiro.)

film may exhibit markedly different behavior. In Figure 12 we show the velocity as a function of the field gradient across the bubble obtained by the bubble velocity method. The two curves were obtained from two regions of a YEuYb epitaxial garnet film.[65] As seen in the figure, in one region v varies as H and in another v varies approximately as $\sqrt{H}$. Since the composition is expected to be uniform over the total area, the very different behavior of the mobility in the two regions demonstrates that the dynamic domain behavior in films of this type is in large part determined by some as yet unspecified collection of defects. It is not possible to pursue this problem further due to the lack, at this time, of more extensive data pertinent to magnetic films of this type.

3.3. Acoustic Loss

Much of the recent work dealing with magnetoelastic interactions in ferromagnets has been concerned with the coupling of elastic waves and spin waves. These interactions do not directly depend on material imperfections[66] and we shall not consider them here, but rather we will limit ourselves to material imperfections that produce acoustic attenuation. In this connection, the most prominant feature is the large acoustic loss caused by the presence of domain walls (somewhat loosely, the presence of domain walls is considered to be an imperfection in this context). The stresses induced by the acoustic vibrations cause domain wall motion which is apparently a lossier process than pure mechanical vibration. The additional acoustic loss produced by domain walls is shown in a striking way in Figure 13, where Q^{-1} of a single-crystal YIG sphere is plotted as a function of the applied field. [If y is the amplitude of the damped harmonic motion, Q is defined by the relation $y = y_0(\sin \omega t)e^{-\omega t/2Q}$.] A sphere is single domain if the applied field is both parallel to the easy axis of magnetization and is larger than the demagnetizing field ($H > 4\pi M_s/3 = 590$ Oe). Domain walls will form for smaller fields. In Ref. 66 it is shown that a reversal of only 2×10^{-3} part of the magnetization of the sphere changes Q by several orders of magnitude.

The mechanisms by which domain walls couple the mechanical energy to the lattice have not been investigated in high-resistivity oxides like YIG. It is uncertain whether the intrinsic loss associated with wall motion (Section 3.2) is sufficient to produce such high damping or whether hysteresis loss caused by domain walls being driven irreversibly over imperfections must also be present. It is interesting to note that domain walls in antiferromagnets also cause acoustic loss since an applied elastic stress can move this type of wall.[67]

The causes of acoustic loss in metals are better understood and can be divided into three categories[68]: (1) macro eddy currents, (2) micro eddy currents, and (3) hysteresis. The first arises from the eddy currents due to the time variation of the spatial average of the magnetization over the sample. The second is produced by the local eddy currents caused by wall motion between individual domains. The hysteresis loss is generally attributed to domain walls moving irreversibly through regions of nonuniform stress. The damping (Q^{-1}) due to this mechanism is proportional to $\lambda E/S_i$, where λ is the mag-

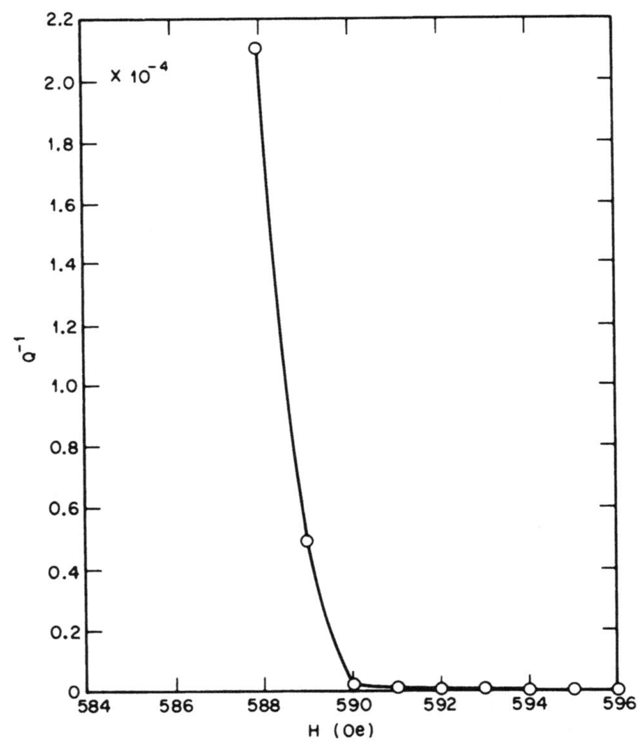

Fig. 13. The acoustic loss Q^{-1} of a YIG single-crystal sphere as a function of the applied field. The measurements were made at room temperature with a frequency of 2.8×10^7 Hz. The applied field was along the easy axis (⟨111⟩). (After Ref. 66.)

netostrictive constant, E Young's modulus, and S_i the average internal stress opposing wall motion.[69]

In magnetically saturated high-resistivity oxides there has been some preliminary work concerning the interaction of elastic waves with the rare earth ions in iron garnets.[66] This study is by no means as complete as the equivalent work concerning the microwave losses produced by these ions (Section 3.1) and it will not be considered here. The investigations using oxides have been primarily centered on the acoustic attenuation caused by valence exchange (electron hopping). Here neither the loss mechanism nor the coupling of energy from the acoustic vibration to the lattice depends explicitly on the presence of the magnetization. This type of loss is of interest because valence exchange does contribute to magnetic processes such as microwave loss (Section 3.1), damping of domain wall motion (Section 3.2), and induced anisotropy at low temperatures (Section 5.1).

Silicon-doped YIG has two microwave loss peaks, one at about 40°K and one, which was discussed in Section 3.1, above room temperature.[70] Both of these loss peaks presumably arise from the presence of Fe^{2+}. The high-temperature peak has an activation energy of 0.25 eV, which is in very good agreement with the activation energy found for conductivity in a sample with a similar Si content.[27] This same valence exchange loss peak is observed acoustically in the 10^6-Hz frequency range at 200°K.[66] The low-temperature loss peak has been attributed to an intraionic relaxation process (slow relaxation) in the Fe^{2+} ions present.[71] However, this interpretation is not consistent with the calculations of the Fe^{2+} energy levels given in Ref. 70 and the origin of the low-temperature peak is still in doubt. It is interesting to note that acoustically there is no indication of a loss peak that would correspond to the low-temperature microwave loss, indicating that in this case the acoustic energy is not coupled to the Fe^{2+} ion.[72]

In nominally pure polycrystalline YIG, Gibbons and Chirba,[73] using frequencies of the order of 10^5 Hz, observed an acoustic loss peak with an activation energy of 0.12 eV. The same activation energy was obtained from measurements carried out at 10^7 Hz.[74] While it is clear that the loss is caused by impurities or by nonstoichiometry, the exact origin of this loss peak has not been identified. The peak cannot be attributed to Si contamination since a study of the activation energy as a function of Si content shows that the activation energy varies from 0.23 eV for high Si content to more than 1.0 eV for very low Si levels.[75,76]

Loss peaks attributed to Fe^{2+}–Fe^{3+} valence exchange have also been reported for Ni ferrite.[77] The activation energy is 0.03 eV, which is identical to the activation energy found by microwave resonance measurements.

Two loss peaks are observed in a Mn ferrite of somewhat uncertain composition.[77] Presumably the composition is near $MnFe_2O_4$. At this composition Mn and Fe appear in both the divalent and trivalent state.[78] The low-temperature peak (40°K at 1.5×10^5 Hz) has an activation energy of 0.03 eV and is associated with Fe valence exchange. The high-temperature peak is due to valence exchange between Mn^{2+} and Mn^{3+}. A study of this peak was carried out in the system $Mn_xFe_{3-x}O_4$, with $1.0 \leq x \leq 1.6$.[79] The maximum loss increases linearly with x accompanied by an increase in activation energy from 0.3 eV for $x = 1.16$ to 0.5 eV for $x = 1.6$. These activation energies are in good agreement with those obtained from electrical conductivity measurements. For $1.2 \leq x \leq 2.0$ the valence states may be represented by [78] $Mn^{2+}Mn^{3+}_{x-1}Fe^{3+}_{3-x}O_4$, and so in this compositional range conductivity results from electron hopping between Mn ions. Furthermore, the observed increase in acoustic attenuation with x shows that the loss is proportional to the amount of Mn^{3+} present.

Even without the possibility of valence exchange the presence of Mn^{3+} may produce a large acoustic loss. In YIG with 0.15 wt. % Mn additions a large low-temperature acoustic loss is observed.[72] The activation energy is in the range of 0.01–0.03 eV.[80] This loss is attributed to the interaction of acoustic vibrations with the Jahn–Teller distortion of the octahedral sites containing Mn^{3+}.

The simple valence exchange loss due to Fe^{2+}–Fe^{3+} ions which is observed in Ni and Mn ferrite is not observed in $Fe^{2+}Fe^{3+}_2O_4$ (magnetite). At about 100°K the Fe^{2+} and Fe^{3+} ions in magnetite order, and in the low-temperature ordered state the activation energy for electron hopping is relatively high.[81] However, the presence of cation vacancies does permit valence exchange,[82] and a loss peak, the amplitude of which is proportional to the vacancy content, is observed at 40°K with a frequency of 10^6 Hz.[83] Another peak associated with the order–disorder transition occurs near 100°K. A third loss peak at about 690°K (1 eV) also has an amplitude that depends on vacancy content.[83] The mechanism that produces this peak is also responsible for the induced anisotropy and the disaccommodation observed in magnetite at room temperature. It will be shown in Section 5.1 that cation vacancy migration plays an important role here.

In ferromagnetic metals obviously valence exchange need not be considered. Here interstitial atoms are the important impurity leading to acoustic loss. Interstitial carbon in iron, which produces the disaccommodation discussed in Section 4.1, is a classical example of such a loss.[84] The acoustic stress causes one or two of the three types of sites that can be occupied by the carbon to become energetically favorable. The result is a relaxation loss that is maximum when $\omega\tau = 1$, where ω is the angular acoustic frequency and τ (the relaxation time) is the time taken for the carbon atoms to go from one site to another. Similar results are obtained with interstitial nitrogen.

4. Structure-Sensitive Properties

4.1. Permeability

A major use of magnetic materials is for inductor and transformer cores. In this application a large permeability with low loss is usually essential.

The susceptibility χ of polycrystalline magnetic materials can arise from domain wall displacement, rotation, or a combination of both. To a good approximation the effects are additive so that $\chi = \chi_{rot} + \chi_{dw}$. The role of each process is most conveniently seen by considering the susceptibility as a function of frequency, with grain size as a parameter. In Figure 14 the permeability $(4\pi\chi + 1)$ for samples having different grain size is plotted versus frequency. For the lowest curve the diameter of the grains is less than 10^{-5} cm, which is too small to support a domain wall; thus for this sample the susceptibility is due to rotation only. The rotational susceptibility is not expected to depend strongly on grain size and therefore is taken the same for all samples. At higher frequencies where χ_{dw} is small the independence of the rotational resonant frequency on grain size is clearly demonstrated in Figure 14. It can also be seen that at low frequencies samples with large grain size have a χ_{dw} an order of magnitude larger than χ_{rot}. This predominance of χ_{dw} appears to be true in most cases of practical importance. It should be noted that Figure 14 was chosen because it clearly shows the contribution of the two processes. The data obtained usually do not show such a clear separation.

Rigorous analysis of χ is not possible since in a polycrystal the domain configuration cannot be experimentally determined and calculations concerning the domain pattern are prohibitively difficult.

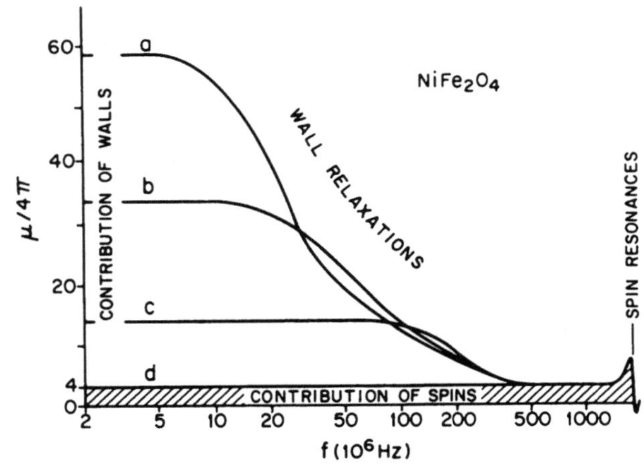

Fig. 14. The permeability of $NiFe_2O_4$ as a function of frequency for samples with different grain sizes and porosity. (a) $D = 11\ \mu m$, $p = 0.05$; (b) $D = 5\ \mu m$, $p = 0.05$; (c) $D = 2\ \mu m$, p $= 0.21$, (d) $D < 0.2\ \mu m$, $p = 0.32$. (After Ref. 85.) The original data on which this figure is based is given in Ref. 86.

It is worthwhile, however, to consider rather crude models for the susceptibility, since even highly oversimplified calculations do set forth the pertinent material parameters.

For the rotational process we will take the material as uniaxial and we will neglect any internal demagnetizing fields. The latter assumption is fairly good if the crystallites are large enough to support domain walls and if the frequency of the applied field is well below the resonant frequency γH_K. In the absence of an applied field, sufficient domain walls form so that the demagnetizing field caused by the discontinuity of **M** at grain boundaries is small.[87] With a small applied field ($H \ll H_K$) at a frequency well below resonance the magnetization in all the grains will move in unison and therefore no appreciable additional internal demagnetizing fields will be produced. If **H** is applied at an angle θ_0 to the easy axis, it can readily be shown[88] that the quasistatic susceptibility is given by $M_s^2(\sin^2\theta_0)/2K_u$. Any demagnetizing effects present will tend to raise the effective value of K_u. For a random distribution of grains the average of $\sin^2\theta_0$ must be used and $\chi_{rot} = M_s^2/4K_u$. In the case of cubic anisotropy and a random distribution of crystallites $\chi_{rot} \sim M_s^2/K_1$; the exact form depends on whether K_1 is positive or negative. As shown in Section 3.1, the losses due to rotational processes are small away from resonance and in the vicinity of resonance they are determined by the considerations given in that section. The preceding comments concerning the losses tacitly imply high-resistivity material. If the resistivity is low, there will be eddy current losses. These losses per unit volume are proportional to $\chi^2\omega^2H^2L^2/10\rho c^2$, where ρ is the resistivity, L the dimension of the sample, and the constant of proportionality is approximately equal to one.* For sheets L is the thickness, for wires L is the radius. This calculation of the eddy current loss is only valid if the applied field can penetrate the sample; i.e., L is less than the skin depth. (For iron with a permeability of 10^3, $\rho = 10^{-5}$ ohm-cm and a frequency of 100 Hz the skin depth is about 1 mm.)

The domain wall contribution to the susceptibility χ_{dw} can be deduced from (16) if we assume a domain configuration where each domain wall has an area d^2 and the domain walls are in a cubic array with a spacing d. This model approximates a polycrystal with a grain size of d and with one wall per grain. With this model $\chi_{dw} = 2M_s y/Hd$, where the wall displacement y is given by (16). Initially only the low-frequency response with large damping will be considered and thus the term $m\ddot{y}$ is neglected. Then we have for the real and imaginary parts of χ_{dw}

$$\chi'_{dw} = \frac{4M_s{}^2}{d}\frac{\kappa}{\kappa^2 + (\omega\beta)^2}, \qquad \chi''_{dw} = \frac{4M_s{}^2}{d}\frac{\omega\beta}{\kappa^2 + (\omega\beta)^2} \tag{21}$$

The quality factor Q (the ratio of the stored energy to the energy lost per cycle) if $4\pi\chi' \gg 1$, is given by,

$$Q = \chi'_{dw}/\chi''_{dw} = \kappa/\omega\beta \tag{22}$$

These two equations set forth some of the general materials requirements needed to meet specific goals. As seen from Eq. (21), the low-frequency susceptibility ($\omega \sim 0$) is proportional to $1/\kappa$, and the relaxation frequency at which χ_{dw} has fallen to half value is κ/β. So a large susceptibility requires a small κ, while a high relaxation frequency requires a large κ. The damping β should be small but not less than $1.4(m\kappa)^{1/2}$ since smaller values of β will give rise to domain wall resonance. Thus the exact materials requirements are a compromise determined by the permeability and frequency range needed.

In order that κ not be zero in our model, it is necessary to consider the domain wall to be fixed at some points. The spring constant κ then arises from the added energy resulting from stretching the wall

* All equations are in the Gaussian system of units.

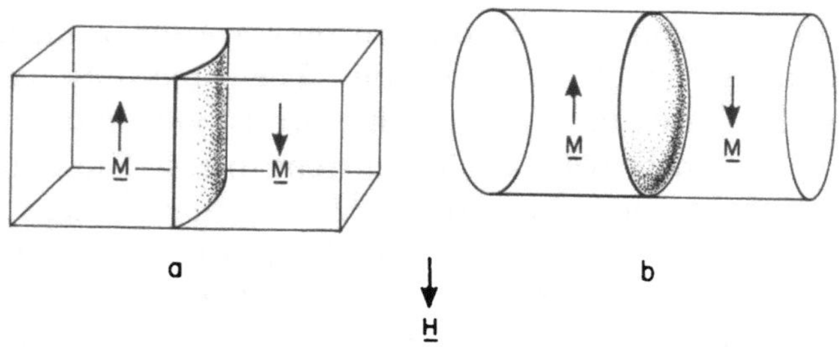

Fig. 15. Two possible types of domain wall pinning. (a) The wall is pinned only along two edges. (b) The wall is pinned along the whole circumference.

and from additional demagnetizing energy produced as the applied field changes the domain wall configuration. In Figure 15(a) a situation where only additional wall energy is important is schematically shown. A 180° wall assumed fixed along both ends bows out under the influence of the applied field in such a way that no additional discontinuities in **M** are produced. The original calculation of this case by Kersten is reviewed in Ref. 88 and we find $\kappa = 18\sigma_w/d^2$. Here σ_w is the wall energy. In Figure 15(b) a circular wall fixed around the circumference is shown. As this wall bows out, discontinuities in **M** are introduced and demagnetizing effects become important. In fact as shown in Ref. 24, the demagnetizing energy is much larger than the additional wall energy and so the wall energy may be neglected. For this case $\kappa = K_1^2/5M_s^2d$. The anisotropy constant appears in this expression since **M** tends to rotate somewhat so as to decrease the demagnetizing energy. The low-frequency susceptibility χ_0 for a wall with limited pinning then is

$$\chi_0 = 2M_s^2d/9\sigma_w \sim M_s^2d/9(AK_1)^{1/2} \tag{23}$$

and for a wall pinned on all sides is

$$\chi_0 = 5M_s^4/\pi K_1^2 \tag{24}$$

From (23) and (24) it is obvious that for a large χ_0 the magnetization should be as large as possible and K_1 as small as possible. The exchange constant A cannot be varied substantially if T_C is to remain well above room temperature. So far the influence of internal strains has been neglected. If internal strains are present, the magnetostrictive energy adds to the anisotropy energy, and in Eqs. (23) and (24), λT should be added to K_1, where T is the tensile stress. For simplicity an isotropic material with $\lambda_{100} = \lambda_{111} = \lambda$ has been assumed. An additional requirement for a large χ_0 then is that λT be small.

The model that leads to (24) gives a χ_0 that is not explicitly structure sensitive. This lack of structure sensitivity is contrary to almost all experimental observations and so the model that leads to Eq. (23) appears to be a better description of the actual domain configuration present even though the details of the interaction between grains cannot be specified. In any case, the models should be considered illustrative rather than an accurate description of the domain configuration present.

If the resistivity of the material is large, the damping β is, in principle at least, determined by the considerations given in Section 3.2. If eddy current damping is important, this additional loss must be considered. The calculation of the losses due to eddy currents is extremely difficult since they depend strongly on the exact domain pattern present. We will only consider two cases. First we make the somewhat artificial assumption that in our model the eddy currents are limited to each grain, that is, the grain boundaries are insulating. For this condition[89]

$$\beta \sim 10^2 M_s^2 d/\rho c^2 \tag{25}$$

A typical value for domain wall damping in the absence of eddy currents is $\beta = 0.1$ dyn sec/cm^3 (Section 3.2). For $d = 0.1$ cm from Eq. (25) we obtain that a resistivity of 4×10^{-3} ohm-cm will give the same β due to eddy currents if M_s is taken as 200 G. Second we assume that there are so many walls that the magnetization change may be taken as uniform. This assumption underestimates the loss and gives[89]

$$Q = 12(\delta/L)^2 \tag{26}$$

where L is the sample dimension and δ the skin depth. For both (25) and (26) it was assumed that $\delta \gg L$. For $L = 1$ cm a frequency of 10^6 Hz, and $\chi = 10^2$ we obtain $Q = 100$ if $\rho = 40$ ohm-cm.

The preceding, somewhat abstract discussion will now be applied to actual materials. A vast number of compositions have been investigated but we will only consider some of the more technically important materials. At low frequencies where eddy current damping is not too large, iron has a distinct advantage. The moment of iron ($4\pi M_s = 22{,}000$) is about a factor of five greater than that available with ferrimagnetic oxides. From Eq. (23) it can be seen that for an equal χ_0 the ferrimagnetic oxides would have to have K_1 (or more accurately $K_1 + \lambda T$) three orders smaller than K_1 of iron. For iron[68] $K_1 = 4.1 \times 10^5$ ergs/cm^3. Effective anisotropy constants of the order of 400 ergs/cm^3 are in general not obtainable since second-order effects neglected in our discussion usually keep the effective anisotropy larger than 10^2–10^3 ergs/cm^3.

In deriving (23) it was assumed that the interior of each grain was free of precipitates, voids, or any other imperfection that would impede wall motion. If this is not the case, d in (23) is a measure of the free wall length and is of course smaller than the grain diameter. For a material to show a larger χ_0, it must have large crystallites and they must be free of imperfections. Iron commonly contains carbon as an impurity and this will precipitate out as Fe_3C for concentrations less than 0.025%.[90] Cioffi[68] eliminated the precipitates by using a hydrogen anneal to remove impurities. This treatment also results in a relatively stress-free sample and yields $\chi_0 = 2.1 \times 10^3$. This value of χ_0 is consistent with Eq. (23) ($\sigma_w = 1.6$ ergs/cm^2) if d is taken as 4×10^{-3} cm, which is not an unrealistic value.

Of more commercial importance than the hydrogen anneal has been the addition of about 3% Si. According to an investigation by Yensen quoted in Ref. 68, the Si causes the carbon impurity to precipitate as graphite rather than Fe_3C and the graphite has a relatively small influence on the domain wall displacement. Furthermore, Si increases the resistivity: from 10^{-5} ohm-cm for pure iron to 4.5×10^{-5} ohm-cm for 3% Si addition. This change of course decreases the eddy current loss. A decrease in K_1 of 26% also tends to increase χ_0. In addition, the Si suppresses the γ phase of iron. With 3% Si present, iron is single phase up to the melting point and this property allows a more flexible rolling and annealing procedure. Further improvements of χ_0 are expected with increasing Si content, but high-Si alloys are generally brittle. The effect of grain orientation produced by rolling is beyond the scope of this chapter and is discussed in Ref. 68.

A further increase in χ_0 is obtained with the use of Ni–Fe alloys (Permalloys). For 80% Ni–20% Fe both K_1 and λ are quite small. By slight changes in Ni/Fe ratio, one of these constants, but not both, can be made essentially zero. The value of M_s is only reduced to 8×10^2 G.[68] In this compositional range there is an order–disorder transition which influences the value of K_1 and λ. In order to suppress this transition and to increase the resistivity, about 4% Mo is added. The presence of the Mo also makes it possible to have both K_1 and λ simultaneously near zero[91] (Figure 16). For 79% Ni, 15% Fe, 5% Mo, and about 0.5% Mn,[68] $\chi_0 = 10^4$.

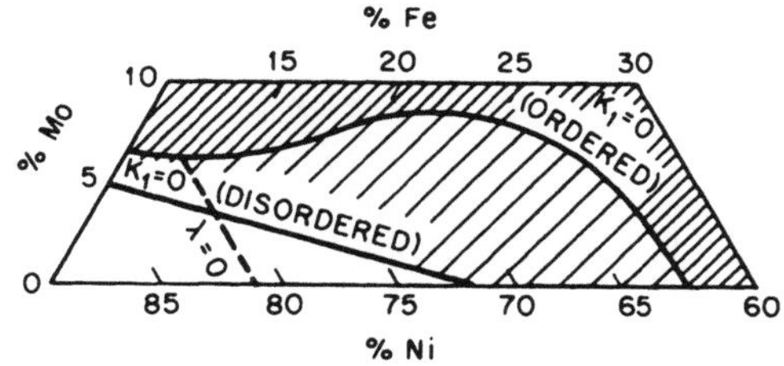

Fig. 16. The anisotropy constant K_1 and the average magnetostrictive constant λ in the Ni–Fe–Mo system. In the doubly shaded region K_1 is positive for the ordered system. In both shaded regions K_1 is positive in the disordered system. The solid lines give $K_1 = 0$ for each case. (After Ref. 91.)

To suppress eddy currents, the materials discussed are used in laminated or powder form. Consequently, the ease of rolling or powdering the alloys is an important materials consideration. Sulfur is added to the Ni–Fe alloys to obtain these in powder form. The sulfur precipitates in the grain boundaries and this yields a brittle material which is readily powdered by rolling.

At higher frequencies the large conductivity of the metallic materials just discussed would require impractically thin laminated sheets (for Fe at 10^6 Hz the sheet thickness would have to be about 10^{-4} cm). Consequently, in this frequency range ferrites with their typically higher resistivity are generally used. As discussed at the beginning of this section, the frequency range required determines some of the materials parameters. For frequencies up to about 10^6 Hz, MnZn ferrite is the favored material.

The dependence of the permeability on the mean grain size given by Ross[(92)] is shown in Figure 17, where the linear dependence of χ given by Eq. (23) is clearly demonstrated. It should be noted that Ross attributed the increase in χ to the decrease in strain and demagnetizing effects with increasing crystallite size rather than to the model that leads to Eq. (23). There are insufficient data to definitely choose between the two explanations. However, subsequently it will be seen that Eq. (23) gives a fairly good description of the experimental data and we will continue to rely on it as a framework for our discussion. The linear dependence shown in Figure 17 is only obtained if the interior of the grains is essentially free of voids. A deviation from linearity occurs at small grain sizes where the grains can no longer support domain walls and rotation becomes the dominant process.

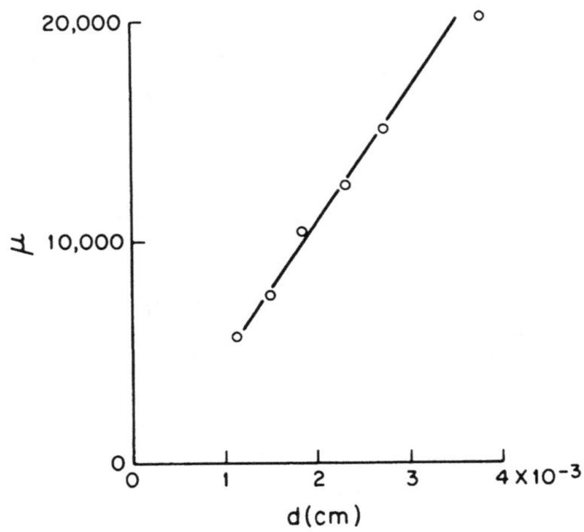

Fig. 17. The dependence of the permeability on the mean grain size for $Fe_{2.1}Mn_{0.5}Zn_{0.4}O_4$. (After Ref. 92.)

Most applications require that χ be temperature independent for some specified temperature range. The temperature dependence of χ in ferrites is generally larger than that in metals since T_C is lower in ferrites and $|K_1|$ decreases rapidly with temperature as T_C is approached. However, for a range of compositions of MnZn ferrite,[93,94] K_1 is also zero at room temperature, as is shown in Figure 18. Fortunately, as seen in the figure, for compositions near 53 mole % Fe_2O_3, 25 mole % MnO, and 22 mole % ZnO ($Fe_{2.1}Mn_{0.5}Zn_{0.4}O_4$), λ is also zero. The boundaries at which $K_1 = 0$ shift with temperature and so by adjusting the composition, the zero in anisotropy can be made to occur substantially below room temperature. In Figure 19 we show both K_1 and the permeability as a function of temperature. As expected, when K_1 is zero the permeability peaks. In between the two peaks χ is relatively temperature insensitive. The compositions that yield a region of positive K_1 (Figure 17) are all iron rich and therefore contain Fe^{2+}. Typically the resistivity of a commercial MnZn ferrite may be about 1 ohm-cm and so without further additions would lead to an unacceptably low Q. A procedure for limiting the effect of the eddy currents is to add Ca to the ferrite raw material.[95] The Ca segregates at the grain boundaries to form an insulating layer. These insulating layers yield an overall resistivity of about 10^3 ohm-cm. From the previous discussion of eddy current damping it can be seen that for grains less than about 10^{-2} cm in diameter it is the

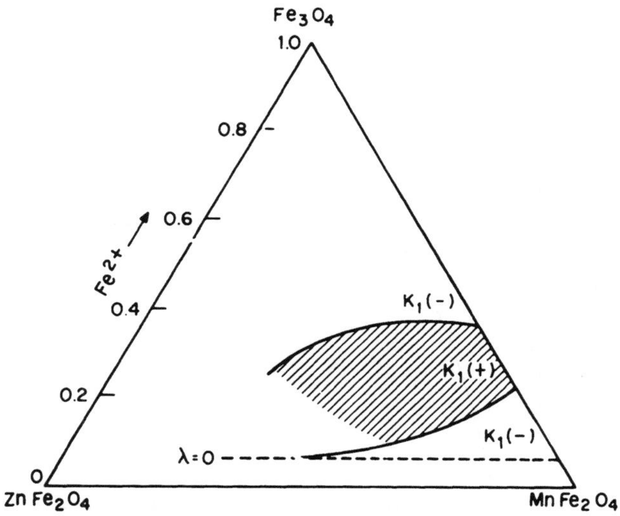

Fig. 18. The anisotropy constant K_1 and average magnetostrictive constant λ in the $(Mn\text{–}Zn\text{–}Fe)O_4$ system. (After Refs. 93 and 94.)

overall resistivity and sample dimensions rather than the resistivity of the individual grains that determine the Q at 10^6 Hz.

Typical values[96] of MnZn ferrite now in use are $M_s = 5 \times 10^2$, $\chi = 3 \times 10^2$, and $Q = 10^2$ at a frequency of 10^5 Hz. If the wall energy σ_w is taken as 0.1 erg/cm^2, which is a reasonable value for this material, Eq. (23) gives $d = 5 \times 10^{-4}$ cm, which is in good agreement with the average grain size of 7×10^{-4} cm observed. The product $\chi Q \sim 4M_s^2/d\omega\beta = 3 \times 10^4$ gives $\beta \sim 0.1$ dyn sec/cm^3. This corresponds to a domain wall mobility of approximately 10^4 cm/sec-Oe, which may be compared to a mobility[54] of 2×10^3 cm/sec-Oe obtained at 195°K with a $Mn_{1.4}Fe_{1.6}O_4$ single crystal. Considering that MnZn ferrite has a smaller K_1 and hence should tend to have a higher mobility [Eq. (11)] and also that the mobility in general increases with temperature, this agreement is quite good.

The best MnZn ferrite available to date[97] has $\chi = 10^2$ and $Q = 10^3$ at 10^5 Hz. A similar analysis yields $d = 2 \times 10^{-4}$ cm and β still approximately 0.1 dyn sec/cm^3. Since the width of a domain wall is between 10^{-4} and 10^{-5} cm in a material of this type, it appears that domain walls take up a relatively large fraction of the total volume. Thus in this case the distinction between rotation and wall displacement becomes rather small.

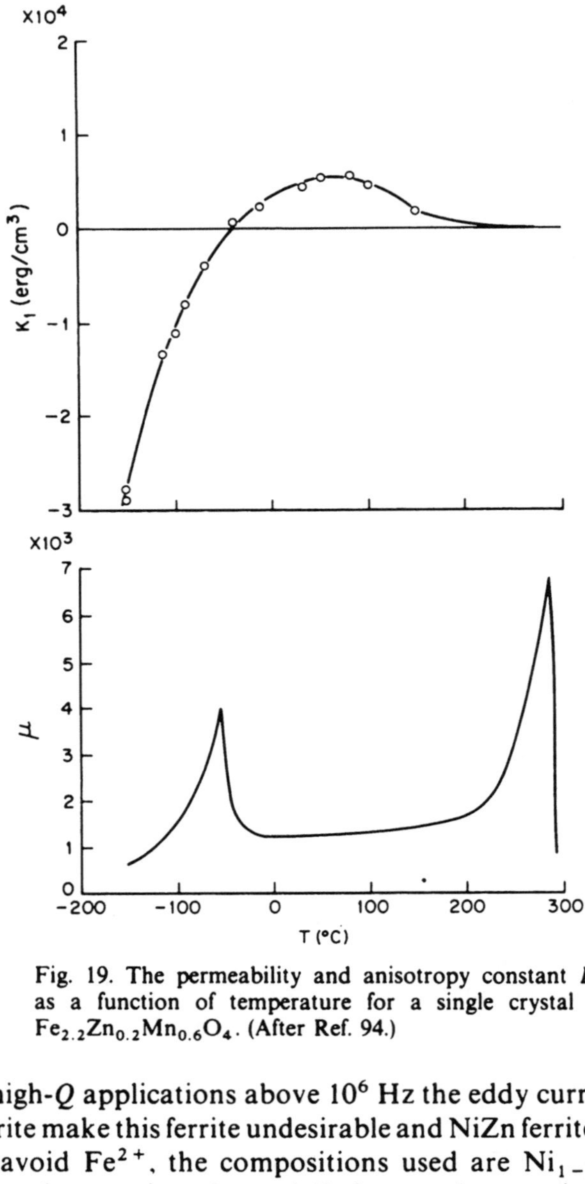

Fig. 19. The permeability and anisotropy constant K_1 as a function of temperature for a single crystal of $Fe_{2.2}Zn_{0.2}Mn_{0.6}O_4$. (After Ref. 94.)

For high-Q applications above 10^6 Hz the eddy current losses of MnZn ferrite make this ferrite undesirable and NiZn ferrite is generally used. To avoid Fe^{2+}, the compositions used are $Ni_{1-x}Zn_xFe_2O_4$. Since in practice κ varies substantially from grain to grain, the domain wall relaxation losses appear over a wide frequency range. Thus at the higher frequencies it is necessary to suppress domain wall displacement to avoid the relaxation losses associated with this process. A small-grained material such as illustrated in Figure 14(d) is desired. As stated earlier, for a rotational process resonance occurs for

$\omega_r = \gamma H_K$ and χ is given approximately by M_s/H_K. So, as was first pointed out by Snoek,[24] $\omega_r \chi \sim \gamma M_s$. This relation shows the desirability of a large M_s. The value of T_C and hence H_K decreases with increasing Zn but since M_s at 0°K increases with Zn (Section 2.1), the value of M_s at room temperature is relatively independent of the Zn content for $0.20 < x < 0.65$. The Zn content can thus be chosen to obtain the maximum χ for the frequency range required. The small grain size required (about 10^{-4} cm) may be obtained by ceramic techniques such as hot pressing.[41]

Another material of interest is $Y_3Fe_5O_{12}$. The question here is why this material does not have useful characteristics as an inductor material. YIG has very low microwave losses and consequently might be expected to be a low-loss material for frequencies in the 10^6-Hz range. It has already been pointed out in Section 3.2 that the domain wall mobility μ_0 in single crystals is three orders of magnitude less than that predicted from the microwave loss data. In fact, the measured mobility of YIG is no better than the mobility obtained with Mn ferrite. However, even the losses predicted using the measured domain wall mobility in a single crystal are much smaller than the losses observed in permeability measurements using polycrystals. This latter difference is not so puzzling, for the following reason. The activation energy for electron hopping deduced from the microwave loss peak of Si-doped YIG is 0.25 eV.[27] In the range of 10^6–10^7 Hz this results in a relaxation loss peak ($\omega\tau = 1$) at about room temperature. Thus it appears plausible that the unexpectedly high loss observed in polycrystalline YIG is due to the presence of Fe^{2+} ions. This loss due to Fe^{2+} is definitely present if the Fe^{2+} is introduced by the addition of 0.05 Si^{4+} atom per formula unit.[98] The activation energy for electron hopping in spinels is much lower and thus does not lead to a room-temperature relaxation loss.

Almost all applications require that, in addition to temperature stability, χ be stable with respect to time. However, a number of materials exhibit a change of χ as a function of time (disaccommodation). For example, Fe with C impurities shows a decrease of χ with time of approximately[88] 20–80%. This change in χ is attributed to a redistribution of the C atoms in the iron lattice. Carbon goes into interstitial sites that are located halfway between the Fe atoms at the corners of the cubic unit cell. There are three types of sites specified by the three $\langle 001 \rangle$ directions. In the absence of a magnetic vector these sites are equivalent. However in the presence of magnetization, pseudodipolar coupling tends to make the site whose $\langle 001 \rangle$ direction is along **M** energetically unfavorable and the C atoms will tend to migrate to the other two types of sites.[99] Thus the material becomes somewhat uniaxial with the ⟨001⟩ axis parallel to **M** the easiest direction of magnetization. The change in the anisotropy is too small to directly account for the observed change in χ. Typically the anisotropy may change by only 0.1 % while χ changes by 50 %. Rather it is the fact that the anisotropy energy is no longer spatially uniform that accounts for the disaccommodation. In a domain wall the local direction of **M** becomes the easy axis. For the wall shown schematically in Figure 2 (with $\phi = 0$), well away from the wall the easy axis is in the z direction, in the center of the wall ($y = 0$) x is the easy axis. In this way the wall becomes stabilized at $y = 0$ and χ decreases. The rate at which χ decreases depends on the rate at which the C atoms relax over the potential barrier between the three types of sites.

Analogous mechanisms result in disaccommodation in spinels. For these materials at room temperature it appears that cation vacancies diffuse among the four nonequivalent octahedral sites. The presence of Fe^{2+} seems to facilitate the cation migration.[100] There is at present, however, no independent evidence to show that vacancy migration does indeed have the activation energy observed for disaccommodation.

4.2. Coercive Force

As was discussed in the introduction, the susceptibility is determined by only the small reversible changes in the magnetization from the demagnetized state. The coercive force H_c is the field required to reduce the magnetization to zero from the saturated state and this implies large irreversible motion of the magnetization. At this point two cases may be defined: (1) for $H = 0$ (point A in Figure 1) there are no walls present; (2) for $H = 0$ there are walls present. For case 1 the coercive force is determined by the field required to nucleate walls since the nucleation field is in general larger than the field required to move walls. The problem of nucleation of domain walls is discussed in Ref. 101 and experimentally is observed with good single crystals of yttrium orthoferrite.[102] If the walls are already present (case 2) H_c is determined by various impedances to wall motion. The large wall displacements result in very complex domain configurations and it is not profitable to develop a simple model such as was used to discuss χ in the preceding section. Instead more general arguments must be employed.

In a perfect crystal the wall (if already nucleated) would be free to move and so H_c would be zero. However, almost all imperfections will impede the motion of the wall. The interaction of walls and crystal defects is most easily demonstrated in single crystals that have few imperfections and a simple domain configuration. Some of these interactions are shown in Figure 20. Using a similar technique it was found that in orthoferrites H_c due to dislocations is 1 Oe; due to growth striations, flux inclusions, voids, and precipitates H_c ranges from 0 to 50 Oe; intersections of growth surfaces give H_c between 1 and 50 Oe; grain boundaries with an angle θ give $H_c = 100\theta^2$ Oe; and $\langle 110 \rangle$ twin planes lead to $H_c > 10^3$ Oe.[104,105]

Unfortunately, the data obtained with single crystals do not give a great deal of insight into the polycrystalline case since in polycrystals the domain configuration and the types and distribution of defects are not known. Furthermore, polycrystals may have a large,

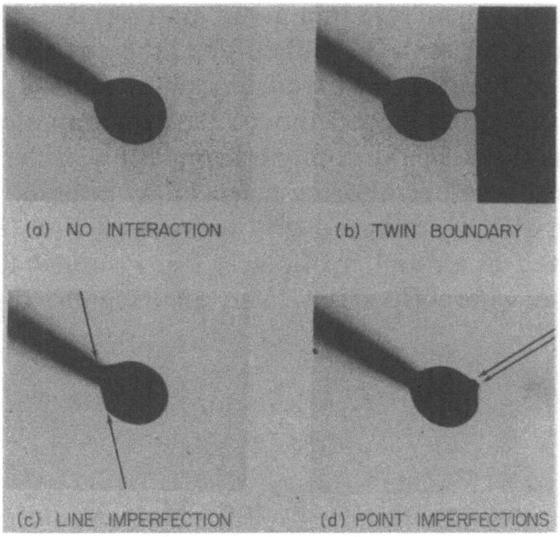

Fig. 20. A cylindrical domain (3×10^{-2} cm in diameter) in a thin crystal of $(SmTb)FeO_3$ stabilized at the tip of a magnetized Vicalloy probe. The light and dark regions, observed by the Faraday effect, correspond to the magnetization directed into and out of the plane of the figure. The domain is being moved from right to left. (a) The domain wall is free of defect interactions. (b) The wall has been pinned at a twin boundary. (c) The wall is impeded as it encounters a linear imperfection, such as a scratch. (d) The wall has been pinned by two point imperfections, such as inclusions. (After Ref. 103.)

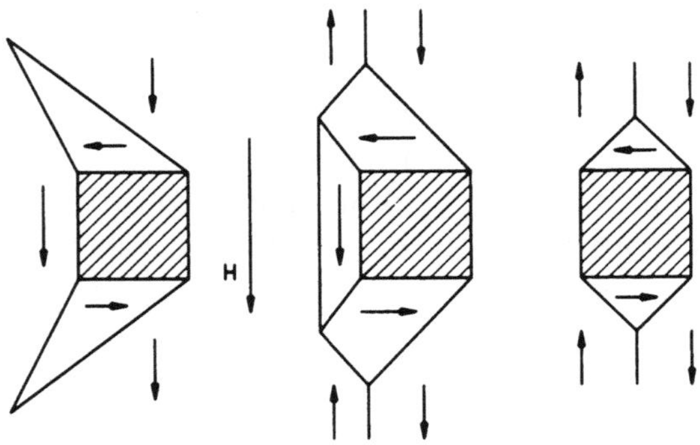

Fig. 21. Possible domain configurations around a defect as a domain wall passes through. (After Ref. 107).

nonuniform internal stress present. Thus a few idealized imperfections are considered and their role in a given material is only deduced by relatively indirect means. The difficulties inherent in predicting H_c are illustrated in Figure 21 where a possible series of domain configurations is shown as a wall moves past an imperfection. For a differently shaped defect, the configuration will be different. Also the closure domains from neighboring defects may interact. Much of the theory dealing with H_c is based on the work of Kersten and Néel, which is reviewed in a number of places (for instance, Ref. 106).

In the case where H_c arises from an irregular distribution of internal stress

$$H_c \sim \tfrac{1}{2}v(\lambda T)^2/K_1 M_s \qquad \text{for} \qquad \lambda T < K_1 \tag{27}$$

and

$$H_c \sim v\lambda T/M_s \qquad \text{for} \quad \lambda T > K_1 \tag{28}$$

where v is the fraction of the volume influenced by the stress. The coercive force caused by nonmagnetic inclusions is

$$H_c \sim K_1 p/M_s \tag{29}$$

where p is the volume fraction of the nonmagnetic material. So for materials where $K_1 > \lambda T$ or $(\lambda T)^2/K_1$, inclusions are expected to play a more important role in determining H_c than is the nonuniform stress. For example, if $T = 3 \times 10^9$ dyn/cm^2 (the approximate yield stress) and $v = p$, Eqs. (27)–(29) indicate that in Fe, inclusions determine H_c, in Ni, the stress contribution predominates, and in $MnFe_2O_4$ ($K_1 = 3 \times 10^4$ ergs/cm^3, $\lambda = 5 \times 10^{-6}$, $M_s = 400$ G), both mechanisms contribute about equally. Of course, the relative importance of stress or inclusions depends on the value of T chosen and this value is in practice only very approximately known. Equations (27)–(29) all show that H_c increases as M_s decreases. This result is obvious since the driving force on a wall is the magnetostatic energy ($\mathbf{M} \cdot \mathbf{H}$) while the forces retarding the wall depend on both λ and K_1 in addition to M_s. The dependence on M_s is dramatically demonstrated by materials that have a compensation point.[108] At the compensation temperature M_s goes to zero but K_1 and λ are not very temperature sensitive. So for an ideally uniform material and neglecting second-order effects, H_c is expected to approach infinity at the compensation temperature. In rare earth iron garnets, if H_c is plotted as a function of temperature, the peak at the compensation point is typically 50 Oe high and the width at the half-maximum points is 2–7°C wide. For these crystals the finite value of H_c is attributed to the susceptibility of the rare earth lattice.[109]

Equation (29) is incomplete inasmuch as the size of the inclusions is not specified. Quite complicated analysis shows that, for constant p, if the diameter of the inclusions is equal to the wall width, the effect on H_c is a maximum.[110] This behavior is shown in Figure 22, where H_c is plotted as a function of the diameter of the Fe_3C precipitate in iron. (The precipitate size is determined in part from a knowledge of the precipitation process rather than from actual observation.) The

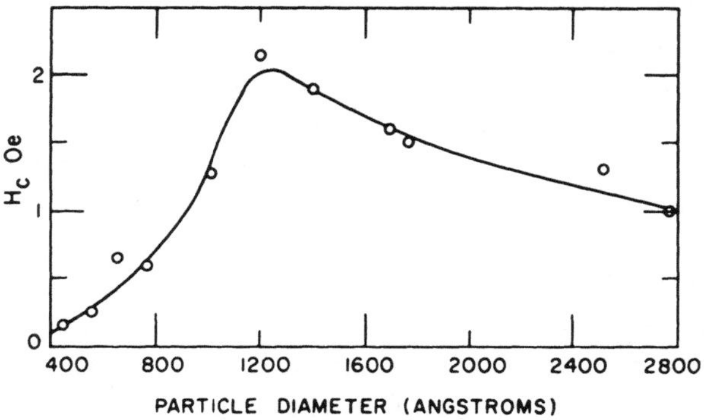

Fig. 22. The coercive force as a function of the diameter of the Fe_3C precipitate in iron. The volume fraction of the precipitate p was 3×10^{-3}. (After Ref. 110.)

clearly demarked peak at 1200 Å agrees well with the calculated wall width of iron. Control of H_c by precipitation has been employed in a number of cases in addition to C in Fe. For example, Au precipitates have been used to control H_c in Permalloy[111] and Zr, Ti, Be, and Au have been used to raise H_c of Co–Fe alloys.[91] While some investigations with Co ferrite have been carried out,[112] the precipitation technique has not been extensively used with the ferrimagnetic oxides. Instead, by careful control of the sintering temperature, voids are left in the grains and at the grain boundaries. To eliminate the rather sensitive underfiring required to obtain the desired void distribution, either palladium or thoria particles about 10^3 Å in diameter are added to the ferrite raw material.[113,114] With these materials the desired H_c (1–4 Oe) is obtained by adjusting the particle content [p in Eq. (29)] and H_c is quite insensitive to the sintering time.

So far we have neglected the effect of grain size. For reasons very similar to the ones given in Section 4.1 showing that χ decreases with decreasing grain size d, H_c increases with decreasing d. For example, in $(YGd)_3Fe_5O_{12}$, H_c increases from 0.8 to 2.8 Oe as d decreases from 15×10^{-4} to 1×10^{-4} cm.[115] As a rule, it is impractical to obtain dense samples with d less than 10^{-4}–10^{-5} cm and consequently the effect of grain size cannot be investigated in this size range independent of porosity.

The discussion so far has tacitly assumed that the porosity p is relatively small and that the material can support a sufficient number of domain walls so that the magnetization can be reversed by wall motion. The coercive force in this regime will vary roughly between tenths of Oe to a few hundred Oe. For higher values of H_c it is convenient to first consider separate fine particles rather than a polycrystalline sample. Sufficiently small particles cannot support domain walls and the value of H_c that results from rotational reversal of the magnetization is substantially simpler to calculate than the H_c due to wall motion. Also, the rotational process sets the upper limit on H_c, the coercive force due to wall motion always being lower.

Before continuing further it is worthwhile to consider very briefly the application of high-H_c materials. High-H_c magnets (permanent magnets) are used almost exclusively to produce an external magnetic field, for instance, the field in the air gap of a "horseshoe" magnet. The continuity requirement on **B** at the magnet–air boundary shows that a large field in the air gap requires a large M_s. Also the magnetic field in the gap is proportional to the internal demagnetizing field in the magnet. Since the demagnetizing field cannot be larger than H_c

without decreasing M to zero (Figure 1), it is clear that a large H_c is also required. The details of the materials requirements in practice are only approximated by the preceding argument and in detail are somewhat complex.[116]

For simplicity only spherical and cylindrical fine particles will be considered. Simple analysis gives that for uniform rotation of **M** with the field along the long axis of the cylinder

$$H_c = (2K_u/M_s) + (2\pi M_s)D + (3\lambda T/M_s) \qquad (30)$$

where $D = 1$ for a cylinder and $D = 0$ for a sphere. To derive (30), it was assumed that the material was uniaxial with the easy axis of magnetization parallel to the long axis. Also, the stress T is assumed to be along the long axis. With only minor changes in constants (30) is valid for materials having other types of anisotropy (cubic, hexagonal, etc.). The first term in (30) arises from the anisotropy energy, the second from demagnetizing energy, and the third from the stress energy. Since typically the maximum values of λ and T are about 10^{-5} and 10^9 dyn/cm^2 respectively, the third term is in general small compared to the first and will henceforth be neglected.

The shape anisotropy has been exploited in the development of elongated single-domain (ESD) magnets.[117] Fine, elongated particles of Fe or Fe–Co alloys are formed by electrodeposition from an aqueous electrolyte into a mercury cathode. If, as is the case here, the demagnetizing effects control H_c, the particle diameter must be even smaller than the diameter at which the particle can no longer support a domain wall. The reason for this is that above a critical diameter even a single-domain particle will not reverse by coherent rotation but instead by some nonuniform rotational mode which minimizes the demagnetizing energy and hence H_c. ESD magnets having particle diameters between 100 and 300 Å have been produced which have a coercive force of about 10^3 Oe.[117]

Permanent magnets that depend on the anisotropy energy must still be composed of single-domain particles if the high H_c associated with the rotational process is to be exploited. However, in this case H_c is not as dependent on particle shape or packing density as was the case with ESD magnets since even nonuniform rotation must overcome the anisotropy energy. The required fine particles are obtained either by grinding or by precipitation of the magnetic material into a nonmagnetic matrix. The use of powder as a starting material affords added flexibility since the particles may be oriented in a magnetic field before compacting. Orienting the powder allows adjustment of the B–H loop to meet specific device applications. Two important examples of powdered magnets where the anisotropy energy predominates are hexagonal ferrites ($BaFe_{12}O_{19}$) and Co–rare earth alloys, the first having the lower value of K_1 ($K_1 = 3.3 \times 10^6$ ergs/cm^3, $H_k = 1.7 \times 10^4$ Oe). In practice finely ground oxide particles are sintered so as to maintain the smallest possible grain size consistent with the needed density.(24) A high density or low porosity p is desired since the effective magnetization of the material is approximately given by $(1 - p)M_s$. The alloy $SmCo_5$ has a very large anisotropy field ($H_k = 2.5 \times 10^5$ Oe) and yields one of the most powerful permanent magnets produced to date. The powdered starting material is pressed to the highest possible density using a number of different techniques.(118) To avoid the necessity of grinding the very fine particles required, Nesbitt(119) used the alloy Co_3Cu_2Sm to precipitate nonmagnetic inclusions in the magnetic material. With nonmagnetic inclusions to impede wall motion a coarser powder still has a high coercive force.

So far we have not considered how well the single-domain picture describes the magnets just considered. As a matter of fact, the measured H_c is substantially lower than the predicted by Eq. (30). For $SmCo_5$, $H_c = 2.5 \times 10^4$ Oe, almost an order of magnitude less than is expected from Eq. (30).(118) Crystal oriented $BaFe_{12}O_{19}$ has a coercive force of about 1.5×10^3 Oe, also well below that expected for rotational magnetization reversal.(24) These deviations are attributed to the fact that the particles are not truly single domain.(120) In some cases the grain diameter ($\sim 10^{-4}$ cm) is too large for single-domain behavior. Even for smaller particles local deviations of the value of K_1, sharp edges, surface imperfections, and particle interaction may all act to form domain walls in particles that might otherwise be expected to be single domain.

Up till now we have only discussed permanent magnets where the fine particles are distinct and fairly well defined. In many cases where the particles are produced by precipitation the situation is not so simple since the shape and composition of the precipitate depend critically on the heat treatment employed. Also the precipitate and the matrix are often both magnetic. For example, Alnico 5 (24% Co, 14% Ni, 8% Al, 3% Cu) consists of two phases whose size, shape, and magnetic properties can be altered by various annealing treatment. Torque(121) measurements on single crystals of Alnico show that the second term (shape effect) in Eq. (30) is the major contribution to H_c. Thus, as might be expected, there is evidence that the highest H_c is reached when the matrix is nonmagnetic.[107] In cobalt–platinum (CoPt) the large H_c is attributed to the high anisotropy of a face-centered tetragonal ordered phase precipitated in a disordered face-centered cubic matrix.[122]

The preceding two compositions are just two examples of the large number of permanent magnets now in use produced by precipitation. The general principles outlined here allow one to interpret the behavior of these materials. For the details concerning composition, heat, and mechanical treatment a number of sources, such as Refs. 107, 123, and 124, may be consulted.

4.3. *B–H* Loop

In the preceding section we discussed the origin of the coercive force H_c. In this section we will consider the shape of the whole *B–H* loop. (While the plot of magnetization versus applied field is usually obtained as *B–H* loop, it is more convenient to consider an *M–H* loop; unless H is comparable to M, there is, of course, only a negligible difference.) A large variety of *B–H* loops are observed and to keep this section within reasonable bounds we will describe only three types. These three types were chosen either for their practical importance or because they demonstrate points of special interest.

We will start our discussion of the *M–H* loop by considering the polycrystalline sample to be a macroscopically closed toroid so that there are no demagnetizing effects due to the shape of the sample. The field is applied circumferentially and is taken as large and negative (point *B* in Figure 1). At this point, if H is larger than the effective anisotropy field, all the magnetization in the various grains is parallel to H and $M = M_s$. As the absolute value of H is decreased, the magnetization in each grain will tend to rotate toward the nearest easy axis. At $H = 0$ (point *A* in Figure 1), the magnetization will be along that easy axes. The value of M at point *A* is denoted by M_r, the remanent magnetization. If the material is uniaxial, the appropriate average over all possible orientations gives $M_r/M_s = 0.5$. For cubic material with $K_1 < 0$, $M_r/M_s = 0.87$ and for $K_1 > 0$, $M_r/M_s = 0.83$.[88] The preceding statements rest upon the assumption that each grain is single domain and in the case of cubic materials that the effects due to magnetostriction are much smaller than the effects due to the crystalline anisotropy. In a cubic crystal, if λT is larger than K_1, the material will take on a uniaxial character.

In real materials the grains are probably not all single domain since demagnetizing effects at grain boundaries and at pores or non-magnetic second phases may result in closure domains. (Figure 21 shows a possible closure configuration around a pore.) These closure domains, of course, reduce M_r from the calculation based on a single domain per grain. Calculation shows that for typical ferrites (8×10^4 $cm^{-1} \leq (K_1/A)^{1/2} \leq 2 \times 10^5$ cm^{-1}) closure domains do not form if the pore size is less than 4–11 $\times 10^{-4}$ cm and for metals (6×10^5 $cm^{-1} \leq (K_1/A) \leq 10^6$ cm^{-1}) the minimum pore size for a closure domain is 0.5–2.0 $\times 10^{-4}$ cm.[125] For a porosity p of 0.03 and all the pores of minimum size, the ratio of M_r/M_s is decreased by 10%. Similar calculations lead to a critical grain size below which closure domains will not form.[126] For $K_1 \gg M_s^2$ this critical size is approximately $3(AK_1)^{1/2}/M_s^2$ and for $M_s^2 > K_1$ it is about $40(A/K_1)^{1/2}$. (We have assumed $\lambda T \ll K_1$.) For a representative ferrite ($A = 10^{-6}$ erg/cm, $K_1 = 10^4$ ergs/cm^3) the critical radius is 4×10^{-4} cm, which is of the order of magnitude of the grain size commonly obtained. In part due to the difficulty of controlling porosity, grain size, and, with the ferrites, composition, there are no experiments that systematically investigate the effect of grain size and porosity on the M_r/M_s ratio. Tables of M_r/M_s for various ferrites, such as given in Ref. 24, do not give both grain size and porosity. However, it does appear to be well established that, as suggested previously, a large M_r/M_s ratio in a cubic material requires that $K_1 > \lambda T$. In more detail, this requirement, for K_1 negative, is that $|K_1| > \lambda_{111} T$ since for a negative anisotropy constant the magnetization lies along the $\langle 111 \rangle$ direction and so λ_{111} is the pertinent magnetostrictive constant.[127] With $Ni_{0.7}Fe_{2.2+x}\square_{0.1-x}O_4$, where $\square$ represents a cation vacancy, Wijn[127] used controlled atmospheres during firing to vary x and thereby obtained a series in which λ_{111} passed through zero. The samples with the lowest λ_{111} had the largest ratio of M_r/M_s. In addition, the Mg–Mn ferrite system is known to yield high values of M_r/M_s[24] and λ_{111} goes through zero in this system when the Mn concentration is approximately equal to the Mg concentration (the exact composition is not given).[128]

To continue, we will neglect the complication of closure domains and furthermore assume each individual grain has a coercive force equal to H_{cg}. Magnetostatic interaction between grains will also be neglected. An individual grain then will reverse if the component of the applied field **H** antiparallel to **M** in the grain is equal to H_{cg}. With increasing field from point A (Figure 23) there is a slight change in M

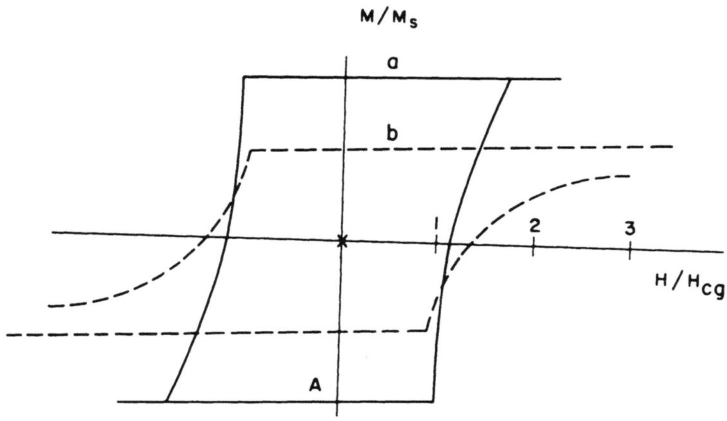

Fig. 23. A M–H loop derived from the model in which each individual grain has a coercive force of H_{cg}: (a) cubic with $K_1 < 0$; (b) uniaxial.

as the magnetization in each grain rotates in the anisotropy field toward **H**. (H_{cg} is taken as much smaller than K_1/M_s.) At $H = H_{cg}$ the grains whose easy axes are parallel to H reverse. For cubic material ($K_1 < 0$) the largest possible angle between the easy axes ($\langle 111 \rangle$) and the field is 55°. Therefore when $H = H_{cg}/\cos 55° = 1.7H_{cg}$ the domain wall reversal is complete. For uniaxial materials some grains have their easy axes nearly perpendicular to **H** and domain wall reversal and rotation will merge under these conditions (Figure 23).

At this point it is convenient to consider the computer application of square-loop material. Small toroids of square-loop material are used as memory elements. Since this application requires flux reversal times of the order of 10^{-6} sec, mostly ferrite cores with their high resistivity are used. For Permalloy tapes 10^{-3} cm thick or thicker the reversal time is dominated by eddy current effects, and thus only tapes thinner than this have found much application.(129) Since the use of Permalloy is quite limited, we will consider only ferrite cores. Basically the requirement on the material is that there exist a field H_m that produces no irreversible flux change and that a field of $2H_m$ reverses the toroid. This feature permits coincident current selection of a specific core in an array(130) (Figure 24). Since, as was mentioned earlier, even a cubic material in which strain predominates behaves as a uniaxial material, it is obvious from Figure 23 that a crystal with $K_1 > \lambda T$ (or specifically $|K_1| > \lambda_{111}T$ for $K_1 < 0$) is required for square-loop operation. This condition is fulfilled in the Mg–Mn

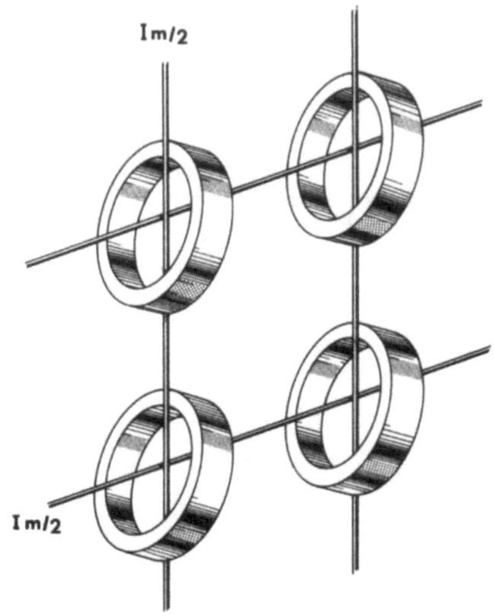

Fig. 24. An array of cores for coincident current select. Only the core with both the wires carrying a current $I_m/2$ can be reversed. The current I_m produces a field H_m; see text.

system, which is a commonly used square-loop composition. In fact, in a number of systems the addition of Mn decreases λ_{111}. This is true in ytterium iron garnet,[131] lithium ferrites,[132] and, of course, Mg ferrite. In view of this it is not surprising that almost all square-loop compositions contain Mn. Besides the requirement on λ_{111}, the microstructure must be such that the spread in H_{cg} is as small as possible. It appears that the presence of a large number of pores which have a diameter below the critical diameter for closure domains leads to a small spread in H_{cg}.[133] Also, while not included in the calculation in Ref. 133, it seems likely that interaction between grains will favor a square loop if the grain size is smaller than that required for closure domains. We conclude our discussion of square-loop ferrites by remarking that most of the preceding guidelines for obtaining square loops were developed after the fact and did not have much impact on the practical development of these materials. Instead, a suitable composition and an appropriate sintering treatment were optimized by empirical means and in general neither the exact composition nor the details of the microstructure are available.

Values of M_r/M_s larger than 0.87 can be obtained if the easy axes are not randomly distributed. As mentioned in Section 4.2, single crystals of $BaFe_{12}O_{19}$ or $SmCo_5$ may be oriented in a magnetic field before sintering. Care must be taken so that the preferred orientation is not lost during the sintering process. For barium ferrite the preferred orientation is actually improved during grain growth as the correctly oriented crystallites grow at the expense of the misoriented ones.[(24)] For oriented $BaFe_{12}O_{19}$ values of M_r/M_s equal to 0.94 have been obtained[(24)] and for $SmCo_5$ values up to 0.97 have been reported.[(134)]

Grain orientation in metals can be obtained by the appropriate rolling or wire-drawing procedures which produce a deformation texture, or by annealing of the cold-worked material to obtain a recrystallization texture. The details are available in Volume 1, Chapter 4. This type of control of the squareness has found use in some computer applications where thin tapes of Permalloy are used as a memory element.[(91)]

With randomly oriented grains an M_r/M_s ratio approaching one can be obtained if an anisotropy is induced along the circumference of the toroid. Induced anisotropies (see Section 5.1) comparable to the crystalline anisotropy are possible if ionic migration can take place below the Curie temperature T_C. At some elevated temperature, but below T_C, a field is applied to the sample that aligns the magnetization, and ionic rearrangement occurs which tends to stabilize this direction of magnetization. Important examples are ferrites with Co additions,[(24)] Ni–Fe alloys (Permalloy),[88] and Ni–Fe–Co alloys (Permimvar).[(68)] In Ni–Fe alloys an anisotropy can also be induced by rolling. This anisotropy arises from a preferential ionic arrangement and not the crystal texture just discussed (see Volume 1, Chapter 4 and Ref. 88).

An interesting consequence of large induced easy axes occurs when the sample is cooled in the demagnetized state. The induced easy axes then stabilize the domain walls present in the same manner as that which leads to disaccommodation (Section 4.1). To alter this stabilized domain configuration requires a relatively large field and consequently a constricted loop such as shown in Figure 25 is obtained. This loop occurs only for small changes in M, and is a minor loop. If M is taken to saturation, the original stabilized domain configuration is lost and a normal loop is obtained.[(135)] This method of suppressing domain wall motion has been suggested as a way to eliminate wall losses in high-frequency inductor ferrites (Section 4.1).[(136)]

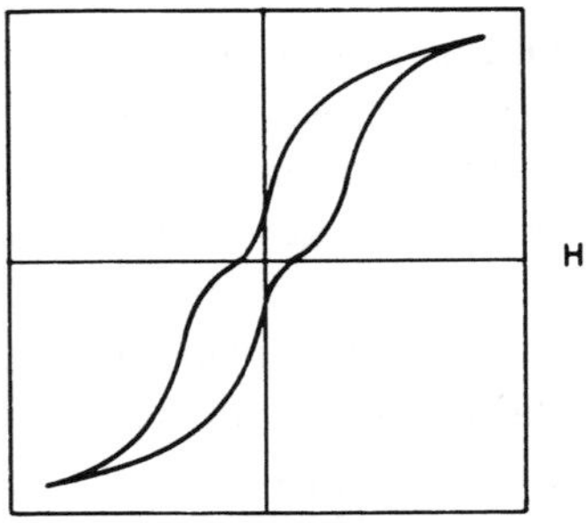

Fig. 25. A constricted M–H loop.

5. Lowered Symmetry

5.1. Field- and Growth-Induced Anisotropy

For a number of magnetic materials a field-induced anisotropy is produced if a magnetic field is applied to the sample in a temperature range in which either atomic or electron migration is possible. Field-induced anisotropy has already been mentioned in connection with disaccommodation (Section 4.1) and constricted M–H loops (Section 4.3). Usually field-induced anisotropy is investigated by applying a magnetic field at some elevated temperature T' at which migration is fairly rapid and subsequently measuring the anisotropy at some lower temperature T where migration is slow. In general, electron migration is rapid at room temperature and therefore the induced anisotropy due to electron migration is only of interest at low temperatures. A theory based only on symmetry arguments and limited to cubic crystals is reviewed by Slonczewski.[137] In this development of the theory the induced anisotropy is assumed to arise from N atoms (or pairs) per unit volume. Each of these atoms (or pairs) can be situated on a number of possible distinct sites or orientations. The ith site or orientation has an energy E_i and an axis of local symmetry $\mathbf{L}_i$. For individual atoms $\mathbf{L}_i$ is the axis of local crystal field symmetry and for pairs $\mathbf{L}_i$ is parallel to the pair axis.

To continue, we let the applied field hold the magnetization $\mathbf{M}$ at T' parallel to a unit vector $\boldsymbol{\beta}$. The material is then quenched to the lower temperature T after which $\mathbf{M}$ has an arbitrary orientation specified by the unit vector $\boldsymbol{\alpha}$. Then it can be shown that the leading terms in the energy due to the heat treatment at T' are given by

$$E = -F(\alpha_1^2\beta_1^2 + \alpha_2^2\beta_2^2 + \alpha_3^2\beta_3^2) - \\ -G(\alpha_1\alpha_2\beta_1\beta_2 + \alpha_2\alpha_3\beta_2\beta_3 + \alpha_1\alpha_3\beta_1\beta_3) \quad (31)$$

where α_p and β_p are the components of $\boldsymbol{\alpha}$ and $\boldsymbol{\beta}$, respectively, with respect to the cube axes. Since the symmetry arguments in both cases are identical, it is not surprising that Eq. (31) has the same form as that given for the elongation of a crystal due to magnetostriction (Section 2.4). Equation (31) represents an orthorhombic energy surface and therefore after magnetic anneal a cubic crystal is no longer magnetically cubic. The values of F and G are in part determined by $\mathbf{L}_i$. Tables of these constants for various $\mathbf{L}_i$ are given in Refs. 137 and 138. For the present purposes it is sufficient to note that $G = 0$ for $\mathbf{L}_i$ parallel to $\langle 100 \rangle$, $F = 0$ for $\mathbf{L}_i$ along $\langle 111 \rangle$, and $F/G = 1/4$ for $\mathbf{L}_i$ along $\langle 110 \rangle$. For a polycrystal, appropriate averaging of Eq. (31) gives $F/G = 1/2$.

Much of the impetus for the development of Eq. (31) came from an effort to understand the field-induced anisotropy in Ni–Fe alloys. The important magnetic alloys in this system, known as Permalloys, contain more than 40% Ni and form a continuous series of solid solutions with a face-centered cubic structure.[68] The induced anisotropy of these alloys is attributed to the directional order of like-atom pairs.[88] Primary evidence for this conclusion is the dependence of K_u, the magnitude of the induced anisotropy, on composition. Calculation of the theoretical dependence on composition is not totally straightforward since the Curie temperature also varies with composition and the number of pairs formed at T' depends on $T_C - T'$. However, after correcting for the change in T_C the value of K_u is fairly well represented by $C_{Ni}^2(1 - C_{Ni})^2$, which is the expected form for like-pair ordering in an ideal solution.[139] Here C_{Ni} is the concentration of Ni, which was kept larger than 0.4 in this series of experiments. The maximum K_u for $C_{Ni} \sim 0.5$ is approximately[88,139] 5×10^3 ergs/cm^3. To obtain this agreement of K_u with concentration, T' must be above the critical temperature for long-range superstructure formation (i.e., ordering of the Ni and Fe). Obviously the tendency to form long-range order will decrease the possibility for directional ordering of like-atom pairs. If long-range order is possible, a relatively complicated dependence of K_u on composition may be observed. For example, 40% Co–Fe forms a superlattice and the induced anisotropy for this composition is low compared to compositions that do not form long-range order.[137] Another effect of long-range order is seen in the role of oxygen in the magnetic anneal of Ni–Fe alloys.[140] Oxygen-free samples do not develop a uniaxial anisotropy, but they do order very readily. Oxygen-contaminated samples do not order as readily, but they do develop a uniaxial anisotropy. If these observations are correct, the induced anisotropy is not really intrinsic to the pure alloys.

If the simple pair ordering model were strictly correct, the expected F/G value for Ni–Fe would be 1/4 since the nearest-neighbor pair direction in a fcc crystal is $\langle 110 \rangle$. Actually, the measured F/G ratio is about 1/9, indicating that more complicated configurations than pairs must be considered.[88] Similar observations can be made for other alloy systems.[88,137]

In the fcc crystals just considered, preferential occupation by single ions cannot give rise to an induced anisotropy since the local symmetry of each site is cubic and $\mathbf{L}_i$ cannot be defined. However, in spinel and garnets the octahedral sites have a trigonal symmetry with the symmetry axis along one of the four $\langle 111 \rangle$ directions and therefore single-ion preferential occupation of these sites can lead to an induced anisotropy. In this case ($\mathbf{L}_i$ along $\langle 111 \rangle$) F will be zero. Perhaps the clearest demonstration of this effect is shown by Fe_3O_4 (magnetite)

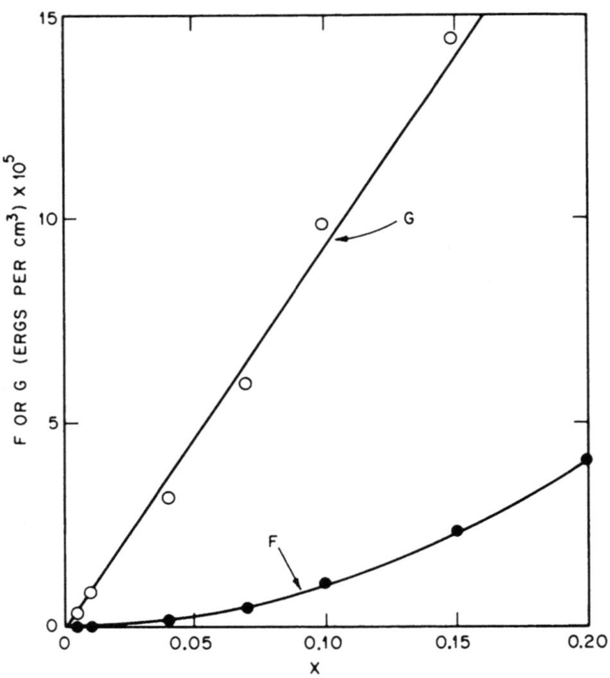

Fig. 26. The anisotropy coefficients F and G measured at room temperature in $Co_xFe_{3-x}O_4$. The heat treatment temperature was 102°C. (After Ref. 141.)

with Co additions. Cobalt enters the spinel lattice as Co^{2+} on the octahedral sites.[24] The experimental data[141] are shown in Figure 26, where F and G are plotted as a function of the cobalt concentration x. The purely quadratic behavior of F with x and the linear dependence of G indicates that, as predicted, only G results from single-ion effects. The quadratic dependence of F shows that this term results from a pair interaction. The symmetry axis observed for this pair interaction ($G = 0$) is $\langle 100 \rangle$. This direction does not correspond to the pair direction between neighboring octahedral sites, which is $\langle 110 \rangle$. Consequently, it is necessary to discard the earlier assumption that $\mathbf{L}_i$ for a pair is parallel to the bond direction. (It is assumed in this discussion that $\mathbf{E}_i$ is uniaxial. The more general case is discussed in Ref. 142). Smit and Wijn[24] justify $\langle 100 \rangle$ as the symmetry axis in this case by considering that the pertinent interaction is not direct between Co ions, but rather through adjacent oxygens. In any case, if $\mathbf{L}_i$ cannot be determined from simple geometric arguments, the usefulness of Eq. (31) is greatly diminished.

The kinetics of the formation and decay of the induced anisotropy at large concentrations is relatively complicated.[142,143] At low concentrations the situation is simpler and it is observed[142,143] that the values of F and G are independent of the cation vacancy content, but that the relaxation time is inversely proportional to the vacancy density. Thus it is clear that the migration of the Co ions takes place by means of a cation vacancy mechanism.

In a number of ferrites of commercial importance an anisotropy can be induced at room temperature. These ferrites also, of course, exhibit disaccommodation. (It should be remembered that disaccommodation results from the induced anisotropy stabilizing the domain walls which exist in the sample.) The origin of this anisotropy is not clear. Some of the difficulties in understanding the induced anisotropy in these materials stem from the fact that different mechanisms appear to be important at different temperatures. For example, disaccommodation data show that at room temperature and above, three distinct mechanisms are needed to explain the results obtained with iron-rich Mn ferrite.[100] One is important near room temperature, one predominates at about 100°C. The high-temperature ($\sim$400°C) mechanism is presumably due to ionic diffusion similar to that observed in Co-doped Fe_3O_4. It is not certain, however, what ionic configuration is formed to produce the anisotropy. The activation energy observed for this high-temperature process is consistent with the activation energy found for long-range diffusion.[144]

The amplitude of the room-temperature induced anisotropy in Fe-rich Mn and Ni ferrites appears to be proportional to the product of the Fe^{2+} and cation vacancy contents.[144–146] Thus, the effect appears to be due to the preferential occupations by Fe^{2+} ions of the octahedral sites near vacancies.[144,147] However, since many of the details of this model are still vague and since the observed activation energy is too high for electron hopping and too low for simple ionic diffusion,[144] it appears that a complete explanation of the room-temperature induced anisotropy is not yet available.

A phenomenon analogous to field-induced anisotropy may occur during crystal growth. This growth-induced anisotropy has been extensively studied in connection with the mixed garnets used for bubble domain devices.[148,149] The composition used may be represented by $(\mathbf{R}_1\mathbf{R}_2)_3(FeGa)_5O_{12}$, where $\mathbf{R}_1$ and $\mathbf{R}_2$ are two rare earths (or Y). During growth, preferential occupation of $\mathbf{R}_1$ and $\mathbf{R}_2$ on the dodecahedral sites accounts for the induced anisotropy. The Fe and Ga contents are adjusted to yield the desired magnetization (Section 2.1). In this manner a material which magnetically is essentially uniaxial (or more accurately, orthorhombic) can be obtained from a cubic crystal. Figure 5 shows the domain structure of a plate of such a uniaxial mixed garnet with the easy axis normal to the plate. As discussed earlier in connection with Figure 5, the domain pattern shown is typical of a uniaxial material.

With the assumption that a growth direction can be defined, the symmetry of the growth-induced anisotropy is identical to that of the field-induced anisotropy.[138] In the field-induced case E_i is assumed quadratic, which is quite plausible.[137] A similar, but less justifiable, assumption concerning the quadratic behavior of the site preference is made in the development of the growth-induced theory.[138] Since the symmetry and assumptions involved for the two types of anisotropy are formally identical, it then follows that Eq. (31) must describe the growth-induced anisotropy energy if $\boldsymbol{\beta}$ is taken as the growth direction. For simplicity $\boldsymbol{\beta}$ is taken as the growth facet normal [to indicate the different assumptions involved, in the literature A and B replace F and G in Eq. (31)].

Flux-grown garnets have (110) and (211) natural faces. A consequence of Eq. (31) (with A and B for F and G) is that the induced anisotropy in both growth facets is determined by only two constants, A and B. This observation is quite well verified by experiment.[150] In fact, a knowledge of A and B obtained from flux-grown crystals appears to be of value in predicting the behavior of films grown epitaxially in a number of different crystal directions.[151]

The hypothesis of preferential occupation of the dodecahedral sites rather than pair interaction is supported by direct observation with spin resonance techniques in Nd- or Yb-doped aluminum garnet.[152] In most cases of interest where the induced anisotropy is large the amount of preferential occupation increases with increasing mismatch of the ionic radii of the two rare earth ions.[153]

In conclusion we remark that in magnetic alloys which exhibit long-range order (e.g., Ni_3Fe) plastic deformation may also produce an induced anisotropy (Section 4.3 and Volume 1, Chapter 4). In this case ordered like-atom pairs are produced by the slip process accompanying the deformation. However, since slip formation is not a smooth function of angle, the argument that led to Eq. (31) and hence Eq. (31) itself do not apply in this case. The details are given in Ref. 154.

5.2. Photoinduced Anisotropy

Photoinduced effects are changes in magnetic properties which occur when the material is irradiated with light. Only a small number of substances have shown this effect and of these Si-doped YIG is the most important. Work in this field started with the discovery by Teale and Temple[155] that illuminating a YIG:Si sample annealed out the low-temperature field-induced anisotropy. This discovery was followed by the observation that polarized light could lower the symmetry of the material in a fashion analogous to annealing in a magnetic field. The recent work in this field is reviewed in Refs. 156 and 157. The symmetry arguments for the photoinduced effects are more involved than for the field- or growth-induced effects and will not be reproduced here. This added complication arises from the fact that in the last two cases the symmetry of the induced energy surface is determined only by the direction of the applied field or the crystal growth. However, in the photoinduced case the polarization direction in addition to the direction of the applied field must be considered.

For simplicity we will only consider the irradiation of a (001) plate at YIG:Si with polarized light. Charge neutrality requires that the presence of Si^{4+} produces Fe^{2+} ions. The Fe^{2+} ions occupy the four magnetically nonequivalent octahedral sites. The experiments on the photoinduced effects are carried out at sufficiently low temperatures (generally at or below 77°K) so that the Fe^{2+} ions may be considered localized. In other words, the temperature is below that at which electron hopping readily takes place. As was discussed in Section 5.1, the octahedral sites have a trigonal symmetry with the $\langle 111 \rangle$ directions the symmetry axis. For the (001) plane only two types of sites need be considered: those whose trigonal axes project on [110] (type 1) and those whose trigonal axes project on $[1\bar{1}0]$ (type 2). We let n_1 and n_2 be the number of Fe^{2+} ions per unit volume on sites of type 1 and type 2, respectively. The applied field is taken along [100] so that the field does not influence the occupation of the octahedral sites. {From Eq. (31), for $\mathbf{L}_i$ along $\langle 111 \rangle$ we have $F = 0$ and the induced energy vanishes if $\boldsymbol{\beta}$ is parallel to [100].} Incident light will excite the additional electron on the Fe^{2+} ions to a higher excited state. In this excited state the electron is more mobile and it may decay to any one of the other octahedral sites.

The Fe^{2+} wave functions have as their axis of symmetry the local axis of the site they occupy. Therefore the probability of exciting a particular Fe^{2+} ion depends on the orientation of the polarization of the light with respect to the site symmetry. Thus irradiation with light propagating parallel to [001] and polarized along [110] will unbalance the population of the two types of sites. Specifically, in this case n_2 will grow at the expense of n_1. (For the polarization along $[1\bar{1}0]$, n_1 will be greater than n_2.) This preferential occupation by the Fe^{2+} ions leads to an uniaxial anisotropy. If $n_1 > n_2$, the [110] axis is the easy axis. The effect is relatively large. For YIG with 0.028 Si atom per formula unit the light-induced anisotropy is about 2×10^4 ergs/cm^3. The wavelength of the light used in these experiments is about 1 μm. This wavelength is at an absorption peak of the Fe^{2+} ion, while pure YIG is quite transparent at this wavelength.

5.3. Jahn–Teller Effect

Ions with an orbitally degenerate ground state are a special kind of impurity since these ions tend to distort the sites they occupy. The origin of this distortion can be pictured by considering first-order perturbation theory. In first-order perturbation theory the center of gravity of a degenerate level stays unchanged as the level is split by the perturbation, which in this case is the distortion of the site containing the ion. The energy of the ground state then decreases linearly with the distortion. The elastic energy of the distortion increases as the square of the distortion and as a result there is some value of the distortion that minimizes the total (electronic and elastic) energy.

This type of distortion was first discussed by Jahn and Teller and is reviewed in Ref. 158.

In most of the magnetic oxides of interest, ions that show a Jahn–Teller effect occupy octahedral sites. In these sites the largest effects arise from Mn^{3+} (weak field, one electron in the e_g level) and by Cu^{2+} (a simple hole in the *d* shell).

If the concentration of the Jahn–Teller ions is low, the distortions of the sites will be independent of each other. The distortion may be along either the *x*, *y*, or *z* direction of the local axis of the octahedral site (Figure 27). The three directions of distortion are separated by a potential barrier of height V_0. Thus for $kT < V_0$ the distortion is frozen into a specific direction and for $kT > V_0$ the distortion moves among the three possible directions. If an acoustic stress is applied, say a compression along the *z* axis in Figure 27, that direction of distortion will be energetically favored. The result is a relaxation process which produces a large acoustic loss. From ultrasonic measurements using Mn-doped YIG, V_0 is found to be in the range 0.01–0.03 eV. The distortion deduced from these experiments is about 15–30%, which is probably too large. A likely explanation for this high estimate is that the octahedron is taken as isolated from the crystal and therefore interactions with neighboring oxygens are neglected.[80]

With high concentrations of Jahn–Teller ions the individual distortions are no longer independent. Instead the crystal as a whole changes from cubic to tetragonal. A large number of spinels containing

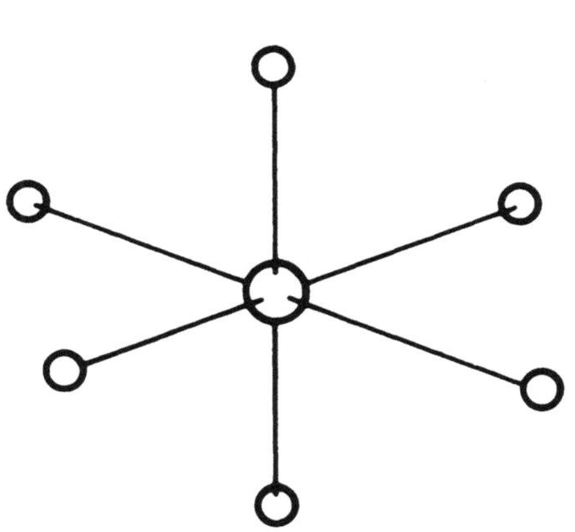

Fig. 27. An octahedral site tetragonally distorted along the *z* axis. The six outer ions represent oxygens and the center one a Jahn–Teller ion.

either Cu, Mn, or both have been studied.[159] Typical *c/a* ratios are about 1.15. A plot of the *c/a* ratio at room temperature as a function of Mn content in $Mn_xFe_{3-x}O_4$ is shown in Figure 28, where the cubic to tetragonal transformation at $x \sim 1.9$ is clearly demonstrated. With $x \sim 1.9$, roughly half of the available octahedral sites are occupied by Mn^{3+}. For a large number of spinels this is approximately the critical fraction of octahedral sites that must be occupied by either Cu^{2+} or Mn^{3+} to produce the tetragonal phase at room temperature.[159] Curves very similar to Figure 28 are obtained if the *c/a* ratio for a fixed composition is plotted as a function of temperature.[160] Both plots show a rather abrupt change in the *c/a* ratio as the critical composition or temperature is reached. This behavior is a consequence of the fact that in the spinel lattice the transition is first order.[160]

The mechanism by which this first-order phase transformation is accomplished has been studied in $CuFe_2O_4$.[161] Microscopic investigations show that the transformation is diffusionless and occurs by a double shear mechanism. The first shear occurs in a (101) plane along a [10$\bar{1}$] direction. The second occurs in the (011) plane in a [0$\bar{1}$1] direction. These shears result in a lamellar microstructure

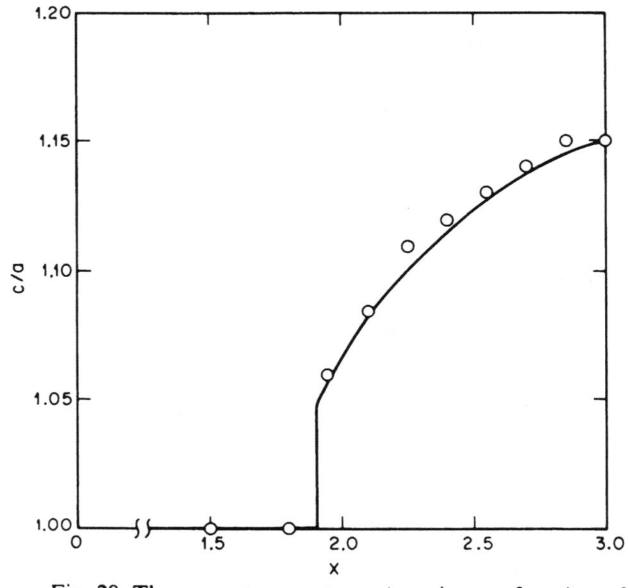

Fig. 28. The room-temperature *c/a* ratio as a function of Mn content in $Mn_xFe_{3-x}O_4$. The data for this plot are from Ref. 160.

in which the orientation difference between adjacent lamellae are functions of the degree of tetragonality.

Since the equilibrium distribution of Cu ions over the octahedral and tetrahedral sites is temperature dependent, the transition temperature is determined in part by the thermal history of the sample. However, this problem is further complicated by the fact that this equilibrium is influenced by the oxygen stoichiometry. Samples quenched from 900°C and from 500°C but with different oxygen stoichiometry have the same fraction of Cu ions on octahedral sites and thus the same transition temperature.[162]

So far we have not considered the behavior of cubic material near the transition point. There is some evidence that local tetrogonal distortions appear in the cubic phase as either the critical composition or temperature is approached. The Mössbauer spectra[163] of $FeCr_2O_4$ (the measurements were made above the Curie temperature) shows quadrupole splittings both above and below the transition temperature. The observed splittings in the cubic phase are associated with the tetragonal distortion as it reorients dynamically through the three equivalent axes. The activation energy for this reorientation is 0.1 eV, which is two to three times the activation energy found for dilute Mn^{3+} in YIG. It should be remembered that the time scale of a Mössbauer observation is of the order of 10^{-7} sec and reorientation times comparable to this will appear as essentially static in this experiment. For measurements made with a time scale large compared to the reorientation time the material will, of course, appear cubic. The details of how the Mössbauer spectra are altered as the relaxation time changes are actually somewhat more complicated than given here and a discussion may be found in Ref. 164.

The investigation of whether truly static local distortions are present in the cubic phase has been centered on Mn ferrite. Unfortunately, the discussion of Mn ferrite with a composition near the critical one is difficult because the valence states of the Fe and Mn ions are uncertain. Futhermore, clustering of octahedral Mn^{3+} may occur. The Jahn–Teller distortions associated with the Mn^{3+} ions, it is supposed, provided an increase in elastic energy which can be released if the distorting ions group themselves together.[165] Evidence for this effect is seen in X-ray data which show that certain Mn ferrites after special heat treatments have patterns with broadened lines.[166] Indirect evidence from various electrical and magnetic measurements is also taken to support cluster formation.[167] Alternate explanations have been sought in lattice strain and in chemical inhomogeneities caused by oxidation.[168] Thus, this question is yet to be finally resolved.

Acknowledgments

I am grateful for permission to use the unpublished data supplied by R. C. Le Craw, G. P. Vella-Caleiro, and R. Wolfe. In addition, I greatly profited from discussions with G. Y. Chin, J. F. Dillon, Jr., M. Dixon, F. B. Hagedorn, H. M. O'Bryan, E. A. Nesbitt, F. J. Schnettler, M. D. Sturge, L. R. Walker, and J. H. Wernick.

References

1. C. Kittel, Physical theory at ferromagnetic domains, *Rev. Mod. Phys.* **21**, 541–583 (1949).
2. C. Kittel, *Introduction to Solid State Physics*, 3rd ed., Wiley, New York (1966).
3. F. Brailsford, *Physical Principles of Magnetism*, Van Nostrand, London (1966).
4. H. Cohen and R. A. Chegwidden, Control of the electromagnetic properties of the polycrystalline garnet $Y_3Fe_{3.75}Al_{1.25}O_{12}$, *J. Appl. Phys.* **37**, 1081–1082 (1966).
5. R. C. LeCraw and R. Wolfe, Personal communication.
6. C. Herring and C. Kittel, On the theory of spin waves in ferromagnetic media, *Phys. Rev.* **81**, 869–880 (1951).
7. J. E. Kunzler, L. R. Walker, and J. K. Galt, Adiabatic demagnetization and specific heat in ferrimagnets, *Phys. Rev.* **119**, 1609–1614 (1960).
8. K. Yosida, The status of the theories of magnetic anisotropy, *J. Appl. Phys.* **39**, 511–518 (1968).
9. M. D. Sturge, E. M. Gyorgy, R. C. LeCraw, and J. P. Remeika, Magnetic behavior of cobalt in garnets, II. Magnetocrystalline anisotropy and ferrimagnetic resonance of cobalt-doped yttrium iron garnet, *Phys. Rev.* **180**, 413–423 (1969).
10. P. Hansen, Contribution of some 4*d* and 5*d* transition-metal ions on octahedral sites to the anisotropy of ferrites and garnets, *Phys. Rev.* **B3**, 862–870 (1971).
11. L. R. Bickford, J. M. Brownlow, and R. F. Penoyer, Magnetocrystalline anisotropy in cobalt-substituted magnetite single crystals, *Proc. IEE B* **104**(Suppl.), 238–244 (1957).
12. E. M. Gyorgy and H. M. O'Bryan, Jr., The effect of oxygen stoichiometry on the magneto-crystalline anisotropy of magnetite, *Phys. Lett.* **23**, 513–514 (1966).
13. J. Kanamori, in *Magnetism* (G. T. Rado and H. Suhl, eds.), Vol. I, pp. 127–203, Academic, New York (1963).
14. A. H. Bobeck, D. H. Smith, E. G. Spencer, L. G. Van Uitert, and E. M. Walters, Magnetic properties of flux grown uniaxial garnets, *IEEE Trans. Mag.* **MAG-7**, 461–463 (1971).
15. L. Landau and E. Lifshitz, On the theory of the dispersion of magnetic permeability in ferromagnetic bodies, *Phys. Z. Soviet Un.* **8**, 153–169 (1935).
16. B. Lax and K. J. Button, *Microwave Ferrites and Ferrimagnetics*, McGraw-Hill, New York (1962).
17. A. M. Clogston, Relaxation phenomena in ferrites, *Bell System Tech. J.* **34**, 739–760 (1955).

18. B. H. Clarke, K. Tweedale, and R. W. Teale, Rare-earth relaxation time and *G* tensor in rare-earth-doped yttrium iron garnet. I. Ytterbium, *Phys. Rev.* **139**, A1933–A1943 (1965).
19. M. Sparks, Effect of impurities on the microwave properties of ytterium iron garnet, *J. Appl. Phys.* **38**, 1031–1038 (1967).
20. W. H. Von Aulock, *Handbook of Microwave Ferrite Materials*, Academic, New York (1965).
21. E. Schlomann, J. J. Green, and U. Milano, Recent developments in ferromagnetic resonance at high power levels, *J. Appl. Phys.* **31**(Suppl.), 286S–395S (1960).
22. R. W. Teale and K. Tweedale, Ytterbium-ion relaxation in ferrimagnetic resonance, *Phys. Lett.* **1**, 298–300 (1962).
23. M. Sparks, Theory of giant-anisotropy ferromagnetic-resonance linewidth, *Phys. Rev.* **161**, 497–498 (1967).
24. J. Smit and H. P. J. Wijn, *Ferrites*, Wiley, New York (1959).
25. J. K. Galt and E. G. Spencer, Loss mechanisms in spinel ferrites, *Phys. Rev.* **127**, 1572–1576 (1962).
26. W. A. Yager, J. K. Galt, and F. R. Merritt, Ferromagnetic resonance in two nickel–iron ferrites, *Phys. Rev.* **99**, 1203–1210 (1955).
27. D. J. Epstein and L. Tocci, High temperature resonance losses in silicon-doped yttrium-iron garnet (YIG) *Appl. Phys. Lett.* **11**, 55–57 (1967).
28. R. C. LeCraw and R. D. Pierce, unpublished.
29. G. F. Dionne, Effects of surface roughness on ferrimagnetic resonance line width measurements, *J. Appl. Phys.* **43**, 1221–1224 (1972).
30. C. W. Haas and H. B. Callen, Ferromagnetic line width in cobalt-substituted ferrites, *Phys. Rev.* **122**, 59–68 (1961).
31. M. D. Sturge, E. M. Gyorgy, R. C. LeCraw, and J. P. Remeika, Magnetic behavior of cobalt in garnets. II. Magnetocrystalline anisotropy and ferrimagnetic resonance at cobalt-doped yttrium iron garnet, *Phys. Rev.* **180**, 413–423 (1969).
32. A. M. Clogston, H. Suhl, L. R. Walker, and P. W. Anderson, Ferromagnetic resonance line width in insulating materials, *J. Phys. Chem. Solids* **1**, 129–136 (1956).
33. H. B. Callen and E. Pittelli, Anisotropic ferromagnetic resonance line width in ferrites, *Phys. Rev.* **119**, 1523–1531 (1960).
34. R. C. LeCraw, unpublished.
35. E. G. Spencer, D. A. Lepore, and J. W. Nielsen, Measurement on lithium ferrite crystals having near-zero defect concentrations, *J. Appl. Phys.* **39**, 732–733 (1968).
36. R. C. LeCraw, E. G. Spencer, and C. S. Porter, Ferromagnetic resonance line width in yttrium iron garnet single crystals, *Phys. Rev.* **110**, 1311–1313 (1958).
37. E. Schlomann, Inhomogeneous broadening of ferromagnetic resonance lines, *J. Appl. Phys.* **40**, 1199–1201 (1969).
38. E. Schlomann, The microwave susceptibility of polycrystalline ferrites in strong dc fields and the influence of nonmagnetic inclusions on the microwave susceptibility, Conference on Magnetism and Magnetic Materials, AIEE Spec. Publ. T-91, 600–609 (1958).
39. P. E. Seiden and J. G. Grunberg, Ferrimagnetic resonance line width in dense polycrystalline ferrites, *J. Appl. Phys.* **34**, 1696–1698 (1963).
40. R. C. Le Craw, Personal communication.
41. R. M. Fulrath and J. M. Pask (eds.), *Ceramic Microstructures*, Wiley, New York (1968).

42. C. E. Patton and H. J. Van Hook, in *AIP Conference Proceedings* (C. D. Graham, Jr. and J. J. Rhyne, eds.), No. 5, pp. 1565–1567, American Institute of Physics, New York (1972).
43. J. K. Galt, Motion of a ferromagnetic domain wall in Fe_3O_4, *Phys. Rev.* **85**, 664–669 (1952).
44. J. F. Dillon, Jr. and H. E. Earl, Jr., Domain wall motion and ferrimagnetic resonance in a manganese ferrite, *J. Appl. Phys.* **30**, 202–213 (1959).
45. F. B. Hagedorn and E. M. Gyorgy, Domain wall mobility in single-crystal yttrium iron garnet, *J. Appl. Phys.* **32**, 285S–283S (1961).
46. M. A. Wanas, Domain-wall motion in yttrium-iron garnets, *J. Appl. Phys.* **38**, 1019–1021 (1967).
47. U. Enz and H. van der Heide, Domain-wall mobility in Si-doped YIG, *J. Appl. Phys.* **39**, 435–437 (1968).
48. H. Harper and R. W. Teale, Damping of magnetic domain wall motion in pure and ytterbium-doped yttrium iron garnet, *J. Phys. C.* (*Solid State Phys.*) **2**, 1926–1930 (1969).
49. A. H. Bobeck *et al.*, Uniaxial magnetic garnets for domain wall "bubble" devices, *Appl. Phys. Lett.* **17**, 131–134 (1970).
50. P. W. Shumate, Jr., Magnetooptic measurement techniques for magnetic-bubble materials, *IEEE Trans. Mag.* **MAG-7**, 586–590 (1971).
51. G. P. Vella-Coleiro, D. H. Smith, and L. G. Van Uitert, Resonant motion of domain walls in yttrium-gadolinium iron garnets, *J. Appl. Phys.* **43**, 2428–2430 (1972).
52. C. Kittel and J. Galt, in *Solid State Physics* (F. Seitz and D. Turnbull, eds.), Vol. 3, pp. 439–564, Academic, New York (1956).
53. F. B. Hagedorn, in *AIP Conference Proceedings* (C. D. Graham, Jr. and J. J. Rhyne, eds.), No. 5, pp. 72–90, American Institute of Physics, New York (1972).
54. J. F. Dillon, Jr., in *Magnetism* (G. T. Rado and H. Suhl, eds.), Vol. 3, pp. 415–464, Academic, New York (1963).
55. H. Harper and R. W. Teale, Damping of magnetic domain wall motion by manganic ions in managanese ferrite, *Solid State Commun.* **5**, 327–330 (1967).
56. H. Harper and R. W. Teale, Magnetic domain wall mobility limitations by anisotropic impurity ions in ferrimagnetic insulators, *J. Phys. Chem. Solids* **28**, 1781–1787 (1967).
57. H. Harper and R. W. Teale, The influence of erbium doping on the velocity of magnetic domain-wall propagation in yttrium iron garnet, *J. Phys. C.* (*Proc. Phys. Soc.*) **1**, 1046–1051 (1968).
58. G. P. Vella-Coleiro, D. H. Smith, and L. G. Van Uitert, Domain-wall mobility in some rare-earth iron garnets, *IEEE Trans. Mag.* **MAG-7**, 745–747 (1971).
59. G. P. Vella-Coleiro, D. H. Smith, and L. G. Van Uitert, Damping of domain wall motion in rare earth iron garnets, *Appl. Phys. Lett.* **21**, 36–37 (1972).
60. E. G. Spencer, R. C. Le Craw, and A. M. Clogston, Low temperature line width maximum in yttrium iron garnet, *Phys. Rev. Lett.* **3**, 32–33 (1959).
61. H. Callen and R. M. Josephs, Dynamics of magnetic bubble domains with an application to wall mobilities, *J. Appl. Phys.* **42**, 1977–1982 (1971).
62. G. P. Vella-Coleiro and W. J. Tabor, Measurement of magnetic bubble mobility in epitaxial garnet films, *Appl. Phys. Lett.* **21**, 7–8 (1972).

63. B. A. Calhoun, E. A. Giess, and L. L. Rosier, Dynamic behaviour of domain walls in low-moment yttrium-gallium-iron garnet crystals, *Appl. Phys. Lett.* **18**, 287–289 (1971).
64. B. E. Argyle, J. C. Slonczewski, and A. F. Mayadas, in *AIP Conference Proceedings* (C. D. Graham, Jr. and J. J. Rhyne, eds.), No. 5, pp. 175–179, American Institute of Physics, New York (1972).
65. G. P. Vella-Coleiro, unpublished.
66. R. C. Le Craw and R. L. Comstock, in *Physical Acoustics* (W. P. Mason, ed.), Vol. III B, pp. 127–199, Academic, New York (1965).
67. Y. Shapira, Ultrasonic behavior near the spin–flop transitions of hematite, *Phys. Rev.* **184**, 589–600 (1969).
68. R. M. Bozorth, *Ferromagnetism*, Van Nostrand, Princeton, N.J. (1951).
69. G. W. Smith and J. R. Birchak, Effect of internal stress distribution on magneto-mechanical damping, *J. Appl. Phys.* **39**, 2311–2316 (1968).
70. T. S. Hartwick and J. Smit, Ferromagnetic resonance in Si–doped YIG, *J. Appl. Phys.* **40**, 3995–4001 (1969).
71. D. I. Tchernev, Effect of low-temperature magnetic anneal on the linewidth of garnets containing Fe^{2+} ions, *J. Appl. Phys.* **38**, 1046–1047 (1967).
72. E. M. Gyorgy and R. C. Le Craw, Influence of Mn and Si on the low temperature acoustic properties of YIG single crystals, *Appl. Phys. Lett.* **6**, 32–33 (1965).
73. D. F. Gibbons and V. G. Chirba, Acoustical loss and Young's modulus of yttrium iron garnet, *Phys. Rev.* **110**, 770–771 (1958).
74. N. G. Einspruch, Ultrasonic loss in yttrium iron garnet, *Phys. Chem. Solids* **23**, 1341–1342 (1962).
75. J. F. Graczyk, Electrical conduction in pure and doped single crystal yttrium iron garnet, MS Thesis, Mass. Inst. of Tech. (1965).
76. J. Verweel and B. J. M. Roovers, in *Solid State Physics in Electronic and Telecommunications* (M. Desirant and J. L. Michiels, eds.), Vol. 3, pp. 475–487, Academic, New York (1960).
77. D. F. Gibbons, Acoustic relaxations in ferrite single crystals, *J. Appl. Phys.* **28**, 810–814 (1957).
78. W. Palmer, Low temperature anisotropy of manganese-iron ferrites, *J. Appl. Phys.* **33**(Suppl.), 1201–1202 (1962).
79. J. H. Hendriks and F. J. Dijkstra, Acoustic relaxation in manganese ferrite ($Mn_xFe_{3-x}O_4$) at temperatures between 90 and 400°K for $1.0 < x < 1.6$, *J. Phys. Chem. Solids* **31**, 2793–2795 (1970).
80. E. M. Gyorgy, R. C. Le Craw, and M. D. Sturge, Influence of Jahn–Teller ions on the acoustic and magnetic properties of YIG, *J. Appl. Phys.* **37**, 1303–1309 (1966).
81. E. J. W. Verwey and P. W. Haayman, Electrical conductivity and transition point of magnetite (Fe_3O_4), *Physica* **8**, 979–987 (1941).
82. S. Iida, Detailed electronic mechanism for the low temperature acoustic loss in magnetite, *J. Phys. Soc. Japan* **22**, 1233–1235 (1967).
83. W. B. Suiter, Jr. and R. Blair, Acoustic properties of magnetite as a function of temperature and cation vacancies, *J. Appl. Phys.* **36**, 1156–1158 (1965).
84. W. P. Mason, *Physical Acoustics and the Properties of Solids*, Van Nostrand, Princeton, N.J. (1958).

85. M. A. Globus, P. Duplex, and M. Guyat, Determination of initial magnetization curve for crystallites size and effective anisotropy field, *IEEE Trans. Mag.* **MAG-7**, 617–622 (1971).
86. M. A. Globus, Influence des dimensions des parois sur la perméabilité initiale, *Compt. Rend.* **255**, 1709–1711 (1962).
87. J. E. Knowles, The magnetostatic energy associated with a polycrystalline ferrite, *Brit. J. Appl. Phys.* **1**, 987–997 (1968).
88. S. Chikazumi, *Physics of Magnetism*, Wiley, New York (1964).
89. H. J. Williams, W. Shockley, and C. Kittel, Velocity of a ferromagnetic domain boundary, *Phys. Rev.* **80**, 1090–1094 (1950).
90. M. Hansen, *Constitution of Binary Alloys*, McGraw-Hill, New York (1958).
91. G. Y. Chin, Review of magnetic properties of Fe–Ni alloys, *IEEE Trans. Mag.* **MAG-7**, 102–113 (1971).
92. E. Ross, I. Hanke, and E. Moser, Mangan-Zink-Ferrite mit Anfangspermeabilitaten bis uber 20,000 und ihr Gefuge, *Z. Angew. Phys.* **17**, 504–508 (1964).
93. K. Ohta and N. Kobayashi, Magnetostriction constants of Mn–Zn–Fe ferrites, *Japan J. Appl. Phys.* **3**, 576–580 (1964).
94. K. Ohta, Magnetocrystalline anisotropy and magnetic permeability of Mn–Zn–Fe ferrites, *J. Phys. Soc. Japan* **18**, 685–690 (1963).
95. C. Guillaud, N. Paulus, and R. Vautier, La diminution des pertes par courant Foucault dans le ferrites de manganese-zinc par addition de calcium, *Compt. Rend.* **242**, 2712–2715 (1956).
96. M. Dixon, Personal communication.
97. T. Akashi, Low loss manganese zinc ferrite, *NEC Res. & Dev.* **19**, 98–101 (1970).
98. J. Verweel, Permeability and ferromagnetic resonance line width of some ferrites with garnet structure, *Proc. IEEE* **109**, Part *B* (Suppl.), 95–98 (1962).
99. G. De Vries, D. Van Geest, R. Gersdorf, and G. Rathenau, Determination of the magnetic anisotropy energy caused by interstitial carbon or nitrogen in iron, *Physica* **25**, 1131–1138 (1959).
100. S. Krupicka and K. Zaveta, Magnetic aftereffects in ferrimagnetic oxidic spinels, *J. Appl. Phys.* **39**, 930–938 (1968).
101. W. F. Brown, Jr., *Micromagnetics*, Interscience, New York (1963).
102. R. C. Sherwood, J. P. Remeika, and H. J. Williams, Domain behavior in some transparent magnetic oxides, *J. Appl. Phys.* **30**, 217–225 (1959).
103. P. W. Shumate, Jr., Operational method for locating and mapping imperfections in single-crystal platelets of rare earth orthoferrite, *J. Appl. Phys.* **42**, 1274–1275 (1971).
104. L. J. Varnerin, in *Ferrites: Proc. of the Int. Conf. 1970 Japan* (Y. Hoshino, S. Iida, and M. Sugimota, eds.), University Park Press, Tokyo (1970).
105. A. J. Kurtzig, Interactions of magnetic domain walls with twin and grain boundaries in orthoferrites, *IEEE Trans. Mag.* **MAG-6**, 497–500 (1970).
106. K. H. Steward, *Ferromagnetic Domains*, University Press, Cambridge, England (1954).
107. R. S. Tebble and D. J. Craik, *Magnetic Materials*, Wiley–Interscience, London (1969).
108. K. P. Belov and Av. Ped'ko, Anomalies in the temperature dependence of the coercive force in rare earth element garnet ferrites in the neighborhood of the compensation point, *Soviet. Phys.—JETP* **12**, 666–668 (1961).

109. J. P. Hanton, Intrinsic coercive force of rare earth iron garnets near the compensation temperature, *IEEE Trans. Mag.* **MAG-3**, 505–509 (1967).
110. L. J. Dijkstra and C. Wert, Effect of inclusions on coercive force of iron, *Phys. Rev.* **79**, 979–985 (1950).
111. E. A. Nesbitt and E. M. Gyorgy, Two-phase permalloy for high-speed switching, *J. Appl. Phys.* **32**, 1305–1308 (1961).
112. J. B. Labroue, M. Takahashi, and M. E. Fine, in *AIP Conference Proceedings* (C. D. Graham, Jr. and J. J. Rhysne, eds.), No. 5, pp. 1098–1072, American Institute of Physics, New York (1972).
113. P. D. Baba, E. M. Gyorgy, and F. J. Schnettler, Two-phase ferrite for high-speed switching, *J. Appl. Phys.* **34**, 1125–1126 (1963).
114. P. D. Baba, Nonmagnetic inclusions in ferrites for high speed switching, *J. Am. Ceram. Soc.* **48**, 305–309 (1965).
115. H. J. Shaw, B. J. Elliott, K. J. Harker, and A. Karp, Microwave generation in pulsed ferrites, *J. Appl. Phys.* **37**, 1060–1066 (1966).
116. R. J. Parker and R. J. Studders, *Permanent Magnets and Their Application*, Wiley, New York (1962).
117. F. E. Luborsky, Development of elongated particle magnets, *J. Appl. Phys.* **32**(Suppl.), 171S–183S (1961).
118. F. Hofer, Dauermagnetwerkstoffe aus Legierungen des Kobalts mit Seltenen Erdmetallen, *Z. Angew. Physik* **30**, 26–32 (1970).
119. E. A. Nesbitt, G. Y. Chin, R. C. Sherwood, and J. H. Wernick, Effect of processing on permanent magnet materials containing rare-earth metals, *J. Appl. Phys.* **40**, 4006–4009 (1969).
120. H. Zijlstra, Coercivity and wall motion, *IEEE Trans. Mag.* **MAG-6**, 179–181 (1970).
121. E. A. Nesbitt and H. J. Williams, Shape and crystal anisotropy of Alnico 5, *J. Appl. Phys.* **26**, 1217–1221 (1955).
122. D. L. Martin, Processing and properties of cobalt–platinum permanent magnet alloys, in *Conf. Mag. and Mag. Materials Boston 1956*.
123. H. Stablen, in *AIP Conference Proceedings* (C. D. Graham, Jr. and J. J. Rhyne, eds.), No. 5, pp. 950–919, American Institute of Physics, New York (1972).
124. F. E. Luborsky, Permanent magnets in use today, *J. Appl. Phys.* **37**, 1091–1094 (1966).
125. G. F. Dionne, Effect of pore size on remanence ratios of magnetic materials. *J. Appl. Phys.* **40**, 431–432 (1969).
126. Y. I. Il'ijashenko and Y. KH. Krivorutsky, Dependence of the spontaneous appearance of reverse domains on anisotropy in polycrystal ferrite materials, *IEEE Trans. Mag.* **MAG-7**, 609–612 (1971).
127. H. P. J. Wijn, E. W. Gorter, C. J. Esveldt, and P. Geldermans, Conditions for square hysteresis loops in ferrites, *Philips Tech. Rev.* **16**, 49–58 (1954).
128. G. F. Dionne, Magnetic anisotropy and magnetostriction constants of Mg-Mn ferrites at 300°K, *J. Appl. Phys.* **41**, 831–832 (1970).
129. N. Menyuk, Magnetic materials for digital computer components. II. Magnetic characteristics of ultra-thin molybdenum-permalloy cores, *J. Appl. Phys.* **26**, 692–697 (1955).
130. R. S. Weiss, in *Magnetic Properties of Materials* (J. Smit, ed.), pp. 205–237, McGraw-Hill, New York (1971).

131. E. M. Gyorgy, J. T. Krause, R. C. Le Craw, L. R. Testardi, and L. G. Van Uitert, Variation of sound velocity with magnetic field in Mn-doped YIG, *J. Appl. Phys.* **38**, 1226–1227 (1967).
132. G. F. Dionne, Magnetic-anisotropy and magnetostriction constants of substituted lithium ferrites at 300°K, *J. Appl. Phys.* **40**, 4486–4490 (1969).
133. J. E. Knowles, The simulation of domain wall motion in square loop ferrites, and other aspects of wall behavior, *Brit. J. Appl. Phys.* **1**, 821–830 (1968).
134. D. L. Martin and M. G. Benz, Cobalt–rare earth permanent magnet alloys, *Cobalt* **5**, 11–15 (1971).
135. M. Kornetzki, J. Brackmann, and J. Frei, Magnetische Ferrite mit Perminvarschleife, *Naturwiss.* **42**, 482 (1955).
136. J. G. M. DeLau and A. L. Stuijts, Chemical composition and high-frequency properties of Ni-Zn-Co ferrites, *Phillips Res. Repts* **21**, 104–112 (1966).
137. J. C. Slonczewski, in *Magnetism* (G. T. Rado and H. Suhl, eds.), Vol. 1, pp. 205–241, Academic, New York (1963).
138. E. M. Gyorgy, A. Rosencwaig, E. I. Blount, W. J. Tabor, and M. E. Lines, General conditions for growth-induced anisotropy in garnets, *Appl. Phys. Lett.* **18**, 479–480 (1971).
139. E. T. Ferguson, Uniaxial magnetic anisotropy induced in Fe–Ni alloys by magnetic anneal, *J. Appl. Phys.* **29**, 252–253 (1958).
140. E. A. Nesbitt, B. W. Batterman, L. D. Fullerton, and A. J. Williams, Effect of oxygen on the magnetic and long-range ordering properties of Ni_3Fe, *J. Appl. Phys.* **36**, 1235–1236 (1965).
141. R. F. Penoyer and L. R. Bickford, Jr., Magnetic annealing effect in cobalt-substituted magnetite single crystals, *Phys. Rev.* **108**, 271–277 (1957).
142. W. Palmer, Kinetics of magnetic annealing in cobalt-substituted magnetite, *Phys. Rev.* **120**, 342–352 (1960).
143. S. Iida and H. Miwa, Theory of the relaxation of the uniaxial anisotropy induced by ionic migration in iron-cobalt ferrites, *J. Phys. Soc. Japan* **21**, 2505–2539 (1966).
144. S. Ogawa, T. Nakajima, T. Sasaki, and M. Takahashi, Ion diffusion and disaccommodation in ferrites, *Japanese J. Appl. Phys.* **7**, 899–903 (1968).
145. P. D. Baba and E. M. Gyorgy, Induced anisotropy in manganese ferrite, *J. Appl. Phys.* **38**, 909–910 (1967).
146. S. Krupicka, Study of relaxation spectrum of $MnFe_3O_4$ ferrites, *J. Phys. Soc. Japan* **1**(7B-1), 304–308 (1962).
147. A. Braginski, Magnetic aftereffects in iron rich ferrites containing vacancies, *Phys. Stat. Sol.* **11**, 603–615 (1965).
148. H. Callen, On Growth-induced anisotropy in garnet crystals, *Mat. Res. Bull.* **7**, 931–938 (1971).
149. A. Rosencwaig and W. J. Tabor, in *AIP Conference Proceedings* (C. D. Graham, Jr. and J. J. Rhyne, eds.), No. 5, pp. 57–70, American Institute of Physics, New York (1972).
150. F. B. Hagedorn, W. J. Tabor, and L. G. Van Uitert, Growth induced magnetic anisotropy in seven different mixed rare earth ion garnets, *J. Appl. Phys.*, To be published.
151. A. Rosencwaig and W. J. Tabor, Growth-induced magnetic anisotropy in epitaxial garnet films, *Appl. Phys. Lett.* **20**, 40–42 (1972).

152. R. Wolfe, M. D. Sturge, F. R. Merritt, and L. G. Van Uitert, Facet-related site selectivity for rare-earth ions in yttrium aluminum garnet, *Phys. Rev. Lett.* **26**, 1570–1573 (1971).
153. E. Gyorgy, M. D. Sturge, L. G. Van Uitert, E. J. Heilner, and W. H. Grockiewicz, Growth-induced anisotropy of some mixed rare earth ion garnets, *J. Appl. Phys.*, To be published.
154. G. Y. Chin, Effect of plastic deformation on the magnetic properties of metals and alloys, *Adv. Mat. Res.* **5**, 217–280 (1971).
155. R. W. Teale and D. W. Temple, Photomagnetic anneal, a new magneto-optic effect, in Si-doped yttrium iron garnet, *Phys. Rev. Lett.* **19**, 904–905 (1967).
156. J. F. B. Hawkes and R. W. Teale, Spontaneous and photo-induced linear dichroism in silicon doped yttrium iron garnet, *J. Phys. C, Solid State Phys.* **5**, 481–492 (1972).
157. R. Alben, E. M. Gyorgy, J. F. Dillon, Jr., and J. P. Remeika, Polarization-dependent photo-induced effects in silicon-doped yttrium iron garnet, *Phys. Rev. B* **5**, 2560–2577 (1972).
158. M. D. Sturge, in *Solid State Physics* (F. Seitz, D. Turnbull, and H. Ehrenreich, eds.), Vol. 20, pp. 91–211, Academic Press, New York (1967).
159. J. B. Goodenough, *Magnetism and the Chemical Bond*, Interscience, New York (1963).
160. P. J. Wojtowicz, Theoretical model for tetragonal-to-cubic phase transformations in transition metal spinels, *Phys. Rev.* **116**, 32–45 (1959).
161. H. J. Levinstein, F. J. Schnettler, and E. M. Gyorgy, Magnetic and crystallographic study of the phase transition in copper ferrite, *J. Appl. Phys.* **36**, 1163–1164 (1965).
162. H. M. O'Bryan, Jr., H. J. Levinstein, and R. C. Sherwood, Effect of stoichiometry on the copper ferrite phase transition, *J. Appl. Phys.* **37**, 1438–1439 (1966).
163. M. Tanka, T. Tokora, and Y. Aiyama, Jahn–Teller effects on Mössbauer spectra of Fe_{57} in $FeCr_2O_4$ and FeV_2O_4, *J. Phys. Soc. Japan* **21**, 262–267 (1966).
169. H. H. Wickman, in *Mössbauer Effect Methodology* (I. J. Gruvermen, ed.), Vol. 2, pp. 39–66, Plenum, New York (1966).
165. V. A. M. Brabers, Cation migration, cation valencies and the cubic-tetragonal transition in $Mn_xFe_{3-x}O_4$, *J. Phys. Chem. Solids* **32**, 2181–2191 (1971).
166. L. Cervinka and D. Vetterkind, The influence of low temperature (293–4.5°K) on the structure of manganese rich manganese ferrite, *J. Phys. Chem. Solids* **29**, 171–179 (1968).
167. S. Krupicka, Z. Simsa, and Z. Smetana, The influence of Mn^{3+} clusters on the physical properties of Mn–Fe spinel, *Czech. J. Phys.* **18**, 1016–1025 (1968).
168. H. M. O'Bryan, Jr. and H. J. Levinstein, The structure of $Mn_{1.88}Fe_{1.12}O_4$, *J. Phys. Chem. Solids* **30**, 1719–1720 (1969).

7

The Imperfect Solid—Mechanical Properties

Eckhard Nembach
Institut für Metallphysik
Universität Göttingen
Göttingen, Germany

1. Introduction to Plasticity

The application of stress to a solid causes a change in its dimensions. If the stress is small, the solid will regain its former shape with removal of the stress. In this case the deformation is called elastic. Higher stresses, however, will cause permanent deformations which do not disappear upon removal of the stress. This type of deformation is called plastic. The capability of metals to undergo extensive plastic deformation without fracturing makes them most favored construction materials. Nearly all crystalline materials can be plastically deformed, at least at elevated temperatures.

The scientific investigation of plasticity started at the beginning of this century. It had been noticed that the surface of a metal polycrystal showed parallel lines after plastic deformation. Figure 1 gives an example for brass. These lines are called slip lines. Their straightness led Ewing and Rosenhain(1) to the assumption that during the deformation of metals, layers of the specimen slide on one another. This process is demonstrated in Figure 2. This type of plastic flow is completely different from viscous flow, which occurs in amorphous condensed phases like hot glass or tar.

On the basis of Ewing and Rosenhain's idea, one expects plastic deformation to start when the force τ acting per unit area in the glide plane in the glide direction reaches a critical value τ_0, called the

Fig. 1. Slip lines on a brass polycrystal.

critical shear stress. The force τ is related to the tensile stress σ acting on the ends of a tensile specimen by a simple geometric relation:

$$\tau = \sigma \cos \varphi \cos \psi \qquad (1)$$

Equation (1) can easily be derived with reference to Figure 3; φ is the angle between the axis of the specimen and the normal of the

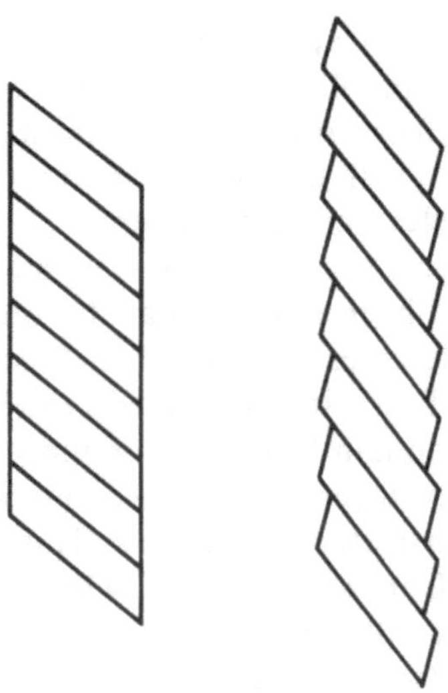

Fig. 2. Plastic deformation of a single crystal by slip (schematic).

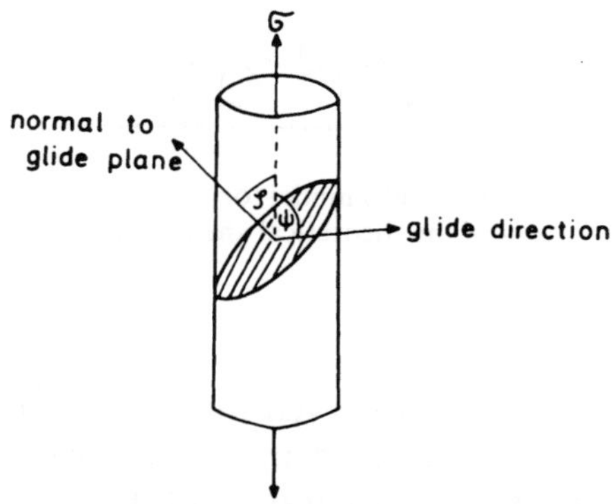

Fig. 3. Geometric relations of slip; the slip plane is shaded.

glide plane, and ψ is the angle between the axis and the direction of glide. The factor $\cos\varphi\cos\psi$ is called the Schmid factor. During glide φ and ψ change; in a tensile test the direction of glide approaches the specimen axis.

The minimum tensile stress σ_0 to cause plastic flow has been measured by Schmid and co-workers(2–4) and Elam(5) for many crystal orientations and various materials. They found that τ_0 calculated from σ_0 with Eq. (1) was nearly independent of φ and ψ (Schmid's law), i.e., the stress component normal to the glide plane has no influence on τ_0. For fcc and hexagonal metals τ_0 was about 100 g/mm^2 and for α-iron about 10 kg/mm^2; for today's high-purity materials the corresponding values are lower by a factor of three to ten.

In the theory of plasticity one uses the glide strain a instead of the tensile strain ε, $\varepsilon = (l - l_0)/l_0$, where l_0 is the length of the specimen before deformation and l is the length after deformation. The glide strain a is the mean slip distance on a slip plane divided by the mean separation of such slip planes:

$$a = (\cos\varphi_0)^{-1}\{[(1 + \varepsilon)^2 - \sin^2\psi_0]^{1/2} - \cos\psi_0\} \tag{2}$$

here φ_0 and ψ_0 are the respective angles before the onset of plastic deformation. For the derivation of Eq. (2) see, for instance, Ref. 6. Obviously Eq. (2) is analogous to Eq. (1). In the following τ and a will always denote quantities resolved onto the slip system and σ and ε the tensile or unresolved quantities.

Frenkel(7) tried to calculate τ_0 from the known binding forces in the crystal. He found that the calculated values $\tau_{s,th}$ were larger by a

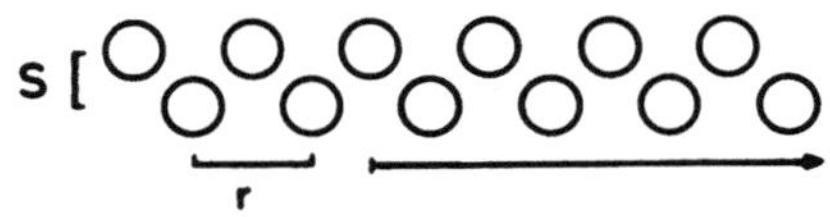

Fig. 4. Frenkel's model for the calculation of the theoretical shear stress.

factor of 10^4 than the ones observed for τ_0. A simplified version of Frenkel's calculation is presented in the following. Two planes of atoms are considered. Let the separation of the atoms in the plane be r and the separation of the planes be s (Figure 4). Frenkel assumed that the shear stress $\tau(x)$ necessary to slide the planes past one another varies sinusoidally with the amount of slip x:

$$\tau(x) = \alpha \sin(2\pi x/r) \tag{3}$$

The constant α has to be determined from the elastic limit $x \to 0$. For this one gets $\tau(x) = -Gx/s$, where G is the appropriate shear modulus. Thus $\alpha = -rG/2\pi s$ and

$$\tau(x) = -(r/2\pi s)G \sin(2\pi x/r) \tag{4}$$

$\tau_{s,th}$ is the maximum of $\tau(x)$; roughly, it equals $G/10$. More accurate calculations which have been reviewed by Seeger[8] gave very similar results. The experimentally observed values for τ_0 are between $10^{-4}\,G$ and $10^{-5}\,G$.

There had been several attempts to solve this discrepancy, but all were unsatisfactory until in 1934 Taylor,[9] Orowan,[10] and Polanyi[11] introduced the concept of the edge dislocation. Five years later Burgers[12] discovered the screw dislocation (Chapter 5 of Volume 1). It is easy to see how an edge dislocation lowers the critical shear stress of a crystal. In Frenkel's model the whole slip plane simultaneously goes through an energetically unfavorable condition (in Figure 4, $x = \frac{1}{2}r$). If a crystal contains an edge dislocation, only its environment is disturbed; the remainder of the slip plane is in perfect order. Figure 5 shows how an edge dislocation is moved by a shear stress. When the dislocation has swept out the whole plane from left to right, the upper half of the crystal has been displaced one lattice constant to the right. The dislocation breaks down the slip of a whole plane into many small steps. The effect of a screw dislocation is exactly the same, except that its direction of propagation and the direction of the shear produced by it are at right angles.

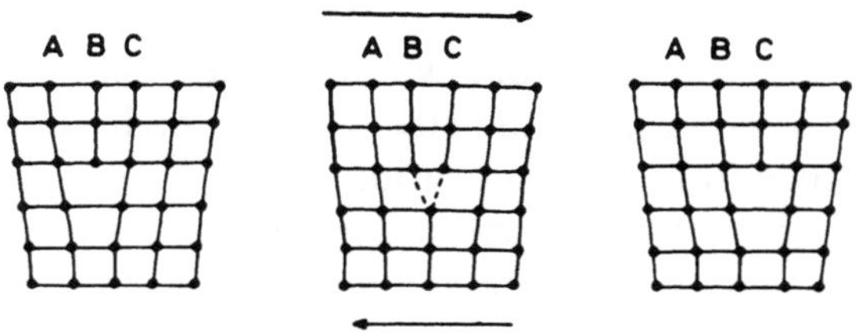

Fig. 5. Movement of an edge dislocation under an applied shear stress.

Whiskers are very thin crystals having a diameter of 10–50 μm. They are so thin that they do not contain dislocations which can produce slip. Therefore τ_0 of whiskers equals $\tau_{s,th}$.[13]

Besides the plastic deformation by slip discussed so far, there is another mechanism: twinning. This process can be understood with reference to Figure 6: Part of the crystal is sheared by the stress into a position which has mirror symmetry with respect to the unchanged part. Twinning plays only a minor role in the mechanical behavior of solids. But it should be mentioned that twinning, together with slip, was proposed as a possible mechanism of plastic deformation of ionic crystals by Reusch[14] as early as 1867.

2. Dislocations

The properties of dislocations have been extensively discussed in Chapter 5 of Volume 1. It will suffice to give a short summary of those properties which are important for the plastic deformation of solids.

Dislocations are one-dimensional crystal lattice defects. They are characterized by their Burgers vector **b** and a vector **e** of unit length which is parallel to the dislocation line. For edge dislocations $\mathbf{b} \cdot \mathbf{e} = 0$ holds and for screw dislocations $\mathbf{b} \cdot \mathbf{e} = |\mathbf{b}|$. Edge dislocations can move conservatively only in planes which contain **b** and **e**. An edge dislocation which moves normal to its glide plane is said to climb.

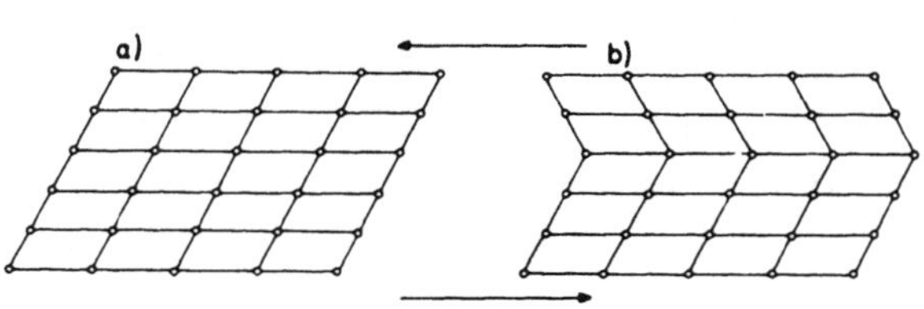

Fig. 6. Mechanical twinning: (a) before, (b) after. The arrows symbolize the shear stress.

This movement creates or absorbs point defects. Screw dislocations can glide conservatively in any plane which contains **b**. Dislocations are surrounded by a stress field, the strength of which varies as $1/r$, where r is the distance from the dislocation.

The force $d\mathbf{K}$ which a stress exerts on a dislocation element of length dl has been calculated by Peach and Koehler:[15]

$$d\mathbf{K} = (\mathbf{b} \cdot \sigma) \times \mathbf{e}\, dl \tag{5}$$

σ is the stress tensor. Dislocations will interact with one another via their stress fields. The force between parallel dislocations is proportional to the inverse of their separation. Energy is stored in the strain field of a dislocation. This line energy is proportional to Gb^2; therefore dislocations have always the smallest possible Burgers vector. On the other hand, **b** has to be a translation vector of the lattice, otherwise the passage of a dislocation would destroy the crystal structure. A dislocation whose Burgers vector is a translation vector may dissociate into two partial dislocations whose Burgers vectors are not translation vectors. Between these two partials which move together is an areal defect: a stacking fault.

There are three possible ways for dislocations to get into the crystal: by accidents during the growth of the crystal, from the surface, and by multiplication. The most important multiplication mechanism is the Frank–Read source. Most of the other mechanisms which have been proposed so far are similar to it, differing only in the process which produces the original dislocation segment.

3. Plastic Deformation of Pure Single Crystal by Slip

Figure 7 gives the resolved shear stress τ versus the shear strain a measured for a copper single crystal deformed in tension. The strain rate $\dot{a}$ is kept constant. Using Eq. (1), one can calculate the shear stress τ which acts on dislocations from the force K which is delivered by the testing machine. The force K equals the tensile stress σ times the specimen's cross section. The strain rate $\dot{a}$ determines the velocity at which dislocations have to move. Normally the flow stress τ increases with $\dot{a}$. The stress τ_0 at which plastic deformation starts is defined as the stress that causes a plastic resolved strain of 0.002. At τ_0 the dislocations in the crystal start moving over macroscopic distances. To continue plastic deformation, increasing stresses have to be applied; this phenomenon is called work hardening. At high strains ductile fracture occurs. Work hardening prevents the specimen

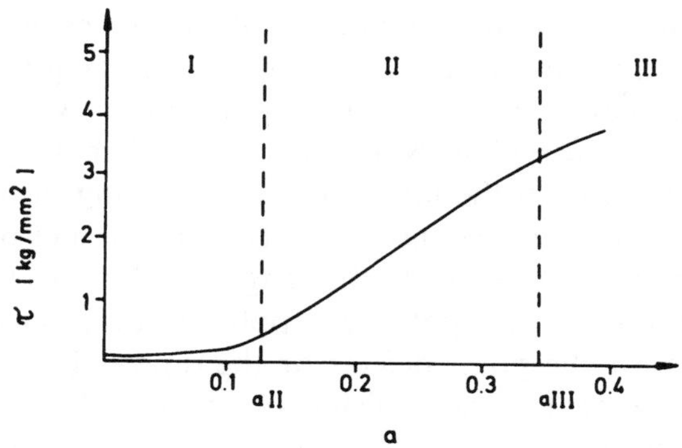

Fig. 7. $\tau(a)$ for a copper single crystal deformed at room temperature (after Diehl[16]).

from fracturing when τ_0 is reached. If there were no work hardening, plastic deformation would start at some weak spot of the specimen. There the cross section would be reduced, so that τ would be locally increased. This in turn would accelerate plastic flow at this spot until the specimen fractured. If work hardening is strong enough to offset the local reduction in cross section, no necking will occur.

In most cases more than one glide plane and direction are possible. But the resolved shear stress calculated with Eq. (1) will be different for various glide systems. Normally the one with the highest stress is activated. It is called the primary glide system; the others are the secondary ones.

As soon as dislocations start moving, they also multiply. Processes of dislocation multiplication have been described in Chapter 5 of Volume 1. Some of the basic mechanisms of work hardening are discussed below.

1. Dislocations interact with one another via their stress fields. Let N be the density of dislocations, measured in cm length per cm^3. Then the mean separation between dislocations is $1/\sqrt{N}$. Thus the mean stress τ_G exerted on a dislocation by its fellow dislocations is given by

$$\tau_G = pbG\sqrt{N} \tag{6}$$

p is a constant equal to about 0.1, b is the Burgers vector, and G is the shear modulus. The stress which has to be applied to the specimen in order to continue plastic flow must be at least τ_G. If N increases by dislocation multiplication, the flow stress increases, too.

2. Moving dislocations encounter others which lie in different glide planes. If dislocations cut one another, they acquire jogs. A jog in a screw dislocation is a short edge dislocation which can move conservatively only in a plane which is different from the glide plane of the screw dislocation. If the latter continues gliding in its own glide plane, the jog cannot move conservatively, but it creates point defects. If the jog does not move at all, the length of the screw dislocation increases by pulling out a dipole. The applied stress has to provide the energy needed for the point defects or the dipole. In any case, jogs impede the movement of screw dislocations seriously.

3. Dislocation dipoles left behind by dislocations hinder the movement of the following ones.

4. Dislocations react with one another and form stable configurations; thus they become immobile.

The flow stress τ can be divided into two parts:

$$\tau = \tau_G + \tau_s \tag{7}$$

τ_G depends only weakly on temperature through the shear modulus, whereas τ_s is strongly temperature dependent. The term τ_G describes the long-range elastic interaction between dislocations; τ_s is the stress which is needed to move dislocations past small local obstacles like other dislocations which cut the glide plane, or small impurity particles. These short-range obstacles are overcome by thermal activation. Therefore their influence on the flow stress is strongly temperature dependent.

In order to understand the stress–strain curve of a material, one has to take into consideration the various possible mechanisms which determine the mobility of dislocations and their multiplication. Applying statistical methods, one has then to calculate the relation between stress and strain. So far this has been impossible because of complications. The only thing which can be done is to single out one or two of the most important processes and base the theory on them. These methods have been quite successful in understanding plastic flow in crystalline solids. Since plastic deformation is an irreversible process, $\tau(a)$ is not a function of state, but depends on how the strain a has been reached.

The most important means of investigation in the field of plasticity are as follows.

1. Single crystals are deformed in tension or compression. Either the strain rate $\dot{a}$ is kept constant and τ is measured as a function

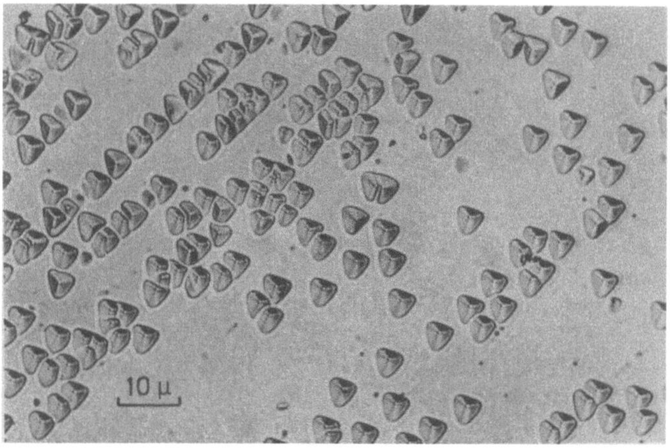

Fig. 8. Dislocation etch pits on a {111} plane of a slightly deformed germanium single crystal.

of a, or τ is kept constant and a is measured as a function of time. The latter experiment is called a creep test.

2. The microstructure of the deformed specimen is studied. (a) From slip lines like those in Figure 1, one gets the following information: Miller indices of the slip plane, slip direction, amount of slip, mean free path of dislocations, and positions at which slip occurs. (b) On certain crystallographic planes the spots where dislocations pierce the surface can be made visible by etching (Figure 8). The etching solution attacks the material preferentially at dislocations, where it produces small pits. (c) In thin slices (typically 1000 Å thick) which have been prepared from the deformed specimen the dislocation arrangement can be studied by electron transmission microscopy. Figure 9 gives two examples. Similar transmission pictures can be obtained by X-ray diffraction, but at a much lower magnification.[18]

Since the behavior of dislocations depends strongly on the crystal structure of the material, each structure will be discussed separately in the following sections. The conditions which the Burgers vector has to meet have been discussed in Section 2. The glide plane must contain **b** and must be densely packed. The latter condition is equivalent to a large separation of adjacent glide planes. Already from Frenkel's model of slip it can be seen that this will result in a low flow stress because it makes the ratio r/s in Eq. (4) small.

(a)

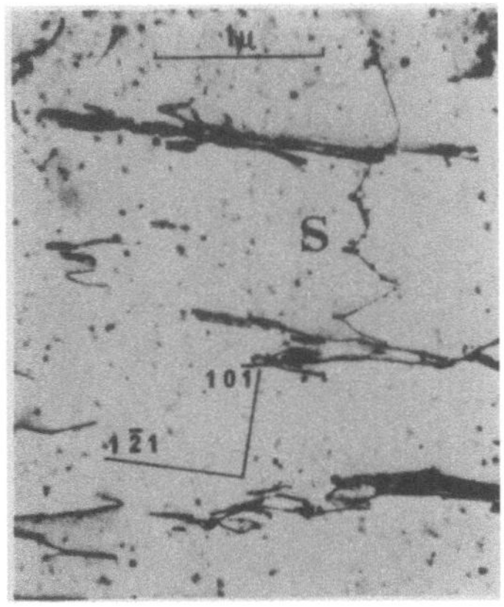

(b)

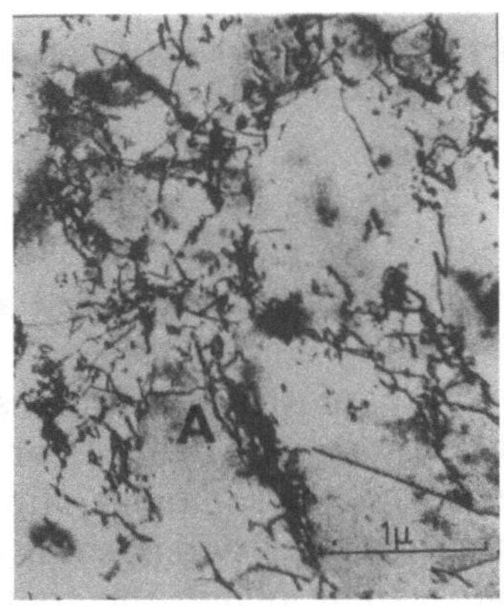

Fig. 9. Electron transmission photos of the dislocation arrangement in a deformed copper single crystal (Hirsch and Steeds[17]). (a) Stage I, $a = 0.03$; (b) stage II, $a = 0.23$.

3.1. Plastic Deformation of FCC Crystals

The mechanical behavior of fcc metals has been studied most extensively. The glide plane is the densely packed $\{111\}$ plane. The Burgers vector is $\frac{1}{2}s\langle 110\rangle$, where s is the lattice constant. It has to be kept in mind that this Burgers vector normally splits up:

$$\tfrac{1}{2}s[\bar{1}01] = \tfrac{1}{6}s[\bar{1}\bar{1}2] + \tfrac{1}{6}s[\bar{2}11] \tag{8}$$

Each dislocation consists of two partials which are nearly parallel to one another and lie in the same plane. Between them there is a stacking fault. Its width depends on the stacking fault energy; typically it will be 5–20s.

Figure 7 shows the stress–strain curve $\tau(a)$ of a copper single crystal deformed at room temperature in tension. Diehl[16] has pointed out that—besides the elastic part—three sections of the curve can be distinguished: In stages I and II the curve is straight. The slope $d\tau/da$ is $10^{-4}\,G$ and $4 \times 10^{-3}\,G$, respectively, where G is the shear modulus. Stage III is curved; often it is parabolic.

3.1.1. Critical Shear Stress

At room temperature τ_0 of fcc metals ranges from 20 to 1000 g/mm^2; except for nickel, it is less than 100 g/mm^2.[16,19–21] But Young[22] has demonstrated by an etch pit technique that already at stresses below 10 g/mm^2 dislocations may move in perfect copper single crystals. At τ_0 many dislocations move over macroscopic distances.

The dependence of τ_0 on crystal orientation is only weak; τ_0 is smallest in the middle of the stereographic triangle (87 g/mm^2) and largest for orientations close to $\langle 111 \rangle$ (135 g/mm^2).[16] Evidently Schmid's law is not exactly followed. The variation of τ_0 with temperature is shown in Figure 10 for silver. After eliminating the temperature dependence of the shear modulus, τ_0 of oxygen-free crystals varies only negligibly with temperature. The τ_0 value of specimens containing 4 ppm oxygen increases at low temperatures. Only for aluminum does τ_s, the temperature-dependent part of τ_0, not vanish. For the other fcc metals τ_0/G is independent of temperature; evidently they do not contain small obstacles which can be overcome by thermal activation. Because of the low stacking fault energy, dislocations are split up widely. Therefore a forest dislocation, i.e., a dislocation which lies in another, nonparallel glide plane, does not represent a small localized obstacle. Cutting of such a forest dislocation is not thermally activated. Aluminum, however, has a high stacking fault energy. Its forest dislocations are overcome by thermal activation; thus the temperature dependence of τ_0 is strong.

Seeger[8] has calculated τ_s as a function of temperature T. Let U_0 be the energy necessary to overcome a local obstacle and d its width, measured in the direction of glide. Two different sources contribute to U_0: the thermal energy of the crystal and the work done by the applied stress τ. Not the whole amount of τ acts on the dislocation, because part of it is compensated by the stress τ_G which is produced

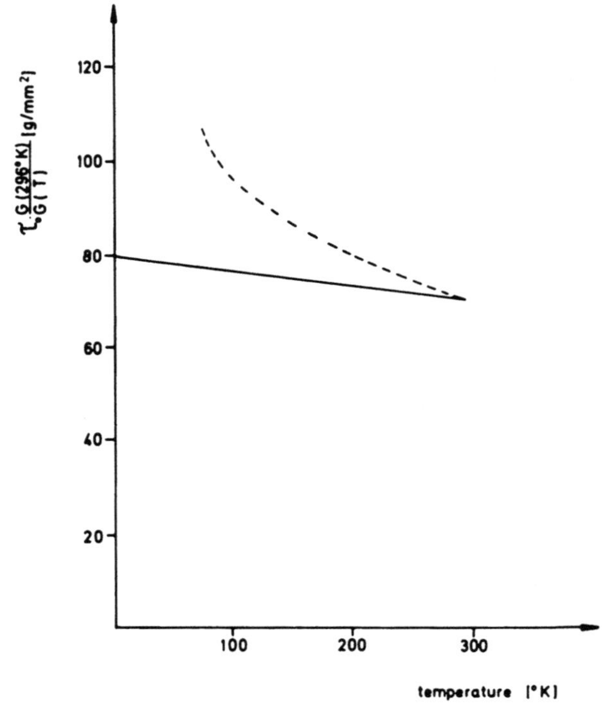

Fig. 10. Temperature dependence of τ_0 for silver (after Ahlers[19]). (- - -) 4 w. ppm oxygen, (——) 1 w. ppm oxygen. The temperature dependence of the shear modulus has been eliminated by multiplying by $G(296°\mathrm{K})/G(T)$.

by other dislocations. The work done by $\tau_s = \tau - \tau_G$ in cutting through the obstacle can be calculated with reference to Figure 11. The mean free length of the dislocation is l_0, and the force acting on the obstacle is $2l_0\tau_s b$. The path on which this force acts is on an average $(1/2)d$. Thus the work done by τ_s equals $\tau_s b l_0 d$. Using Boltzmann statistics the mean waiting time t before the obstacle is overcome can be calculated:

$$t = (1/\nu_0)\exp[(U_0 - v\tau_s)/kT] \tag{9}$$

where ν_0 equals the Debye frequency times the ratio b/l_0 and $v = bl_0 d$. The term v is called the activation volume. If there are N' dislocation segments in the crystal and the area swept out by each of them in one glide step is F (Figure 11), one gets for the shear strain a

$$a = N'(F/q)(b/h) \tag{10}$$

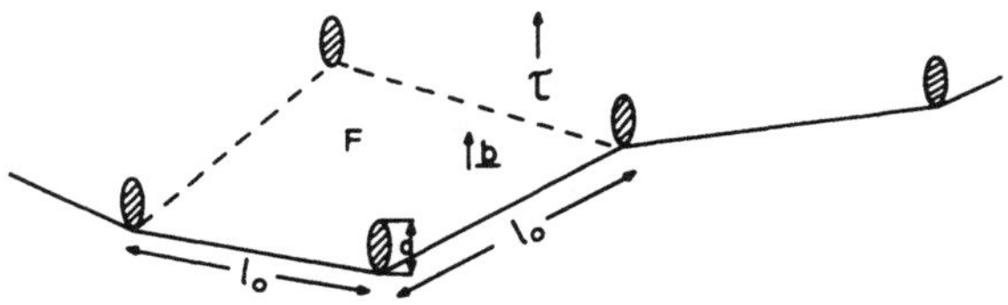

Fig. 11. Dislocation held up by an obstacle (after Seeger[8]). F is the area swept out by the dislocation once it has cut through the obstacle. The obstacle is assumed to be a dissociated forest dislocation. Other obstacles will have more circular cross sections.

where q is the area of the glide plane and h the specimen length, measured normal to the glide plane. N' is related to the dislocation density N by

$$N = N'l_0/qh \tag{11}$$

Equation (10) generalizes the definition of a given in Section 1 for cases where dislocations do not sweep out the whole glide plane. This adds the factor F/q. The strain rate $\dot{a}$ is derived from Eqs. (9) and (10):

$$\dot{a} = a/t = (NFb/l_0)\nu_0 \exp[-(U_0 - v\tau_s)/kT] \tag{12}$$

Here it is assumed that the time a dislocation spends moving between two obstacles is small compared with the time it spends waiting in front of them. In a tensile test $\dot{a}$ is given by the machine and τ_s is measured. From Eq. (12) one gets

$$\tau_s = (1/v)[U_0 - kT\ln(NFb\nu_0/l_0\dot{a})] \tag{13}$$

This yields for the total critical shear stress τ_0

$$\tau_0 = \tau_G + (1/v)[U_0 - kT\ln(NFb\nu_0/l_0\dot{a})] \tag{14}$$

One expects a linear decrease of τ_s with T until τ_s equals zero; from then on τ_0/G should be constant. Equation (14) does not only hold for τ_0, i.e., a nearly zero, but also for higher strains. Seeger[8] has shown that the temperature dependence of the flow stress of aluminum single crystals deformed to a strain of 0.25 followed Eq. (14). In this case the obstacles were forest dislocations. In Figure 10 the obstacles in the impure silver crystals must have been impurity atoms.

Since dislocations are flexible, they will bend out between the obstacles when a stress is applied. This reduces their mean free length

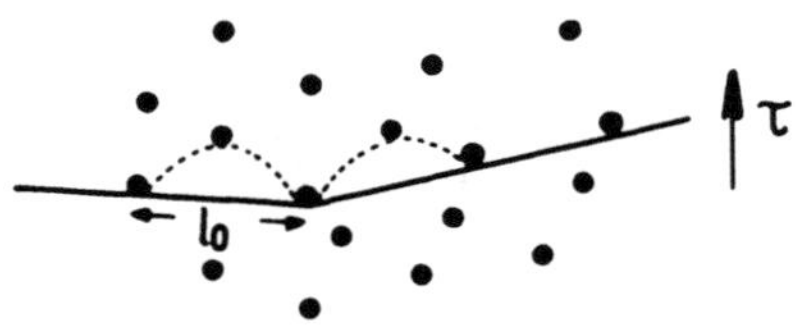

Fig. 12. A flexible dislocation (- - -) bends out between obstacles.

l, as can be seen from Figure 12. Friedel[23] and Labusch[24] have calculated l as a function of τ; Labusch's result for high stresses is

$$l = (2E_l/n_F b\tau)^{1/3} \tag{15}$$

E_l is the line tension of the dislocation and n_F the number of obstacles per unit area. The variation of l with τ causes a corresponding variation of the activation volume v.

Seeger explains the orientation dependence of τ_0 in the following way: If the orientation of the crystal is close to the edges of the stereographic triangle, the resolved shear stress in secondary glide systems will be high enough to cause already at τ_0 some dislocation movement and multiplication in these systems. This means a decrease in the ratio F/l_0 in Eq. (14) and an increase in τ_0.

3.1.2. Stage I

Diehl[16] found for copper single crystals that the slope $\vartheta_{\text{I}} = d\tau/da$ and the extent a_{II} of stage I (Figure 7) depend on the orientation of the specimen: ϑ_{I} is largest for orientations close to ⟨111⟩ and smallest for orientations close to ⟨123⟩; the orientation dependence of a_{II} is just opposite to that of ϑ_{I}, i.e., crystals with orientations close to ⟨123⟩ are softest. Using electron transmission microscopy, Essmann[25] found that nearly all dislocations have edge character. Figure 9(a) shows the dislocation arrangement. The Burgers vector is $\frac{1}{2}s[10\bar{1}]$ and the direction of most of the dislocations is parallel to $[1\bar{2}1]$, i.e., they are edge dislocations. Because of the low flow stress and the small work-hardening rate ϑ_{I}, stage I is called the stage of easy glide.

Argon and East[26] picture plastic flow in this stage in the following way. Sources emit dislocations. Dislocations which lie in parallel glide planes and have different signs attract one another. The screw segments cross-slip (Section 3.1.4) and annihilate one another; the edge segments form stable dislocation dipoles. These dipoles capture other passing dislocations and thus produce multipoles, which limit the mean free path of the following dislocations. The dipoles are clearly visible in Figure 9(a). The velocity of moving dislocations is mainly governed by two mechanisms: cutting forest dislocations and interactions with phonons. In a statistical analysis Argon and East calculated the characteristic parameters of stage I as a function of temperature and orientation. The higher the resolved shear stress in secondary glide systems is, the more glide occurs in them. This increases the density of forest dislocations and thus impedes glide in the primary system. Thus the orientation dependence of ϑ_{I} and a_{II} mentioned above can be explained.

3.1.3. Stage II

In stage II the work-hardening curve is steepest; its slope ϑ_{II} depends on temperature only through the shear modulus. This means work hardening is only due to elastic interactions between dislocations; τ_s of Eq. (7) is negligible. ϑ_{II} has been found to be roughly $3 \times 10^{-3}\ G$. Seeger *et al.*[27] have presented a model for work hardening in stage II. In this stage secondary slip becomes noticeable. The mean free path of primary dislocations is limited by Lomer–Cottrell dislocations lying in the glide plane. A Lomer–Cottrell dislocation is an immobile dislocation which is formed by a reaction between a dissociated primary dislocation and a secondary one. The probability of the reaction increases with the number of secondary dislocations. From slip line studies Seeger *et al.* determined the mean slip distances of edge and screw dislocations, l_1 and l_2, respectively, as a function of a:

$$l_i = \Lambda_i/(a - a^*) \tag{16}$$

with $\Lambda_1 = 8 \times 10^{-4}$ cm, $\Lambda_2 = 4 \times 10^{-4}$ cm, and $a^* \approx a_{II}$.

The distances l_i depend on a because with increasing strain, more and more Lomer–Cottrell dislocations are formed. Seeger assumed that a source, once it becomes activated, emits n dislocation loops in rapid succession and afterward remains inactive. For an increase da in strain, dm new sources per unit volume have to be activated:

$$da = nbl_1 l_2\, \mathrm{dm} \tag{17}$$

Equation (17) is similar to Eq. (10). Inserting l_i from Eq. (16) and integrating gives

$$\tfrac{1}{3}(a - a^*)^3 = (1/2)nb\Lambda_1{}^2 m \tag{18}$$

m is related to the mean internal stress τ_G by Eq. (6):

$$\tau_G = 0.1nbG[2m(l_1 + l_2)]^{1/2} \tag{19}$$

The n dislocations which are emitted by a source and are piled up in front of a Lomer–Cottrell dislocation act like a giant dislocation of Burgers vector nb. Since the temperature-dependent part τ_s of the shear stress is zero, the flow stress τ equals τ_G. After elimination of m and l_i from Eqs. (16), (18), and (19) one gets

$$\vartheta_{II} = d\tau/da = 0.1G(2nb/\Lambda_1)^{1/2} \tag{20}$$

Using the experimentally observed value of 20 for n, one gets $\vartheta_{II} = 3.2 \times 10^{-3}\,G$, which is in good agreement with the experimental result. Of course, the numerical values can be calculated more accurately.

Equations (17) and (19) are quite general, but $d\tau/da$ cannot be calculated from them without knowledge about the dependence of l_i on a or τ. In contrast to Seeger, who derived this relation from experiments, Kuhlmann-Wilsdorf[28] gave a purely theoretical estimate for $d\tau/da$.

Figure 9(b) is an electron transmission picture of a copper crystal deformed into stage II. There are areas of high dislocation density which enclose others which are nearly free of dislocations. The formation of such an inhomogeneous dislocation distribution becomes more distinct for higher deformations.

Measuring the dislocation density N by an etch pit technique, one can check whether Eq. (6) holds. This has been done by Livingston[29] for copper and by Levinstein and Robinson[30] for silver. In both cases the flow stress was proportional to $\sqrt{N}$.

3.1.4. Stage III

The decrease of slope of the work-hardening curve at high stresses has been explained by Diehl *et al.*[31] in terms of the cross-slip of screw dislocations. Once the stress on a screw dislocation which is pressed against a Lomer–Cottrell barrier becomes high enough to recombine the two partial dislocations, the screw dislocation may leave its glide plane and change to another one. In Figure 13 a segment of a screw dislocation moves from a (111) plane to a $(11\bar{1})$ plane. In both planes it can glide because its Burgers vector is $\frac{1}{2}s[1\bar{1}0]$. If it returns later to another (111) plane, it can continue in its former glide direction. The stress at which cross-slip occurs is higher for metals with a low

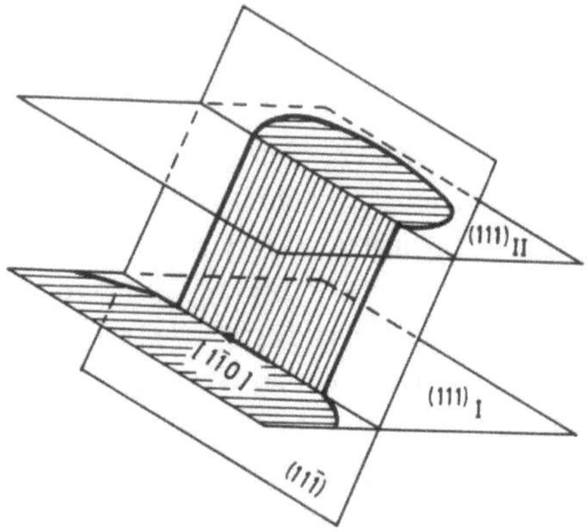

Fig. 13. Cross-slip of a screw dislocation (after Diehl *et al.*[31]).

stacking fault energy γ than for metals with a high γ value, because in the latter metals, the two partial dislocations stay close together: aluminum (high γ) cross-slips more easily than copper or nickel (low γ).[31] After a screw dislocation has been sufficiently constricted by the applied stress, cross-slip is accomplished by thermal activation. Therefore τ_{III}, the stress at which cross-slip becomes possible, varies strongly with temperature. Haasen[21] has measured τ_{III} of nickel single

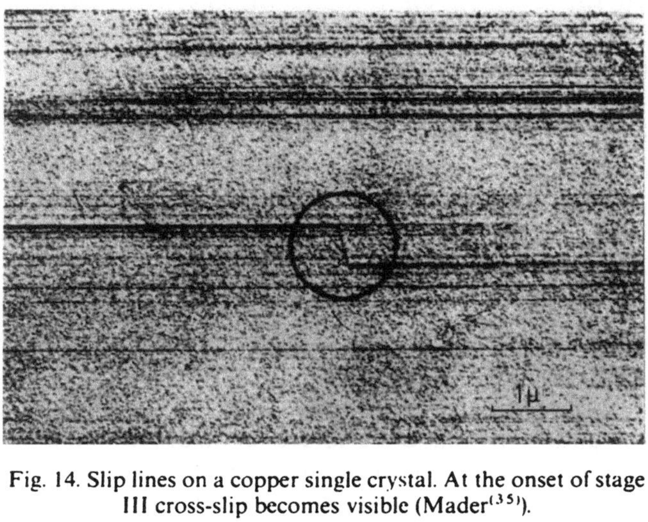

Fig. 14. Slip lines on a copper single crystal. At the onset of stage III cross-slip becomes visible (Mader[35]).

crystals as a function of temperature, finding

$$\tau_{III} = 18.1 - 0.039T \tag{21}$$

with τ_{III} in kg/mm^2 and T in °K. Schoeck and Seeger[32] and Seeger *et al.*[33] have given numerical relations between τ_{III} and γ. The γ values calculated from τ_{III} agree fairly well with those derived from other experiments.[34] The view that τ_{III} is governed by the onset of cross-slip is also supported by Mader's[35] slip-line studies (Figure 14).

Recently the theory for the onset of stage III has been generalized by Mecking and Lücke[36] for the whole stage III. They interpret work hardening in stage III as a combination of two processes: the hardening process of stage II and dynamic recovery by cross-slip. In this way $d\tau/da$ can be calculated as a function of τ.

3.2. Plastic Deformation of Hexagonal Crystals

Because of the lower symmetry of hexagonal crystals, they have a great number of glide systems which are crystallographically not equivalent. The glide systems which have been observed so far have been tabulated by Hirth and Lothe.[37] The most important glide plane is the basal plane; the Burgers vectors lying in it are parallel to $\langle 11\bar{2}0\rangle$. For materials with a small axial ratio, $\{10\bar{1}0\}$ planes are the preferred glide planes.

Comprehensive studies of the plasticity of hexagonal metals have been published by Lücke *et al.*[38] and Seeger and Träuble[39] on zinc and Hirsch and Lally[40] on magnesium. The work-hardening curve can be divided into three stages A, B, and C, which correspond to the stages I, II, and III, respectively, of fcc metals. The extent of stage A is very large at temperatures above 0°C; it may exceed shear strains of 1.0. In stage B Hirsch and Lally observed twinning. If the specimen is oriented for basal slip, Seeger's theory of τ_0 of fcc metals (Section 3.1.1) applies also to hexagonal metals. They behave like fcc metals with a high stacking fault energy. The linear decrease of τ_0 with increasing temperature which is predicted by Eq. (14) has been observed by Seeger and Träuble for zinc and by Boas and Schmid[41] for cadmium. For basal glide τ_0 is of the same order of magnitude as τ_0 of fcc metals. For prismatic glide it is much higher. For zinc at 256°C Gilman[42] found 1275 g/mm^2 for prismatic glide and 10 g/mm^2 for basal glide.

Hirsch and Lally have advanced a theory for slip in stage A which is very similar to Argon and East's[26] theory for stage I of fcc metals. Stage B is again governed by the elastic interaction of dislocations, and in stage C a thermally activated process starts operating. Seeger and Träuble believe that it is climb of edge dislocations.

3.3. Plastic Deformation of Crystals Having the Diamond Structure (DS)

At temperatures which are higher than about two-thirds of their absolute melting temperature germanium, silicon, and similar materials can be plastically deformed. Figure 15 shows a work-hardening curve of germanium. Again three stages can be distinguished. But before plastic flow starts, a pronounced yield point occurs. Plastic flow of DS crystals has been extensively studied by Haasen and co-workers; their results have been recently reviewed.[44] Slip occurs on $\{111\}$ planes, the Burgers vector is $\frac{1}{2}s\langle 110\rangle$, where s is the lattice constant. The dislocation mechanisms in DS crystals are similar to the ones in fcc metals. This is to be expected from the similarity of the two crystal structures. The temperature-dependent dynamic recovery process in stage III is probably a combination of cross-slip and climb, at least at high temperatures.[45]

The yield point is caused by the low dislocation density N_0 of the undeformed DS crystals. Typically N_0 is less than $10^4\ \mathrm{cm}^{-2}$ and may even be zero. In common metal crystals N_0 is $10^6\ \mathrm{cm}^{-2}$. In a tensile test the specimen is forced to change its length at a rate which is given by the machine. The specimen will deform partially elastically and partially plastically:

$$\dot{a} = \dot{a}_{\mathrm{elast}} + \dot{a}_{\mathrm{plast}} \tag{22}$$

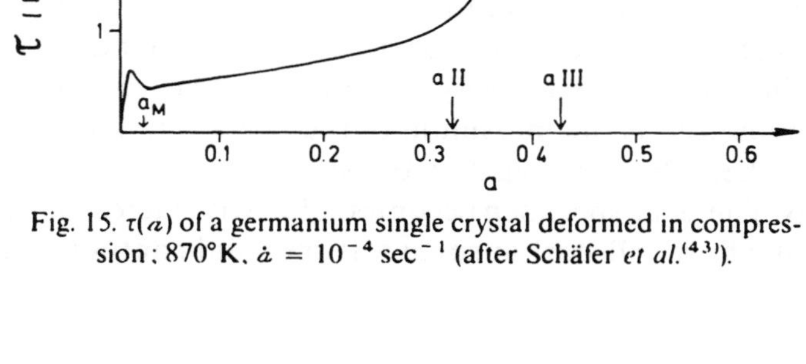

Fig. 15. $\tau(a)$ of a germanium single crystal deformed in compression; 870°K, $\dot{a} = 10^{-4}\ \mathrm{sec}^{-1}$ (after Schäfer *et al.*[43]).

The "elastic glide strain rate" $\dot{a}_{\text{elast}}$ is the elastic technical strain rate $\dot{\varepsilon}_{\text{elast}}$ after transformation according to Eq. (2). The rate $\dot{a}_{\text{plast}}$ is related to the density N and velocity v of dislocations by an equation similar to Eq. (12):

$$\dot{a}_{\text{plast}} = Nbv \tag{23}$$

where v depends on the applied stress τ:

$$v = (\tau - A\sqrt{N})^Z Be^{-U/kT} \tag{24}$$

Equation (24) is an empirical result. The terms Z, A, B, and U have been measured by Schäfer[46]: $Z = 1.1$, $A \approx 10^{-3}$ kg/mm, $U = 1.6$ eV. Using $(\tau - A\sqrt{N})$ instead of τ in Eq. (24) means that a dislocation does not feel the whole applied stress because part of it is counterbalanced by the stress produced by the other dislocations. The term $A\sqrt{N}$ represents τ_G of Eq. (7). Dislocation multiplication is governed by another empirical law:

$$\dot{N} = KNv(\tau - A\sqrt{N}) \tag{25}$$

where K is a constant. Equation (25) implies that every dislocation may act as a dislocation source and the probability of multiplication is proportional to the slip path. Equations (24) and (25) have been confirmed by independent experiments; Eq. (24) seems to be somewhat questionable for small stresses, but v does increase strongly with increasing stress. Putting Eq. (22)–(25) together, one gets a complete description of the work-hardening curve up to strains slightly beyond the minimum. Unfortunately, these differential equations cannot be solved analytically. Therefore only a qualitative view is presented here. At the start of the deformation only very few dislocations are in the specimen; therefore $\dot{a}_{\text{plast}}$ is small and most of the deformation is elastic. This causes a fast increase in τ, which increases the velocity of the few dislocations [Eq. (24)] and this in turn increases N [Eq. (25)], which rises steeply. Gradually $\dot{a}_{\text{plast}}$ becomes equal to $\dot{a}$. Since so many dislocations have been created, they can move more slowly [Eq. (23)]. This reduces τ [Eq. (24)]. Afterwards, i.e., beyond the stress minimum, N becomes so large that the dislocation interaction makes a new increase in stress necessary. Schröter *et al.*[47] found that a plastic glide strain of 0.02 increases N by a factor of 100. The smaller N_0 is, the more pronounced is the maximum of the stress–strain curve. In that part of it where $d\tau/da$ is negative the specimen will deform inhomogeneously. There the specimen becomes softer with increasing deformation. Plastic flow will start at some soft spot and continue there until this region has been deformed beyond the minimum of the stress–strain curve. Then the stress field of the dislocations will initiate plastic flow in adjacent parts of the crystal. This wavelike propagation of plastic flow is by so-called Lüders bands. Because the crystal does not deform homogeneously, the calculations presented above have to be applied locally, not to the whole crystal.

3.4. Plastic Deformation of BCC Crystals

The investigation of bcc metals has long been hampered by the difficulty of preparing high-purity single crystals. Since interstitially dissolved elements like oxygen, nitrogen, and carbon give rise to strong tetragonal lattice distortions, their total content has to be reduced below 1 ppm. A very detailed study of niobium has been published by Duesbery and Foxall.[48,49] The Burgers vector is always $\frac{1}{2}s\langle 111\rangle$ and slip planes may be either $\{110\}$ or $\{112\}$. Above 400°K slip seems to occur on noncrystallographic planes, i.e., different slip planes are activated simultaneously. The stress–strain curves depend strongly on temperature and orientation. For room temperature, again three stages are found; the first and second ones are straight. For temperatures below 200°K, however, $d\tau/d\alpha$ decreases continuously with increasing α. The value of τ_0 varies strongly with temperature: At 78°K it is 13 kg/mm^2 and above 373°K it is 0.3 kg/mm^2. It should be remembered that τ_0/G of high-purity silver is nearly independent of temperature.

The following description applies only to room temperature and crystal orientations in the middle of the stereographic triangle. The qualitative appearance of the stress–strain curves is the same as for fcc crystals. Stage I is rather long; it may go up to shear strains of 0.2. The value of $d\tau/d\alpha$ is close to zero, and hardly any work hardening is visible. Slip lines are straight and only one glide system is activated. The end of stage I, as determined from the stress–strain curve, coincides with the occurrence of slip lines of a secondary glide system. In stage II, $d\tau/d\alpha$ is $2.5 \times 10^{-3}\,G$[48] and this is very close to the corresponding values for fcc metals. Taylor and Christian[50] measured the density of secondary dislocations by electron transmission microscopy and found it to increase strongly in stage II. Work hardening in this stage seems to be due to the difficulty of forcing primary dislocations to cut secondary ones. The arrangement of dislocations is similar to the one of fcc metals deformed into stage II. In stage III a dynamic recovery process operates; so far it is not clear which one it is.

A very interesting discovery has been made by Šesták and Zárubová.[51] Single crystals of Fe–Si were deformed in bending at room temperature. It was observed that the slip planes in the tension and compression regions were different. Duesbery and Foxall[48] found that Nb single crystals, which slipped on {112} planes, could be deformed more easily in compression than in tension. The difference in τ_0 was 20%. Their explanation for the asymmetries is that the dislocations may split into partial dislocations and the stacking faults lie in different planes. Before a dislocation can move, its partials have to be brought together. Only in the case of compression is the direction of the applied stress such that it helps to recombine the partial dislocations. Body-centered cubic metals are the only materials for which $\tau(a)$ curves measured in compression differ from the ones measured in tension.

Plastic flow below room temperature seems to be governed by a combination of intrinsic and impurity effects. Below 100°K the Peierls potential may determine the flow stress.

3.5. Plastic Deformation of Crystals Having the NaCl Structure

Ionic crystals like NaCl and LiF can be plastically deformed at all temperatures; the glide strains achieved may exceed 0.2 at room temperature.[52] The only perfect Burgers vector which has been observed for this crystal structure is $\frac{1}{2}s\langle 110\rangle$, where s is the lattice constant. But there are three glide planes: {110}, {100}, and {111}. In evaluating the probabilities of activating the different glide systems, one has to take the special features of ionic bonding into consideration, which have been discussed by Wernick in Chapter 4 of Volume 1 of this treatise. Glide planes must be chosen so that during slip, ions of equal sign do not pass one another. Moreover, one has to avoid creating charged glide steps on the surface of the specimen. At least two glide systems of the type $\langle 110\rangle$, {110} are always equally stressed; this follows immediately from Eq. (1). Mendelson[53] and Hesse[52] found for NaCl that in reality only one of the equally stressed slip systems is activated. The $\tau(a)$ curves measured by these authors look exactly like the ones of fcc metals. At room temperature τ_0 is 40 g/mm^2; τ_0/G decreases linearly with increasing temperature and is constant above 250°K. In stage II, $d\tau/da$ is 2.7×10^{-3} G. Matucha[54] has published a detailed study of slip lines on NaCl. According to him, secondary slip is important for work hardening in stage II. For specimens of $\langle 100\rangle$ orientation another {110} plane is activated and for specimens of $\langle 110 \rangle$ orientation secondary slip occurs on $\{100\}$ planes. Dislocations of secondary systems limit the mean free path l of primary dislocations; l varies with strain in the same way as it does for fcc metals [Eq. (16)]. Matucha believes that the model which Seeger developed for work hardening in stage II of fcc metals (Section 3.1.3) can be applied to NaCl, too. The onset of stage III of NaCl crystals coincides with the beginning of cross-slip. Screw dislocations escape from their $\{110\}$ glide plane first on a $\{111\}$ plane, then change to a $\{100\}$ plane, and finally return to another $\{110\}$ plane which is parallel to their original glide plane. Fontaine(55) has calculated that dislocations in NaCl may dissociate. Therefore the theory developed for cross-slip in fcc metals is at least qualitatively applicable to NaCl.

3.6. General Features of Stress–Strain Curves of Pure Single Crystals

For all five crystal structures which have been discussed the stress–strain curve can be divided into three stages. In stage II the movement of dislocations is mainly governed by their elastic interaction; the slope $d\tau/da$ is roughly $3 \times 10^{-3}\, G$. The contribution of the elastic interaction between dislocations to the work-hardening rate becomes noticeable when their density exceeds $10^8\ \text{cm}^{-2}$. The transition to such a high density from the low grown-in one corresponds to stage I. In stage III a thermally activated process operates: cross-slip of screw dislocations or climb of edge dislocations. These mechanisms reduce the work-hardening rate.

4. Mechanical Twinning

Twinning as a mechanism of plastic deformation has already been mentioned in Section 1. It occurs in nearly all metals. However, bcc(56–58) and fcc(21,59,60) metals twin only at low temperatures. For hexagonal metals twinning is very common if the specimen orientation is unfavorable for basal slip. Figure 16 demonstrates the effect of twinning on the deformation curve. Each load drop corresponds to the creation of one or more twin lamellae. Such a lamella has mirror symmetry with respect to the rest of the crystal, and its thickness is typically between 1 and 10 μm. The formation of a twin is often accompanied by an audible click. Twinning spreads over the whole specimen in a similar manner as slip does during the propagation of a Lüders band. Usually, rather high stresses are needed for twinning. These can only be built up if slip by dislocations is difficult. Twin

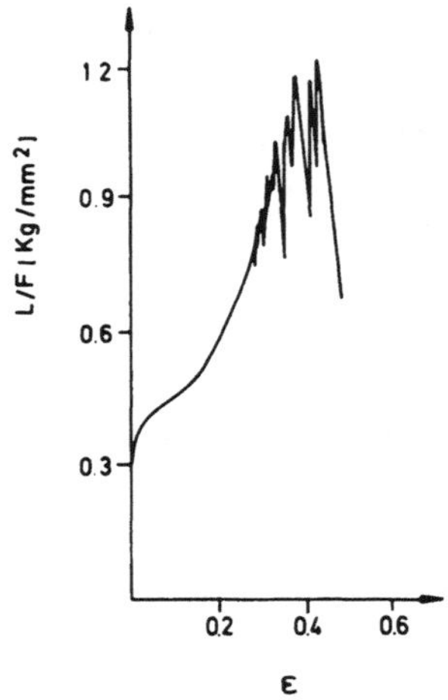

Fig. 16. Deformation of a cadmium single crystal by twinning. L is the load, F the cross section before deformation, ε the technical strain (after Boas and Schmid[31]).

planes, i.e., mirror planes in Figure 6, are $\{111\}$ for fcc, $\{112\}$ for bcc, and $\{10\bar{1}2\}$ for hexagonal crystals.

Models which explain the formation of twins on the basis of dislocation mechanisms have been suggested by Cottrell and Bilby[61] and Venables.[62] Here a simplified version for fcc metals will be presented. The stacking sequence of $\{111\}$ planes is *ABCABC*. A $\frac{1}{2}s[01\bar{1}]$ dislocation can split up into a Frank (FD) and a Shockley (SD) partial dislocation:

$$\underset{}{\tfrac{1}{2}s[01\bar{1}]} = \underset{\text{FD}}{\tfrac{1}{3}s[11\bar{1}]} + \underset{\text{SD}}{\tfrac{1}{6}s[\bar{2}1\bar{1}]} \qquad (26)$$

The FD and the SD lie in the $(11\bar{1})$ plane. The FD is a sessile dislocation; it cannot glide. Between FD and SD there is a stacking fault. There the stacking sequence is *ABCABABC* (Figure 17b). By sweeping out the whole $(11\bar{1})$ plane the SD creates a twin which is three atomic layers thick. These twin lamellae can grow in thickness if other SD's pass on adjacent planes. But then the question arises of how all these dislocations are to be created. The solution to this problem is that only one SD is needed if it is lifted to the next glide plane after it has swept out one. Let there be another dislocation P which intersects the $(11\bar{1})$ plane. The component of P's Burgers vector parallel to $[11\bar{1}]$ must equal the separation of $\{111\}$ planes. Then P makes a spiral ramp out of the

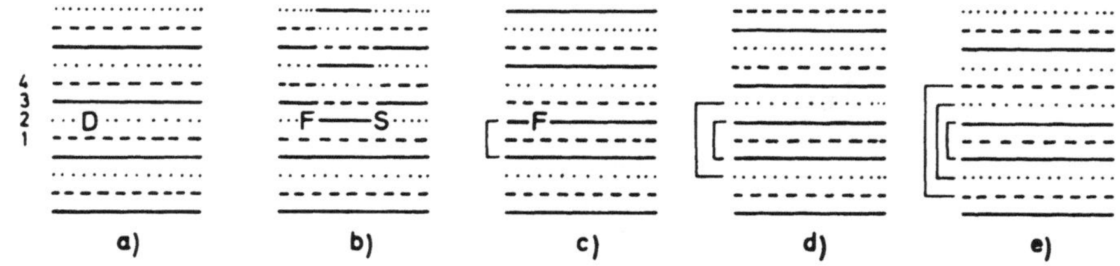

Fig. 17. Formation of a twin lamella by the movement of a Shockley partial dislocation. (——) *A* plane, (- - -) *B* plane, (. . .) *C* plane. (a) *D* marks a perfect dislocation in the $(11\bar{1})$ plane. (b) *D* splits into a Frank (F) and a Shockley (S) partial dislocation. (c) *S* has swept out the whole of plane 2; thus the stacking fault extends through the whole specimen. (d) Another Shockley partial dislocation has swept out plane 3. (e) Same as (d) for plane 4. A twin lamella has been created. Corresponding planes are linked by brackets. Plane 1 is the mirror plane.

$(11\bar{1})$ planes. If the SD meets *P* and is unable to cut it, it will swing around it. As can be seen from Figure 18, each arm of the SD moves into another direction, i.e., one spirals up and the other one down. Thus a set of twin lamellae is created. Because of its function, *P* is called a pole dislocation.

The model described above suggests a close connection between stacking fault energy γ and twinning. For small values of γ the FD and the SD are far from one another; they can easily be separated. This should make twinning easy. Indeed, Venables[62] and Peissker[63] observed a decrease of the stress necessary to cause twinning if γ of copper was reduced by alloying.

5. Solid-Solution Hardening

5.1. Experimental Observations

Work hardening is not the only process by which the flow stress of a material can be increased; this can also be achieved by alloying.

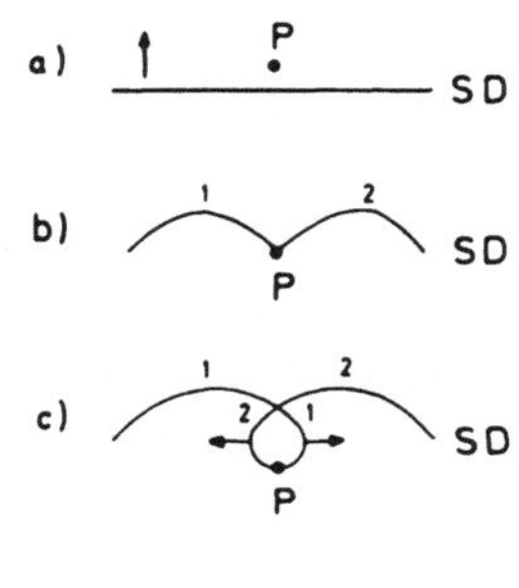

Fig. 18. Shockley partial dislocation (SD) encountering a pole dislocation *P*. The arrows mark the direction of the movement of SD. In (c) the two arms of SD move in opposite directions.

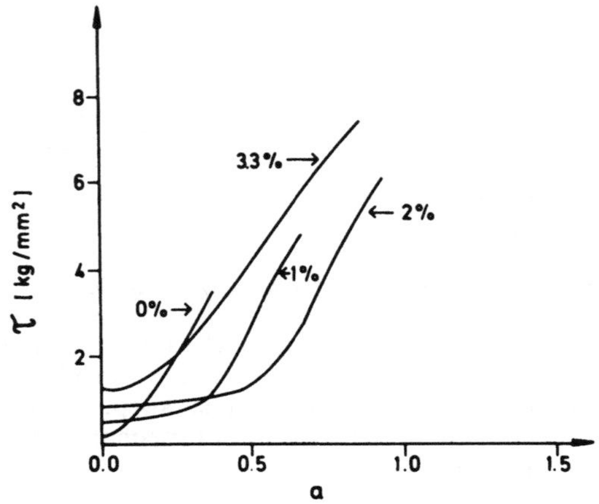

Fig. 19. Room-temperature work-hardening curves of copper–germanium solid-solution single crystals (after Peissker[64]). The germanium content is given in atomic percent, the arrows mark τ_{III}.

Fig. 19 shows work-hardening curves of copper–germanium solid-solution single crystals. The addition of germanium increases τ_0, τ_{III}, and, for low concentrations, also a_{II}. Germanium hardens copper, i.e., reduces the mobility of dislocations. The longer easy glide stage is no contradiction. a_{II} is reached after some glide in secondary systems has happened (Section 3.1.3). Since these, too, are hardened by alloying, a_{II} increases. The decrease of a_{II} for high germanium concentrations seems to be due to another mechanism which does not depend on the type of the alloying element but only its concentration.[64] The temperature dependence of τ_0 is shown in Figure 20 for various copper-base solid solutions. For temperatures T below 550°K, τ_0 decreases with increasing temperature; above 550°K, τ_0 is constant. This type of $\tau_0(T)$ curve has been found for many other solid solutions: for fcc,[66,67] hexagonal,[42] and even bcc[68–70] materials. For temperatures approaching the melting point τ_0 decreases again and finally reaches the value for the pure metal.[42] In bcc metals a solution-softening effect has been found. At low temperatures τ_0 has a minimum as a function of solute concentration. Mitchell and Raffo[71] measured τ_0 of various tantalum–rhenium alloys at 77°K. For pure tantalum τ_0 was 37 kg/mm^2; 2 at. % rhenium brought τ_0 down to 25 kg/mm^2. Beyond 2 at. %, τ_0 increased monotonically with the rhenium content. If the temperature is raised, solution softening becomes less pronounced.

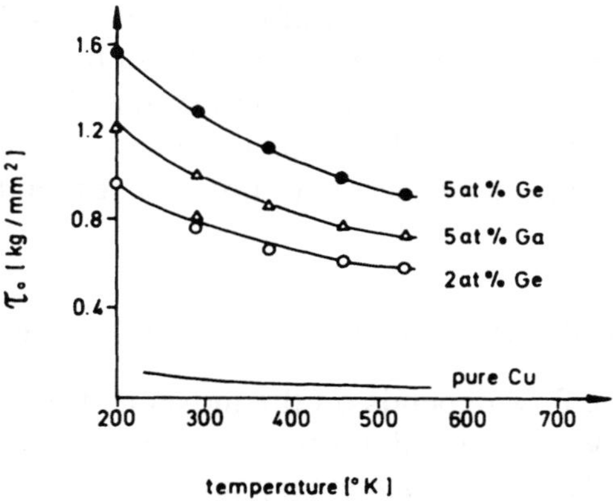

Fig. 20. Temperature dependence of τ_0 of various copper solid solutions, after a compilation by Kostorz and Haasen.[65]

Iron has long been known to exhibit a yield point similar to that of germanium.[72,73] Paxton and Churchman[72] have demonstrated that the yield point in iron single crystals occurs only if the specimen contains carbon. This becomes clear from Figure 21. If the deformation of a specimen which does contain carbon is interrupted beyond the stress minimum and continued after a short heat treatment at 60°C, a new yield point is observed. This is in contrast to the findings for germanium, where aging does not produce a new yield point.

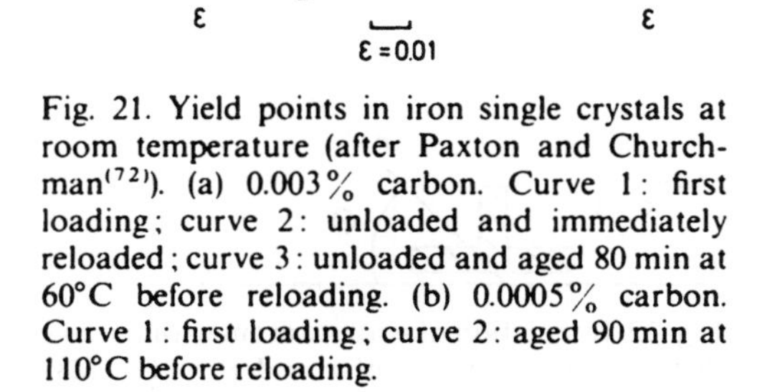

Fig. 21. Yield points in iron single crystals at room temperature (after Paxton and Churchman[72]). (a) 0.003% carbon. Curve 1: first loading; curve 2: unloaded and immediately reloaded; curve 3: unloaded and aged 80 min at 60°C before reloading. (b) 0.0005% carbon. Curve 1: first loading; curve 2: aged 90 min at 110°C before reloading.

At temperatures where the diffusion rates of solute atoms are appreciable many alloy crystals have discontinuous stress–strain curves which look similar to the ones observed in the case of twinning, but the amplitude of the yield drops is much smaller[(74)] (Portevin–Le Chatelier effect). Such serrated flow corresponds to repeated yielding.

5.2. Interaction Mechanisms between Dislocations and Solute Atoms

Since plastic flow depends on the movement of dislocations, the mechanical properties of solid solutions cannot be understood without detailed knowledge of the elementary interaction processes between dislocations and solute atoms. In the following sections a compilation of these mechanisms is given.

5.2.1. Size Effect

If the solute atoms differ in size from the matrix atoms, they will give rise to a stress field by which they interact with dislocations. The interaction energy E_R between an edge dislocation and a solute atom has been calculated by Cottrell.[(75)] The difference in atomic radius is characterized by ε_R, the relative change in lattice constant s with concentration c:

$$\varepsilon_R = (1/s)(ds/dc) \tag{27}$$

$$E_R(r) = [4(1 + \nu)/3(1 - \nu)][GbR^3(\sin\theta)/r]\varepsilon_R \tag{28}$$

where ν is Poisson's ratio and R is the atomic radius of the solute atoms; for r and θ see Figure 22. Equation (28) holds as long as r is

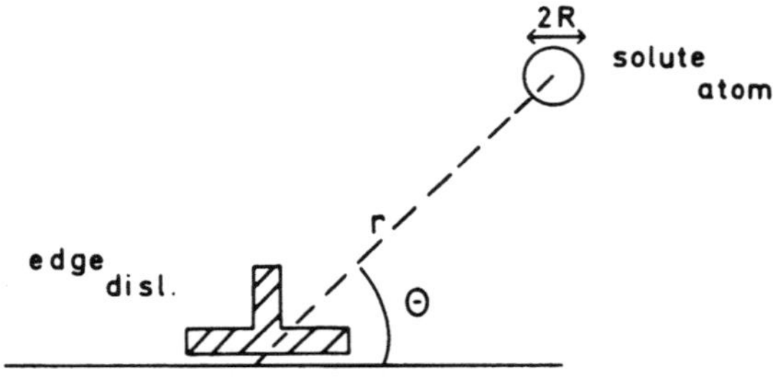

Fig. 22. Interaction of a solute atom with an edge dislocation; definition of r and θ of Eq. (28).

several times larger than s. One notices that large solute atoms, for which ε_R is positive, are attracted by the dilation center of the dislocation ($\pi < \theta < 2\pi$); small solute atoms have lower energies in the compression center ($0 < \theta < \pi$).

To a first approximation, no dilation is produced by a screw dislocation; therefore it does not interact with misfitting solute atoms. A second-order approximation yields an interaction energy which is much smaller than that of an edge dislocation.[76] If the distortion caused by the solute atom has only tetragonal symmetry, the interaction with both types of dislocations is strong. That is the reason why bcc metals are strongly hardened by interstitially solved oxygen, nitrogen, or carbon.

5.2.2. Modulus Effect

A solute atom changes the elastic properties of its environment.[77] The strain energy density $D(r)$ of a screw dislocation varies with the distance r as (Chapter 5 of Volume 1)

$$D(r) = Gb^2/8\pi^2r^2 \tag{29}$$

Thus the energy change $E_G(r)$ caused by one solute atom of radius R at a distance r from the screw dislocation is

$$E_G(r) = [(G_s - G_M)b^2/8\pi^2r^2](4\pi/3)R^3 \tag{30}$$

where G_M is the modulus of the matrix and G_s is that of the solute atom. Equation (30) holds if r is much larger than the lattice constant. For edge dislocations a rough estimate gives a value for $E_G(r)$ which is $[1/(1 - \nu)]^2$ times as large as that of screw dislocations. Fleischer[77] related the difference $G_s - G_M$ to the macroscopic change η of the modulus with solute concentration c:

$$\varepsilon_G = (G_s - G_M)/G_M = \eta/(1 + \tfrac{1}{2}|\eta|) \tag{31}$$

$$\eta = (1/G)(dG/dc) \tag{32}$$

This yields, instead of Eq. (30),

$$E_G(r) = G_M\varepsilon_G b^2(R^3/6\pi r^2) \tag{33}$$

5.2.3. Electric Charge Effect

In ionic crystals a dislocation may become charged because of the presence of elementary jogs.[78] In metals dislocations are charged because the specific volume at a dislocation differs from the one in the remainder of the material. Thus an edge dislocation becomes a line dipole. In most cases solute atoms carry a charge, too. Friedel[23] has pointed out that these charges give rise to a very short-ranging interaction force in metals.

5.2.4. Effect of Short-Range Order

The influence of short-range order on plastic flow has been discussed by Fisher.[79] The reason for short-range order is that the binding forces between unlike atoms are larger than the ones between identical atoms. By the movement of dislocations, strong bonds are cut and replaced by weak ones. Since the energy difference has to be supplied by the applied stress, an increase in flow stress is the consequence. This mechanism is of importance only for rather concentrated solid solutions.

5.2.5. Suzuki Effect

Suzuki[80] realized that the free energy of a solute atom which sits on the stacking fault of a dissociated dislocation may be lower than that of a solute atom which is in the rest of the material. In this case solute atoms will gather on stacking faults and thus lock dislocations. This is of course only possible if the dislocations are at rest. The interactions of Sections 5.2.1–5.2.4, however, are felt by moving dislocations, too.

5.2.6. Effect of Changes in Stacking Fault Energy

The stacking fault energy γ of fcc metals depends on their electron concentration. If solute and matrix atoms differ in their valence, γ depends on the composition of the alloy. Since γ determines the width of the stacking fault of dissociated dislocations, their mobility is affected by alloying. In contrast to the dislocation–solute atom interactions of Sections 5.2.1–5.2.5, this is not a local interaction. The energy of a dislocation does not depend on its position relative to the solute atoms.

5.3. Interpretation of the Experiments on Solid-Solution Hardening

The effects of solute atoms on plastic flow can be grouped into two classes: (1) the solute atoms segregate to the dislocations; (2) the solute atoms are homogeneously distributed throughout the material.

5.3.1. Hardening by Segregated Solute Atoms

The diffusion coefficient of carbon in iron is so high at room temperature that carbon atoms can segregate to dislocations. The driving force for diffusion is the elastic interaction between interstitially solved carbon atoms and dislocations. Carbon atoms gather in a cloud around dislocations and make them immobile.[75] Therefore only a few of them are available for plastic flow. It is the same situation as for germanium; there the low grown-in dislocation density of the crystals causes a yield point. Hahn[73] explained the yield point in iron–carbon crystals in the same way as Haasen and co-workers did in the case of germanium. A strong yield point will always occur if the following three conditions are fulfilled: low density of mobile dislocations, strong dependence of their velocity on stress, rapid dislocation multiplication. If an already deformed iron–carbon specimen is aged at temperatures between 0 and 200°C, carbon atoms migrate to the dislocations and lock them. Therefore the specimen will show another yield point if it is deformed again (Figure 21a).

If a stress is exerted upon a dislocation which is surrounded by an atmosphere of solute atoms, the dislocation will be slightly shifted from the center of this atmosphere. The solute atoms will try to follow this shift by diffusion. In this way the dislocation can slowly move, dragging its cloud behind. Cottrell and Jaswon[81] have calculated the maximum velocity v_m at which solute atoms can follow a moving dislocation. Let F be the interaction force between a solute atom and a dislocation; then one gets from Einstein's formula

$$v_m = (D/kT)F \tag{34}$$

where k is Boltzmann's constant, T is the absolute temperature, and D is the diffusion coefficient of the solute atoms. The radius ρ of the solute atmosphere is related to F by the following two equations:

$$F = A/r^2 \tag{35}$$

$$\rho = A/kT \tag{36}$$

where A is a constant and r is the distance between the solute atom and the dislocation. A reasonable estimate for the mean value of r is $\rho/2$. Thus one gets for v_m

$$v_m = 4D/\rho \tag{37}$$

As long as the dislocation moves at a velocity below v_m, its atmosphere can follow. The dislocation can move faster only if it breaks away from its cloud. A dislocation whose average velocity approaches v_m cannot move smoothly. At certain times it will move faster than v_m and break away from its cloud, whereas at other times it will move so slowly that the cloud can catch up again and slow it down. The movement of the dislocation becomes jerky. This is the explanation for serrated flow (Portevin–Le Chatelier effect).

5.3.2. Hardening by Homogeneously Distributed Solute Atoms

At high temperatures, where the interactions between dislocations and solute atoms are easily overcome by thermal activation, they will be homogeneously distributed throughout the specimen. The same holds if it is quenched from high temperatures and tested at temperatures at which diffusion is negligible. Except for the yield point in iron and the Portevin–Le Chatelier effect, all solid-solution hardening effects are produced by homogeneously distributed solute atoms.

For fcc solid solutions Fleischer[76] has presented a theory which relates the increase of the critical shear stress $\Delta\tau_0$ due to solid-solution hardening to the elementary interactions between solute atoms and dislocations. This theory does not allow for thermal activation; therefore it can be directly applied only at 0°K. Fleischer identifies $\Delta\tau_0$ with the stress needed to force dislocations past solute atoms. He considers only those atoms that lie immediately above or below the glide plane. From Eqs. (28) and (33) the maximum interaction force F between a solute atom and a dislocation can be derived. Let l be the mean distance between two solute atoms with which a dislocation interacts. The distance l depends on the flexibility of the dislocation (Figure 12). It is related to the line tension E_l, the concentration c of solute atoms, and the applied stress τ by Eq. (15):

$$l = (2E_l s^2/cb\tau)^{1/3} \tag{15'}$$

Here c/s^2 has replaced the number of solute atoms n_F per unit area, and s is the lattice constant. The dislocation can be moved past a solute atom if the force exerted by $\Delta\tau_0$ on the two dislocation segments of length l equals F:

$$2(\Delta\tau_0)bl = F \tag{38}$$

Eliminating l from Eqs. (15′) and (38) yields

$$\Delta\tau_0 = 0.25c^{1/2}F^{3/2}(1/bsE_l^{1/2}) \tag{39}$$

The numerical constant in Fleischer's original equation was slightly different because he used another relation between l and c.

Labusch[82] has done a more careful statistical analysis of this problem; he allows for a statistical distribution of forces. His result at 0°K is

$$\Delta\tau_0 = zc^{2/3}F^{4/3}(w^{1/3}/bs^{4/3}E_l^{1/3}) \tag{40}$$

Here F is the maximum interaction force between solute atoms and dislocations, w is the mean range of these forces, and z is a numerical factor of about 0.19. Labusch[83] took also fluctuations in the distribution of solute atoms into consideration. This changes z but leaves the relation between $\Delta\tau_0$, c, and F unaffected. Labusch[83] has calculated the dependence of $\Delta\tau_0$ on temperature T. The function $\Delta\tau_0(T)$ decreases with increasing T and becomes constant at higher temperatures. This is in agreement with the experimental findings shown in Figure 20. In the temperature range where $\Delta\tau_0$ is independent of T it is still proportional to $c^{2/3}$. Therefore Eqs. (39) and (40) have to be only slightly modified for high temperatures.

In calculating F from ε_R, the misfit parameter, and ε_G, the parameter which describes the difference in modulus, one has to take into account that the interactions with screw and edge dislocations are different. If neither ε_R nor ε_G vanishes, a suitable combination of both of them has to be used.[66,70,76,82,84,85] Kan and Haasen[84] proposed for the total ε_t

for screw dislocations $\quad \varepsilon_t = |\varepsilon_G - 3\varepsilon_R|$

for edge dislocations $\quad \varepsilon_t = (\varepsilon_G{}^2 + 256\varepsilon_R{}^2)^{1/2} \qquad (41)$

The authors replaced Eq. (40) by

$$\Delta\tau_0 = (G/550)\varepsilon_t^{4/3}c^{2/3} \tag{42}$$

and found for Au–Al and Au–Ge[85] solid solutions reasonable agreement between theory and experiment if they used ε_t of edge dislocations. For $c \geq 3 \times 10^{-3}$ Eqs. (39) and (40) describe the experimentally observed variation of $\Delta\tau_0$ with c equally well, but only Eq. (40) holds for c approaching zero. This has been demonstrated by Jax *et al.*[66] for various fcc solid solutions.

Additions of germanium or gallium to copper decrease its stacking fault energy (Section 5.2.6). Therefore dislocations dissociate more widely and cross-slip becomes more difficult.[64,86] This is the reason why τ_{III}, the stress necessary for cross-slip, increases with increasing germanium content (Figure 19).

Solution softening of bcc alloys has so far not been satisfactorily explained. Harris[87] suggested that it may be due to an anomalous dependence of the shear modulus on solute concentration.

If a solid is irradiated by fast neutrons or electrons, point defects and point-defect agglomerates are created. All of these defects increase τ_0 in a similar way as do solute atoms. For high strains the stress–strain curves of irradiated specimens do not differ from those of unirradiated specimens, because the passage of dislocations helps vacancies and interstitials to annihilate one another; the same holds for the agglomerates of these defects.[88]

6. Hardening by Particles of a Second Phase

The most efficient way to strengthen a material is to introduce small particles of a second phase. This can be seen from the following comparison: At room temperature τ_0 of pure copper is about 0.05 kg/mm^2; dissolving 2 at. % germanium in copper raises τ_0 to 0.8 kg/mm^2 (Figure 20), but introducing 1 vol. % of silicon oxide particles of 910 Å diameter increases τ_0 to 1.8 kg/mm^2.[89]

When a dislocation meets a particle in its glide plane it has two principally different possibilities to overcome it: (1) The dislocation may cut through the particle. Then the applied stress has to provide the energy which is needed for creating new surfaces on entering and leaving the particle. These surfaces are marked A_2 in Figure 23a. Figure 23b is an electron transmission picture of particles which have been cut by dislocations. (2) The dislocation may bow out between two obstacles and finally leave a loop around each of them (Figure 24) or it may overcome the obstacle by cross-slip. The first mechanism has been proposed by Orowan[91] and the second one by Ashby.[92]

Whether particles are cut or overcome by one of the two other mechanisms depends on their structure, size, and spacing. Moreover, in most cases the particle will be surrounded by a stress field which also hinders the movement of dislocations.

6.1. Cutting of Particles

At the beginning of a precipitation process the particles are often coherent with the matrix. They have a very similar or even the same crystal structure and orientation as the matrix, but their lattice constant, solute concentration, or degree of order is different. If a dislocation encounters such a particle in its glide plane, it can continue

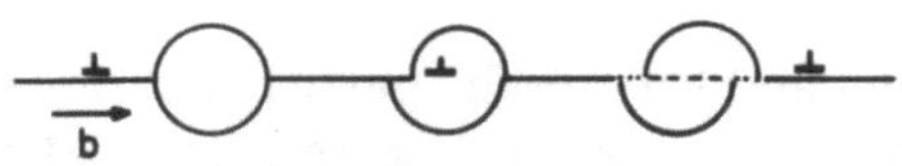

Fig. 23a. An edge dislocation cuts through a particle; area A_1 (- - -); area A_2 (· · ·).

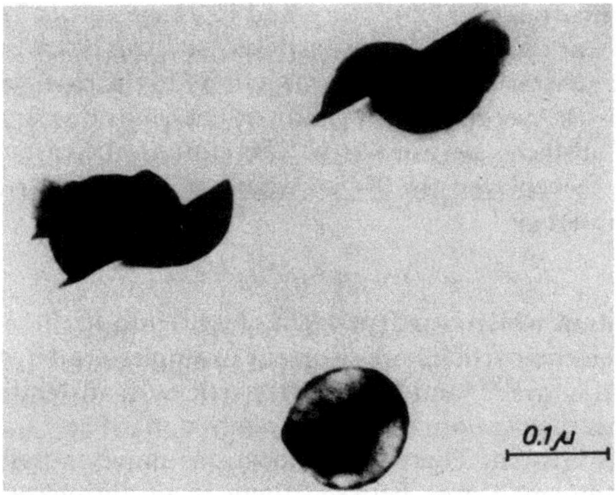

Fig. 23b. Cut particles in a Ni–Cr–Al alloy (after Gleiter and Hornbogen[90]).

its motion in the particle, because there the slip system and the Burgers vectors differ only slightly from the ones in the matrix. Three contributions to the hardening effect of such particles can be distinguished. The particles are assumed to be spherical.

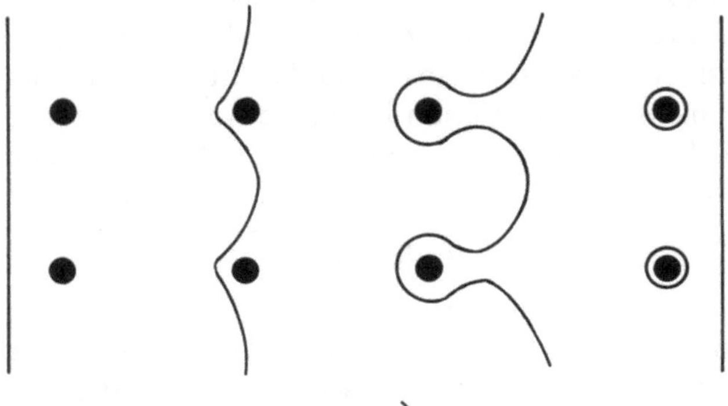

Fig. 24. Orowan[91] mechanism. The dislocation bows out between the particles which cannot be cut, and finally leaves a loop around each of them. The image plane is the glide plane; the arrow marks the direction of glide.

1. On entering and leaving the particle a dislocation creates new surfaces; they are marked A_2 in Figure 23(a). Let γ_i be the surface energy involved. Since the dislocation will generally not pass through the center of a particle, the radius of the area swept out by it equals $(2/3)^{1/2}R$, where R is the radius of the particle. The maximum interaction force is roughly $(2/3)^{1/2}R\pi\gamma_i$ and its range w is b. The increase in critical shear stress due to the particles, $\Delta\tau_0$, can be calculated from Eq. (40) as a function of the volume fraction f of the particles and their mean radius R. According to Eq. (40), $\Delta\tau_0$ depends on c/s^2, i.e., on the number of obstacles per unit area. Therefore in the case of particles c/s^2 has to be replaced by Bf/R^2, where B is a geometric constant. Thus one obtains

$$\Delta\tau_0 = A_i\gamma_i^{4/3}f^{2/3} \tag{43}$$

A_i is a constant which is independent of γ_i, f, and R.

2. Often the particles are ordered; examples are Ni_3Al particles in Ni–Cr–Al alloys[90] and $Ni_3(Al, Ti)$ particles in austenitic steels.[93] If only one dislocation passes through a particle, an antiphase boundary is created. Therefore dislocations move in pairs, because then only the small area between them is an antiphase boundary. This problem has been treated by Gleiter and Hornbogen,[94] who find

$$\Delta\tau_0 = A_0\gamma_0^{3/2}f^{1/3}R^{1/2} \tag{44}$$

where γ_0 is the energy which is connected with the antiphase boundary and A_0 is a constant which is independent of γ_0, f, and R.

3. The difference in specific volume between the matrix and the particles gives rise to a stress field which can be treated similarly to the stress of individual solute atoms in solid solutions (Sections 5.2.1 and 5.3.2). The maximum interaction force F between a dislocation and the stress field of a particle is proportional to R,[76] and the same holds for the range of F. Instead of Eq. (42), one gets

$$\Delta\tau_0 = A_s\varepsilon_R^{4/3}f^{2/3}R^{1/3} \tag{45}$$

where A_s is a constant independent of ε_R, f, and R, ε_R describes the misfit of the particle. The modulus effect can be treated similarly.

Generally all three processes will be superimposed. Then $\Delta\tau_0$ will increase if the particles grow but keep their volume fraction constant. This is in agreement with many experiments.[94,95,93] A typical example is shown in Figure 25 for $R < 70$ Å. For larger radii the Orowan process starts operating. The hardening mechanisms

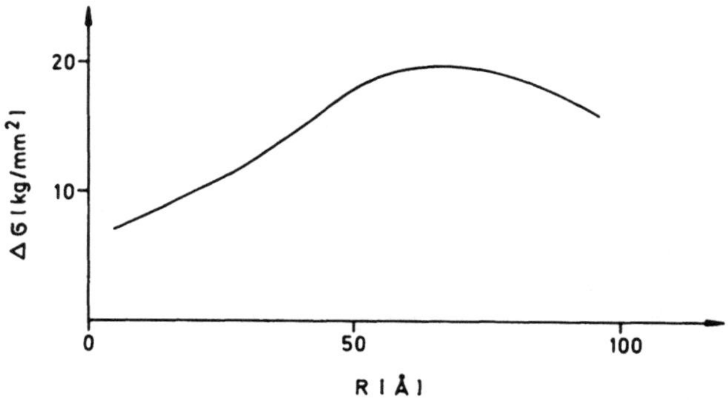

Fig. 25. Relation between the shear proof-stress increase $\Delta\sigma$ and the radius R of spherical $Ni_3(Al, Ti)$ particles in a Fe–Cr–Ni–Mo matrix (after Raynor and Silcock[93]). The volume fraction of the particles is kept constant.

which have been discussed above are of great technical interest. Two examples are: Aluminum is strengthened by fine copper- or magnesium-rich particles and austenitic steels by martensite plates or $Ni_3(Al, Ti)$ particles.

6.2. Orowan Mechanism

If the particles are widely spaced, dislocations may bow out between them according to Figure 24 or they may overcome the particles by cross-slip. If the particles are not ductile, i.e., no dislocation movement is possible in them, these two mechanisms are the only ones by which plastic flow is possible. Such a material has been investigated by Ashby and Ebeling[89,96,97]: copper containing particles of amorphous silicon oxide, which cannot be sheared.

Orowan[91] has calculated the stress τ_B which is necessary to make dislocations pass between particles:

$$\tau_B = E_l/(bd/2) \tag{46}$$

where E_l is the line tension of the dislocations and d is the mean particle spacing in the slip plane. The τ_B of Eq. (46) can be identified with the increase in critical shear stress $\Delta\tau_0$ due to the particles. Ashby[96] has given a more careful analysis of this problem by taking into account that E_l is different for screw and edge dislocations. In Eq. (46) $d/2$ can be replaced by the volume fraction f of the particles and their mean radius R. Then one obtains for $\Delta\tau_0$

$$\Delta\tau_0 = \Omega G f^{1/2} R^{-1} \tag{47}$$

where Ω is a constant independent of f and R. Here $\Delta\tau_0$ depends on temperature only through G.

If ductile particles coarsen while f stays constant, $\Delta\tau_0$ increases first with increasing R [Eqs. (44) and (45)], but once the spacing between the particles becomes wide enough, the dislocations bow out between them. From then on $\Delta\tau_0$ decreases for further coarsening [Eq. (47)]. The plot of $\Delta\tau_0(R)$ shows a maximum (Figure 25). If dislocations overcome the particles by cross-slip, dislocation loops are left behind, but they do not lie in the original glide plane of the dislocations.(92) The work-hardening rate has been found to increase strongly as soon as the Orowan or the cross-slip mechanism starts operating.(93) Each dislocation which passes a particle leaves one or more loops around it and thus makes it more difficult for the following dislocations to pass, too.

7. Mechanical Properties of Polycrystals

Grain boundaries affect plastic deformation in two ways. Since they are obstacles for dislocations, they harden the material. On the other hand, they render a new deformation mechanism possible: The grains may slip on one another. In this section only tensile stress and strain, σ and ε, will be considered. Relating σ and ε to the τ and α of each grain is a complex problem, beyond the scope of this book.

7.1. Plastic Deformation of Polycrystals by Dislocation Mechanisms

For temperatures below two-thirds of the absolute melting temperature plastic deformation of polycrystals is only possible by the movement of dislocations. In a polycrystal the grains cannot deform independently from one another; each of them is restrained by its neighbors. If the grains are not to lose contact with one another, i.e., no pores or even cracks are to be created, each grain must be able to deform proportional to the shape of the specimen. v. Mises(98) and Taylor(99) proved that if this has to be achieved by dislocation glide, five independent slip systems are needed. The condition is fulfilled for fcc and bcc metals, but for hexagonal materials only if slip on non-basal planes is easy. This explains the poor cold-workability of many hexagonal materials.

The shape of the stress–strain curves depends on the grain size, as has been shown by Fleischer and Hosford[100] for aluminum and Carreker and Hibbard[101] for copper. In the following Zankl[102] and Schwink's[103] results for nickel are presented. The stress–strain curve can be divided into three stages: stage 1 goes up to $\varepsilon_2 = 10^{-2}$ and stage 2 up to $\varepsilon_3 = 4 \times 10^{-2}$. Zankl has observed that in stage 1 the length of slip lines equals the grain size, whereas in stage 2 their length decreases with increasing deformation. Schwink[103] believes that work hardening in stage 2 is governed by the same mechanism which governs stage II of fcc single crystals. In stage 3 cross-slip has been observed.[102] There is a close connection between dislocation mechanisms in single crystals and polycrystals of fcc metals.

The dependence of the critical shear stress σ_0 on the grain diameter D has been extensively investigated.[104–107] The result is

$$\sigma_0 = \sigma_{00} + kD^{-1/2} \tag{48}$$

where σ_{00} and k are constants. Equation (48) can be understood from the following. It is assumed that all grains have the same size, but since they have different orientations and thus different Schmid factors, slip will start in each grain at a different stress. Macroscopic yielding occurs when all grains can be deformed. If in a favorably oriented grain g_f a source starts emitting dislocations, they can only glide up to the grain boundary, where they are piled up. This bundle of dislocations increases the stress which acts on a source in a neighboring, unfavorably oriented grain g_u. The number n of dislocations which have been emitted by the source in g_f before their back-stress stops it is proportional to $D\sigma$,[108] where σ is the applied stress. These n dislocations act like a giant dislocation of Burgers vector nb. The stress exerted by them upon the source in g_u is proportional to $nb\sigma$. This source starts emitting dislocations once the force exerted on it reaches a critical value f_0. If σ is strong enough to activate sources in all grains, the critical shear stress σ_0 is reached:

$$f_0 \propto (D\sigma_0)(b\sigma_0) \tag{49}$$

or

$$\sigma_0 \propto D^{-1/2} \tag{50}$$

The friction stress σ_{00} has to be added to the right side of Eq. (50) in order to get Eq. (48).

It has been mentioned in Section 1 that single crystals change their orientation during plastic flow. In polycrystals this causes the grains to assume special orientations with respect to the direction of the applied stress. In a cold-rolled fcc metal sheet a {110} plane of the grains is parallel to the plane of the sheet and one of their ⟨112⟩ directions is parallel to the rolling direction. For bcc metals the planes and directions are {100} and ⟨110⟩, respectively. Polycrystals whose grains have such prefered orientations are said to have a texture.[109]

7.2. Plastic Deformation of Polycrystals by Grain Boundary Sliding

Since binding across grain boundaries is rather weak, it should be easy for two grains to slide upon one another. However, incompatibilities arise where the grain boundaries deviate from a plane. In a polycrystal, compression and dilatation centers are created. There are three ways in which these can be relieved: (a) elastic deformation, (2) dislocation movement, (3) diffusion, either grain boundary or bulk diffusion. The contribution of elastic deformation is negligible. The following discussion will be limited to cases where no dislocation motion is possible. Diffusion in connection with plastic flow in polycrystals has been considered first by Nabarro[110] and Herring[111] and recently by Raj and Ashby.[112] The latter authors have calculated the strain rate $\dot{\varepsilon}$ as a function of the applied stress σ, temperature T, grain diameter D, and volume diffusion coefficient D_V:

$$\dot{\varepsilon} = 42 D_V \Omega \sigma / D^2 kT \tag{51}$$

where k is Boltzmann's constant and Ω the atomic volume. The velocity at which grains slide on one another is governed by the velocity at which atoms can migrate from the compression to the dilatation areas. The proportionality between $\dot{\varepsilon}$ and σ means that in this case the material behaves like a viscous fluid. This is no surprise, since now the atoms are no longer fixed to their lattice sites but are free to move by diffusion. In the case of grain boundary diffusion a formula similar to Eq. (51) has been obtained by Raj and Ashby. Folweiler[113] has observed for alumina $\dot{\varepsilon} \propto \sigma/D^2$; this agrees with Eq. (51). It should be noted that grain boundary sliding does not produce any work hardening.

8. Fracture

8.1. Fracture by Unidirectional Deformation

In everyday life a material is called ductile or brittle depending on whether it can be plastically deformed or not. In ductile single crystals fracture happens after large plastic strains. These strains have reduced the cross section of the specimen so much that a small neck which has been formed accidentally causes a large relative change in cross section and thus a stress concentration, due to which plastic flow will proceed at the neck. Since at high strains work hardening is small, it cannot compensate for the stress concentration. The specimen fractures at the neck. In the following, brittle fracture, i.e. fracture which is preceded by very little, if any, plastic flow, will be discussed. The theoretical fractures stress $\sigma_{f,\text{th}}$ can be calculated in a similar manner as the theoretical shear strength in Section 1. If the two parts I and II of the crystal shown in Figure 26(a) are to be separated along plane A, a tensile stress has to be applied to overcome the atomic binding forces. The necessary stress σ varies with the separation x as shown in Figure 26(b); it reaches $\sigma_{f,\text{th}}$ at $x = m$. The equilibrium value of x is the lattice constant s. The function $\sigma(x)$ can be approximated by

$$\sigma(x) = \sigma_{f,\text{th}} \sin\{[\pi/2(m - s)](x - s)\} \tag{52}$$

For $x \to s$, Eq. (52) reduces to Hooke's law:

$$\sigma(x) = [(x - s)/s]E \tag{53}$$

where E is Young's modulus. From Eq. (52) and (53) m can be calculated:

$$m = s + (\sigma_{f,\text{th}}\pi s/2E) \tag{54}$$

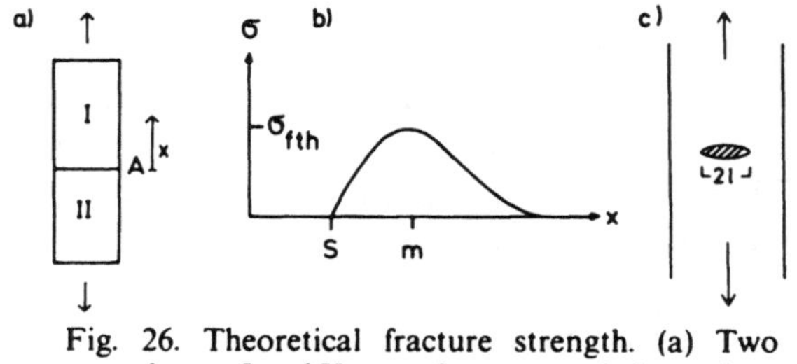

Fig. 26. Theoretical fracture strength. (a) Two crystal parts I and II are to be separated along plane A. (b) Variation of the stress needed to separate I from II as a function of the separation x. (c) Crack nucleus (after Griffith[114]). Normal to the image plane the crack extends through the whole specimen.

The work done in pulling I and II apart equals the energy $2F\gamma$ which is needed to create two new surfaces of area F, where γ is the surface energy:

$$2F\gamma = F\int_{s}^{2m-s} \sigma(x)\,dx = 2\sigma_{f,\mathrm{th}}^2 sF/E \tag{55}$$

This yields for $\sigma_{f,\mathrm{th}}$

$$\sigma_{f,\mathrm{th}} = (E\gamma/s)^{1/2} \tag{56}$$

In reality materials fail at much lower stresses σ_f. Griffith[114] tried to explain the discrepancy between $\sigma_{f,\mathrm{th}}$ and σ_f of glass by assuming that there are small cracks in the material which have elliptical cross sections (Figure 26c). The stress σ_f is the stress needed for the propagation of such cracks. They will expand if the elastic energy released by their propagation equals the energy needed for creating new surfaces:

$$\sigma_f = (2\gamma E/\pi l)^{1/2} \tag{57}$$

where $2l$ is the length of the crack. In Eq. (57) l has replaced the lattice constant s of Eq. (56). Griffith's model is in agreement with the experimental observations of glass. It cannot be directly applied to crystalline solids in which stress concentrations can be relieved by dislocation glide. If a crack grows, energy is needed not only for creating new surfaces but also for the dislocation processes at the tip of the crack. This has been considered by Orowan.[115] In Eq. (57) γ has to be replaced by $\gamma + \gamma_D$, where γ_D is the energy needed for the dislocation mechanisms involved. If one calculates l from the fracture strength σ_f, one obtains values that are much too high. One is sure that good metal samples do not contain crack nuclei this large. It has been found that brittle fracture is always preceded by some plastic flow. Cottrell[116] has suggested a dislocation mechanism by which a crack nucleus can be formed in bcc crystals. Parallel edge dislocations which lie in different slip planes glide toward the intersection of these planes and react there with one another, forming immobile edge dislocations. Their missing half-planes are equivalent to a crack nucleus. The necessary dislocation arrangement can only be produced by very localized slip. This is, e.g., the case if in iron at some place dislocations are freed from their impurity cloud and multiply rapidly.

Many bcc metals are ductile at room temperature and brittle at low temperatures. This transition is caused by the strong temperature dependence of the dislocation mobility in these materials (Section 3.3).

Brittle fracture will occur if it is easier to relieve stress by crack propagation than by dislocation glide.

8.2. Fatigue Fracture

If a material is subjected to alternating stresses, it will fail at much lower stresses than in unidirectional tests. The number N of cycles which can be endured without failure depends on the amplitude τ_e of the alternating stress. The experimental results on fcc single crystals can be described by the following relation between N and τ_e:

$$\tau_e = \tau_f - \tfrac{1}{6}(\tau_f - \tau_{\text{III}}) \log N \qquad (58)$$

τ_f is the fracture strength in a unidirectional test and τ_{III} is the stress at which cross-slip starts (Section 3.1.4). The applied tensile stress $\sigma \sin \omega t$ has to be resolved to the glide system with the highest Schmid factor. If the resolved stress amplitude is smaller than τ_{III}, no failure will occur within 10^6 cycles. At the start of a fatigue test the stress amplitude has to be slowly increased up to τ_e, otherwise large plastic strains would occur. Fatigue fracture is rather reproducible. For a given τ_e the scatter in N will be less than 5%, that is, about the same accuracy as that with which $\tau(a)$ curves can be determined.

Alternating stresses move dislocations to and fro. Dislocations of opposite sign which glide on parallel slip planes attract one another and form dipoles or higher multipoles. Neumann[117] has shown that unstable arrangements of multipoles are built up in this way. If such an arrangement breaks down, many dislocations are freed which locally produce extensive slip. In Section 8.1 it was shown that such localized slip leads to the creation of crack nuclei. In fatigue fracture these are normally formed at the surface of the specimen. Therefore the endurance limit depends on the conditions of the surface.

9. Anelasticity

So far two kinds of deformations have been distinguished: elastic and plastic. In the case of elastic deformations ε is proportional to σ, whereas for plastic deformations $\varepsilon(\sigma)$ is a complicated function which depends on how a certain σ has been reached. Plastic deformation is not reversible. Anelastic deformations are characterized by their reversibility and their time dependence. An example is presented in Figure 27. At t_1 a constant stress is applied and the elastic strain ε_e is observed immediately. Then ε increases slowly and reaches the value

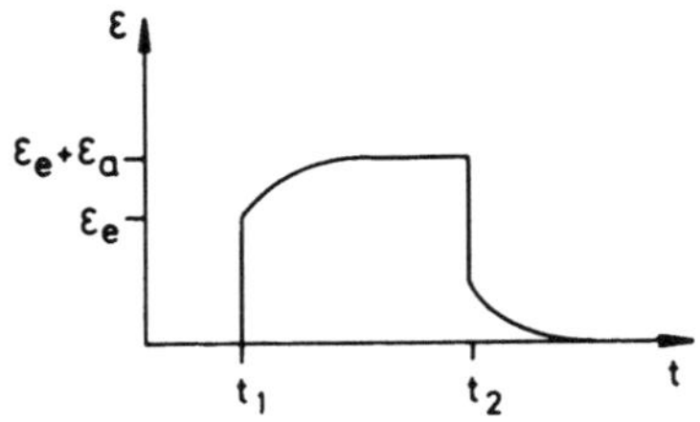

Fig. 27. Anelastic deformation. At t_1 stress is applied and at t_2 removed.

$\varepsilon_e + \varepsilon_a$ asymptotically. At t_2 the stress is removed and ε decreases immediately and approaches zero for long times. The asymptotic values of ε are unique functions of σ. It can be assumed that ε, $\dot{\varepsilon}$, σ, and $\dot{\sigma}$ are connected to one another by a linear relation[118]

$$M_a\sigma + t_0 M_e\dot{\sigma} = \varepsilon + t_0\dot{\varepsilon} \tag{59}$$

where M_a, M_e, and t_0 are constants which depend on the properties of the material. Using Eq. (59), one can calculate ε_e and ε_a of Figure 27. Let the applied stress be σ_z; then

$$\varepsilon(t) = \sigma_z M_e + \sigma_z(M_a - M_e)\{1 - \exp -[(t - t_1)/t_0]\} \tag{60}$$

The elastic strain ε_e is obtained from Eq. (60) for $t = t_1$; the anelastic strain ε_a is the total strain for $t - t_1 \gg t_0$ minus ε_e: $\varepsilon_e = \sigma_z M_e$, $\varepsilon_a = \sigma_z(M_a - M_e)$. Normally ε_e is much larger than ε_a. If a material which can be described by Eq. (60) is subjected to an alternating stress, $\sigma(t)$ and $\varepsilon(t)$ are not in phase, but there is a phase difference θ:

$$\sigma(t) = \sigma_m \sin \omega t, \qquad \varepsilon(t) = \varepsilon_m \sin(\omega t - \theta) \tag{61}$$

where σ_m and ε_m are the amplitudes of stress and strain, respectively, and ω is the circular frequency. The phase difference θ can be calculated by inserting Eq. (61) into Eq. (60):

$$\theta = \tan^{-1}[(1/M_e)(M_a - M_e)\omega t_0/(1 + \omega^2 t_0^{\,2})] \tag{62}$$

Equation (62) holds for $M_e \gg (M_a - M_e)$, i.e., for $\varepsilon_e \gg \varepsilon_a$. The phase difference θ is a measure of the attenuation of the vibrations. It is strongest for $\omega t_0 = 1$. Many anelastic effects are caused by lattice defects. In the following two sections it will be shown how point defects and dislocations can produce anelastic behavior of solids.

9.1. Anelasticity Due to Point Defects

If there are crystallographic positions for point defects (impurity atoms, vacancies, interstitial atoms) which are energetically equivalent only in the unstrained crystal, but not in the strained one, stress will force point defects to move to those places which are energetically most favorable. This movement gives rise to an additional strain, i.e., an anelastic effect. This process is thermally activated; t_0 of Eq. (59) varies with temperature:

$$t_0 = t_{00} e^{E/kT} \tag{63}$$

where t_{00} is a constant and E the activation energy. An example for such a process is the Snoek effect.[119,120] Snoek investigated the damping of mechanical vibrations in α-iron containing interstitially solved carbon. Carbon atoms occupy positions of the type (1/2, 0, 0) and (1/2, 1/2, 0); they cause tetragonal lattice distortions. In unstressed specimens the carbon atoms are equally distributed over all crystallographically equivalent positions. But if a specimen is stretched along one axis, carbon atoms will prefer it, because there the energy of the tetragonal distortion is lower. The time constant with which carbon atoms adjust to a change in σ equals t_0 of Eq. (60). Snoek used a torsion pendulum for his investigations, the frequency of which was kept constant. However, t_0 was varied by varying the temperature. The phase difference θ was found to obey Eq. (62).

9.2. Anelasticity Due to Dislocations

Only two of the many dislocation mechanisms which give rise to anelasticity will be discussed. They are normally investigated by observing the internal friction, i.e., the attenuation of sound waves, in the material. Bordoni[121] measured the internal friction of cold-worked fcc metals in the temperature range 20–300°K and at frequencies between 10 and 40 kHz. He found that damping was strongest at about 100°K. Seeger[122] gave the following interpretation for this Bordoni peak. Because of the Peierls potential the line energy of a dislocation varies periodically with its position. In the same manner as carbon atoms in α-iron jump under the combined influence of thermal energy and stress into the most favorable position, a dislocation segment can jump from one Peierls minimum to the next. Such a segment, which is one lattice constant ahead of the rest of the dislocation, is called a kink. The relaxation time of the kink movement depends on temperature. Therefore the internal friction caused by this process is temperature dependent, too.

A dislocation mechanism whose damping depends only slightly on temperature is the following. A dislocation segment which is fixed at its ends may vibrate like a string.[123] Since any dislocation movement consumes energy, this mechanism causes damping of mechanical vibrations; it is strongest for frequencies above 1 MHz.

When measuring elastic constants in a static experiment care must be taken to avoid errors because of anelastic effects. If dislocations bow out under stress, the strain is overestimated. This can be avoided by measuring elastic constants not for pure materials, but for solid solutions and extrapolating to a solute concentration of zero.

Acknowledgments

The author wishes to express his thanks to Prof. P. Haasen for reading and commenting on the manuscript.

The following publishers and institutions have kindly granted permission to use copyright material: Akademie-Verlag (Figures 15 and 23b), the Institute of Metals (Figure 25), Pergamon Publishing Company (Figures 19 and 21), the National Physical Laboratory (Figure 9), Dr. Riederer-Verlag (Figures 7, 10, 13, and 20), and Springer-Verlag (Figures 11, 14, and 16).

References

1. J. A. Ewing and W. Rosenhain, The crystalline structure of metals, *Phil. Trans. Roy. Soc.* (*London*) *A* **193**, 353–375 (1900).
2. P. Rosbaud and E. Schmid, Über Verfestigung von Einkristallen durch Legierung und Kaltreckung, *Z. Physik* **32**, 197–225 (1925).
3. W. Boas and E. Schmid, Über die Dehnung von Cadmiumkristallen, *Z. Physik* **54**, 16–45 (1929).
4. W. Fahrenhorst and E. Schmid, Über die plastische Dehnung von α-Eisenkristallen, *Z. Physik* **78**, 383–394 (1932).
5. C. F. Elam, Tensile tests of large gold, silver and copper crystals, *Proc. Roy. Soc.* (*London*) *A* **112**, 289–296 (1926).
6. J. J. Gilman, *Micromechanics of Flow in Solids*, McGraw-Hill, New York (1969).
7. J. Frenkel, Zur Theorie der Elastizitätsgrenze und der Festigkeit kristalliner Körper, *Z. Physik* **37**, 572–609 (1926).
8. A. Seeger, in *Handbuch der Physik* (S. Flügge, ed.), Vol. VII/2, pp. 1–210, Springer-Verlag, Berlin (1958).
9. G. I. Taylor, The mechanism of plastic deformation of crystals, *Proc. Roy. Soc.* (*London*) *A* **145**, 362–387 (1934).

10. E. Orowan, Zur Kristallplastizität III, *Z. Physik*. **89**, 634–659 (1934).
11. M. Polanyi, Über eine Art Gitterstörung, die einen Kristall plastisch machen könnte, *Z. Physik* **89**, 660–664 (1934).
12. J. M. Burgers, Some considerations on the fields of stress connected with dislocations in a regular crystal lattice, *Koninkl. Ned. Akad. Wetenschappen, Proc. Section of Sciences* **42**, 293–325 (1939).
13. S. S. Brenner, in *Growth and Perfections of Crystals* (R. H. Doremus, B. W. Roberts, and D. Turnbull, eds.), pp. 157–188, Wiley, New York (1958).
14. E. Reusch, Über eine besondere Gattung von Durchgängen im Steinsalz und Kalkspath, *Ann. Phys. Chem.* **5/12**, 441–451 (1867).
15. M. Peach and J. S. Koehler, The forces exerted on dislocations and the stress fields produced by them, *Phys. Rev.* **80**, 436–439 (1950).
16. J. Diehl, Zugverformung von Kupfereinkristallen, *Z. Metallk.* **47**, 331–343 (1956).
17. P. B. Hirsch and J. W. Steeds, in *The Relation between the Structure and Mechanical Properties of Metals* (N. P. Allen, ed.), Vol. I, pp. 48–82, H.M.S.O., London (1963).
18. A. R. Lang, Studies of individual dislocations by x-ray diffraction microradiography, *J. Appl. Phys.* **30**, 1748–1755 (1959).
19. M. Ahlers, Der Einfluß kleinster Sauerstoffgehalte auf die Plastizität von Silbereinkristallen, *Z. Metallk.* **56**, 741–750 (1965).
20. S. Bühler and K. Lücke, Plastische Verformung von Silbereinkristallen bei hohen Temperaturen, *Z. Metallk.* **55**, 331–342 (1964).
21. P. Haasen, Plastic deformation of nickel single crystals at low temperatures, *Phil. Mag.* **3**, 384–418 (1958).
22. F. W. Young, Jr., Elastic–plastic transition in copper crystals as determined by an etch-pitch technique, *J. Appl. Phys.* **32**, 1815–1820 (1961).
23. J. Friedel, *Les Dislocations*, Gauthier-Villars, Paris (1956).
24. R. Labusch, Die "Aktivierungslänge" bei der thermischen Versetzungbewegung über Hindernisse auf der Gleitebene, *Z. Phys.* **167**, 452–460 (1962).
25. U. Essmann, Elektronenmikroskopische Untersuchung der Versetzungsanordnung verformter Kupfereinkristalle, *Phys. Stat. Sol.* **12**, 707–722 (1965).
26. A. S. Argon and G. East, in Proc. Int. Conf. on the Strength of Metals and Alloys (T. Hirone, ed.), *Trans. Japan Inst. Metals* **9**(Suppl.), pp. 756–767 (1968).
27. A. Seeger, J. Diehl, S. Mader, and H. Rebstock, Work-hardening and work-softening of face-centered cubic metal crystals, *Phil. Mag.* **2**, 323–350 (1957).
28. D. Kuhlmann-Wilsdorf, A new theory of work-hardening, *Trans. Met. Soc. AIME* **224**, 1047–1061 (1962).
29. J. D. Livingston, The density and distribution of dislocations in deformed copper crystals, *Acta Met.* **10**, 229–239 (1962).
30. H. J. Levinstein and W. H. Robinson, in *The Relation between the Structure and Mechanical Properties of Metals* (N. P. Allan, ed.), Vol. I, pp. 179–203, HMSO, London (1963).
31. J. Diehl, S. Mader, and A. Seeger, Gleitmechanismus und Oberflächenerscheinungen bei kubisch flächenzentrierten Metallen, *Z. Metallk.* **46**, 650–657 (1955).
32. G. Schoeck and A. Seeger, in *Report of the Conference on Defects in Crystalline Solids* (N. F. Mott, ed.), pp. 340–346, The Physical Society, London (1955).

33. A. Seeger, R. Berner, and H. Wolf, Die experimentelle Bestimmung von Stapelfehlerenergien kubisch-flächenzentrierter Metalle, *Z. Physik* **155**, 247–262 (1959).
34. K. D. Rogausch, Die Stapelfehlerenergie von Legierungen nach der Knoten- und der τ_{III}-Methode, *Z. Metallk.* **58**, 50–56 (1967).
35. S. Mader, Elektronenmikroskopische Untersuchung der Gleitlinienbildung auf Kupfereinkristallen, *Z. Physik* **149**, 73–102 (1957).
36. H. Mecking and K. Lücke, Quantitative Analyse der Bereich—III—Verfestigung von Silber-Einkristallen, *Acta Met.* **17**, 279–289 (1969).
37. J. P. Hirth and J. Lothe, *Theory of Dislocations*, McGraw-Hill, New York (1968).
38. K. Lücke, G. Masing, and K. Schröder, Dehnversuche an Zink-Einkristallen, *Z. Metallk.* **46**, 792–800 (1955).
39. A. Seeger and H. Träuble, Die plastische Verformung von Zinkeinkristallen, *Z. Metallk.* **51**, 435–456 (1960).
40. P. B. Hirsch and J. S. Lally, The deformation of magnesium single crystals, *Phil. Mag.* **12**, 595–648 (1965).
41. W. Boas and E. Schmid, Über die Temperaturabhängigkeit der kritischen Schubspannung von Cadmiumkristallen, *Z. Physik* **57**, 575–581 (1929).
42. J. J. Gilman, Plastic anisotropy of zinc monocrystals, *Trans. Met. Soc. AIME* **206**, 1326–1336 (1956).
43. S. Schäfer, H. Alexander, und P. Haasen, Plastische Verformung von Germanium und Indiumantimonid im dynamischen Druckversuch, *Phys. Stat. Sol.* **5**, 247–264 (1964).
44. H. Alexander and P. Haasen, in *Solid State Physics* (F. Seitz, D. Turnbull, and H. Ehrenreich, eds.) Vol. 22, pp. 27–158, Academic, New York and London (1968).
45. L. Weiss, Zur Versetzungserzeugung und dynamischen Erholung in Germanium, Diploma Thesis, University of Göttingen (1968).
46. S. Schäfer, Messung von Versetzungsgeschwindigkeiten in Germanium, *Phys. Stat. Sol.* **19**, 297–312 (1967).
47. W. Schröter, H. Alexander, and P. Haasen, Die Inhomogenität der Verformung von Germanium im Streckgrenzenbereich, *Phys. Stat. Sol.* **7**, 983–998 (1964).
48. M. S. Duesbery and R. A. Foxall, in *Realstruktur und Eigenschaften von Reinststoffen* (E. Rexer, ed.), Part 3, pp. 439–448, Akademie-Verlag, Berlin (1967).
49. M. S. Duesbery and R. A. Foxall, A detailed study of the deformation of high purity niobium single crystals, *Phil. Mag.* **20**, 719–751 (1969).
50. G. Taylor and J. W. Christian, Experiments on the deformation of niobium single crystals, *Phil. Mag.* **15**, 893–929 (1967).
51. B. Sěsták and N. Zárubová, Asymmetry of slip in Fe–Si alloy single crystals, *Phys. Stat. Sol.* **10**, 239–250 (1965).
52. J. Hesse, Die plastische Verformung von Natriumchlorid, *Phys. Stat. Sol.* **9**, 209–230 (1965).
53. S. Mendelson, Dislocations and plastic flow in NaCl single crystals I, *J. Appl. Phys.* **33**, 2175–2181 (1962).
54. K. H. Matucha, Elektronenmikroskopie von Gleitlinien auf verformtem NaCl, *Phys. Stat. Sol.* **26**, 291–310 (1968).
55. G. Fontaine, Dissociation des dislocations sur les plans (110) dans des cristaux ioniques du type NaCl, *J. Phys. Chem. Sol.* **29**, 209–214 (1968).

56. C. S. Barrett and R. Bakish, Twinning and cleavage in tantalum, *Trans. Met. Soc. AIME* **212**, 122–123 (1958).
57. H. W. Schadler, Deformation behavior of zone-melted tungsten single crystals, *Trans. Met. Soc. AIME* **218**, 649–655 (1960).
58. C. J. McHargue, Twinning in columbium, *Trans. Met. Soc. AIME* **224**, 334–339 (1962).
59. T. H. Blewitt, R. R. Coltman and J. K. Redman, Low-temperature deformation of copper single crystals, *J. Appl. Phys.* **28**, 651–660 (1957).
60. H. Suzuki and C. S. Barrett, Deformation twinning in silver–gold alloys, *Acta Met.* **6**, 156–165 (1958).
61. A. H. Cottrell and B. A. Bilby, A mechanism for the growth of deformation twins in crystals, *Phil. Mag.* **42**, 573–581 (1951).
62. J. A. Venables, in *Deformation Twinning* (Metallurgical Society Conferences, Vol. 25, R. E. Reed-Hill, J. P. Hirth, and H. C. Rogers, eds.), pp. 77–111, Gordon and Breach, New York (1964).
63. E. Peissker, Mechanische Zwillingsbildung in kupferreichen Mischkristallen, *Z. Metallk.* **56**, 155–164 (1965).
64. E. Peissker, Quergleitspannung und Stapelfehlerenergie von kupferreichen Mischkristallen, *Acta Met.* **13**, 419–431 (1965).
65. G. Kostorz and P. Haasen, Mischkristallhärtung von Kupfer, *Z. Metallk.* **60**, 26–28 (1969).
66. P. Jax, P. Kratochvil, and P. Haasen, Solid solution hardening of gold and other f.c.c. single crystals, *Acta Met.* **18**, 237–245 (1970).
67. A. A. Hendrickson and M. E. Fine, Solid solution strengthening of Ag by Al, *Trans. Met. Soc. AIME* **221**, 967–974 (1961).
68. H. H. Kranzlein, M. S. Burton, and G. V. Smith, Solid solution strengthening in iron–nickel and iron–platinum alloys, *Trans. Met. Soc. AIME* **233**, 64–70 (1965).
69. G. Kostorz, Mischkristallhärtung von Niob, *Z. Metallk.* **59**, 941–952 (1968).
70. P. Jax, Mischkristallhärtung von Niob-Einkristallen, *Z. Metallk.* **62**, 279–291 (1971).
71. T. E. Mitchell and P. L. Raffo, Mechanical properties of some tantalum alloys, *Can. J. Phys.* **45**, 1047–1061 (1967).
72. H. W. Paxton and A. T. Churchman, The effect of temperature and composition on the deformation of single crystals of iron, *Acta Met.* **1**, 473–479 (1953).
73. G. T. Hahn, A model for yielding with special reference to the yield-point phenomena of iron and related bcc metals, *Acta Met.* **10**, 727–738 (1962).
74. A. Wijler and J. Schade van Westrum, Strain rate experiments and the Portevin–Le Chatelier effect in Au (14 at. % Cu), *Scripta Met.* **5**, 531–535 (1971).
75. A. H. Cottrell, in *Report of a Conference on Strength of Solids* (N. F. Mott, ed.), pp. 30–36, The Physical Society, London (1948).
76. R. L. Fleischer, in *The Strengthening of Metals* (D. Peckner, ed.), pp. 93–140, Reinhold, New York (1964).
77. R. L. Fleischer, Solution hardening, *Acta Met.* **9**, 996–1000 (1961).
78. F. Seitz, Speculations on the properties of the silver halide crystals, *Rev. Mod. Phys.* **23**, 328–352 (1951).
79. J. C. Fisher, On the strength of solid solution alloys, *Acta Met.* **2**, 9–10 (1954).
80. H. Suzuki, Chemical interaction of solute atoms with dislocations, *Sci. Rep. Res. Inst. Tohoku Univ. A* **4**, 455–463 (1952).

81. A. H. Cottrell and M. A. Jaswon, Distribution of solute atoms around a slow dislocation, *Proc. Roy. Soc. A* **199**, 104–114 (1949).
82. R. Labusch, A statistical theory of solid solution hardening, *Phys. Stat. Sol.* **41**, 659–669 (1970).
83. R. Labusch, to be published.
84. T. Kan and P. Haasen, Solid solution hardening of silver single crystals by cadmium, *Mater. Sci. Eng.* **5**, 176–178 (1969/70).
85. T. Kan and P. Haasen, Solid solution hardening of gold single crystals by aluminum and by germanium, *Z. Metallk.* **61**, 355–356 (1970).
86. A. Howie and P. R. Swann, Direct measurements of stacking-fault energies from observations of dislocation nodes, *Phil. Mag.* **6**, 1215–1226 (1961).
87. B. Harris, The influence of some solutes on Young's modulus of niobium, *J. Less-Common Metals* **12**, 247–250 (1967).
88. J. Diehl, in *Vacancies and Interstitials in Metals* (A. Seeger, D. Schumacher, W. Schilling, and J. Diehl, eds.) pp. 739–783, North-Holland, Amsterdam (1970).
89. R. Ebeling and M. F. Ashby, Dispersion hardening of copper single crystals, *Phil. Mag.* **13**, 805–834 (1966).
90. H. Gleiter and E. Hornbogen, Beobachtung der Wechselwirkund von Versetzungen mit kohärenten geordneten Zonen (II), *Phys. Stat. Sol.* **12**, 251–264 (1965).
91. E. Orowan, in *Symposium on Internal Stresses in Metals and Alloys*, pp. 451–453, The Institute of Metals, London (1948).
92. M. F. Ashby, in *Electron Microscopy and Strength of Crystals* (G. Thomas and J. Washburn, eds.), pp. 891–897, Interscience, New York (1963).
93. D. Raynor and J. M. Silcock, Strengthening mechanisms in γ'-precipitating alloys, *Metal Sci. J.* **4**, 121–130 (1970).
94. H. Gleiter and E. Hornbogen, Precipitation hardening by coherent particles, *Mater. Sci. Eng.* **2**, 285–302 (1967/68).
95. M. Witt and V. Gerold, Die Plastizität von Kupfer-Kobalt-Einkristallen mit und ohne Ausscheidung, *Z. Metallk.* **60**, 482–487 (1969).
96. M. F. Ashby, in *Oxide Dispersion Strengthening* (Metallurgical Society Conferences, Vol. 47); (G. S. Ansell, T. D. Cooper, and F. V. Lenel, eds.), pp. 143-205, Gordon and Breach, New York (1968).
97. R. Ebeling, Elektronenmikroskopische Gleitlinienuntersuchungen an dispersionsgehärteten Kupfereinkristallen, *Z. Metallk.* **61**, 42–52 (1970).
98. R. v. Mises, Mechanik der plastischen Formänderung von Kristallen, *Z. Angew. Math. Mech.* **8**, 161–185 (1928).
99. G. I. Taylor, Plastic strain in metals, *J. Inst. Metals* **62**, 307–324 (1938).
100. R. L. Fleischer and W. F. Hosford, Jr., Easy glide and grain boundary effects in polycrystalline aluminum, *Trans. Met. Soc. AIME* **221**, 244–247 (1961).
101. R. P. Carreker, Jr. and W. R. Hibbard, Jr., Tensile deformation of high-purity copper as a function of temperature, strain rate, and grain size, *Acta Met.* **1**, 654–663 (1953).
102. G. Zankl, Magnetische und elektronenmikroskopische Untersuchungen zum plastischen Verhalten von vielkristallinem Nickel, *Z. Naturforsch.* **18a**, 795–809 (1963).

103. C. Schwink, Über die Verfestigung homogener kubisch flächenzentrierter Vielkristalle bis zum Beginn dynamischer Erholungsvorgänge, *Phys. Stat. Sol.* **8**, 457–474 (1965).
104. A. Cracknell and N. J. Petch, Frictional forces on dislocation arrays at the lower yield point in iron, *Acta. Met.* **3**, 186–189 (1955).
105. E. O. Hall, The deformation and aging of mild steel III, *Proc. Phys. Soc. B* **64**, 747–753 (1951).
106. R. Armstrong, I. Codd, R. M. Douthwaite, and N. J. Petch, The plastic deformation of polycrystalline aggregates, *Phil. Mag.* **7**, 45–58 (1962).
107. H. H. Tjerkstra, The effect of grain size on the stress–strain curve of α-iron and the connection with the plastic deformation of the grain boundaries, *Acta Met.* **9**, 259–263 (1961).
108. G. Leibfried, Verteilung von Versetzungen im statischen Gleichgewicht, *Z. Physik* **130**, 214–226 (1951).
109. G. Wassermann and J. Grewen, *Texturen metallischer Werkstoffe*, Springer-Verlag, Berlin (1962).
110. F. R. N. Nabarro, in *Report of a Conference on Strength of Solids* (N. F. Mott, ed.), pp. 75–90, The Physical Society, London (1948).
111. C. Herring, Diffusional viscosity of a polycrystalline solid, *J. Appl. Phys.* **21**, 437–445 (1950).
112. R. Raj and M. F. Ashby, On grain boundary sliding and diffusional creep, *Met. Trans.* **2**, 1113–1127 (1971).
113. R. C. Folweiler, Creep behavior of pore-free polycrystalline aluminum oxide, *J. Appl. Phys.* **32**, 773–778 (1961).
114. A. A. Griffith, The phenomena of rupture and flow in solids, *Phil. Trans. Roy. Soc. (London) A* **221**, 163–198 (1921).
115. E. Orowan, Fracture and strength of solids, *Rep. Prog. Phys.* **12**, 185–232 (1948–1949).
116. A. H. Cottrell, Theory of brittle fracture in steel and similar metals, *Trans. Met. Soc. AIME* **212**, 192–203 (1958).
117. P. Neumann, Coarse slip model of fatigue, *Acta Met.* **17**, 1219–1225 (1969).
118. B. S. Berry and A. S. Nowick, in *Physical Acoustics* (W. P. Mason, ed.), Vol. III A, pp. 1–42, Academic, New York and London (1966).
119. J. L. Snoek, Effect of small quantities of carbon and nitrogen on the elastic and plastic properties of iron, *Physica* **8**, 711–733 (1941).
120. J. L. Snoek, Tetragonal martensite and elastic aftereffect in iron, *Physica* **9**, 862–864 (1942).
121. P. G. Bordoni, Teoria della dissipazione elastica nei monocristalli secondo la meccanica quantistica: un nuovo effetto di rilassamento, *Ric. Sci.* **19**, 851–862 (1949).
122. A. Seeger, On the theory of the low-temperature internal friction peak observed in metals, *Phil. Mag.* **1**, 651–662 (1956).
123. A. Granato and K. Lücke, Theory of mechanical damping due to dislocations, *J. Appl. Phys.* **27**, 583–593 (1956).

Index

MIX
Papier aus verantwortungsvollen Quellen
Paper from responsible sources
FSC® C105338

If you have any concerns about our products,
you can contact us on
ProductSafety@springernature.com

In case Publisher is established outside the EU,
the EU authorized representative is:
Springer Nature Customer Service Center GmbH
Europaplatz 3, 69115 Heidelberg, Germany

Printed by Libri Plureos GmbH
in Hamburg, Germany